RARE EARTH INTERMETALLICS

MATERIALS SCIENCE AND TECHNOLOGY

A. S. Nowick and B. S. Berry, ANELASTIC RELAXATION IN CRYSTALLINE SOLIDS, 1972

E. A. Nesbitt and J. H. Wernick, RARE EARTH PERMANENT MAGNETS, 1973

W. E. Wallace, RARE EARTH INTERMETALLICS, 1973

J. C. Phillips, BONDS AND BANDS IN SEMICONDUCTORS, 1973

In preparation

J. H. Richardson and R. V. Peterson (editors), SYSTEMATIC MATERIALS ANALYSIS, VOLUMES I AND II

Rare Earth Intermetallics

W. E. WALLACE

Department of Chemistry
University of Pittsburgh

With an Appendix Prepared in Collaboration with

E. SEGAL

Ministry of Defense
Scientific Department
Tel Aviv, Israel

ACADEMIC PRESS New York and London 1973
A Subsidiary of Harcourt Brace Jovanovich, Publishers

ACADEMIC PRESS, INC.
111 Fifth Avenue, New York, New York 10003

United Kingdom Edition published by
ACADEMIC PRESS, INC. (LONDON) LTD.
24/28 Oval Road, London NW1

LIBRARY OF CONGRESS CATALOG CARD NUMBER: 72-82651

PRINTED IN THE UNITED STATES OF AMERICA

CONTENTS

PART III RARE EARTH–TRANSITION METAL SYSTEMS

Chapter 9 **Nickel Compounds**

Chapter 10 **Cobalt Compounds**

Chapter 11 **Iron and Manganese Compounds**

Chapter 12 **Compounds with 4d and 5d Transition Metals**

Appendix **Energies, Eigenfunctions, and Magnetic Moments of Rare Earth Ions in a Hexagonal Field** 211

PREFACE

Whatever else may be said of the decade of the sixties, they were wonderful years for those individuals working with rare earths and their compounds. During that decade the subject "grew up." The many bizarre characteristics of the elements were revealed and elucidated, at least in general terms, and very many intermetallic compounds containing rare earths were synthesized and characterized. Developments came, at least in the first two-thirds of this era, with almost overwhelming rapidity, and the subject enlarged from a pure research activity to the threshold of major applications. Workers in the field, "old hands" as well as the many neophytes attracted to this area because of either the odd characteristics of or applications prospects for rare earth intermetallics, are in danger of being submerged by the volume of information becoming available.

The information explosion in the rare intermetallic field is primarily in the form of individual research reports. These accounts of individual research which involve many workers and many journals and many countries and which often tell only a small fraction of a total story are in obvious need of collation and summary. This has been done in part in the excellent compilation of M. Hansen and K. Anderko, and in the supplement by R. P. Elliott, entitled "Constitution of Binary Alloys," and in W. B. Pearson's compendium, "Handbook of Lattice Spacings and Structures." These works have provided information concerning the structural and constitutional features of rare earth (and other) intermetallics in highly concise and readable form. To date there has been no comparable work dealing with magnetic and thermal characteristics of intermetallics. This monograph has been prepared to meet the obvious need for such a compilation.

The writing of this monograph is a logical, if not inevitable, consequence of my involvement with rare earth intermetallics over the past fifteen years. In 1968, in response to the pressure of emerging results, I prepared a review dealing with the magnetic characteristics of these intermetallics. This review (Reference 4, Chapter 1) provided a reasonably complete status report for the field as of that date; it has subsequently been enlarged and updated (Reference 5, Chapter 1). This monograph constitutes a further progression in the series.

The review of the literature is complete to August, 1971.

ACKNOWLEDGMENTS

Although the groundwork for this compilation was laid over a period of years, preparation of the written account began in earnest only in April, 1971, during sabbatical leave from the University of Pittsburgh and with the assumption of my position as Guest Professor in the Institute of Experimental Physics IV at the Ruhr University, Bochum, Germany. I am deeply indebted to the two institutions for the essential help which they gave in furthering this work—to my home institution for granting me leave to undertake this project and to the Ruhr University for providing me a delightful "home-away-from-home" in which to carry it out. I am also indebted to many friends and colleagues in Pittsburgh and Bochum for equally essential help. The debt which I owe in this regard is distributed over many individuals, more than I can individually cite. At the Ruhr University I am particularly indebted to Professors S. Methfessel and H. Kirchmayer for arranging my appointment, tending to myriads of details for me, and meeting my needs within their rapidly developing and awesomely big new university in such an efficient and effective manner. I am equally indebted to numerous colleagues in Pittsburgh who assisted my efforts. The lively interest of my graduate students and postdoctorals in rare earth systems has played a vital role in my efforts to grow with the developing subject area. Their contribution to the generation of this manuscript, while indirect, is nonetheless very substantial. I am particulary indebted to Ms. Nancy Chen Yang for her contribution; she, as a part of her graduate program, carried the major burden of the literature searches. Without her able, dedicated, and conscientious services it is doubtful that this monograph would have come into being—at least at this time. I am equally indebted to my secretary, Ms. Patricia Buddemeyer, for her skill and effectiveness in transforming a rather checkered manuscript into beautiful typewritten copy. Special thanks are also due to Dr. E. Segal for his contribution to my understanding of the quantum mechanical details of the crystal field interaction and for his careful and conscientious calculation of the effects of this interaction, some of which are tabulated in the Appendix. Lastly, and most importantly, I must draw attention to the essential role played by my friend and professional colleague, Professor Raymond S. Craig. He and I have been collaboratively involved in the study of metallic systems (rare earths and non-rare earths) for the past two decades. His keen insight and unflagging interest have represented indispensable ingredients in our joint efforts to stay abreast of developments in the rare earth field. Without the stimulation provided by this long-standing collaboration I fear that the writing of this monograph would never have passed beyond the idea stage.

In addition, it is a pleasure to call attention to the interest and support of various government agencies for this and other of my activities involving rare earths—the Air Force Office of Scientific Research, the National Science Foundation, the National Aeronautics and Space Agency, and particularly the U.S. Atomic Energy Commission and the Army Research Office—Durham.

Part I

INTRODUCTORY AND BACKGROUND MATERIAL

Chapter 1

INTRODUCTION AND SCOPE

Prior to World War II intermetallic compounds received little attention from physical scientists. They were regarded as laboratory curiosities of little interest except to those individuals concerned with phase equilibria in polycomponent metallic systems. More recently interest in intermetallics, particularly those containing rare earths, has quickened so that they are now engaging the attention of large numbers of material scientists, chemists, physicists, engineers, and, very recently, production specialists.

The rare earth intermetallics exhibit a number of features which make them of interest. First, the rare earths are prolific compound formers, entering into chemical union with metals of almost every group in the Periodic Table. Second, the numerous compounds show great range and diversity in respect to their structures and stoichiometries. Third, the systems display a number of esoteric characteristics—magnetic ordering to produce in some cases rather exotic magnetic structures, a wide range of crystal field effects, unusual transport properties, such as resistance minima, etc. Lastly, some of the rare earth intermetallics have magnetic characteristics which make it likely that they will soon become of use in the fabrication of a wide variety of electromechanical devices. In addition to these several points of interest, the rare earth intermetallics possess the feature, which they share with other kinds of rare earth systems, of affording the investigator the opportunity of studying a series of closely related isostructural compounds which differ only in the nature of the rare earth constituent.

A few scattered studies of the magnetic properties of rare earth intermetallics predate the decade of the sixties. However, the present era of intensive investigation of these materials began with the work of Nesbitt *et al.* [1] in 1959 and that of Hubbard *et al.* [2] and Nassau *et al.* [3] in 1960, on systems involving the rare earths in chemical union with a 3d-transition metal. Since these seminal studies the fundamental properties of these and other rare earth intermetallics have been very exhaustively investigated until, as noted above, developments have advanced to a stage that applications seem imminent. Although the latter remains at present more a bright hope than a tangible reality, the prospects of commercial utility of these materials are drawing into the field numerous individuals who lack experience with the rare earths. This and the many facets which these intermetallics display have made it appropriate at this time to review the field and set forth the accomplishments to date in summary form. This has been done on two preceding occasions [4,5] in journal articles of somewhat more limited scope. The present work, which may be regarded as an outgrowth of the earlier reviews, follows essentially the same format as employed earlier and is arranged with the same principal objective in mind—namely, to provide an up-to-date account of the magnetic characteristics of rare earth intermetallics in a form which is as concise as possible, consistent with reasonable clarity. In addition, in this review a summary of the (relatively small amount of) work which has been performed dealing with their thermal properties is also included.

Most of the investigations which form the substance of the present monograph were carried out since 1960. They have been largely focused on two aspects of these materials—their structural and magnetic features. A few studies have been concerned with their thermal properties (low temperature heat capacities, third law entropies, etc.). Reasonably current and comprehensive accounts of the structural and constitutional characteristics of rare earth intermetallics [6–8] are available. However, there has been no correspondingly comprehensive treatment of their magnetic and thermal behavior, which is, of course, needed if their full technological capabilities are to be realized. The scope of the present monograph has been defined so as to meet this need. In the main two kinds of information are included: (1) bulk magnetic characteristics (temperature dependence of susceptibility, saturation magnetization, and the nature of the cooperative magnetic phase) and (2) low temperature specific heats and related thermal properties. We decided to include as well a small amount of structural information, namely, the structural type of the compound, to spare the reader the inconvenience of continually having to refer to other sources for these data. Results obtained through the use of NMR, ESR, or Mössbauer spectroscopy are not included per se. They are included only if they serve a useful function in regard to interpretations of the overt magnetic characteristics of the system—supportive or corroborative evidence, etc.

In view of the burgeoning commercial significance of the RCo_5 compounds as materials for use in permanent magnet fabrication a section has been included in Chapter 10 giving the desiderata for a substance to be useful for these purposes. This information is, of course, to be found elsewhere, but it is scattered and, moreover, lies mainly within the province of the material or electrical engineer. The pure scientists, chemist or physicist, is often unaware of the properties required of a material if it is to assume technological significance.

In the further definition of the scope of the work we must address ourselves to the troublesome question of what is to be included under the term "intermetallic compound." The choice that we have made is arbitrary. It excludes those systems in which the second component is boron, carbon, hydrogen, a halogen, or an element of groups VA or VIA. We are thus not including the rather extensive work on the chalcogenides and pnictides. Apart from these exclusions we take a rather liberal view of systems to be regarded as intermetallics. Some of the inclusions may refer to disordered or only partially ordered substitutional solid solutions. These are a minority. Most of the substances which we have included are real compounds in the sense that they have a characteristic structure, separate and distinct from that of the components, and a stoichiometry which, although not as sharply defined as for salts, is still reasonably sharp; that is, the system is stable only over a fairly restricted range of composition.

In the ensuing chapters we present first a brief description of the interactions experienced by a rare earth ion as a result of its neighbors in the crystal (Chapters 2 and 3) and then (Chapters 4–12) a résumé of the results obtained for various materials. Perhaps the most useful and most significant aspect of this monograph will be the tables in the last nine chapters. In these the essential thermal and magnetic features are summarized—third law entropies, electronic specific heat coefficients, ordering temperatures, Weiss constants, paramagnetic and ferromagnetic moments. Notation is included to convey in concise form information as to the magnetic nature of the compounds in the paramagnetic and ordered states. As regards the latter, compounds will be identified as being ferromagnetic, ferrimagnetic, or antiferromagnetic. Description of their paramagnetic behavior is amplified to the extent of indicating whether the susceptibility (χ) varies inversely with temperature to the lowest measured temperatures (Curie or Curie–Weiss behavior) or shows deviations from linearity at low temperatures, and if so the probable nature of the phenomena producing the deviations. As indicated above, the description of magnetic behavior provided by these tabulations is intended to provide workers in the field with a highly concise status report for a given compound; it is not an exhaustive recapitulation of all the studies that have been made. For the latter purpose the reader is referred to the original literature to which extensive references are made.

The nomenclature used in tabulating the data is given in Table 1-1. μ_{or}, given in the tables in Chapters 4 through 12, is the moment for the material in the magnetically ordered state. The value of μ_{or} is for 4.2°K, unless another temperature is specified.

TABLE 1-1

Nomenclature Used to Magnetically Type the Rare Earth Intermetallics

Symbol	Ordered form	Symbol	Paramagnetic form
F	Ferromagnetic	P	Pauli paramagnetic
F_i	Ferrimagnetic	C	Curie law behavior
AF	Antiferromagnetic	C–W	Curie–Weiss behavior
AF(M)	Antiferromagnetic, exhibits meta magnetism[a]	N,x	χ^{-1} nonlinear with temperature over entire range studied
		C–W,x[b]	Curie–Weiss (or Curie) behavior at high temperature but deviating at low temperatures

[a] If the critical field, $H_c > 25$ kOe, the material is regarded as an antiferromagnetic material and the type is merely given as "AF."

[b] $x =$ M (multiplets narrowly spaced); $x =$ VC (valence change); $x =$ VV (Van Vleck paramagnetism(; x = CF (crystal field effect other than Van Vleck paramagnetism); x = K (Kondo effect); x = IEL (incomplete electron localization); x = ? (cause of the deviation either is unknown or many causes are probably involved); x = ?? (behavior at low temperatures has not been determined).

REFERENCES

1. E. A. Nesbitt, J. H. Wernick, and E. Corenzwit, *J. Appl. Phys.* **30,** 365 (1959).
2. W. M. Hubbard, E. Adams, and J. V. Gilfrich, *J. Appl. Phys.* **31S,** 368, (1960).
3. K. Nassau, L. V. Cherry, and W. E. Wallace, *J. Phys. Chem. Solids* **16,** 131 (1960).
4. W. E. Wallace, *Prog. Rare Earth Sci. Technol.* **3,** 1 (1968).
5. W. E. Wallace, *Prog. Solid State Chem.* **6,** 155 (1971).
6. W. B. Pearson, "Handbook of Lattice Spacings and Structures." Pergamon, Oxford, 1967.
7. M. Hansen and K. Anderko, "Constitution of Binary Alloys." McGraw-Hill, New York, 1958.
8. R. P. Elliott, "Constitution of Binary Alloys." McGraw-Hill, New York, 1965.

Chapter 2

THE MAGNETIC INTERACTIONS

The magnetic characteristics of rare earth intermetallics presented in Chapters 4–12 can not be fully appreciated without at least a rudimentary understanding of the nature of the magnetic interactions. The following is a very brief summary of the currently accepted concept of the nature of these interactions. Fuller accounts are available elsewhere [1].

At sufficiently high temperatures all rare earth intermetallics are in the paramagnetic state. Upon cooling the magnetic entropy is extracted, in accordance with the requirements of the third law of thermodynamics. Usually the magnetic entropy is removed by cooperative establishment of magnetic order.* In exceptional cases removal of magnetic disorder is accomplished by the development of Van Vleck paramagnetism through the mechanism which is briefly described in the next chapter. The formation of magnetic structures implies the existence of magnetic interactions.

In discussing the nature of the magnetic interactions we shall find it convenient to separate the intermetallics into two categories—those in which the rare earth is combined with a nonmagnetic partner (NM-type) and those in which the partner is magnetic (M-type), the partner in this case almost always being a 3d-transition metal. Interactions in the M-type compounds are of three kinds: R–R, R–M, and M–M. (M here denotes the magnetic partner.) In the NM-type compounds we

*These remarks, of course, pertain to systems containing local moments. Compounds such as $LaAl_2$, $LaNi_2$, etc. exhibit only the Pauli paramagnetism of the conduction electrons and remain paramagnetic to the lowest temperatures. However, these Pauli paramagnets also comply with the third law.

have only R–R interactions. Evidence to be presented later (Chapter 10) suggests that there is no important difference in the character of the R–R interactions in the two types of compounds; hence they can be discussed together. The nature of the R–M interactions has not been fully clarified but it seems likely that they closely resemble the R–R interactions. Thus conceptually there are only two kinds of interactions involved in the rare earth compounds—the M–M and R–R interactions. Since the M–M separation in the elements and compounds is almost identical, we presume the M–M interactions to be little different from those in the 3d elements. The exact nature of these interactions represents a major unsolved problem in the solid state field and will not be considered further in this monograph. We thus confine our attention in the ensuring discussion to the nature of the R–R interactions.

The temperature at which the formation of the cooperative magnetic phase begins in the NM compounds ($\sim 10^2$ °K) excludes the possibility that the dipole–dipole interaction is the dominant influence.* If so, ordering would occur in the range 0.1–1°K. Interatomic distances between rare earth ions in the elements and in intermetallics are about ten times larger than the radius of the 4f shell, ~ 0.4 Å. Thus, the 4f electrons are so localized that overlap between f orbitals centered on adjacent atoms is very small or negligible. Hence dipole–dipole interactions and direct exchange normally constitute only a small portion of the total magnetic interaction between rare earth ions in the elements or in intermetallics. The dominant influence is indirect exchange via the conduction electrons.

The particular mechanism involved was initially conceived and developed by Ruderman and Kittel [2] as a basis for interpreting NMR results obtained for certain metallic systems (specifically Knight shift measurements on Cu–Mn alloys). A short time later Kasuya [3] drew attention to the inadequacy of direct exchange as a coupling mechanism in elemental gadolinium and to the appropriateness of the Ruderman–Kittel treatment as a basis for describing this interaction. This mechanism is now referred to as the Ruderman–Kittel–Kasuya–Yoshida (RKKY) interaction. The essence of this treatment is the following: There is an exchange interaction between the localized 4f electrons and the itinerant conduction electrons. If the ion cores are nonmagnetic, there is equivalence of electrons with spin up and spin down in a metal and the net spin density vanishes everywhere. Exchange between the 4f and conduction electrons unbalances the spin. The imbalance is large in the vicinity of the magnetic ion but decreases in an oscillatory fashion

*In exceptional cases, e.g., $CeAl_2$ (see Chapter 4), the dipole–dipole interaction may be the predominant contribution to the magnetic interactions.

(*vide infra*) as distance from the magnetic ion increases. The spin disturbance is rather long range, the polarization being appreciable at distances ten or more times the radius of the 4f shell. This polarization provides coupling between the magnetic ions in the crystal.

We will now outline the mathematical formulation of the RKKY interaction. The interaction between spins of the conduction electrons (**s**) and localized electrons (**S**) is given by the Hamiltonian

$$\mathcal{H} = -\Gamma \mathbf{S} \cdot \mathbf{s}, \tag{2-1}$$

where Γ is the effective exchange integral. By use of perturbation theory, taken to second order, one obtains the expression [2–4] for the net spin density, ρ:

$$\rho = -[9\pi Z^2 \Gamma S/4E_F]\, F(x). \tag{2-2}$$

In this equation Z is the number of conduction electrons per atom, $E_F = k^2\hbar^2/2m$ is the free electron Fermi energy, and $F(x) = x^{-4}(x \cos x - \sin x)$, where $x = 2k_F r$. Here k_F is the radius of the Fermi sphere and r is the distance from the magnetic ion. Equation (2-2) is obtained by invoking free electron behavior for the conduction electrons. As will be indicated in Chapter 4, Γ is frequently negative (*vide infra*), in which case ρ appears as in Fig. 2-1.

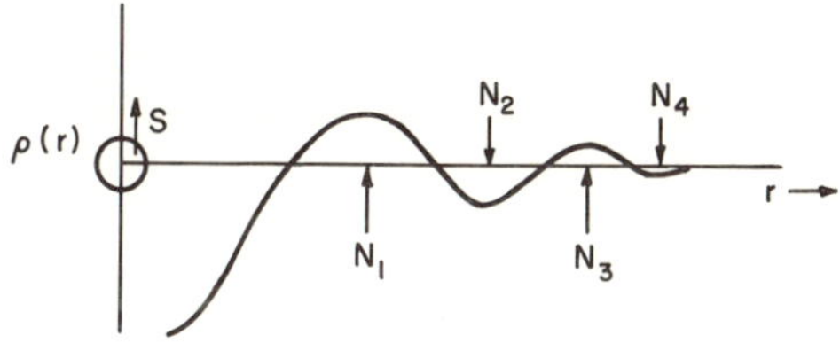

Fig. 2-1. Spin density, $\rho(r)$, versus distance from the magnetic ion for the case $\Gamma < 0$.

We see that the polarization of the conduction electrons decreases as the distance from the magnetic ion increases in a damped oscillatory fashion. A second magnetic ion with $\Gamma < 0$ will experience positive coupling (i.e., ferromagnetic alignment) if it is situated at distances such that the amplitude of the spin density wave is negative, i.e., at positions N_2 and N_4 in Fig. 2-1. Negative or antiferromagnetic coupling will result if the interionic separation corresponds to positive amplitudes of the spin density wave (N_1 and N_3 in Fig. 2-1).

This is the situation for pairwise interactions. In the real case the interaction experienced by a given ion is obtained by an appropriate sum involving all the surrounding magnetic ions. Each ion tends to generate a spin-density wave such as shown in Fig. 2-1. Thus the central ion experiences a composite interaction arising out of all the surrounding ions. In the molecular field approximation the Weiss constant, θ, is a measure of the interaction. De Gennes [5] has treated an assemblage of ions interacting via the RKKY mechanism and obtained an expression for the Weiss constant:

$$\theta = -\{[3\pi Z\Gamma^2(g-1)^2 J(J+1)]/4kE_F\} \Sigma F(x). \tag{2-3}$$

Here $\Sigma F(x)$ is the quantity obtained by summing the RKKY function, $F(x)$, over all the surrounding ions with the origin at the site of one of the rare earth ions. In the molecular field approximation the magnitude of the Weiss constant is identical with the ordering temperature. Thus in this case the Néel temperature (T_N) or Curie temperature (T_C) is also given by Eq. (2-3). According to the treatment of de Gennes the kind of order which develops depends on the sign of $\Sigma F(x)$: antiferromagnetic if this sum is positive and ferromagnetic if it is negative.

The ordering temperatures of an isostructural series of rare earth compounds are of special interest in the light of Eq. (2-3). If the rare earth has the same valence throughout the series (so that Z, E_F, and $F(x)$ are constant) and if Γ is constant, the ordering temperature is expected to vary linearly with the quantity $(g-1)^2J(J+1)$, the so-called de Gennes function. Close approach to the expected linearity is indeed observed for many intermetallics as is brought out in subsequent chapters. This behavior, which is, of course, confined to systems in which the R–R interactions are dominant, affords rather strong support for the general concept, if not the details, of the RKKY mechanism.

The utility of the RKKY formalism and the results obtained from it, such as Eq. (2-3), are unquestioned. However, we must emphasize the limitations of the model on which it is based, the free electron description of the conduction electrons. Validity of conclusions reached using the RKKY treatment must always be assessed in this context. In this regard it is perhaps worth mentioning that use of the RKKY treatment entails difficulties over and above those stemming from the simplified model. For example, it is frequently difficult if not impossible to establish the number of "free" electrons (Z) and hence k_F. Since $\Sigma F(x)$ is sensitively dependent on k_F, the utility of Eq. (2-3) is consequently somewhat restricted. Nevertheless, for lack of a more realistic treatment, considerable use is made of it as is brought out, for example, in the discussion of rare earth–aluminum compounds (Chapter 4).

It is significant to note that the sign of θ, as given by Eq. (2-3), is electron concentration (e.c.) dependent. As e.c. is increased, x increases and θ oscillates back and forth from positive to negative values. Thus on the e.c. scale there are regions in which ferromagnetic interactions prevail interspersed with those in which antiferromagnetic interactions predominate. This is a feature of very considerable importance in connection with efforts to produce rare earth-containing magnetic materials of controlled magnetic structure (see, for example, Chapter 10).

There is another interesting, but often unappreciated, facet of Eq. (2-3)—namely, that while θ is electron concentration dependent, it is unaffected by lattice expansions or contractions which leave the structure of the material unchanged. This follows since θ depends only on the $k_F r$ product and not on these two quantities individually. k_F and r bear inverse and direct relationships, respectively, to lattice size so that the product is independent of the dimensions of the lattice. Hence, if the RKKY mechanism is dominant and is the correct description of the magnetic interactions, efforts to influence the sign of the magnetic coupling by altering the lattice dimensions will be ineffective.

That Γ in Eq. (2-1) can be negative has been alluded to earlier. This is frequently, but not universally, the case. If Γ were a true exchange integral, it would of course be positive. From Knight shift measurements, such as are described in Chapter 4, both positive and negative values of Γ are deduced. Clearly Γ is not a pure exchange integral but instead is a composite parameter. Values of Γ for several intermetallics are given in Table 2-1. The dependence of Γ on the chemical nature of the system is ascribed to variations in band structure from one system to another, the negative contribution to Γ originating with interband mixing [6].

In summary, it is generally accepted that the R–R interactions take place via the conduction electrons by means of the RKKY mechanism. Presumably the R–M interactions, where M is a 3d-transition metal, originate in a similar fashion, although this remains to be demonstrated unequivocally. M–M interactions are probably the same as those occurring in the 3d-transition elements.

TABLE 2-1

The Effective Exchange Integral (Γ) *for Various Intermetallics*

Compound	(eV)	Reference
$GdAl_3$	-1.7	[7]
$GdAl_2$	-0.9	[8]
GdAl	-0.39	[9]
GdCu	$+0.72$	[10]
$NdSn_3$	-0.35	[11]
$NdPt_2$	$+0.56$	[12]
$NdPt_5$	$+0.34$	[12]

REFERENCES

1. See, for example, T. Kasuya *in* "Magnetism" (G. T. Rado and H. Suhl, eds.), Vol. IIB, p. 215. Academic Press, New York, 1966.
2. M. A. Ruderman and C. Kittel, *Phys. Rev.* **96,** 99 (1954).
3. T. Kasuya, *Progr. Theor. Phys., Kyoto.* **16,** 45 (1956).
4. K. Yoshida, *Phys. Rev.* **106,** 893 (1957).
5. P. G. de Gennes, *J. Phys. Radium* **23,** 510 (1962).
6. R. E. Watson, S. Koida, M. Peter, and A. J. Freeman, *Phys. Rev.* **139,** A167 (1965).
7. A. M. Van Diepen, H. W. de Wijn, and K. H. J. Buschow, *J. Chem. Phys.* **46,** 3489 (1967).
8. K. H. J. Buschow, J. F. Fast, A. M. Van Diepen, and H. W. de Wijn, *Phys. Status Solidi* **24,** 715 (1967).
9. H. W. de Wijn, K. H. J. Buschow, and A. M. Van Diepen, *Phys. Status Solidi* **29,** 189 (1968).
10. H. W. de Wijn, K. H. J. Buschow, and A. M. Van Diepen, *Phys. Status Solidi* **30,** 759 (1968).
11. V. Udaya Shankar Rao and R. Vijayaraghavan, *Phys. Lett.* **19,** 168 (1965); F. Borsa, R. G. Barnes, and R. A. Reese, *Phys. Status Solidi* **19,** 359 (1967).
12. R. Vijayaraghavan, S. K. Malik, and V. Udaya Shankar Rao, *Phys. Rev. Lett.* **20,** 106 (1968).

Chapter 3

THE CRYSTAL FIELD INTERACTION AND ITS EFFECT ON THERMAL AND MAGNETIC BEHAVIOR

A. THE CRYSTAL FIELD INTERACTION BY THE OPERATOR EQUIVALENT METHOD

Experiment shows, as will become abundantly evident in the chapters to follow, that the various physical properties of rare earth intermetallics—their magnetic and thermodynamic behavior, as well as their transport properties such as electrical conductivity—are significantly influenced by electrostatic interaction of the rare earth ion with the surrounding ions, the crystal field interaction. Some knowledge of this interaction is essential to the understanding of the behavior of rare earth systems. A brief discussion of the crystal field interaction is presented in this chapter; fuller accounts are to be found elsewhere [1].

The magnetic characteristics of metallic rare earth systems are determined almost exclusively by the electrons in the ion core; the delocalized electrons give rise to the Pauli paramagnetism, which is insignificant in comparison with the magnetism originating with the core electrons. The specific feature of the core electrons which is determinative is the spin and orbital motion of the 4f electrons. The latter is significantly influenced by the perturbing effect of the electrostatic potential (V) originating with the ions in the environment. This is the so-called crystal field interaction. We discuss this interaction in the present chapter in terms of a model in which the 4f electrons are confined to the immediate vicinity of the rare earth nucleus (see Chapter 2) and the surrounding ions are treated as point charges. In this way we obtain the point charge expressions for the crystal field intensity parameters, A and B, given in Eqs. (3-4) and (3-5). These expressions in themselves are of limited utility, except as a convenient way of introducing the subject, since the point charge model seldom applies. However, the general Hamiltonians such as are given in Eqs. (3-6) and (3-7), which have been formulated with the assistance of the point charge approach, are exceedingly useful. In these, the crystal field parameters, the B_n^m's, are quantities to be established from experiment.

Taking the origin of our coordinate system to be at the nucleus of the rare earth ion, we find for the electrostatic potential at a point (r, θ, φ) near the origin due to the surrounding k ions

$$V(r, \theta, \varphi) = \sum_k q_k/(\mathbf{R}_k - \mathbf{r}). \tag{3-1}$$

If our rare earth ion has charge q_i at r_i, θ_i, φ_i, its electrostatic energy due to the perturbing potential, V, is

$$W_c = \sum_i q_i V = \sum_i \sum_k q_i q_k/(\mathbf{R}_k - \mathbf{r}). \tag{3-2}$$

The effect of the perturbing potential on the eigenvectors and eigenvalues of the rare earth ions can be established by the straightforward, but tedious, methods of perturbation theory [2]. However, results are more conveniently obtained using the operator equivalent method introduced by Stevens [3]. This approach, which we now describe briefly, is limited to situations in which W_c is small compared to the spin–orbit interaction separating the different J manifolds, so that J remains a good quantum number.*

The Hamiltonian $\mathcal{H}_c$ is formed from W_c by procedures set forth in detail elsewhere [1–3]. The operator equivalent method consists of finding an operator consisting of standard angular momentum operators which act on the angular part of the wave function of the rare earth ion, which is described in the $|LSJM\rangle$ representation. Formulation of the equivalent operator is accomplished by use of the Wigner–Eckart theorem according to which any tensorial operator can be expressed in a power series of angular momentum operators. In the equivalent operator formalism

$$\mathcal{H}_c = \mathcal{H} = \sum_{n=0}^{n'} \sum_{m=-n}^{n} B_n^m O_n^m (\mathbf{J}) \tag{3-3}$$

with n' the maximal value of n. For f electrons n' is 6 and for d electrons

*In cases where J is not a good quantum number use may be made of the operator equivalent method by expressing the crystal field Hamiltonian as a polynomial in L instead of J.

it is 4. In Eq. (3-3) B_n^m is the so-called crystal field intensity parameter and $O_n^m(J)$ represents polynomials of the angular momentum operators J_z, J^2, J_+, and J_-. The polynomials normally needed have been tabulated by Hutchings [1] and also the matrix elements associated with the several operators for various J's. The matrix elements are, of course, characteristic of the total angular momentum of the rare earth ion (J).

The B_n^m coefficients are determined in part by the surrounding ions and in part by the radial extension of the rare earth ion. B_n^m is obtained from the expression

$$B_n^m = \langle J\|\theta_n\|J\rangle \, \langle r^n\rangle \, (1-\sigma_n) \, K_n^m A_n^m. \tag{3-4}$$

In this expression $\langle J\|\theta_n\|J\rangle$ is a reduced matrix element which is usually designated as α_J, β_J, or γ_J for $n=2$, 4, or 6, respectively. $(1-\sigma_n)$ is a shielding factor. A_n^m is controlled by the electrostatic potential generated by the surrounding ions. Values of the constants K_n^m are given in Table 3-1. A_n^m is given by the expression

$$A_n^m = (-1)^{m+1} \, [4\pi/(2n+1)] \, e^2 \sum_k (Z_k/R_k^{n+1}) \, Y_n^{-m} \, (\theta_k, \varphi_k). \tag{3-5}$$

Here Y_n^{-m} represents the spherical harmonics; Z_k and R_k are the valence number and distance to the kth ion in the environment of the rare earth ion.

TABLE 3-1

Values of the Constants K_n^m in Eq. (3-4)

$K_2^0 = 1/4\ (5/\pi)^{1/2}$,	$K_2^1 = (15/2\pi)^{1/2}$,	$K_2^2 = 1/2\ (15/2\pi)^{1/2}$
$K_4^0 = 3/16\ (\pi)^{1/2}$,	$K_4^1 = 3/4\ (5/\pi)^{1/2}$, $K_4^3 = 3/4\ (35/\pi)^{1/2}$,	$K_4^2 = 3/4\ (5/2\pi)^{1/2}$ $K_4^4 = 3/8\ (35/2\pi)^{1/2}$
$K_6^0 = 1/32\ (13/\pi)^{1/2}$,	$K_6^1 = 1/8\ (13 \cdot 21/2\pi)^{1/2}$, $K_6^3 = 1/16\ (13 \cdot 105/\pi)^{1/2}$, $K_6^5 = 3/16\ (13 \cdot 77/\pi)^{1/2}$,	$K_6^2 = 1/32\ (13 \cdot 105/\pi)^{1/2}$ $K_6^4 = 3/32\ (13 \cdot 14/\pi)^{1/2}$ $K_6^6 = 1/32\ (13 \cdot 21 \cdot 11/\pi)^{1/2}$

The number of terms in the Hamiltonian [Eq. (3-3)] would seem to be rather large even with the limitation on n. However the number is sharply reduced as a result of symmetry. In fact, for high symmetry systems most of the B_n^m's vanish. For example, only four terms remain for systems of cubic or hexagonal symmetry:

$$\mathcal{H}_{\text{cub}} = B_4^0(O_4^0 + 5O_4^4) + B_6^0(O_6^0 - 21O_6^4), \tag{3-6}$$

$$\mathcal{H}_{\text{hex}} = B_2^0O_2^0 + B_4^0O_4^0 + B_6^0O_6^0 + B_6^6O_6^6. \tag{3-7}$$

Assuming the point charge model and an ideal axial ratio, i.e., $c/a = (8/3)^{1/2}$, $\mathcal{H}_{\text{hex}}$ can be further reduced to [4,5]

$$\mathcal{H}_{\text{hex}} = B_4^0O_4^0 + B_6^0\left(O_6^0 + \frac{77}{8}O_6^6\right). \tag{3-8}$$

B_n^m in Eqs. (3-6) to (3-8) can be obtained from Eqs. (3-4) and (3-5). However, the shielding factor and $\langle r^n \rangle$ values are not well known so the B's are usually treated as parameters and are evaluated from experiment.

To give an idea of the procedure involved in establishing the details of the crystal field interaction we cite the essential steps for two simple examples for the case of $J = 5/2$:

$$\mathcal{H} = B_2^0O_2^0 + B_4^0O_4^0 \tag{3-9}$$

and

$$\mathcal{H} = B_2^0O_2^0 + B_2^2O_2^2. \tag{3-10}$$

The several polynomials in the angular momentum operators for O_n^m are

$$O_2^0 = 3J_z^2 - J^2, \tag{3-11}$$

$$O_2^2 = \frac{1}{2}(J_+^2 + J_-^2), \tag{3-12}$$

where

$$\mathbf{J}_\pm = \mathbf{J}_x \pm i\mathbf{J}_y \tag{3-13}$$

and

$$O_4^0 = [35J_z^4 - 30J^2J_z^2 + 25J_z^2 - 6J^2 + 3J^4]. \tag{3-14}$$

In working out the elements for O_2^0, O_2^2, and O_4^0, it is necessary to recall that J_z, J_z^2, and J_z^4 are all diagonal matrices with elements M, M^2, and M^4, respectively. J^2 and J^4 are also diagonal; all elements are equal to $J(J+1)$ for J^2 and to $J^2(J+1)^2$ for J^4. The elements for the raising and lowering operators, J_+ and J_-, are

$$\text{for } J_+: \quad \langle J, M+1\,|J_+|J,M\rangle = [J(J+1) - M(M+1)]^{1/2}; \tag{3-15}$$

$$\text{for } J_-: \quad \langle J, M-1\,|J_-|J,M\rangle = [J(J+1) - M(M-1)]^{1/2}. \tag{3-16}$$

All other matrix elements for $J_\pm$ vanish. For $J=5/2$ we readily obtain the matrices given in Eqs. (3-17) to (3-19) for O_2^0, O_2^2, and O_4^0. To see in detail how these matrices are derived we consider, as an example, the element in the upper left-hand corner of the O_2^0 matrix; it connects $M=-5/2$ with $M'=-5/2$. We have from Eq. (3-11)

$$3[\langle -5/2|J_z^2|-5/2\rangle] - \langle -5/2|J^2|-5/2\rangle. \tag{3-11a}$$

The first term of Eq. (3-11a) is $3\times M^2=75/4$, the second is $J(J+1)=35/4$, and so this matrix element is $(75/4) - (35/4) = 10$, as indicated in Eq. (3-17). The other elements are derived similarly.

$$O_2^0 = \begin{pmatrix} 10 & 0 & 0 & 0 & 0 & 0 \\ 0 & -2 & 0 & 0 & 0 & 0 \\ 0 & 0 & -8 & 0 & 0 & 0 \\ 0 & 0 & 0 & -8 & 0 & 0 \\ 0 & 0 & 0 & 0 & -2 & 0 \\ 0 & 0 & 0 & 0 & 0 & 10 \end{pmatrix}, \tag{3-17}$$

$$O_2^2 = \begin{pmatrix} 0 & 0 & \sqrt{10} & 0 & 0 & 0 \\ 0 & 0 & 0 & 3\sqrt{2} & 0 & 0 \\ \sqrt{10} & 0 & 0 & 0 & 3\sqrt{2} & 0 \\ 0 & 3\sqrt{2} & 0 & 0 & 0 & \sqrt{10} \\ 0 & 0 & 3\sqrt{2} & 0 & 0 & 0 \\ 0 & 0 & 0 & \sqrt{10} & 0 & 0 \end{pmatrix}, \tag{3-18}$$

$$O_4^0 = 60\begin{pmatrix} 1 & 0 & 0 & 0 & 0 & 0 \\ 0 & -3 & 0 & 0 & 0 & 0 \\ 0 & 0 & 2 & 0 & 0 & 0 \\ 0 & 0 & 0 & 2 & 0 & 0 \\ 0 & 0 & 0 & 0 & -3 & 0 \\ 0 & 0 & 0 & 0 & 0 & 1 \end{pmatrix}. \tag{3-19}$$

Using the Hamiltonian in Eq. (3-9) we obtain, after combining the two matrices $B_2{}^0O_2{}^0$ and $B_4{}^0O_4{}^0$ by matrix addition

$$\mathcal{H} = \begin{pmatrix} E_{11} & 0 & 0 & 0 & 0 & 0 \\ 0 & E_{22} & 0 & 0 & 0 & 0 \\ 0 & 0 & E_{33} & 0 & 0 & 0 \\ 0 & 0 & 0 & E_{44} & 0 & 0 \\ 0 & 0 & 0 & 0 & E_{55} & 0 \\ 0 & 0 & 0 & 0 & 0 & E_{66} \end{pmatrix}, \tag{3-20}$$

where $E_{11} = E_{66} = 10B_2{}^0 + 60B_4{}^0$; $E_{22} = E_{55} = -2B_2{}^0 - 180B_4{}^0$; $E_{33} = E_{44} = -8B_2{}^0 + 120B_4{}^0$. Since $\mathcal{H}$ in Eq. (3-20) is diagonal, E_{ii} represents the energies of the several states in a crystal field described by $\mathcal{H} = B_2{}^0O_2{}^0 + B_4{}^0O_4{}^0$. E_{11} corresponds to the state $|-5/2\rangle$, E_{22} to $|-3/2\rangle$, E_{33} to $|-1/2\rangle, \ldots, E_{66}$ to $|5/2\rangle$. For this Hamiltonian there is no mixing of the $|M_i\rangle$ states of the free ion. This is general for Hamiltonians involving only operators of the form $O_n{}^0$. The wave functions (or eigenvectors) are pure M states. When operators of the form $O_n{}^m$ with $m \neq 0$ are involved, the wave function may consist of linear combinations of two or more M states. The operator $O_n{}^m$ mixes M states differing by m.

The crystal field energy levels corresponding to Eq. (3-20) are shown in Fig. 3-1. For $B_2{}^0$ and $B_4{}^0 > 0$ we note that $|\pm 1/2\rangle$ are the lowest

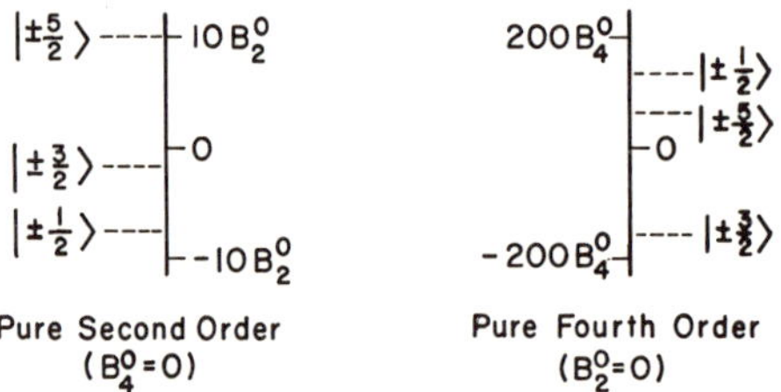

Fig. 3-1. Crystal field splitting for a $J = 5/2$ system described by the Hamiltonian, $\mathcal{H} = B_2{}^0O_2{}^0 + B_4{}^0O_4{}^0$.

lying states for a pure second-order interaction, whereas for a pure fourth-order interaction the states with $M = \pm 3/2$ are lowest in energy. For intermediate situations, that is, when both second- and fourth-order interactions are appreciable, either $|\pm 1/2\rangle$ or $|\pm 3/2\rangle$ is lowest in energy depending on the relative magnitude of $B_2{}^0$ and $B_4{}^0$.

The Hamiltonian given by Eq. (3-10) takes the form

$$\mathcal{H} = \begin{pmatrix} E_{11} & 0 & E_{13} & 0 & 0 & 0 \\ 0 & E_{22} & 0 & E_{24} & 0 & 0 \\ E_{31} & 0 & E_{33} & 0 & E_{35} & 0 \\ 0 & E_{42} & 0 & E_{44} & 0 & E_{46} \\ 0 & 0 & E_{53} & 0 & E_{55} & 0 \\ 0 & 0 & 0 & E_{64} & 0 & E_{66} \end{pmatrix}. \tag{3-21}$$

The off-diagonal elements originate with the $B_2^2O_2^2$ term. In Eq. (3-21), $E_{11} = E_{66} = 10B_2^0$; $E_{22} = E_{55} = -2B_2^0$; $E_{33} = E_{44} = -8B_2^0$; $E_{31} = E_{64} = E_{13} = E_{46} = \sqrt{10}\ B_2^2$, $E_{42} = E_{53} = E_{24} = E_{35} = 3\sqrt{2}\ B_2^2$. Diagonalization of the matrix of Eq. (3-21) is readily accomplished by computer techniques. This rotates the coordinate axes so that the eigenvectors are no longer pure M states but instead are linear combinations of the type $\Sigma\, a_i|M_i\rangle$ involving, in this particular case, states differing in M by 2. The energies are functions of B_2^0 and B_2^2. We do not present the results of diagonalizing matrix (3-21) since it is not clear that the Hamiltonian (3-10) corresponds to any real crystal. However, corresponding treatments of real situations (which are regrettably somewhat more complex) have been carried out in a number of instances. For example, Lea *et al.* [6] in a celebrated paper have provided results for crystals having cubic symmetry using the Hamiltonian in Eq. (3-6) and Segal and Wallace [5] have treated the hexagonal case (point group $D_6 = 622$) using the Hamiltonian in Eq. (3-8). Computations in each case were made in terms of parameters x and W defined by the relationships

$$\frac{B_4^0F_4}{B_6^0F_6} = \frac{x}{1-|x|} \tag{3-22a}$$

and

$$B_4^0F_4 = Wx \tag{3-22b}$$

where F_4 and F_6 are the well-known common factors appearing in the O_4^m and O_6^m matrices, respectively. The parameter x establishes the relative importance of the fourth- and sixth-degree electrostatic interactions.

As examples of the results obtained in these treatments, data are shown for the case of $J = 4$ (Pr^{3+}, Pm^{3+}) in Tables 3-2 and 3-3 and Figs. 3-2 to 3-4. Magnetic moments are obtained from the fundamental ex-

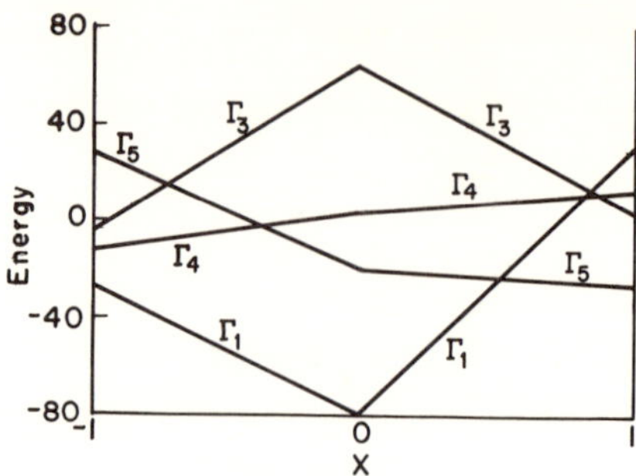

Fig. 3-2. Energy (in arbitrary units) versus x for $J = 4$ and cubic symmetry

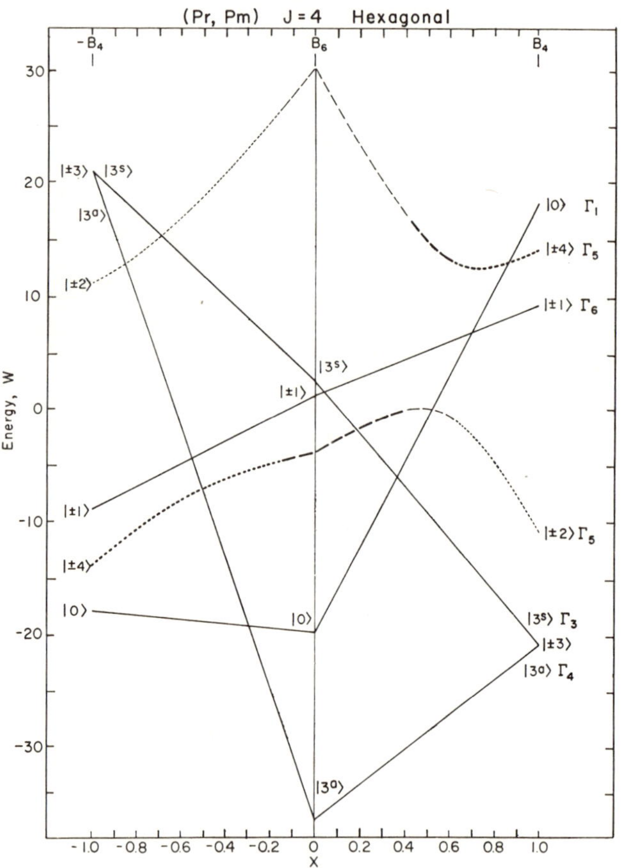

Fig. 3-3. Energy levels and eigenfunctions of an ion with a total angular momentum $J = 4$, as a function of the mixing parameter, x. — The eigenfunctions of the level do not depend on x. - - - The eigenfunctions are over 80% of the eigenfunctions indicated at $x = \pm 1$. – – The eigenfunctions are less than 80% of the eigenfunctions as indicated. The energy scale is in units of W. When $W > 0$, the order of the levels is as shown. If $W < 0$, the order of the levels is inverted.

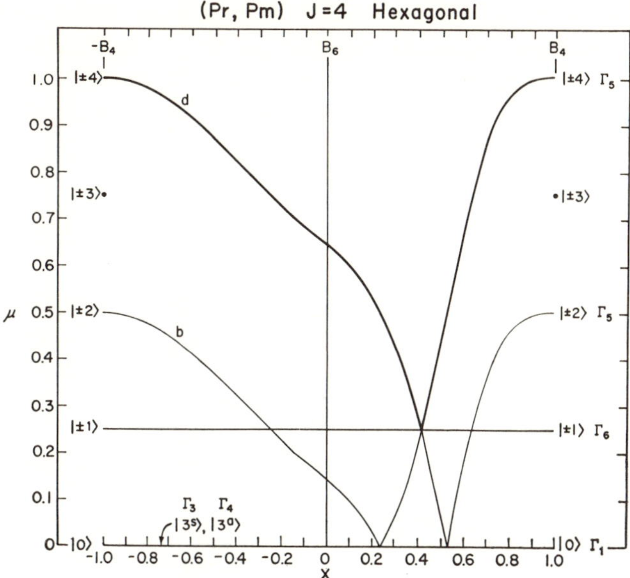

Fig. 3-4. Permanent (parallel) magnetic moments of the levels in Fig. 3-3. The two eigenfunctions of the doublet levels have equal and opposite magnetic moments; in the figure only the positive component is shown. (Note the nonlinear scale of x.)

pression: $\mu = \langle \Gamma_i | \mu_{\text{op}} | \Gamma_i \rangle$ where $\mu_{\text{op}} = gJ$. This reduces for integral values of J to

$$\mu = (1/J) \sum_i a_i^2 M_i \tag{3-23a}$$

where μ is expressed in units of $gJ\,\mu_B$. Equation (3-23a) is also valid for the z component of μ for J nonintegral. However, to obtain the component perpendicular to the z axis we must make use of the expression

$$\mu = \left(\frac{1}{2}J\right) \langle \Gamma_i | (J_+ + J_-) | \Gamma_i \rangle . \tag{3-23b}$$

The moments listed in Table 3-2 were obtained from Eqs. (3-23a) using the eigenvectors given by Lea, Leask, and Wolf; they are identical with the zero field moments, i.e., permanent moments of the crystal field states obtained many years ago by Penney and Schlapp [7] and more recently by Schumacher and Hollingsworth [8].

Results given in Table 3-3 and in the original paper [5] for other J values while of considerable interest are of restricted utility because of the limitation placed on the axial ratio. In a more recent work this restriction has been relaxed so that the results are applicable to systems with nonideal axial ratio. In this case the $B_2^0O_2^0$ term does not vanish from the Hamiltonian. Representative results obtained using the Hamiltonian

$$\mathcal{H} = B_2^0O_2^0 + B_4^0O_4^0 + B_6^0\left(O_6^0 + \frac{77}{8}O_6^6\right) \tag{3-24}$$

have been published [9]. More extensive tables of the results obtained are given in the Appendix, which has been prepared with the assistance of Dr. E. Segal.

TABLE 3-2

Energies,[a] Eigenfunctions,[b] and Magnetic Moments[c] for Cubic Symmetry
$J = 4,\ \mathcal{H} = B_4^0(O_4^0 + 5O_4^4) + B_6^0(O_6^0 - 21O_6^4)$

$J = 4 \rightarrow \Gamma_1 + \Gamma_3 + \Gamma_4 + \Gamma_5$
$F(4) = 60 : F(6) = 1260$

Γ_1:	$0.4564\|4\rangle + 0.7638\|0\rangle + 0.4564\|-4\rangle$		$\mu = 0$
	$x = \pm1$	$E = \pm28$	
	$x = 0$	$E = -80$	
Γ_3:	$0.5401\|4\rangle - 0.6455\|0\rangle + 0.5401\|-4\rangle$		$\mu = 0$
	$x = \pm1$	$E = \pm 4$	
	$x = 0$	$E = 64$	
Γ_4:	$0.3536\|\pm3\rangle + 0.9354\|\mp1\rangle$		$\mu = \pm 0.125$
	$0.7071\|4\rangle + 0.0000\|0\rangle - 0.7071\|-4\rangle$		$\mu = 0$
	$x = \pm1$	$E = \pm14$	
	$x = 0$	$E = 4$	
Γ_5:	$0.9354\|\pm3\rangle - 0.3536\|\mp1\rangle$		$\mu = 0.625$
	$0.7071\|\pm2\rangle - 0.7071\|-2\rangle$		$\mu = 0$
	$x = \pm1$	$E = \mp 26$	
	$x = 0$	$E = -20$	

[a]Energies are in arbitrary units and in all cases are linear with x.
[b]The eigenfunctions are independent of x for $j = 4$ in cubic symmetry. This is not generally true.
[c]Magnetic moments are in units of $gJ\,\mu_B$. Moments in Bohr magnetons are obtained by multiplying by 3.2.

We note that for a number of the states in Tables 3-2 and 3-3 the magnetic moment vanishes. This is the situation for zero strength magnetic field. Although these states lack a moment in zero applied field, they usually acquire a moment by induction when exposed to a magnetic

TABLE 3-3

Energies, Eigenfunctions, and Magnetic Moments for J = 4 and Hexagonal Symmetry, $\mathcal{H} = B_4^0O_4^0 + B_6^0(O_6^0 + 77/8O_6^6)^a$

$J = 4 \rightarrow \Gamma_1$ (singlet) + Γ_3 (singlet) + Γ_4 (singlet) + $2\Gamma_5$ (doublet) + Γ_6 (doublet) (Pr^{3+}, Pm^{3+})

$|\Gamma_1\rangle$: $|0\rangle$ $\mu = 0$

$x = \mp 1$ $E = \mp 18$

$x = 0$ $E = -20$

$|\Gamma_6\rangle$: $|\pm 1\rangle$ $\mu = 0.25$

$x = \mp 1$ $E = \mp 9$

$x = 0$ $E = 1$

$|\Gamma_3\rangle \equiv |3^s\rangle$: $(1/\sqrt{2})\,[\,|-3\rangle + |3\rangle\,]$ $\mu = 0$ for $x \neq \pm 1$

$x = \mp 1$ $E = \pm 21$

$x = 0$ $E = 2.25$

$|\Gamma_4\rangle \equiv |3^a\rangle$: $(1/\sqrt{2})\,[\,|-3\rangle - |3\rangle\,]$ $\mu = 0$ for $x \neq \pm 1$

$x = \mp 1$ $E = \pm 21$

$x = 0$ $E = -36.25$

$|\Gamma_5\rangle$: $a_2\,|\mp 2\rangle + a_4\,|\pm 4\rangle$, $a_2^2 > a_4^2$

x	E	μ	a_2	a_4		
∓1.0	±11.0	0.500	1.000	0.000	1.000	∓14.0
−0.8	13.6	0.478	0.993	0.121	0.978	−10.8
−0.6	16.8	0.414	0.971	0.239	0.914	− 8.2
−0.4	20.8	0.324	0.940	0.342	0.824	− 6.4
−0.2	25.3	0.230	0.906	0.424	0.730	− 5.1
0.0	30.1	0.145	0.873	0.487	0.645	− 4.1
0.2	23.3	0.031	0.829	0.559	0.531	− 1.9
0.4	17.1	0.216	0.723	0.691	0.284	− 0.3
0.6	− 0.9	0.168	0.882	−0.471	0.668	13.1
0.8	− 4.9	0.457	0.985	−0.170	0.957	12.5
x			a_4	a_2	μ	E

$|\Gamma_5\rangle$: $a_4\,|\pm 4\rangle - a_2\,|\pm 2\rangle$, $a_4^2 > a_2^2$

[a]For $|\Gamma_1\rangle$, $|\Gamma_6\rangle$, $|\Gamma_3\rangle$, and $|\Gamma_4\rangle$, E is a linear function of x. When $x = \pm 1$, Γ_3 and Γ_4 converge to a doublet whose eigenfunctions are $|\pm 3\rangle$ and the associated moment is 0.75.

field.* Let us take the Γ_3 state from Table 3-2 as an example. For $H = 0$,

$$\Gamma_3 = (1/\sqrt{2})\,[\,|-3\rangle + |3\rangle\,]. \tag{3-25}$$

*The $\Gamma_1 = |0\rangle$ state is an exception. It has neither a permanent nor an induced moment.

The vanishing moment comes about as a result of the equality of coefficients for $|-3\rangle$ and $|3\rangle$ [see Eq. (3-23a)]. In a magnetic field (either applied or the Weiss molecular field) this equality no longer exists and

$$\Gamma_3 = a_1|-3\rangle + a_2|3\rangle, \quad \text{where } a_1 \neq a_2. \tag{3-26}$$

The difference in a_1 and a_2, and hence the induced moment, is controlled in part by the overall crystal field splitting (E_c) and partly by the strength of the magnetic field, being inversely related to the former and directly to the latter. As an example, calculations by Segal [10] give

$$a_1 = 0.6455 \quad \text{and} \quad a_2 = 0.7637 \tag{3-27}$$

for $E_c = 100°\text{K}$ and $H = 23$ kOe, from which one obtains [Eq. (3-23)] a moment 0.125 $gJ\,\mu_B$. Generally similar calculations were made for ions in a field of cubic symmetry by Penney and Schlapp [7] and Schumacher and Hollingsworth [8]. They expressed their results in the form

$$\mu = \mu_0 + \alpha H \tag{3-28}$$

which is valid only for small H. α in this expression is inversely proportional to E_c, the overall crystal field splitting.

The moments associated with the various crystal field states are needed in calculating the susceptibility (*vide infra*).

B. THE CRYSTAL FIELD PARAMETERS B_n^m

As noted above the quantities B_n^m appearing in the Hamiltonian are usually treated as parameters to be evaluated from experiment. In some instances it is desirable to be able to evaluate them directly. Derivation of the Hamiltonian in Eq. (3-8) from Eq. (3-7) is a case in point. The point charge model is used to evaluate B_6^6 in terms of B_6^0, thus eliminating one parameter and considerably facilitating computations. From Eq. (3-4) we see that the parameter B_n is proportional to the product of $\langle r^n\rangle$ and $(1-\sigma_n)$, neither of which is known very well. However, the ratio B_6^6/B_6^0 is independent of this product and hence it can be determined with greater confidence than either of the quantities alone. For reasons such as this, and many others, it is often necessary to be able to calculate B_n^m. In this section we give two simple examples of the calculation of a crystal field parameter. We shall calculate B_2^0 and B_4^0 for a hexagonal crystal with ideal axial ratio. We know from group theory arguments that $B_2{}^0$ vanishes for such a crystal; this will also be shown numerically.

The spherical harmonics needed [Eqs. (3-4) and (3-5)] in this particular case are as follows:

$$Y_2{}^0 = (1/4)\,(5/\pi)^{1/2}\,(3\cos^2\theta - 1), \tag{3-29}$$

$$Y_4{}^0 = (3/16)\,\pi^{1/2}\,(35\cos^4\theta - 30\cos^2\theta + 3). \tag{3-30}$$

It is convenient to choose the hexagonal axis as the z axis. If we limit consideration to the twelve nearest neighbors surrounding the central ion, there are six ions in a plane which contains the central ion and is perpendicular to the hexagonal axis (layer A) and three ions in the plane above (layer B) and three in a plane below layer A (layer B′). Coordinates of these several ions are as follows:

Layer A: $(a,\pi/2,\pi/6)$; $(a,\pi/2,3\pi/6)$; $(a,\pi/2,5\pi/6)$; $(a,\pi/2,7\pi/6)$; $(a,\pi/2,9\pi/6)$; $(a,\pi/2,11\pi/6)$;

Layer B: $(a,\theta,0)$; $(a,\theta,\pi/3)$; $(a,\theta,2\pi/3)$;

Layer B′: $(a,\theta',0)$; $(a,\theta',\pi/3)$; $(a,\theta',2\pi/3)$;

where $\cos\theta = \sqrt{2/3}$, and $\theta' = 180 - \theta$.

From Eqs. (3-5) and (3-29)

$$\begin{aligned} A_2{}^0 &= -(4\pi/5)(Ze^2/a^3)(1/4)(5/\pi)^{1/2}\,\{6(-1) + 3[3(2/3) - 1] \\ &\quad + 3[3(2/3) - 1]\} \\ &= 0; \end{aligned} \tag{3-31}$$

therefore

$$B_2{}^0 = 0.$$

Also from (3-5) and (3-30)

$$\begin{aligned} A_4{}^0 &= -\frac{4\pi Ze^2}{9a^5}\left(\frac{3}{16}\pi^{1/2}\right)\left\{6(3) + 3\left[35\left(\frac{2}{3}\right)^2 - 30\left(\frac{2}{3}\right) + 3\right]\right. \\ &\quad \left. + 3\left[35\left(\frac{2}{3}\right)^2 - 30\left(\frac{2}{3}\right) + 3\right]\right\} \\ &= -\frac{4\pi Ze^2}{9a^5}\left(-\frac{44}{3}\right) = \frac{176\pi Ze^2}{27a^5}. \end{aligned} \tag{3-32}$$

Hence, since $K_4{}^0 = (3/16)\pi^{1/2}$,

$$B_4{}^0 = -\frac{7\pi^2 Ze^2}{48a^5} \beta_J(1-\theta_j) \langle r^4 \rangle . \tag{3-33}$$

Other values for $B_n{}^m$ are computed in an analogous fashion. The needed spherical harmonics are tabulated in many places. For example, Hutchings [1] gives the ones most commonly involved in crystal field calculations.

C. MAGNETIC AND THERMAL PROPERTIES; VAN VLECK PARAMAGNETISM

In a rare earth system the ions are, of course, distributed over the several states in accordance with the Boltzmann expression. The populations of the states vary with changing temperature. These variations significantly affect a number of macroscopic properties of rare earth systems—their bulk magnetic characteristics, their thermal properties such as the specific heat, and their transport properties such as the electrical conductivity. For example, if, as temperature is lowered, the ions settle into states of low or vanishing moment, electrical conductivity substantially improves due to the disappearance or suppression of spin-disorder, i.e., paramagnetic, scattering [11,12].

The influence of the crystal field interaction on the temperature dependence of susceptibility (χ), discussed many years ago by Penney and Schlapp [7], is made evident from the fundamental Van Vleck equation [13]

$$\chi = N_0/H \sum_i \mu_i \exp(-E_i/kT) \;/\; \sum_i \exp(-E_i/kT) . \tag{3-34}$$

Here μ_i are the moments of the various states of the ions ($\mu_i = -\partial E_i/\partial H$), E_i their energies, and N_0 the number of ions. For $kT > E_i$ Eq. (3-34) leads to reciprocal susceptibility linearly dependent on temperature, but usually with a nonvanishing intercept on the temperature axis—i.e., it leads to Curie–Weiss behavior. At low temperatures such that the condition $kT > E_i$ is no longer fulfilled significant deviations from linearity in the χ^{-1} versus T curve may occur [7]. In the special case that the ground state of the ion is a singlet, i.e., μ_0 in Eq. (3-28) vanishes, expression (3-34) reduces to

$$\chi = N_0\alpha_0 , \tag{3-35}$$

where α_0 is the polarizability of the ion in the singlet ground state.

There is thus the onset of temperature independent paramagnetism at low temperatures and the material is said to become a Van Vleck paramagnet. In this case the paramagnetic entropy disappears on cooling by the ions transforming into nonmagnetic entities rather than by the development of magnetic order. The conditions for this are that (1) the ground state must be nondegenerate (i.e., $\mu_0 = 0$) and (2) the exchange interaction must be weaker than some threshold value. If, however, exchange exceeds a threshold value, magnetic ordering occurs via a mechanism termed by Cooper [14] the "bootstrap" process. This process, which is comparatively rare, is treated in detail elsewhere [14,15]. $PrAl_2$ is discussed briefly in Chapter 4 as an example of bootstrap ferromagnetism.

In the preceding paragraphs we have drawn attention to the manner in which the crystal field interaction significantly influences the low temperature magnetic characteristics of an assemblage of rare earth ions for the case in which the ground state is a singlet. We shall now show that it also has a profound influence on an assemblage of ions in which the ground state is not a singlet. Consider, for example, Nd^{3+} in $NdNi_2$. In the cubic field which exists in this compound the tenfold degenerate ground state multiplet is decomposed into a Γ_6 state, which is doubly degenerate, and two Γ_8 states, which are quadruply degenerate. As is pointed out in Chapter 9, the Γ_6 state is lowest lying. Evidence presented in that chapter indicates that the Γ_8 states are well separated from the ground state so that upon cooling the Nd^{3+} ions settle into the Γ_6 state before magnetic ordering occurs. The moments of the two components of the Γ_6 state are [7] $\pm 1.33\ \mu_B + \alpha H$, where α is the polarizability.* Magnetic ordering then occurs within the Γ_6 state instead of within the entire J manifold. Consequently, the entropy associated with the destruction of magnetic order is $R \ln 2$ instead of $R \ln(2J+1) = R \ln 10$. In this case the remainder of magnetic entropy, $R \ln 5$, is acquired when the sample is heated above the Curie temperature, 10°K for $NdNi_2$. The extra magnetic entropy is associated with the excitation from the Γ_6 to the Γ_8 states and the associated heat capacity anomaly is Schottky-type instead of λ-type.

The effect of the crystal field interaction on the thermal properties of rare earth systems, particularly their heat capacity behavior, is quite simple to visualize for the case in which magnetic ordering does not occur—the Van Vleck paramagnet. Near 0°K the ions populate only the lowest-lying state. Excitation into higher states occurs as temperature is increased introducing a significant contribution to the heat capacity.

* $\alpha = 2.4 \times 10^{-3}\ \mu_B$/kOe for an overall crystal field splitting of 430°K, the value deduced for $NdNi_2$ [16].

This contribution (C_c) is given by the expression

$$C_c = RT^2\, d^2 \ln Q/dT^2 \qquad (3\text{-}36)$$

where Q is the partition function. The process is not cooperative and the heat capacity is then Schottky-type.

We note that the crystal calculations are made in terms of two (or more) parameters. For example, in the work of Lea *et al.*; these parameters were W and x, which are related to B_4^0 and B_6^0 [Eqs. (3-22a) and (3-22b)]. As indicated earlier the crystal field parameters can not be be evaluated sufficiently well from first principles to be useful but are instead obtained from appropriate experiments. Heat capacity and magnetic susceptibility measurements through Eqs. (3-34) and (3-36) can serve to establish the value of these parameters. Numerous examples of this are given in the later chapters. Conductivity measurements also provide useful information in this respect. This approach is outside the main themes of this monograph, however, and will not be discussed, except for the brief account of its use in establishing the ground state of $CeAl_2$ presented in the following chapter.

One final aspect of the crystal field interaction merits comment. Attention in Chapter 10 is drawn to the significance of anisotropy in applications of rare earth intermetallics. Magnetic anisotropy is primarily a consequence of the crystal field interaction.* Since Gd^{3+} and Eu^{2+} are in S states in which the crystal field interaction is vanishingly small, or almost so, we expect systems of these ions to be free, or nearly free, of magnetic anisotropy.

REFERENCES

1. M. T. Hutchings, *Solid State Phys.* **16,** 227 (1966). See this review article for numerous references to many works dealing with the influence of the crystalline electric field on the behavior of systems of paramagnetic ions.
2. B. Bleaney and K. W. H. Stevens, *Rep. Progr. Phys.* **16,** 108 (1953).
3. K. W. H. Stevens, *Proc. Phys. Soc., London.* **A65,** 209 (1952).
4. B. Bleaney, *Proc. Roy. Soc.* **A276,** 39 (1963).
5. E. Segal and W. E. Wallace, *J. Solid State Chem.* **2,** 347 (1970).
6. K. R. Lea, M. J. M. Leask, and W. P. Wolf, *J. Phys. Chem. Solids* **23,** 1381 (1962).
7. W. G. Penney and R. Schlapp, *Phys. Rev.* **41,** 194 (1932).
8. D. P. Schumacher and C. A. Hollingsworth, *J. Phys. Chem. Solids* **27,** 749 (1965).
9. E. Segal and W. E. Wallace, *J. Solid State Chem.* **6,** 99 (1973).
10. K. H. Mader, E. Segal, and W. E. Wallace, *J. Phys. Chem. Solids* **30,** 1 (1969).
11. V. U. S. Rao and W. E. Wallace, *Phys. Rev.* **2,** 4613 (1970).
12. H. J. van Daal and K. H. J. Buschow, *Solid State Commun.* **7,** 217 (1968).

*Anisotropic exchange may contribute slightly to the anisotropy.

13. J. H. Van Vleck, "The Theory of Magnetic and Electrical Susceptibilities," p. 181. Oxford Univ. Press, London and New York, 1932.
14. See Ref. 41, Chapter 4.
15. W. E. Wallace, F. Kissell, E. Segal, and R. S. Craig, *J. Phys. Chem. Solids* **30,** 13 (1969).
16. W. E. Wallace, R. S. Craig, A. Thompson, C. Deenadas, M. Dixon, M. Aoyagi, and N. Marzouk, *Les Éléments des Terres Rares, Coll. Int. C.N.R.S.* No. 180, 427 (1970).

Part II

RARE EARTH–NONTRANSITION METAL SYSTEMS

Chapter 4

COMPOUNDS WITH ALUMINUM

Intermetallics containing the rare earths in chemical union with aluminum have been very extensively studied since they form readily and possess many interesting features. Of all the known rare earth intermetallics they have received the greatest attention and are hence the best characterized with the possible exception of rare earth–cobalt systems.

The known compounds are listed in Table 4-1, together with their structures [1]. Lack of a structural entry indicates that the potential

TABLE 4-1

Structures of Rare Earth–Aluminum Compounds

R	R_3Al	R_2Al	RAl	R_3Al_2	RAl_2	RAl_3	R_3Al_{11}	RAl_4
La	DO_{19}		CeAl		C15	DO_{19}	La_3Al_{11}	
Ce	DO_{19},Ll_2		CeAl		C15	DO_{19}	La_3Al_{11}	
Pr	DO_{19}	C23	ErAl		C15	DO_{19}	La_3Al_{11}	
Nd	DO_{19}	C23	ErAl		C15	DO_{19}	La_3Al_{11}	
Sm		C23	ErAl		C15	DO_{19}		
Eu					C15			Dl_3
Gd		C23	ErAl	Zr_3Al_2	C15	DO_{19}		
Tb		C23	ErAl	Zr_3Al_2	C15	$BaPb_3$		
Dy		C23	ErAl	Zr_3Al_2	C15	$HoAl_3$		
Ho		C23	ErAl	Zr_3Al_2	C15	$HoAl_3$		
Er		C23	ErAl	Zr_3Al_2	C15	Ll_2		
Tm				Zr_2Al_3	C15	Ll_2		
Yb					C15	Ll_2		
Lu					C15	Ll_2		
Ref.	[2,3]	[2–4]	[5,6]	[7]	[8]	[9]	[10]	[10]

compound has not yet been prepared. The ErAl, CeAl, and Zr_3Al_2 structures are characteristic structures which have not as yet received strukturbericht notations. The structures of the two equiatomic compounds contain eight formula units per cell and are of orthorhombic symmetry. The structure of Zr_3Al is tetragonal with four formula units per cell.

A. THE RAl_2 COMPOUNDS

1. Binary Systems

a. General Magnetic Features

All of the lanthanides combine with aluminum to produce this compound. This is the so-called cubic Laves phase structure, named for the Swiss crystallographer who performed early work dealing with the prototype compound $MgCu_2$. $CeAl_2$ was formed and characterized by European crystallographers [11, 12] in the early 1940's. However, the existence of the entire RAl_2 series was not established until two decades later, in the work of Wernick and Geller [8]. Bulk magnetic measurements were made on eleven of the fourteen compounds listed in Table 4-1 by Williams *et al.* [13] in 1962. The remainder of the compounds were studied by Wallace and his associates, who also provided more detailed information on several of the compounds studied earlier [14,15]. The essential aspects of the magnetic behavior of this group of compounds are given in Table 4-2.

Nereson *et al.* [16,17] studied $PrAl_2$, $NdAl_2$, $DyAl_2$, and $ErAl_2$ by neutron diffraction techniques. $ErAl_2$ was also studied by Will [18] by this technique. Results for $ErAl_2$ in Table 4-2 come from these sources: bulk magnetic measurements by Williams *et al.* [13] and Swift and Wallace [15] led to a considerably lower value for the saturation moment ($\sim 7\ \mu_B$/f.u.), probably due to the difficulty of saturating the Er^{+3} ion. Ofer *et al.* [20] used Dy Mössbauer spectroscopy to establish nuclear hyperfine field for Dy in $DyAl_2$. Their measurements led to a Dy moment of 10.2 μ_B, as compared with the free ion value, $gJ = 10\ \mu_B$. Curie temperatures are taken from the work of Nereson *et al.*; bulk measurements by Swift and Wallace [15] made in a 20 kOe field gave results 5 to 10° higher than those listed in Table 4-2.

La is nonmagnetic and hence $LaAl_2$ displays only the Pauli paramagnetism associated with the polarization of the conduction electrons by the applied field. It is of interest to note that, excluding $LaAl_2$ and $CeAl_2$ from consideration, all of the RAl_2 compounds are ferromagnetic except $EuAl_2$. The antiferromagnetism of $EuAl_2$, which was established by Mader and Wallace [14] in 1968, merits special comment.

TABLE 4-2

Magnetic Characteristics of RAl_2

	Paramagnetic state			Ordered states				
R	μ_{eff} (μ_B/f.u.)	θ (°K)	Type	μ_{or} (μ_B/f.u.)	T_C (°K)	T_N (°K)	Type	Ref.
La			P					[14]
Ce	2.54	−33	C–W,?			3	AF	[15,19]
Pr	3.48	32	C–W	2.62,2.94	34		F	[15,16]
Nd	3.59	70	C–W	2.27,2.5	65		F	[15,17]
Sm				0.1	122		F	[13]
Eu	7.88	0	C			15	AF	[14]
Gd	7.94	168	C–W	7.10	182		F	[15]
Tb	9.81	110	C–W	8.60	121		F	[15]
Dy	10.7	64	C–W	9.62,9.5,10.2	62		F	[15,17,20]
Ho	10.7	30	C–W	9.16	42		F	[15]
Er	9.5	14	C–W	8.3,8.8	14		F	[15,16,18]
Tm				4.69	8		F	[13]
Yb			P					[13]
Y			P					[21]

The differing behavior of this compound is a consequence of the fact that Eu in $EuAl_2$ exists in divalent state, whereas the other magnetic rare earths exist in the RAl_2 compounds as trivalent ions. Trivalent europium exists in a $4f^6$ configuration. The special stability of the half filled f shell causes Eu^{+3} to capture an electron from the conduction band to form Eu^{+2}. The divalency of Eu in $EuAl_2$ is clearly indicated by its lattice parameter shown in Fig. 4-1. The lattice parameters for the

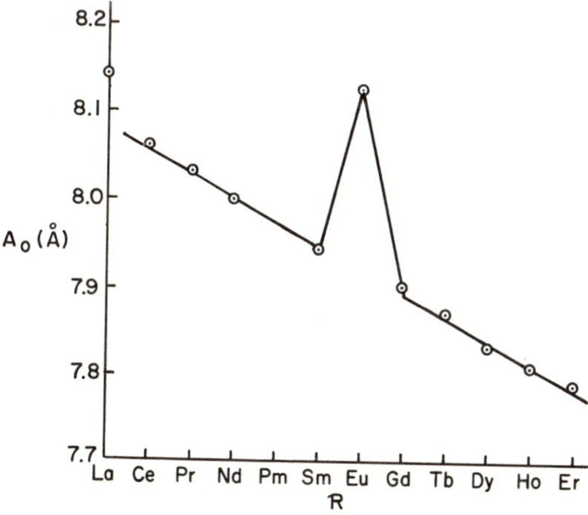

Fig. 4-1. Lattice spacing of RAl_2 compounds.

RAl_2 compounds decrease with increasing atomic number of R because of the well-known lanthanide contraction. The unit cell of $EuAl_2$ is substantially enlarged in comparison with those of other members of the series, strongly suggesting that Eu is in a lower valence state. Divalency is confirmed by ESR measurements by Peter and Matthias [22].

As indicated above in Chapter 2 interactions in lanthanide-containing systems are electron concentration dependent, leading us to the premise that the type of magnetic structure (i.e., ferromagnetic versus antiferromagnetic) which develops is controlled by the electron concentration. This idea was developed theoretically by Mattis [23,24] and has received experimental confirmation by several investigators [14,25–27]. For example, Sekizawa and Yasukochi found that antiferromagnetic GdAg (e.c. = 2) could be transformed into a ferromagnetic material by replacing Ag with In to give $GdAg_{0.5}In_{0.5}$ (e.c. = 2.5). We thus ascribe the antiferromagnetism in $EuAl_2$ to its reduced e.c. as compared to the $R^{3+}Al_2$ materials. Mader and Wallace confirmed [14] the correctness of this postulate by forming the ternaries $Eu_{1-x}La_xAl_2$. They observed that the ternaries were AF for $x < 0.2$ but became ferromagnetic for $x > 0.6$. $Eu_{0.6}La_{0.4}Al_2$ failed to order at 4°K. Swift and Wallace [28] (*vide infra*) have successfully analyzed these experimental results in terms of the RKKY formalism [Eq. (2-3)].

Antiferromagnetism of $CeAl_2$ was first inferred from bulk magnetization measurements by Swift and Wallace [15]. The magnetization-field curve is linear at 4°K or above but below 3°K it becomes concave upward, indicative of the onset of antiferromagnetism. Hill and da Silva [19] shortly after confirmed this from heat capacity measurements which gave a λ-type thermal anomaly at about 3.5°K. However, Buschow and Van Daal [29] have shown that $CeAl_2$ exhibits a resistance minimum at about 12°K, suggesting the establishment of the Kondo state for the Ce^{3+} ion below 10°K. If the resistance minimum in this material is indeed due to the Kondo effect, we have an example in $CeAl_2$ of a concentrated Kondo system. Since the magnetic and thermal properties of such systems are as yet unknown, the conclusion that the transition at 3.5°K is a normal paramagnetic-to-antiferromagnetic transition must be regarded as tentative. The importance of the Kondo phenomenon in this case is possibly great and may even be dominant.

We note that the antiferromagnetism of $CeAl_2$ is not ascribed to electron concentration effects such as described above for $EuAl_2$. It seems instead due to the weakness of the RKKY interaction in this material. The de Gennes factor for Ce^{3+} is only about 1% of that of Gd^{3+}. Hence, ordering via the RKKY interaction might be expected to occur at one-hundredth the $GdAl_2$ ordering temperature or at about 2°K. Other interactions (possibly dipole–dipole) are probably dominant, giving rise to a magnetic structure differing in this case from that for the other $R^{3+}Al_2$ compounds.

b. Nuclear Magnetic Resonance and the Interaction Mechanism

NMR measurements on ^{27}Al have led to useful information about the magnetic behavior of the RAl_2 compounds. In a metal which exhibits Pauli paramagnetism resonance for a magnetic nucleus occurs at a lower field than for the same nucleus in an insulator; this is the so-called Knight (or chemical) shift. The nucleus experiences an internal field which is generated by the Fermi contact interaction with the polarized conduction electrons, the polarization having been produced by the applied external field. The Knight shift (K) in this case is positive and invariant with temperature since the Pauli susceptibility is temperature-independent. In 1960, Jaccarino *et al.* [30] found that K for ^{27}Al in the RAl_2 compounds was an order of magnitude larger than the Knight shift for a Pauli paramagnet, was strongly temperature-dependent, and its sign varied according to whether R was a light or heavy lanthanide. They observed K to be positive for light lanthanides and negative for heavy lanthanides. These observations are consistent with the conduction electron polarization shown in Fig. 4-2.

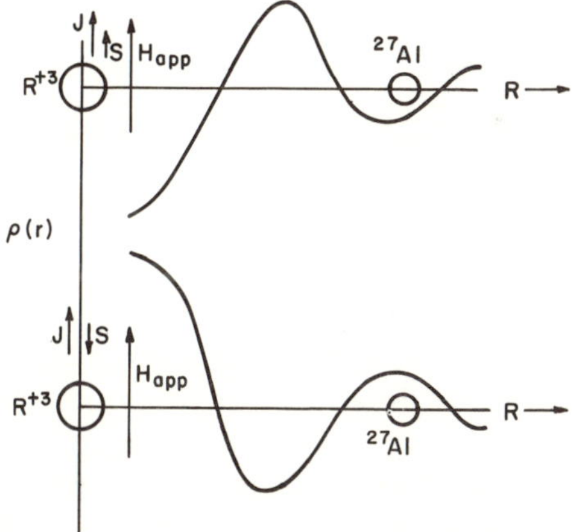

Fig. 4-2. Schematic representation of the conduction electron polarization near a rare earth ion as deduced from NMR measurements on RAl_2 compounds—top, heavy rare earth; bottom, light rare earth. The ^{27}Al nucleus experiences a field due to the conduction electrons which leads to $K < 0$ (top) and $K > 0$ (bottom).

Assuming a Hamiltonian of the form

$$\mathcal{H} = -\Gamma \mathbf{S} \cdot \mathbf{s}, \tag{2-1}$$

several authors were able to show [30–33] that

$$K = K_0[1 + J(g-1)\chi_f/2g\mu_B^2], \tag{4-1}$$

where g is the Landé factor, χ_f is the susceptibility of the 4f shell, K_0 is the contribution to K of the Pauli susceptibility, and J is given as

$$J = 6\pi Z\Gamma \sum F(x), \tag{4-2}$$

in which the symbols have the same significance as in Eq. (2-3), except that in Eq. (2-3) R is measured from the rare earth ion whereas in Eq. (4-2) it is measured from the Al nucleus. The validity of Eq. (4-1) is restricted [34] to cases in which the Van Vleck susceptibility is negligible in comparison with the "normal," i.e., temperature-dependent part of χ_f. It is invalid when these two contributions become comparable or the Van Vleck susceptibility becomes dominant.

Buschow *et al.* [35] have used Eqs. (2-3) and (4-1) to evaluate Γ and k_F, the radius of the effective Fermi sphere. Using k_F^0, the radius of the Fermi sphere computed assuming each atom in the RAl_2 compounds contributes three electrons to the conduction band, θ obtained from Eq. (2-3) is negative. This would imply predominance of antiferromagnetic interactions. Since these RAl_2 compounds, with the exception of $CeAl_2$ and $EuAl_2$, are ferromagnetic, Buschow *et al.* [35] adjusted k_F^0 to k_F where $k_F/k_F^0 \approx 0.9$ to achieve a positive θ. Using Eqs. (2-3) and (4-1) to solve simultaneously for k_F and Γ, they find $k_F = 0.94\, k_F^0$ and $\Gamma = -0.9$ eV for $GdAl_2$. Similar NMR results were obtained and analyses made by Van Diepen, de Wijn, and Buschow for RAl, RAl_3, and R_3Al_{11} compounds [36–38]. Interestingly, two K_{Al} values are observed in R_3Al_{11} corresponding to two crystallographically inequivalent kinds of Al. This observation constitutes rather strong evidence that the spin polarization of conduction electrons around a rare earth ion is, indeed, oscillatory. The NMR work on the RAl_2 compounds, in general, provides substantial support from the coupling mechanism set forth in Chapter 2.

The simplicity of the mechanism leading to the expression for the Weiss constant, θ [Eq. (2-3)] is evident from Fig. 4-2. The rare earth neighbor is only slightly farther from the central ion than that Al nucleus. It (i.e., the neighboring R^{3+}) is situated where the amplitude of the spin density wave is still negative and it couples antiparallel (since Γ is negative) to the net spin of the conduction electrons and hence parallel to the central ion.

It is appropriate at this point to anticipate subsequent discussion of ternary systems of the type $R_{1-x}R_x'Al_2$. We have three possible R,R′ combinations: R and R′ are both heavy, R and R′ are both light, or one is heavy and the other is light. In view of the difference of LS coupling in heavy and light R's we see immediately (from Fig. 4-2 and the comments in the preceding paragraph) that ferromagnetic coupling should occur if both are heavy and both light and ferrimagnetic coupling if one is heavy and the other is light. This is observed experimentally as is pointed out below.

c. $CeAl_2$

The special features of this compound have been adumbrated above. We now should like to consider this compound in greater length. White *et al.* [39] observed that χ^{-1} versus T curves for $CeAl_2$ and the ternaries $Ce_{1-x}La_xAl_2$ were linear above 200°K with a slope close to that expected for an assemblage of free Ce^{3+} ions but below 200°K the slope increased. These observations were confirmed by Swift and Wallace [15]. White *et al.* ascribed the deviations to the influence of the cubic crystal field which decomposes the six-fold degenerate ground state of the Ce^{3+} ion into a Γ_7 (doublet) and a Γ_8 (quartet) state. An assemblage of Ce^{3+} behaves differently, particularly at low temperatures, depending upon whether the Γ_7 or Γ_8 is lower lying. Considerable effort has been expended in an attempt to establish which is the ground state. We now review the experimental evidence—susceptibility [15,39], heat capacity [19], and resistivity [40] behavior—which provides information as to the nature of the ground state. We show that the weight of evidence indicates Γ_7 to be the ground state, which is at variance with expectations based on point charge calculations.

White *et al.* [39] employed the point charge model to obtain the potential (V) of a 4f electron at a position x,y,z near the Ce nucleus. Their expression, except for an unimportant additive constant, is

$$V = (3.89\ Z_R/R_R{}^4 - 3.76\ Z_{Al}/R_{Al}{}^4)\ F(x,y,z), \tag{4-3}$$

where $F(x,y,z) = (x^4 + y^4 + z^4 - 3r^4/5)$. Inasmuch as R_R and R_{Al} are almost equal the contributions from R and Al in Eq. (4-3) nearly cancel. Actually $R_R = 1.045\ R_{Al}$ for $CeAl_2$ and hence if both ions are taken to be tripositive, $V < 0$. With V negative, it can be shown using the calculational procedures outlined in Chapter 3 that Γ_8 is lower lying.

The susceptibility (χ) of an assemblage of cerium ions in a cubic field is given [39] by the expression

$$\chi = \chi_0\ f(y), \tag{4-4}$$

where χ_0 is the Curie law susceptibility for Ce^{3+} ions and $f(y)$ is a function which is dependent on the energy difference (E) of the Γ_7 and Γ_8 states. In the case that Γ_7 is the ground state

$$f(y) = [5/21 + 26/21\, e^{-y} + (32/21\, y)(1 - e^{-y})] \;/\; (1 + 2e^{-y}), \quad (4\text{-}5)$$

where $y = E/kT$. Using this expression, which can readily be derived from Eq. (3-34) employing μ_i values from Schumacher and Hollingsworth [Chap. 3, Ref. 8], a good accounting for the susceptibility behavior was made* with $E = 200°K$. This did not result if Γ_8 was taken as the ground state.

As regards heat capacity results for $CeAl_2$, there is a λ-type thermal anomaly peaking at 3.5°K and a Schottky-type anomaly [42] at 40°K. Hill and da Silva report [19] an entropy associated with the lower anomaly closely approximating $R \ln 2$, which is as expected if disordering occurs within the Γ_7 state. The higher temperature excess† is associated with excitation into the Γ_8 state. The heat capacity data at higher temperatures do not accord well (Fig. 4-3) with either the Γ_7 or Γ_8 state being the ground state. The magnetic entropy is clearly introduced over a wider range of temperature than expected on the "single ion picture," represented by curves Q and D. It is believed that the lack of accord between computed and observed results is due to the interaction between the localized moment and the conduction electrons, the same interaction which gives rise to the Kondo effect in this material. There is great difficulty in separating this effect from the other effects—namely ordering and crystal field excitation.

As noted above the resistivity of $CeAl_2$ is also in accord with the notion that the doublet is the ground state. Rao and Wallace [40] have analyzed the spin-disorder scattering process in the first Born approximation taking due cognizance of the special features introduced by the crystal field interaction. The expression obtained for $\rho_s(T)$, the spin-disorder resistivity, is

$$\rho_s(T) = [(3\pi Nm)/(\hbar^2 e^2 E_F)\, \Gamma^2 (g-1)^2]\, [2a^2(3a^2-1)(P_1+P_2-P_1f_1-P_2f_2) + 9/4\, P_1 + 11\, P_2 + 13/2\, (P_f f_1 + P_2 f_2)]. \quad (4\text{-}6)$$

For the definitions of the quantities appearing in Eq. (4-6) the reader is referred to Rao and Wallace [40]. We only need to note that P_1, P_2, f_1 and f_2 are functions of E/kT, E being the Γ_8, Γ_7 energy difference. The

*The region for $T < 10°K$ did not yield to this analysis presumably because of the onset of ordering or the Kondo process or both.

†The difference between the heat capacity of $CeAl_2$ and $LaAl_2$ is plotted. $LaAl_2$ is taken to represent the lattice and electronic contributions.

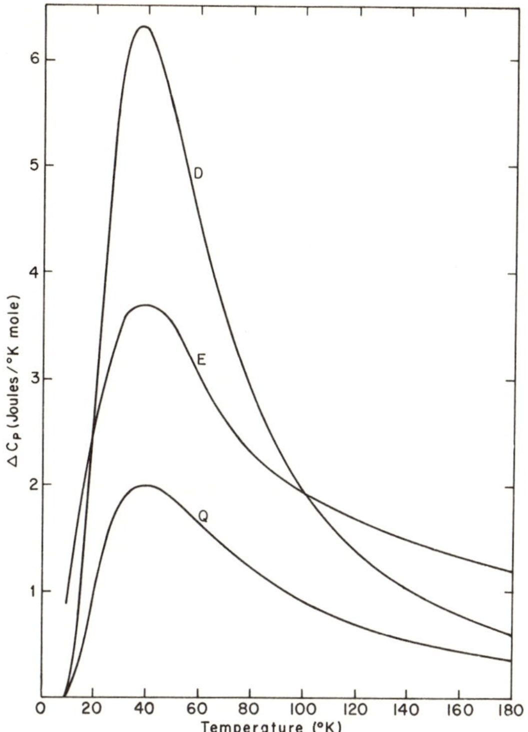

Fig. 4-3. Crystal field heat capacity $[C_{cf} = C(CeAl_2) - C(LaAl_2)]$ versus temperature for $CeAl_2$. D and Q are calculated curves for the ground state being the doublet (Γ_7) and quartet (Γ_8), respectively. *E* represents experiment [42, 43].

qualitative significance of the treatment is that at high temperatures the paramagnetic scattering is controlled by the free ion moment, 2.56 μ_B in case of Ce^{3+}. At low temperatures the ions settle into the ground crystal field state in which there is reduction of the moment and hence of the paramagnetic scattering. The shape of the curve for ρ_s versus T is determined by E and the nature of the ground state. Plots of Eq. (4-6) for Γ_7 and Γ_8 as the ground state together with the results of experiment are shown in Fig. 4-4. Clearly experiment favors a doublet ground state.

The situation for $CeAl_2$ is then as follows: Present experimental evidence clearly favors the Γ_7 state as lowest lying for Ce^{3+} in $CeAl_2$. However, the behavior is not fully in accord with that expected for a system of weakly interacting tripositive cerium ions. There is probably rather strong interaction with the conduction electrons which significantly influences the behavior of χ, C_p and ρ of the system.

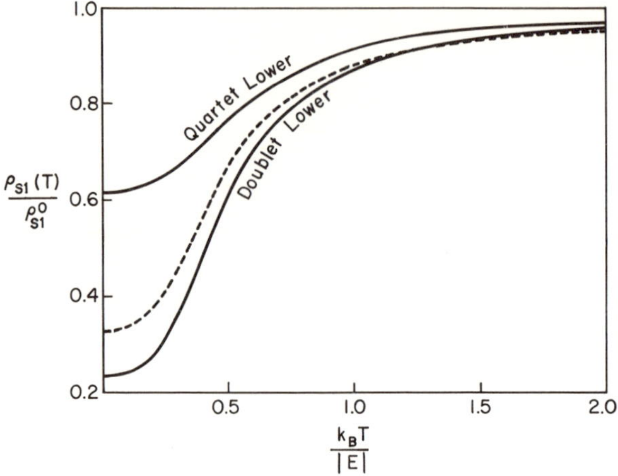

Fig. 4-4. Theoretical curves for spin-disorder resistivity for a cubic crystal field for doublet (Γ_7) and quartet (Γ_8) lower [40]. The dashed curve was estimated from the experimental data for $CeAl_2$. $\rho_s(T)$ is obtained from Eq. (4-6). ρ_0 is the high temperature limiting value of ρ_s and corresponds to the situation in which crystal field effects are unimportant.

If we return to Eq. (4-3), we see that within the framework of the point charge model, $V > 0$ (needed for Γ_7 to be the ground state) can occur only if Al is given a formal charge equal to or less than 2.5. It is obvious that the point charge model is a very poor approximation in a metallic conductor; the conduction electrons act to shield the central ion from the electrostatic influence of its neighbors. However, if the conduction electrons aggregate preferentially in the regions between Ce and Al, they would contribute to the potential in such a way as to reduce the Al contribution perhaps even to the point of reversing its sign. In this way the low (or possibly negative) formal charge of Al could arise. This picture of the conduction electron behavior is consistent with the viewpoint of metallic cohesion advocated by Pauling who has regarded the metallic bond as differing in no fundamental way from the usual covalent bond [44]. The latter occurs because of collection of valence electrons in the intervening space between the bonded nuclei, which is a position of low potential energy.

In conclusion, we regard the potential energy (V) of the 4f electrons as positive and the splitting pattern for Ce^{3+} in $CeAl_2$ to be that shown in Fig. 4-5. The splitting patterns for the other R^{3+} ions for $V > 0$ are also given in Fig. 4-5.

d. $PrAl_2$, an Example of Bootstrap Magnetic Ordering

From the preceding discussion and Fig. 4-5 we see that the ground

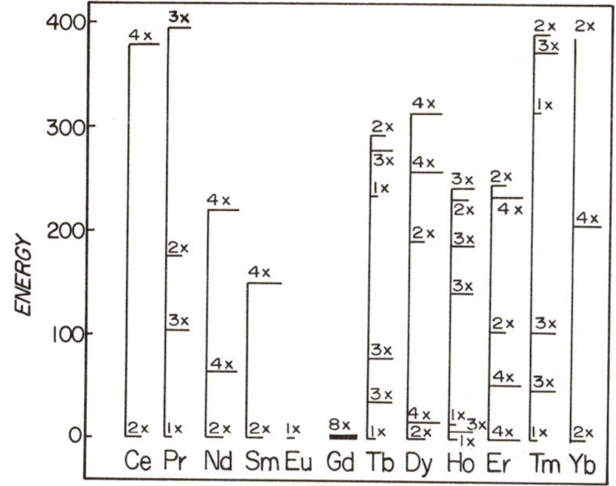

Fig. 4-5. Splitting pattern for R^{3+} ions in a pure fourth-order crystal field with $V > 0$. [V is the potential energy of the 4f electrons and is given by Eq. (3-1).] The diagram is schematic in the sense that the overall splitting is arbitrary.

state of $PrAl_2$ is a nonmagnetic singlet. The wave function of this (Γ_1) state is

$$\Gamma_1 = a|-4\rangle + b|0\rangle + a|4\rangle. \tag{4-7}$$

In the absence of exchange this material would lose its entropy on cooling by deexcitation of the crystal field states; its heat capacity would show a Schottky-type thermal anomaly associated with repopulation of its crystal field states and it would exhibit Van Vleck paramagnetism at low temperatures. However, as remarked in Chapter 3, magnetic order can grow out of a singlet state if exchange exceeds a certain threshold value. This is the case for $PrAl_2$; it exhibits bootstrap ferromagnetism, with a Curie temperature of 34°K. We describe the treatment [21] of the process for this material in this section.

The moments of the crystal field states Γ_i, $i = 1$ to 4, are given in the Table 4-3. These are for $x = 0.80$ and $E_c = 25$°K, which seem to be reasonable values for the two parameters involved in the Hamiltonian. However, the characteristics of the system can be shown to be rather insensitive to the values of x and E_c. The magnetization M in ergs/Oe mole is given by the expression

$$M = \alpha \sum \mu_i \exp(-T_i/T) Q^{-1}, \tag{4-8}$$

where α is a unit conversion constant equal to 1.786×10^4 erg Oe^{-1} $mole^{-1}$, μ_i is the moment in units of gJ, T_i is the energy in °K of the ith

TABLE 4-3

Magnetic Moments[a] *in Units of* gJ *for* Pr^{3+} *in* $PrAl_2$
$\mu_i = a_i + b_i H$

	a_i	$10^6\ b_i$	Energy (°K)
Γ_5	−0.625	−6.20	25
	0.000	−3.98	
	0.625	−7.18	
Γ_3	0.000	−13.44	15.5
	0.000	−13.56	
Γ_4	−0.125	19.90	9.0
	0.000	−15.38	
	0.125	19.78	
Γ_1	0.000	19.40	0

[a] From Mader *et al.* [21].

crystal field state, and Q is the partition function. At low or moderate fields μ_i is linear with H, i.e.,

$$\mu_i = a_i + b_i(H_0 + \lambda M). \tag{4-9}$$

From (4-8) and (4-9) we obtain the expression

$$M = \alpha(A + BH_0)/(Q - \alpha B\lambda), \tag{4-10}$$

where $A = \Sigma_i a_i \exp(-T_i/T)$ and $B = \Sigma b_i \exp(-T_i/T)$. Equation (4-8) may be regarded as the counterpart of

$$M = Ng\mu JB_J(x), \tag{4-11}$$

($x = g\mu_B HJ/kT$) which occurs in the familiar Weiss theory of ferromagnetism. M in this latter treatment, as will be recalled, is evaluated by noting the value of M at the intersection of M versus H from Eq. (4-11) and the linear relationship $M = xkT/\lambda gJ\ \mu_B$; for $T > T_C$ the intersection occurs at the origin and $M = 0$. As temperature falls the slope of the linear relationship decreases and below T_C the intersection occurs at $M > 0$. The condition for T_C is that the slopes of the linear plot and for Eq. (4-11) be equal at $M = 0$. Equation (4-10) is handled in an analogous manner. We represent the quantity $\alpha \Sigma \mu_i \exp(-T_i/T)/Q$ as F and evaluate and plot F versus M for a variety of temperatures. On the same diagram we put a plot of $F = M$. The intersection of this straight line with plots of F versus M gives the equilibrium value of M at the temperature in question. The slope of the linear plot is obviously unity. T_C is found* by invoking the condition $(dF/dM)_{M=0} = 1$. This leads to the expression

$$T_C = (\alpha^2\lambda/Nk)\,\frac{\Sigma a_i^2 \exp(-T_i/T_C) - A^2}{Q - \alpha\lambda B}\,. \tag{4-12}$$

Equation (4-12) is inconvenient to use since T_C is also involved in a complex way in the expression to the right of the equality sign. The temperature may be obtained more conveniently and with sufficient reliability from Eq. (4-10) by ascertaining the temperature at which M begins to diverge (denominator vanishes). T_C is readily obtained in this way using plots of Q and $\alpha B\lambda$ versus temperature and finding the temperature at which these two quantities become equal. Mader *et al.* [21] estimated λ from θ, the Weiss constant for $PrAl_2$, and with Eq. (4-12) calculated T_C to be 39°K, in quite good agreement with experiment, 34°K.

Equation (4-10) shows that there is a cutoff value for λ below which ordering does not occur; $\lambda_c \approx 2.8$ Oe^2 erg^{-1} mole^{-1} for Pr^{3+} ions with an overall crystal field splitting in the range of 25°C. For $PrAl_2$, λ can be decreased by diluting with YAl_2 or $LaAl_2$ and T_C reduced accordingly. Values of T_C computed by Mader, Segal, and Wallace are shown in Fig. 4-6 for two functional dependencies† of λ on x in the formula $Pr_xM_{1-x}Al_2$ (M = Y or La) – $\lambda \propto x$ or $\propto x^{2/3}$. It is to be noted that $Pr_{0.05}Y_{0.95}Al_2$ does not order at low temperatures.

The procedure discussed in this section while conveying the essential features of the bootstrap process suffers from the limitations inherent in the molecular field theory. Cooper [41], who seems to have invented the term, has given more thorough-going treatments of the bootstrap process employing more elegant statistical theories than the molecular field approach—random phase approximation, two site correlation approximation, etc. The reader who wishes to delve deeper into the bootstrap process should consult his numerous publications.

e. *Heat Capacities and Entropies*

In contrast with the extensive work on the magnetic behavior of the RAl_2 compounds comparatively little work has been done on their heat

*The author is indebted to Professor R. S. Craig for drawing attention to this way of handling Eqs. (4-8) and (4-10).

†See Mader *et al.* [21] for the reasons underlying those two postulates for the dependence of λ on x.

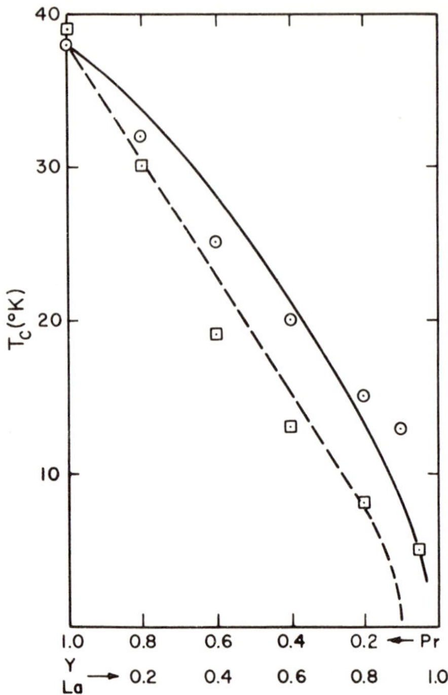

Fig. 4-6. Variation of T_C with x for $Pr_xM_{1-x}Al_2$ ternaries. M = Y (O) and M = La (□). Points are experimental; the full line is computed from Eq. (4-12), assuming λ varies as $x^{2/3}$, the dashed line assuming λ depends linearly on x [21].

capacities and associated thermodynamic characteristics. The work of Deenadas *et al.* [42,43] in the range 8–300°K, that of Hill and da Silva [19] on $CeAl_2$ (0.5–15°K) and of Hungsberg [45] on Y, La, and Lu compounds (2.5–20°K) seem to be the only studies that have been made to date.

Deenadas *et al.* [43] made measurements on the five RAl_2 compounds with R = La, Ce, Pr, Nd, and Gd. Results for $CeAl_2$ [19,43] have been discussed above; those for the other four compounds are shown in Figs. 4-7 to 4-9. The differential heat capacity of $PrAl_2$ and $GdAl_2$ with respect to their nonmagnetic isostructural counterpart, $LaAl_2$, is shown in Figs. 4-10 and 4-11. Results for $NdAl_2$ and $PrAl_2$ are very similar; they are not the simple λ type but instead have a pronounced shoulder. This behavior is probably due to the fact that the exchange and crystal field interactions are in these instances comparable. The simple situation described in Chapter 3 for $NdNi_2$ in which magnetic disordering and the crystal field excitation are clearly resolved appears not to occur in $NdAl_2$ because exchange and the crystal field interaction are of com-

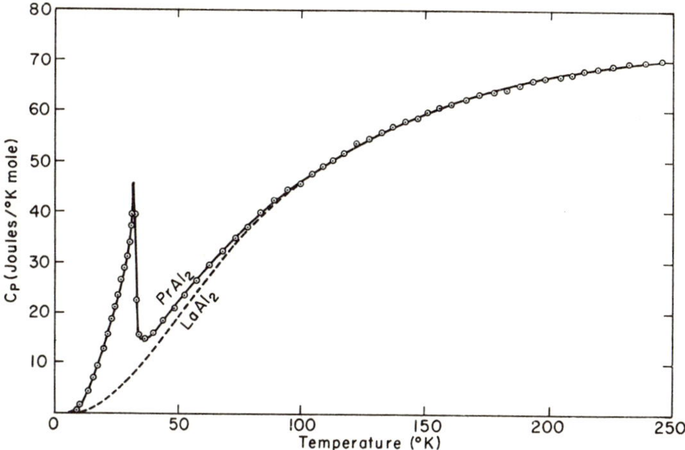

Fig. 4-7. Heat capacities of $PrAl_2$ and $LaAl_2$ versus temperature [42,43].

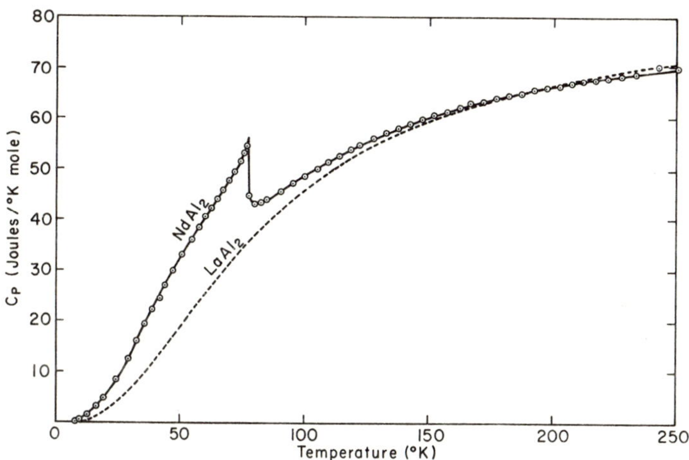

Fig. 4-8. Heat capacities of $NdAl_2$ and $LaAl_2$ versus temperature [42,43].

parable importance; it certainly does not occur in $PrAl_2$, which as noted above, orders by the bootstrap process. Results for $GdAl_2$ are very unusual. The anticipated λ-type thermal anomaly is absent; instead there is an extremely broad heat capacity excess, implying that ferromagnetism is destroyed over an extensive range of temperature. The behavior of $GdAl_2$ is as yet not understood.

Perhaps the only additional feature of these compounds meritorious of comment is that whereas C_p values for $PrAl_2$, $NdAl_2$, and $GdAl_2$ approach that of $LaAl_2$ at temperatures well above T_C, C_p for $CeAl_2$

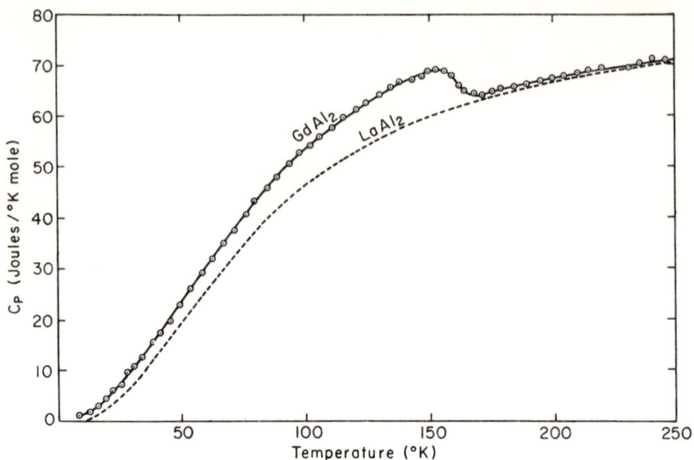

Fig. 4-9. Heat capacities of $GdAl_2$ and $LaAl_2$ versus temperature [42,43].

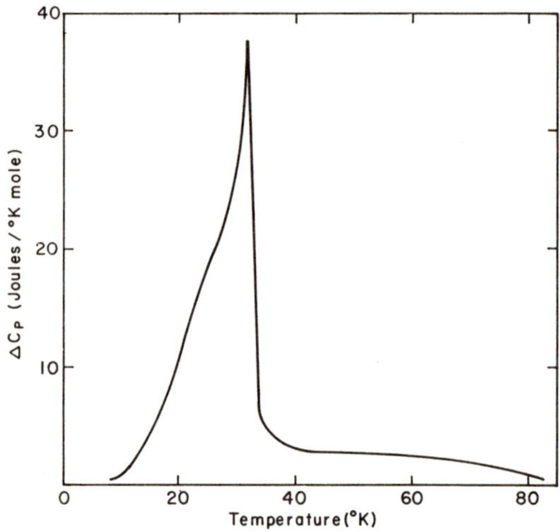

Fig. 4-10. Excess heat capacity versus temperature for $PrAl_2$, i.e., difference in C_p for $PrAl_2$ and $LaAl_2$ [42,43].

does not (see Fig. 4-3). This suggests a stronger interaction between the localized 4f electrons and the conduction electrons for Ce^{3+} than for the other tripositive rare earths.

The measured and calculated magnetic entropies are given in Table 4-4. The entropies, entropies of formation, and other thermodynamic properties are given in Table 4-5. Data for $CeAl_2$ are not given in Table 4-5 because of uncertainties in regard to the extrapolation of the C_p data below 10°K.

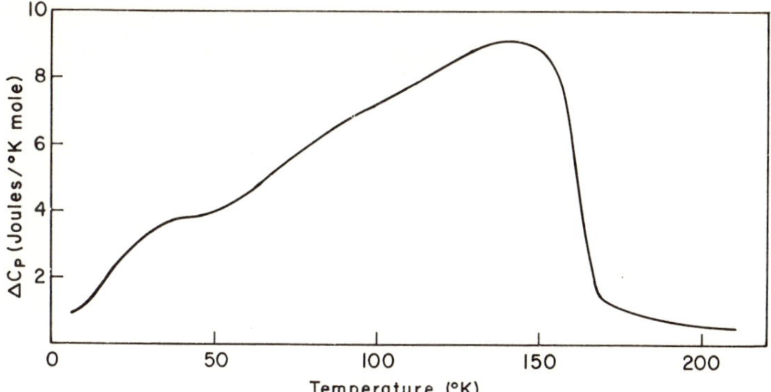

Fig. 4-11. Excess heat capacity versus temperature for $GdAl_2$ [42,43].

The electronic specific coefficients (γ) and Debye temperatures θ_D for the three compounds studied by Hungsberg are listed in Table 4-6. The high γ value for $LaAl_2$ is noteworthy; it undoubtedly arises from the perturbing influence of the vacant 4f levels which are near in energy to E_F, the Fermi energy. Recent theories of superconductivity

TABLE 4-4

Magnetic or Crystal Field Entropies or Both[a]
(J °K^{-1} mole^{-1})

	Theor.	Meas.	Meas./Theor.
$CeAl_2$	$R \ln 6/2 = 9.13$	8.45	0.925
$PrAl_2$	$R \ln 9 = 18.26$	15.94	0.871
$NdAl_2$	$R \ln 10 = 19.11$	16.99	0.886
$GdAl_2$	$R \ln 8 = 17.20$	12.63	0.735

[a]From Deenades *et al.* [43].

TABLE 4-5

Thermodynamic Characteristics of RAl_2 Compounds[a]
(J °K^{-1} mole^{-1} at 25°C)

	C_p	$(H - H_0°)/T$	S	$-(F - H_0°)/T$	$-\Delta S$ (formation)
$LaAl_2$	73.71	49.53	98.76	49.24	14.97
$PrAl_2$	73.81	50.93	114.71	63.78	15.13
$NdAl_2$	73.00	52.43	115.76	63.33	14.29
$GdAl_2$	74.10	52.93	111.40	58.65	11.24

[a]From Deenades *et al.* [43].

associate a high density of states with the development of superconductivity [46]. $LaAl_2$ becomes superconducting below 3.3°K.

TABLE 4-6

Thermal Characteristics[a] of RAl_2

	γ (mJ °K^{-2} mole^{-1})[b]	θ_D (°K)[b]
YAl_2	5.47	249
$LaAl_2$	11.0	331
$LuAl_2$	5.60	264

[a] From Hungsberg and Gschneidner [45].
[b] γ is the electronic specific heat coefficient; θ_D is the Debye temperature.

2. Ternary RAl_2 Systems

a. Systems Involving Substituents on the R Sublattice

The first studies of this nature were carried out by Williams *et al.* [13] in 1962. They studied $R_{1-x}R_x'Al_2$ ternary systems with the following R,R′ combinations: Gd,Pr; Dy,Gd; Gd,Er; Tb,Dy; and Pr,Nd. They observed the coupling mode referred to above, namely ferromagnetic coupling when R and R′ are both heavy or light and antiferromagnetic coupling when one is light and the other is heavy. The same coupling systematics were observed by Swift and Wallace [15] with several additional R,R′ combinations. Typical results are displayed in Tables 4-7 to 4-9. The coupling is such that the spins of R and R′ always couple parallel. The $Ce_{1-x}Eu_xAl_2$ ternaries are an exception to this [15]; most of these fail to order. $Ce_{0.2}Eu_{0.8}Al_2$ seems to order antiferromagnetically at 23°K as compared to 15°K for $EuAl_2$. The Ce–Pr–Al and Ce–Nd–Al ternaries exhibited ferromagnetic coupling. We do not find this unexpected since the weak antiferromagnetism of $CeAl_2$ is unlikely to predominate in the ternaries containing strongly ferromagnetic $PrAl_2$ and $NdAl_2$. The Ce^{3+} moment in these ternaries is 0.64 and 0.66 μ_B for the $PrAl_2$- and $NdAl_2$-based materials, respectively. As noted above the Ce^{3+} ground state multiplet splits into Γ_7 and Γ_8 crystal field states. These have moments 0.71 and 1.5 μ_B, respectively. The moment for Ce^{3+} is in close agreement with that for the Γ_7 state supporting the conclusion reached above that the doublet is lowest lying.

As noted above studies of $Eu_{1-x}La_xAl_2$ alloys showed that antiferromagnetic $EuAl_2$ becomes ferromagnetic when the electron concentration is increased by partial replacement of Eu with La. The change in sign of the coupling occurs at $x = 0.4$ Swift and Wallace [28] analyzed

TABLE 4-7

Magnetic Data for Some Heavy–Heavy $R_{1-x}R_x'Al_2$ Compounds

Compound	$a_0(\text{Å})$	μ_{sat}[a]	gJ[b]	μ_{eff}	$g[J(J+1)]^{1/2}$[b]	θ_W[c] (°K)	Curie temp.[d] (°K)
$DyAl_2$	7.83	9.62	10.0	10.7	10.6	64	70
$Dy_{0.8}Er_{0.2}Al_2$	7.83	8.75	9.8	10.4	10.4	55	61
$Dy_{0.6}Er_{0.4}Al_2$	7.82	8.32	9.6	10.3	10.3	43	49
$Dy_{0.4}Er_{0.6}Al_2$	7.81	7.92	9.4	10.0	10.0	33	39
$Dy_{0.2}Er_{0.8}Al_2$	7.80	7.63	9.2	9.85	9.79	24	31
$ErAl_2$	7.79	7.05	9.0	9.46	9.59	14	24
$Dy_{0.8}Ho_{0.2}Al_2$	7.83	9.51	10.0	10.5	10.6	60	66
$Dy_{0.6}Ho_{0.4}Al_2$	7.82	9.43	10.0	10.7	10.6	48	61
$Dy_{0.4}Ho_{0.6}Al_2$	7.82	9.29	10.0	10.6	10.6	39	54
$Dy_{0.2}Ho_{0.8}Al_2$	7.81	9.20	10.0	10.6	10.6	32	47
$HoAl_2$	7.81	9.16	10.0	10.7	10.6	30	42
$Ho_{0.8}Gd_{0.2}Al_2$	7.82	8.75	9.4	10.0	10.1	65	63
$Ho_{0.6}Gd_{0.4}Al_2$	7.85	8.48	8.8	9.62	9.63	90	87
$Ho_{0.4}Gd_{0.6}Al_2$	7.87	8.10	8.2	8.99	9.10	114	114
$Ho_{0.2}Gd_{0.8}Al_2$	7.89	7.56	9.6	8.45	8.54	142	145
$GdAl_2$	7.90	7.10	7.0	7.94	7.96	168	182
$TbAl_2$	7.87	8.60	9.0	9.81	9.72	110	121
$Tb_{0.8}Gd_{0.2}Al_2$	7.87	8.22	8.6	9.45	9.36	120	126
$Tb_{0.6}Gd_{0.4}Al_2$	7.88	7.85	8.2	9.06	9.06	130	133
$Tb_{0.4}Gd_{0.6}Al_2$	7.88	7.56	7.8	8.69	8.70	142	146
$Tb_{0.2}Gd_{0.8}Al_2$	7.89	7.40	7.4	8.36	8.32	160	168
$Tb_{0.8}Dy_{0.2}Al_2$	7.86	8.52	9.2	9.96	9.94	96	109
$Tb_{0.6}Dy_{0.4}Al_2$	7.85	8.78	9.4	10.3	10.1	88	99
$Tb_{0.4}Dy_{0.6}Al_2$	7.84	9.07	9.6	10.2	10.2	84	94
$Tb_{0.2}Dy_{0.8}Al_2$	7.84	9.10	9.8	10.6	10.5	75	80

[a] Saturation moments in μ_B/f.u. at 4.2°K.
[b] This is an appropriate average for the ternary systems.
[c] θ_W = Weiss constant in °K.
[d] Curie temperatures are for $H = 19$ kOe.

this in terms of the RKKY formalism. The RKKY sums for $EuAl_2$ and $LaAl_2$ are shown in Fig. 4-12. The curve for the ternary alloys progressively shifts from that characteristic of $EuAl_2$ to that for $LaAl_2$ as x increases. From Eq. (2-3) we see that for ferromagnetic interactions $F(x)$ must be negative. Taking k_F/k_F^0 to be 0.92, which is very close to the value used by Buschow *et al.* [35] in analyzing results for $GdAl_2$-based ternaries (*vide infra*), we obtain a change in sign of coupling at $x = 0.4$ in accord with experiment.

Buschow *et al.* have studied [35] not only $Gd_xM_{1-x}Al_2$ systems with M = Y, La, and Th but also $Tb_xY_{1-x}Al_2$ ternaries. As noted above Mader *et al.* [21] have carried out similar studies on the ternaries

TABLE 4-8

Magnetic Data for some Light–Heavy $R_{1-x}R_x'Al_2$ Systems[a]

Compound	a_0(Å)	μ_{sat}	μ_{eff}	$g\,[J(J+1)]^{1/2}$	θ_W (°K)	Curie temp. (°K)
$Ho_{0.75}Pr_{0.25}Al_2$	7.88	6.28	9.31	9.35	24	40
$Ho_{0.5}Pr_{0.5}Al_2$	7.93	3.39	7.98	7.91	15	40
$Ho_{0.5}Pr_{0.75}Al_2$	7.98	0	6.20	6.14	5	—
$Ho_{0.1}Pr_{0.9}Al_2$	8.01	1.80	4.68	4.76	12	39
$PrAl_2$	8.03	2.60	3.46	3.58	30	37
$Ho_{0.75}Nd_{0.75}Al_2$	7.85	6.65	9.25	9.35	33	48
$Ho_{0.5}Nd_{0.5}Al_2$	7.90	3.51	7.83	7.91	12	51
$Ho_{0.25}Nd_{0.75}Al_2$	7.96	0.65	6.25	6.16	0	—
$Ho_{0.1}Nd_{0.9}Al_2$	7.98	1.20	4.71	4.78	23	—
$NdAl_2$	8.00	2.27	3.59	3.62	70	76
$Pr_{0.8}Dy_{0.2}Al_2$	7.98	0.50	5.58	5.73	26	—
$Pr_{0.6}Dy_{0.4}Al_2$	7.93	2.19	7.16	7.25	22	56
$Pr_{0.4}Dy_{0.6}Al_2$	7.90	4.45	8.52	8.52	46	62
$Pr_{0.2}Dy_{0.8}Al_2$	7.87	7.18	9.52	9.61	60	65

[a]Symbols have the same significance as in Table 4-7.

TABLE 4-9

Magnetic Data for some Light–Light $R_{1-x}R_x'Al_2$ Systems[a]

Compound	a_0(Å)	μ_{sat}	gJ	μ_{eff}	$g\,[J(K+1)]^{1/2}$	θ_W (°K)	Curie temp. (°K)
$CeAl_2$	8.06	—	—	2.54	2.54	−33	None
$Ce_{0.8}Pr_{0.2}Al_2$	8.06	0.95	2.35	2.72	2.79	−17	11
$Ce_{0.6}Pr_{0.4}Al_2$	8.05	1.34	2.56	2.92	3.00	0	17
$Ce_{0.4}Pr_{0.6}Al_2$	8.04	1.75	2.78	3.16	2.20	+11	23
$Ce_{0.2}Pr_{0.8}Al_2$	8.03	2.15	2.99	3.28	3.39	32	28
$Ce_{0.2}Nd_{0.8}Al_2$	8.01	1.90	3.04	3.34	3.43	44	59
$Ce_{0.4}Nd_{0.6}Al_2$	8.03	1.56	2.82	3.21	3.23	23	39
$Ce_{0.6}Nd_{0.4}Al_2$	8.04	1.17	2.59	2.97	3.02	4	27
$Ce_{0.8}Nd_{0.2}Al_2$	8.05	0.78	2.37	2.74	2.79	−14	12
$Pr_{0.8}Nd_{0.2}Al_2$	8.02	2.42	3.21	3.48	3.59	36	48
$Pr_{0.6}Nd_{0.4}Al_2$	8.01	2.48	3.23	3.47	3.60	40	55
$Pr_{0.4}Nd_{0.6}Al_2$	8.01	2.34	3.24	3.52	3.60	51	64
$Pr_{0.2}Nd_{0.8}Al_2$	8.00	2.33	3.26	3.49	3.61	65	69

[a]Symbols are the same as in Table 4-7.

$Pr_{1-x}M_xAl_2$ with M = La and Y. The Curie temperature decreases with increasing content of nonmagnetic M in each case. For 20% replacement of Gd by Y or La T_C decreases by about 30°C; the effect of Th is

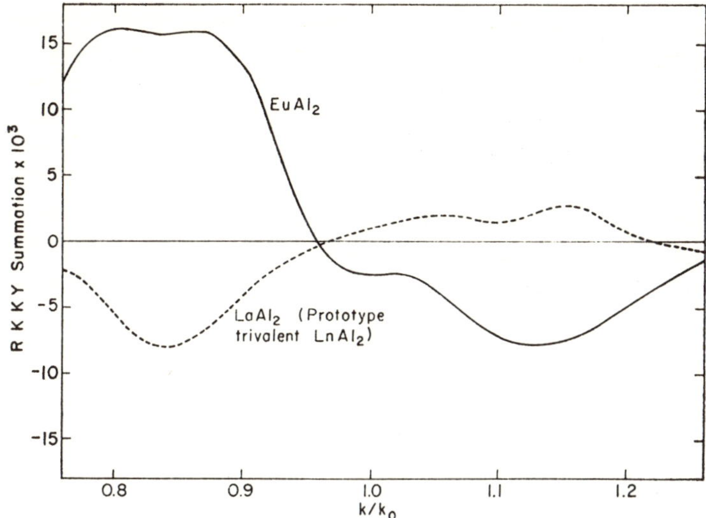

Fig. 4-12. RKKY sums for $EuAl_2$ and an RAl_2 compound containing R^{3+} [28].

about threefold larger. From these observations and using curves such as are shown in Fig. 4-12 Buschow *et al.* concluded that their results could be systematized using $k_F/k_F{}^0 = 0.94$. It will be recalled that this value of k_F was also obtained from NMR measurements on $GdAl_2$.

The decrease in T_C with increasing M content in the $PrAl_2$-based ternaries has been discussed above. Purwins [47] has studied $Er_{1-x}Gd_xAl_2$ by Mössbauer spectroscopy. He finds T_C increases linearly with x and also observes an increase in the Er^{3+} hyperfine field as the Gd content increases.

b. Systems Involving Substituents on the Al Sublattice

Leon *et al.* [48,49] have studied the magnetic behavior of the systems $RAl_{2-x}Ni_x$ with R = Ce, Pr, Nd, Gd, Tb, Dy, Ho, and Er. In all cases except when R = Ce, there are two primary solid solutions, based on RAl_2 and RNi_2, each of appreciable range. The amount of aluminum which can be incorporated into $CeNi_2$ was undetectable, i.e., less than 1%. An intermediate phase of C22 structure (Fe_2P) forms in a composition range about the composition RNiAl. Results obtained [48] are largely summarized in Tables 4-10–4-12. In the majority of cases, a decrease of the Al/Ni ratio in the sample led to a decline in T_C and the saturation magnetization.

Magnetic behavior of CeNiAl and certain of the Pr ternaries is quite unusual. The susceptibility–temperature behavior of $CeNi_{0.98}Al_{1.02}$ is shown in Fig. 4-13 and in a log–log plot in Fig. 4-14. Susceptibility is very low compared to expectation from Curie's law. The influence of the crystal field or spin compensation by the conduction electrons (Kondo phenomenon) or both is undoubtedly responsible. Anderson has shown [50] that if the latter is dominant at low temperatures χ^{-1} varies as $T^{1/2}$ in the low temperature limit. The data in Fig. 4-14 show that at the lowest temperature measured χ^{-1} is linear with $T^{0.57}$ strongly suggesting that CeNiAl is another concentrated Kondo system.

Magnetization–temperature results for $PrAl_{1.96}Ni_{0.04}$ are shown in Fig. 4-15. Other $PrAl_{2-x}Ni_x$ ternaries with x in the range 0.04–0.12

TABLE 4-10

Magnetic Behavior of Ternary $RNi_{2-x}Al_x$ Systems and the Parent Binary Systems

	x in $RNi_{2-x}Al_x$	T_C (°K)[a]	θ (°K)[a]	μ_{eff} (μ_B)[a]	μ_{sat} (μ_B)[a]
Ternaries based on RNi_2					
Ce	DNF[b]	—[b]			
Pr	0.05	VVP[b]	0(4)	3.66(3.57)	
Nd	0.04	16(16)	7(10)	3.46(3.74)	1.73(1.84)
Gd	0.11	92(85)	79(78)	8.23(7.82)	6.88(7.13)
Tb	0.05	53(45)	40(35)	9.91(9.82)	6.59(7.82)
Tb	0.09	56(45)	42(35)	9.78(9.82)	6.45(7.82)
Dy	0.16	42(30)	26(23)	10.8(10.4)	7.92(9.23)
Ho	0.16	30(22)	14(12)	10.6(10.5)	8.02(8.40)
Er	0.17	30(21)	8(11)	9.50(9.37)	5.18(6.75)
Ternaries based on RAl_2					
Ce	1.80	—[c](3.1)[b]	(−41)		
Pr	1.84	—(37)	20(30)	3.44(3.46)	—(2.60)
Nd	1.73	25(76)	19(70)	3.61(3.59)	1.71(2.27)
Gd	1.55	68(182)	58(168)	3.73(7.94)	6.06(7.10)
Tb	1.92	118(121)	102(110)	10.7(9.81)	8.19(8.60)
Tb	1.79	84	72	9.71	7.25
Tb	1.60	59	53	10.1	6.30
Dy	1.80	83(70)	66(64)	10.8(10.7)	8.05(9.62)
Ho	1.55	31(42)	17(30)	10.9(10.7)	8.05(9.16)
Er	1.90	33(24)	16(14)	9.82(9.46)	8.12(9.59)
Er	1.76	27(24)	9(14)	9.87	7.14
Er	1.70	29(24)	6(14)	9.62	6.45

[a] T_C and θ denote the Curie and Weiss temperatures, respectively. μ_{eff} is the paramagnetic moment; μ_{sat} is the saturation moment measured at 4.2°K.

[b] The numbers in parentheses give the ordering temperature for the RNi_2 or RAl_2 parent phase. All except $CeAl_2$ order ferromagnetically. Susceptibility work and heat capacity show that $CeAl_2$ becomes antiferromagnetic. DNF signifies that the ternary does not form. VVP signifies Van Vleck paramagnetism.

[c] $CeNi_{2-x}Al_x$ ternaries remain paramagnetic down to 2.28°K.

TABLE 4-11

Magnetic Properties of the C22 $Ni_{2-x}Al_x$ Alloys

	T_C[a]	θ[a]	μ_{eff}[b]	$g[J(J+1)]^{1/2}$[b]	μ_{sat}[b]	gJ[b]
$CeNi_{0.98}Al_{1.02}$	—[c]	—[d]	—[d]	2.56	—	2.14
$PrNi_{1.02}Al_{0.98}$	—[c]	−10	3.73	3.62	—	3.20
$NdNi_{1.20}Al_{0.80}$	17	5	3.84	3.62	1.64	3.27
$GdNi_{1.05}Al_{0.95}$	61	53	8.90	7.94	6.36	7.00
$TbNi_{1.01}Al_{1.01}$	47	30	11.0	10.6	7.82	10.0
$HoNi_{1.09}Al_{0.91}$	27	12	10.6	10.6	7.25	10.0
$ErNi_{0.90}Al_{1.10}$	16	0	9.85	9.59	7.40	9.00

[a] Curie and Weiss temperatures in °K.
[b] Moments in Bohr magnetons.
[c] These ternaries are paramagnetic at 4.2°K.
[d] Does not exhibit Curie–Weiss behavior.

TABLE 4-12

Magnetic Properties of $RAl_{2-x}Ni_x$ Ternaries

x in $RAl_{2-x}Ni_x$	Weiss constant (°K)	Curie temp. (°K)	Saturation moment (μ_B/f.u.)	Paramagnetic moment (μ_B/f.u.)
		R = Gd		
0.0 ($GdAl_2$)	168	182	7.10	7.94
0.1	184	191	7.21	8.28
0.24	128	98	7.23	8.34
0.45	58	68	6.06	8.73
		R = Tb		
0.0 ($TbAl_2$)	110	121	8.60	9.8
0.08	102	118	8.19	10.7
0.21	72	84	7.25	9.71
0.40	53	59	6.3	10.1
		R = Ho		
0.0 ($HoAl_2$)	30	42	9.16	10.7
0.08	39	46	8.90	10.7
0.24	30	37	8.84	11.5
0.45	17	31	8.05	10.9
		R = Er		
0.0 ($ErAl_2$)	14	24	7.05	9.46
0.03	39	22	7.85	10.3
0.10	16	33	8.12	9.82
0.16	15	31	8.17	10.0
0.23	9	25	7.35	9.73
0.30	6	28.5	6.54	9.62

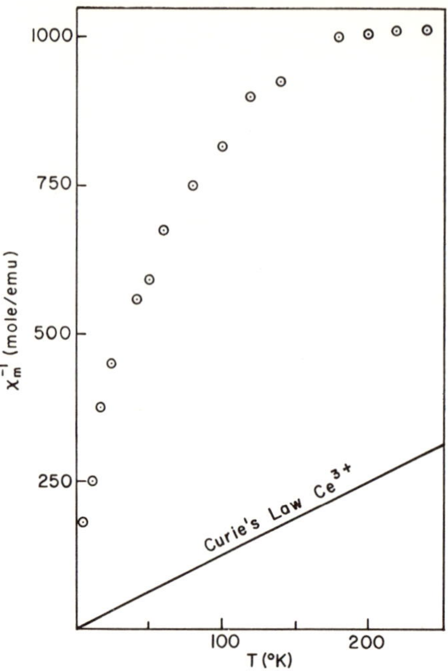

Fig. 4-13. Variation of reciprocal susceptibility with temperature for $CeNi_{0.98}Al_{1.02}$ at 18 kOe [48]. For comparison, Curie's law for Ce^{+3} ions is also shown.

behave similarly. The decline in moment below 20°K was initially taken to indicate the onset of antiferromagnetism, although this seemed rather unlikely in view of the ferromagnetism of $PrAl_2$. Examination of these materials by neutron diffraction techniques [51] showed that they remain ferromagnetic to the lowest temperatures. The decline in moment then must imply a drastic change in magnetic hardness when Al is replaced by nickel. Exchange is weakened when this replacement occurs (as shown by the decline in θ) and the bootstrap process is obviously impaired, leading to a change in hardness by a mechanism which is yet to be clarified.

The results in Table 4-12 indicate two types of behavior as Ni replaces Al (1) by $TbAl_2$-based ternaries in which θ, T_C, and μ_{sat} all decrease monotonically and (2) by the others in which the magnetic properties pass through a maximum as composition is varied. The similarity of results for the Gd-, Ho-, and Er-containing ternaries excludes the possibility that the differing behavior of these ternaries is a crystal field effect or is dependent upon whether the number of f electrons is

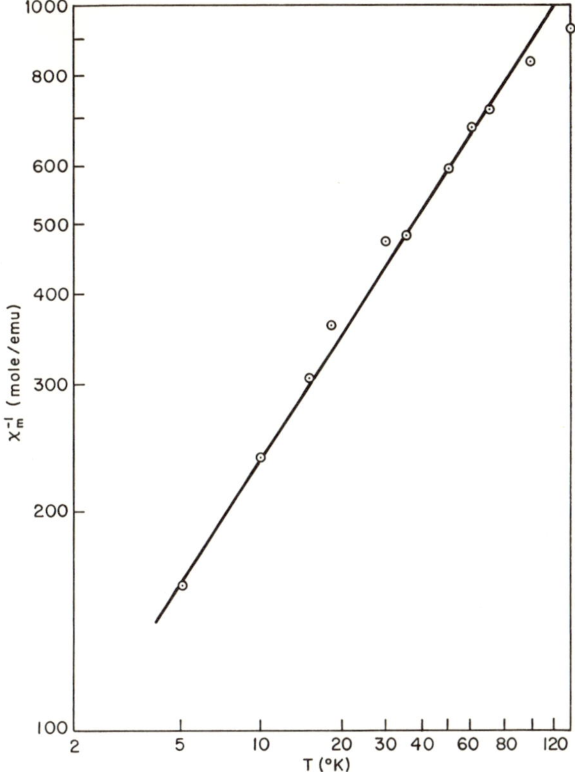

Fig. 4-14. Log–log plot of χ^{-1} versus temperature for $CeNi_{0.98}Al_{1.02}$ [48].

odd or even. The difference, therefore, appears to originate in some subtlety of the RKKY interaction.

Leon *et al.* ascribed the differing behavior of the several ternaries to differing k values for the Tb ternaries on the one hand and the Gd, Ho, and Er ternaries on the other: $\sim 1.3 \times 10^8$ cm^{-1} for the Tb ternaries, and $\sim 1.4 \times 10^8$ cm^{-1} for the others ($k_F{}^0 = 1.63 \times 10^8$ cm^{-1}). In the $RAl_{2-x}Ni_x$ ternaries electron concentration and k_F decrease as x increases; the effect of this on the RKKY sum and hence on the interaction [see Eq. (2-3)] is evident from Fig. 4-16.

With $k_F = 1.4 \times 10^8$ at $x = 0$, the RKKY sum goes through a minimum as x increases whereas with $k_F = 1.3 \times 10^8$ at $x = 0$, the sum diminishes steadily as x increases. Thus the differing behavior can be related to a different k value for the two groups of compounds.

We now inquire as to why there should be a difference in k. The

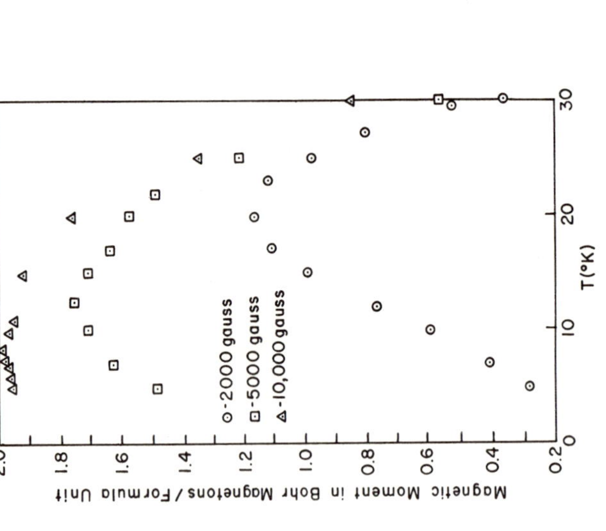

Fig. 4-15. Magnetization versus temperature for $PrNi_{0.04}Al_{1.96}$ [48].

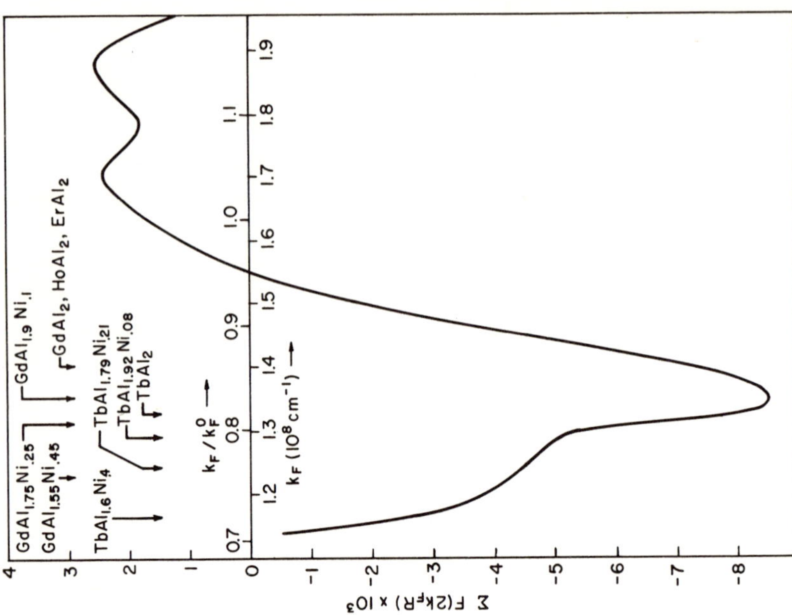

Fig. 4-16. RKKY sum versus k_F, the radius of the Fermi sphere. Arrows indicate the k_F values for the several alloys which are needed to account for the magnetic data [49].

RKKY mechanism, as noted above, is a free electron treatment. The inadequacies are partly taken into account by use of k_F, which differs from k_F^0. The details of the band structure are in some way incorporated into the choice of k_F. Thus a different k_F for the Tb system implies the acceptance of the idea of a different band structure for the Tb system than for the others. We now inquire if this differentiation seems reasonable. Apparently it is. Tb^{3+} has one rather loosely bonded electron beyond the half-filled shell. In some respects Tb^{3+} and Ce^{3+} are similar. The single loosely bonded 4f electron endows Ce compounds with properties substantially different from those of its neighbors in the rare earth sequence. We can imagine Tb compounds to exhibit similar behavior.

B. THE RAl_3 COMPOUNDS

The magnetic behavior of these compounds has been studied by Buschow and Fast [54], Buschow [55], Mader and Swift [56], and Mader *et al.* [21]. Results are summarized in Table 4-13. Only the heavy rare earth compounds show evidence of magnetic ordering. Curie–Weiss behavior is observed at higher temperatures with effective moments close to those expected for an assemblage of free tripositive ions for all except $SmAl_3$ and $YbAl_3$. Nonlinear inverse susceptibility–temperature behavior is ascribed to the narrowness of the multiplet spacings for $SmAl_3$ and to variable valency for $YbAl_3$.

Results for $CeAl_3$ taken from the work of Mader and Swift are shown in Fig. 4-17; they have analyzed these results in detail using the Hamiltonian

$$\mathcal{H} = B_2^0 O_2^0 + B_4^0 O_4^0 - g\mu_B \mathbf{H} \cdot \mathbf{J}. \tag{4-13}$$

The second-order term is important in this case since $c/a < (2/3)^{1/2}$. Diagonalizing the matrix and using the point charge value of

$$B_2^0/B_4^0 = 30,$$

Mader and Swift obtained the eigenvalues

$$E_\parallel = \begin{cases} 12\ B_2^0 \pm 5\ G/2, \\ -4\ B_2^0 \pm G/2, \\ -8\ B_2^0 \pm 3G/2; \end{cases} \tag{4-14a}$$

$$E_{\perp} = \begin{cases} 12\ B_2{}^0 + 1/16\ G^2/B_2{}^0 \text{ (twice)}, \\ -4\ B_2{}^0 \pm 3G/2 + 1/2\ G^2/B_2{}^0, \\ -8\ B_2{}^0 - 9/40\ G^2/B_2{}^0 \text{ (twice)}. \end{cases} \qquad \text{(4-14b)}$$

In these, $G = g\mu_B H$ and $E_{\parallel}$ and $E_{\perp}$ represent the eigenvalues with H parallel and perpendicular to the hexagonal axis, respectively. χ was calculated in terms of $\chi_{\parallel}$ and $\chi_{\perp}$ from the expression

$$\chi = 1/3\ \chi_{\parallel} + 2/3\ \chi_{\perp} \qquad \text{(4-15)}$$

using Eq. (3-34) together with the eigenvalues given above. The susceptibility computed from (4-15) is shown in Fig. 4-17. The calculated χ^{-1} value is lower than experiment by almost a constant amount, suggesting that antiferromagnetic exchange is important. Using the molecular field approach

$$\chi^{-1} = \chi_0{}^{-1} - \lambda, \qquad \text{(4-16)}$$

where χ_0 represents the exchange-free susceptibility. With $\lambda = -26$ $\text{Oe}^2\ \text{erg}^{-1}\ \text{mole}^{-1}$ and $B_2{}^0 = 14°\text{K}$, an excellent accounting for the susceptibility-temperature behavior is achieved (see Fig. 4-17). The order of states is $|\pm 3/2\rangle$ lowest with $|\pm 1/2\rangle$ and $|\pm 5/2\rangle$ progressively higher by 56 and 224°K, respectively. The order of levels is the same as used by van Daal *et al.* [57] and Rao *et al.* [58] in assessing the conductivity behavior of $CeAl_3$.

TABLE 4-13

Magnetic Characteristics of RAl_3

	Paramagnetic state			Ordered state				
R	μ_{eff} (μ_B/f.u.)	θ (°K)	Type	μ_{or} (μ_B/f.u.)	T_C (°K)	T_N (°K)	Type	Ref.
Ce	2.63	−46	C–W,CF					[5]
Pr	3.74	−14	C–W,VV					[21]
Nd	4.11	5	C–W,?					[54]
Sm			N,m					[54]
Gd	8.29	−89	C–W			17	AF	[54]
Tb	10.00	−64	C–W			21	AF	[54]
Dy	10.85	−51	C–W			23	AF	[54]
Ho	10.89	−26	C–W			9	AF	[54]
Er	9.87	6	C–W	6.2	21		F	[54]
Tm	7.88	−19	C–W,VV					[54,55]
Yb	4.62	−300	N,VC					[54]

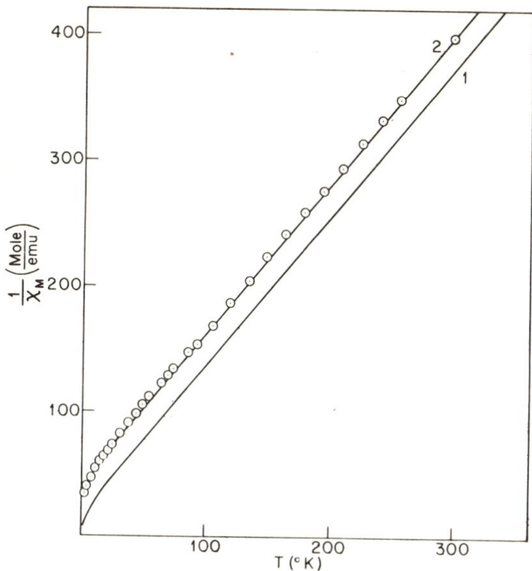

Fig. 4-17. Susceptibility behavior [56] for $CeAl_3$ (○); calculated value using crystal field splitting only (1) and calculated value allowing for exchange with a molecular field paramater $\lambda = -26$ emu/mole (2).

A similar type of analysis of the results for $PrAl_3$ was carried out by Mader *et al.* [21]. This case is more complex because of the presence of a sixth-order term in the Hamiltonian which was used

$$\mathcal{H} = B_2^0\, O_2^0 + B_4^0\, O_4^0 + B_6^0\, [O_6^0 + (77/8)\, O_6^6] - g\, \mu_B\, \mathbf{J} \cdot \mathbf{H}. \qquad (4\text{-}17)$$

By a procedure analogous to that described for $CeAl_3$ we are able to account for the susceptibility behavior of $PrAl_3$. Two sets of parameters lead to accord between calculation and experiment (see Fig. 4-18), each involving an overall crystal field splitting of the order of 100°K. Results of Buschow and Fast [54] show positive deviations from Curie-Weiss behavior for $NdAl_3$ below about 60°K, probably because of the influence of the crystal field interaction. However, detailed calculations such as have been made for the Ce and Pr compounds have not been made for $NdAl_3$.

$TmAl_3$ exhibits Van Vleck paramagnetism at low temperatures indicating that the singlet Γ_1 state is the ground state. Buschow [55] has analyzed results by the classical Penney and Schlapp [59] technique using the eigenvalues and magnetic moments in parametrical form from the work of Schumacher and Hollingsworth [60]. This consists of using

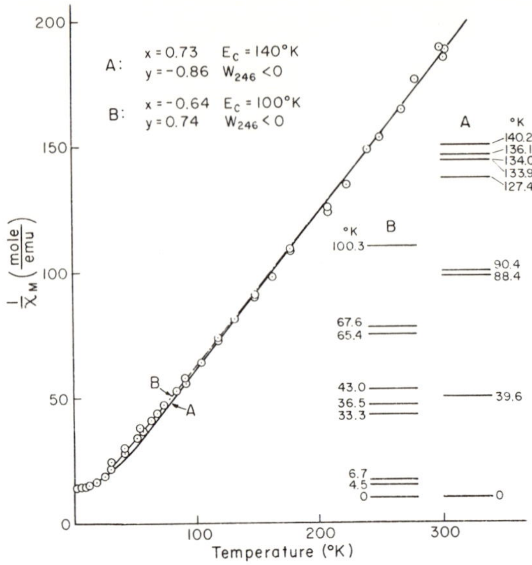

Fig. 4-18. Susceptibility behavior of $PrAl_3$ after Mader *et al.* [21]. Points are experimental. The curve is calculated using the Hamiltonian given in Eq. (4-17).

Eq. (3-34) in parametrical form and adjusting the single parameter (sixth-order interactions were ignored in the treatment of Schumacher and Hollingsworth) for best fit with experiment. Buschow's analysis indicated an overall splitting of 420°K and antiferromagnetic exchange corresponding to $\theta = -8$°K.

Van Diepen *et al.* [36,37] have carried out NMR studies on RAl_3 compounds similar to those described above for the RAl_2 systems. The Knight shift shows the same systematics as in the RAl_2 series, i.e., $K > 0$ for light lanthanides and $K < 0$ for heavy lanthanides. If results are analyzed in terms of the uniform polarization model originally proposed by Jaccarino [30], in which J in Eq. (4-1) is the exchange integral Γ appearing in Eq. (2-1), J (or Γ) is found to be negative, ranging from about ¼ to ½ eV. Using the RKKY approach with $k_F = k_F^0$ Γ is obtained [36] as about −1.3 eV for the RAl_3 compounds. The same procedure for the RAl_2 compounds leads to $\Gamma = -1.7$ eV. However, these results are rather sensitive to the choice of k_F. For example, in the procedure (see above) in which Buschow *et al.* [35] used θ and K to evaluate Γ and k_F for $GdAl_2$ Γ was evaluated to be −0.9 eV with $k_F = 0.94\ k_F^0$. These variations in values established for the parameter Γ emphasize the point that the RKKY formalism even when "renormalized" by adjusting k_F is at best only a crude description of the magnetic interactions in rare earth intermetallics.

C. RAl COMPOUNDS

Fairly extensive studies of these materials—their bulk magnetic characteristics, magnetic structures by neutron diffraction, and their NMR behavior—have been made. Kissell and Wallace established [52] the bulk magnetic characteristics of the RAl compounds with R = La, Ce, Pr, and Nd and Barbara *et al.* investigated [53] the entire series from R = Ce to Tm with the exception of Sm and Eu (Table 4-14). In weak fields all except HoAl behave as antiferromagnetic materials; however, they are actually metamagnetics with critical fields ranging from 5 kOe (for ErAl) to 35 kOe (for GdAl). HoAl orders ferromagnetically. Bécle *et al.* [61] have established the magnetic structure of the members of the RAl series using neutron diffraction. Structures are quite complex and the original papers (Bécle *et al.* [61] and references cited therein) should be consulted for details. It seems appropriate to draw attention here to only two features of this work: (1) none of the structures are collinear; and (2) the R–R nearest neighbor coupling is almost (i.e., allowing for slight noncollinearity) ferromagnetic in all cases except for TmAl.

TABLE 4-14

Magnetic Characteristics of RAl

	Paramagnetic state			Ordered state				
R	μ_{eff} (μ_B/f.u.)	θ (°K)	Type	μ_{or} (μ_B/f.u.)	T_C (°K)	T_N (°K)	Type	Ref.
La			P					[52]
Ce	2.44	−19	C–W,?			10	AF(M)	[52]
Pr	3.58	1	C–W			19	AF	[52]
Nd	3.52	13	C–W,?			25	AF	[52]
Gd	8.15	64	C–W,?			42	AF	[53]
Tb	10.0	10	C–W,?			72	AF	[53]
Dy	10.6	25	C–W,?			20	AF	[53]
Ho	10.7	17	C–W,??	6.2	26		F	[53]
Er	9.65	23	C–W,??			13	AF	[53]
Tm	7.70	−2	C–W,??			10	AF	[53]

Van Diepen *et al.* [38] have examined the RAl series by ^{27}Al NMR spectroscopy for R = Gd through Er. LaAl was also examined. The Knight shift (K) is positive for La compound and negative for the others. Analysis using Eqs. (2-3) and (4-1) led to a Γ value of −0.9 eV and $k_F = 0.96\ k_F^0$.

D. R_3Al_2 COMPOUNDS

This series has been studied by Buschow [62] (Dy_3Al_2) and Barbara *et al.* [63] (R = Gd through Tm). Results are summarized in Table 4-15. Barbara *et al.* observe ferromagnetic behavior for the Gd, Tb, Dy, and Ho compounds and antiferromagnetic behavior for the Er_3Al_2 and Tm_3Al_2 compounds. However, with R = Tb, Dy, and Ho the ferromagnetic form does not persist to the lowest temperatures. Instead, according to Barbara *et al.* [53,63] it transforms into an essentially antiferromagnetic state below a temperature, T_{tr}, which is field-dependent. These authors worked with polycrystalline materials. Buschow working with a "pseudo-single crystal," i.e., an oriented powder, observed essentially the same behavior for Dy_3Al_2, but only if the field was applied perpendicular to the *c* axis.

TABLE 4-15

Magnetic Characteristics of R_3Al_2 Compounds

	Paramagnetic state			Ordered state					
	μ_{eff} (μ_B/R)	θ (°K)	Type	μ_{or} (μ_B/R)	T_C (°K)	T_N (°K)	T_{tr} (°K)[a]	Type	Ref.
Gd	8.3	285	C–W	7.15	282			F	[63]
Tb	10.1	125	C–W		190		10	*a*	[63]
Dy	11.2	31	C–W		76		20	*a*	[63]
Ho	11.1	10	C–W		33		11	*a*	[63]
Er	10.1	−3	C–W			9		AF	[63]
Tm	7.8	−10	C–W			3		AF	[63]

[a] Material transforms at T_{tr} into a weakly magnetic substance.

Barbara *et al.* [64] found the hysteresis loop for Dy_3Al_2 to be quite unusual; it is almost square. They have interpreted these results as implying an unusually thin Bloch wall separating the magnetic domains, possibly the magnetization being reversed in only one interatomic distance. They show how this situation can develop for noncollinear structures in which the exchange coupling between the two sublattices is weak compared to the anisotropy energy. Under these conditions the domain structure is rather like an antiphase structure, leading to the thin Bloch walls and the nearly square hysteresis loop.

The maximum energy product, $(BH)_{max}$, obtained from the hysteresis loop is 73×10^6 G-Oe, indicating this to be an extremely promising material for permanent magnet fabrication. (See Chapter 10 for a discussion of the desired characteristics for permanent magnet materials.)

However, its utility would be limited to very low temperatures since, as seen in Fig. 4-19, its desirable hysteresis characteristics have disappeared by 30°K.

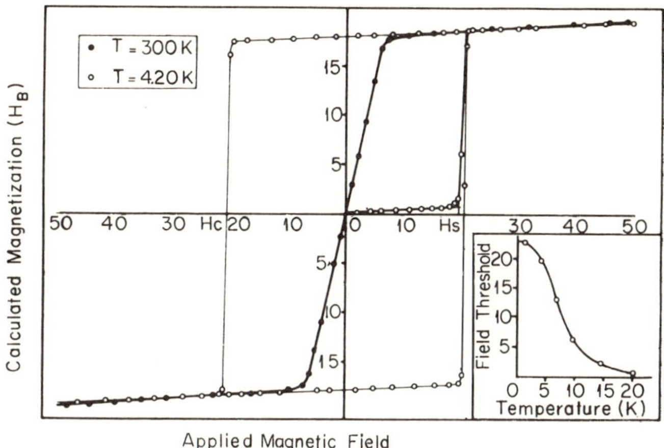

Fig. 4-19. Hysteresis loop for Dy_3Al_2 obtained by Barbara *et al.* [64]. Used with permission of Masson et Cie, Paris.

E. OTHER ALUMINUM COMPOUNDS

We present in this section a summary of magnetic results which have been obtained for other rare earth intermetallics containing aluminum $-R_3Al$, R_2Al, R_3Al_{11}, and $EuAl_4$ (Table 4-16). The R_3Al compounds were investigated by Mader and Wallace [65] for R = Ce and Pr. The cerium compound exists in two allotropic modifications, cubic and hexagonal, with essentially identical behavior. Susceptibility–temperature data for Ce_3Al and Pr_3Al are shown in Fig. 4-20. The Ce_3Al behavior closely resembles that of its antiisomorphous counterpart $CeAl_3$, the analysis of which has been described above. It seems clear that the low temperature deviation from behavior of Ce_3Al also originates with the crystal field interaction. Pr_3Al becomes ferromagnetic below 16°K. The moment is well below $gJ = 3.2$, expected for the free Pr^{3+} ion. The low moment is undoubtedly due to the quenching effect of the crystal field.

Van Diepen *et al.* [66] have made ^{27}Al NMR measurements on the R_3Al_{11} compounds with R = La, Pr, and Nd. These have been alluded to earlier, pointing out that the two crystallographically inequivalent Al sites in the R_3Al_{11} structure can be distinguished by NMR spectroscopy. The conduction electron polarization is different at the two sites, as might be expected, leading to different K values for the two kinds

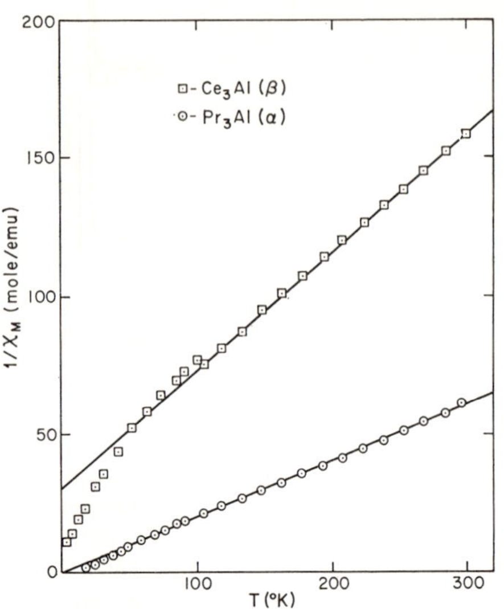

Fig. 4-20. Reciprocal susceptibility versus temperature for Ce_3Al (β) and Pr_3Al [65].

of Al. Van Diepen *et al.* estimate Γ to be -1.4 and -1.1 eV for the Pr and Nd compounds, respectively.

$EuAl_4$ has been studied by Van Viepen *et al.*, by Mader and Wallace [14], and by Wernick *et al.* [67]. The salient magnetic features of this compound are largely summarized in Table 4-16. The moment in the

TABLE 4-16

Magnetic Characteristics of Miscellaneous Rare Earth–Aluminum Compounds

	Paramagnetic state			Ordered state				
	μ_{eff} (μ_B/R)	θ (°K)	Type	μ_{or} (μ_B/R)	T_C (°K)	T_N (°K)	Type	Ref.
Ce_3Al	2.52	−24	C–W,CF					[65]
Pr_3Al	3.62	2	C–W	1.29	16		F	[65]
Pr_2Al	3.59	4	C–W			18	AF	[65]
La_3Al_{11}			P					[66]
Ce_3Al_{11}				0.61	9		F	[56]
Pr_3Al_{11}	3.52	2	C–W			12	AF	[65,66]
Nd_3Al_{11}	3.62	−9	C–W,?					[66]
$EuAl_4$	7.75	13	C–W			13	AF(M)	[14,66,67]

[a] Measurements were actually made on samples of composition RAl_4 in these two cases prior to the knowledge that the true stoichiometry is R_3Al_{11}.

paramagnetic state indicates dipositive europium in this material. This is confirmed by the ESR measurements of Wernick *et al.* who find a g value of 1.995 ± 0.0004. They also observed $EuAl_4$ to be metamagnetic with $H_c \approx 12$ kOe. At higher field the moment rises rapidly and a value of 6.7 μ_B/f.u. is attained at 80 kOe and 4.2°K. This approximately is the value for a $4f^7$ configuration and supports the ESR results as regards the valence state of europium. The valence state of europium is thus the same in $EuAl_4$ and $EuAl_2$ [14].

REFERENCES

1. A complete listing of "strukturbericht" types, e.g., C15 is the $MgCu_2$ structure, etc., is provided by W. B. Pearson, "A Handbook of Lattice Spacings and Structures of Metals and Alloys," pp. 5–78. Pergamon Press, Oxford, 1967.
2. K. H. J. Buschow and J. H. N. Van Vucht, *Z. Metalk.* **56,** 9 (1965); **57,** 162 (1966); K. H. J. Buschow, *J. Less Common Metals* **9,** 452 (1965).
3. C. Bécle and R. Lemaire, *C. R. Acad. Sci. Paris* **264,** 543, 887 (1967).
4. K. H. J. Buschow and A. S. Van der Goot, *J. Less Common Metals* **24,** 117 (1971).
5. K. H. J. Buschow, *J. Less Common Metals* **8,** 209 (1965).
6. C. Bécle and R. Lemaire, *C. R. Acad. Sci. Paris* **264B,** 543, 887 (1967).
7. N. C. Baenziger and J. J. Hegenbarth, *Acta Cryst.* **17,** 620 (1964).
8. J. H. Wernick and S. Geller, *Trans. Met. Soc AIME* **218,** 866 (1960).
9. J. H. N. Van Vucht and K. H. J. Buschow, *J. Less Common Metals* **10,** 98 (1965); *Philips Res. Rep.* **19,** 319 (1964).
10. A. H. Gomes de Mesquita and K. H. J. Buschow, *Acta Cryst.* **22,** 497 (1967).
11. H. J. Wallbaum, *Z. Kristallogr.* **103,** 147 (1941).
12. H. Nowotny, *Z. Metallk.* **34,** 22 (1942).
13. H. J. Williams, J. H. Wernick, E. A. Nesbitt, and R. C. Sherwood, *J. Phys. Soc. Japan Suppl. B1* **17,** 91 (1962).
14. K. H. Mader and W. E. Wallace, *J. Chem. Phys.* **49,** 1521 (1968).
15. W. M. Swift and W. E. Wallace, *J. Phys. Chem. Solids* **29,** 2053 (1968).
16. N. Nereson, C. Olsen, and G. Arnold, *J. Appl. Phys.* **39,** 4605 (1968).
17. N. Nereson, C. Olsen, and G. Arnold, *J. Appl. Phys.* **37,** 4575 (1966).
18. G. Will, *Z. Naturforsch.* **23,** 413 (1968).
19. R. W. Hill and J. M. Machado da Silva, *Phys. Lett.* **30A,** 13 (1969).
20. S. Ofer, M. Rakavy, E. Segal, and B. Khurgin, *Phys. Rev.* **138,** A241 (1965).
21. K. H. Mader, E. Segal, and W. E. Wallace, *J. Phys. Chem. Solids* **30,** 1 (1969).
22. M. Peter and B. T. Matthias, *Phys. Rev. Lett.* **4,** 449 (1960).
23. D. Mattis and W. E. Donath, *Phys. Rev.* **128,** 1618 (1962).
24. D. Mattis, "The Theory of Magnetism," p. 203. Harper and Row, New York, 1965.
25. S. Methfessel, *Z. Angew. Phys.* **18,** 414 (1965).
26. K. Sekizawa and K. Yasukochi, *J. Phys. Soc. Japan* **21,** 684 (1966).
27. G. T. Alfieri, E. Banks, and K. Kanematsu, *J. Appl. Phys.* **37,** 1254 (1966).
28. W. M. Swift and W. E. Wallace, *J. Solid State Chem.* **3,** 180 (1971).
29. K. H. J. Buschow and H. J. van Daal, *Phys. Rev. Lett.* **23,** 408 (1969).
30. V. Jaccarino, B. T. Matthias, M. Peter, H. Suhl, and J. H. Wernick, *Phys. Rev. Lett.* **5,** 251 (1960); V. Jaccarino, *J. Appl. Phys.* **32,** 102S (1961).
31. R. G. Barnes, F. Borsa and D. Peterson, *J. Appl. Phys.* **36,** 940 (1965).

32. V. U. S. Rao and R. Vijayaraghavan, *Phys. Lett.* **19,** 168 (1966).
33. A. M. Van Diepen, H. W. de Wijn, and K. H. J. Buschow, *J. Chem. Phys.* **46,** 3489 (1967).
34. K. H. J. Buschow, A. M. Van Diepen, and H. W. de Wijn, *Phys. Rev.* **161,** 253 (1967).
35. K. H. J. Buschow, J. F. Fast, A. M. Van Diepen, and H. W. de Wijn, *Phys. Status Solidi.* **24,** 715 (1967).
36. A. M. Van Diepen, H. W. de Wijn, and K. H. J. Buschow, *J. Chem. Phys.* **46,** 3489 (1967); **51,** 5259 (1969).
37. A. M. Van Diepen, H. W. de Wijn, and K. H. Buschow, *Phys. Lett.* **26A,** 340 (1968).
38. A. M. Van Diepen, H. W. de Wijn, and K. H. J. Buschow, *Phys. Status Solidi* **29,** 189 (1968).
39. J. A. White, H. J. Williams, J. H. Wernick and R. C. Sherwood, *Phys. Rev.* **131,** 1039 (1963).
40. V. U. S. Rao and W. E. Wallace, *Phys. Rev.* **2B,** 4613 (1970).
41. B. R. Cooper and Y. L. Wang, *J. Appl. Phys.* **40,** 1344 (1969); *Phys. Rev.* **185,** 696 (1969); **163,** 444 (1967); **172,** 539 (1968).
42. W. E. Wallace, R. S. Craig, A. Thompson, C. Deenadas, M. Dixon, M. Aoyagi, and N. Marzouk, *Les Éléments des Terres Rares, Coll. Int. du C.N.R.S.* No. 180, 427 (1970).
43. C. Dennadas, A. W. Thompson, R. S. Craig, and W. E. Wallace, *J. Phys. Chem. Solids* **32,** 1853 (1971).
44. L. Pauling, *J. Amer. Chem. Soc.* **69,** 542 (1947).
45. R. E. Hungsberg and K. A. Gschneidner, Jr., *J. Phys. Chem. Solids* **33,** 401 (1972).
46. D. C. Hamilton and M. A. Jenson, *Phys. Rev. Lett.* **11,** 205 (1963); E. E. Havinga and M. H. Van Maaren, *Phys. Lett.* **31A,** 167 (1970).
47. H. G. Purwins, *Phys. Lett.* **31A,** 523 (1970).
48. B. Leon and W. E. Wallace, *J. Less Common Metals* **22,** 1 (1970).
49. B. Leon, V. U. S. Rao, and W. E. Wallace, *J. Less Common Metals* **24,** 247 (1971).
50. P. W. Anderson, *Phys. Rev.* **164,** 352 (1967).
51. W. C. Koehler, private communication (Oct., 1971).
52. F. Kissell and W. E. Wallace, *J. Less Common Metals* **11,** 417 (1966).
53. B. Barbara, C. Bécle, R. Lemaire, and R. Pauthenet, *J. Appl. Phys.* **39,** 1084 (1968).
54. K. H. J. Buschow and J. F. Fast, *Z. Phys. Chem. ,Frankfurt.* **50,** 1 (1966).
55. K. H. J. Buschow, *Z. Phys. Chem. Neue Folge* **59,** 21 (1968).
56. K. H. Mader and W. M. Swift, *J. Phys. Chem Solids* **29,** 1759 (1968).
57. H. J. van Daal, F. E. Maranzana, and K. H. J. Buschow, *J. Phys. Suppl.* **32,** C 1–424 (1971).
58. V. U. S. Rao, W. Suski, R. S. Craig, and W. E. Wallace, *Proc. Rare Earth Conf., 8th, Reno, Nevada* p. 355 April (1970).
59. W. G. Penney and R. Schlapp, *Phys. Rev.* **41,** 194 (1932).
60. D. P. Schumacher and C. A. Hollingsworth, *J. Phys. Chem. Solids* **27,** 749 (1966).
61. C. Bécle, R. Lemaire, and D. Paccard, *J. Appl. Phys.* **41,** 855 (1970).
62. K. H. J. Buschow, *Phys. Lett.* **29A,** 12 (1969).
63. B. Barbara, C. Bécle, R. Lemaire, and R. Pauthenet, *C. R. Acad. Sci. Paris* **267,** 309 (1968).
64. B. Barbara, C. Bécle, R. Lemaire, and D. Paccard, *J. Phys. Suppl.* **32,** C 1–299 (1971).
65. K. H. Mader and W. E. Wallace, *J. Less Common Metals* **16,** 85 (1968).
66. A. M. Van Diepen, K. H. J. Buschow, and H. W. de Wijn, *J. Chem. Phys.* **51,** 5259 (1969).
67. J. H. Wernick, H. J. Williams, and A. C. Gossard, *J. Phys. Chem. Solids* **28,** 271 (1967).

Chapter 5

GALLIUM, INDIUM, AND THALLIUM COMPOUNDS

In contrast with the compounds with aluminum, which have been intensively investigated, the systems with the other IIIA metals seem to have received the least attention to date of all the metals with which the rare earths enter into chemical union. TbGa was investigated by neutron diffraction techniques by Cable *et al.* [1]. A series of compounds represented by the formula RGa_2 was examined by Barbara *et al.* [2], and Colombo and Olcese [3] studied the several cerium compounds with gallium, indium, and thallium listed in Table 5-4. These appear to be the only studies of R–Ga compounds. The R–Tl systems have received even less attention, the study of $DyTl_3$ by Olsen *et al.* [4] being the only other known investigation dealing with these materials. The compounds involving indium have received somewhat more attention.

A. THE RIn_3 AND R_3In COMPOUNDS

1. Magnetic Properties

The RIn_3 and R_3In compounds have been studied by Buschow *et al.* [5] and Hutchens *et al.* [6], respectively. The single compound, $CeIn_3$, had been investigated earlier by Tsuchida and Wallace [7]. Results obtained are presented in Tables 5-1 and 5-2. Most of the RIn_3 compounds exhibit Curie–Weiss behavior at high temperatures with slopes characteristic of an assemblage of tripositive ions. The La, Ce, Sm, and Yb compounds are exceptions to this. The behavior of $SmIn_3$ (Fig. 5-1) is clearly a consequence of the Sm^{3+} multiplet structure. $LaIn_3$ is a Pauli paramagnet. The behavior of $CeIn_3$ is shown in Fig. 5-2, taken from

Tsuchida's measurements; those of Buschow *et al.* show essentially the same features. These susceptibilities, despite the fact that the slope is "normal" at higher temperatures, must be considered anomalous. The large negative Weiss constant is indicative of the operation of special effects apart from antiferromagnetic exchange. Since van Daal and Buschow [9] observe a resistance minimum for $CeIn_3$, it is probably like $CeAl_2$ in that spin compensation (Kondo phenomenon) and the crystal field interaction both operate to lower the susceptibility. Heat capacities of this compound are also anomalous (see below).

TABLE 5-1

Magnetic Characteristics of RIn_3 Compounds[a]

	Paramagnetic state			Ordered state		
R	μ_{eff} (μ_B/R)	θ (°K)	Type	T_N (°K)	Type	Ref.
La			P			[5]
Ce	2.62[b]	−54[b]	C–W,?	10.4	AF	[5,7,8]
Pr	3.58	−10	C–W,VV			[5]
Nd	3.62	−17	C–W	7	AF	[5]
Sm			N,m	16	AF	[5]
Gd	8.20	−85	C–W	45	AF	[5]
Tb	10.05	−62	C–W	36	AF	[5]
Dy	10.78	−35	C–W	23	AF	[5]
Ho	10.65	−18	C–W	11	AF	[5]
Er	9.75	−10	C–W	6	AF	[5]
Tm	7.6	−6	C–W			
Yb			N,VC			

[a]All compounds occur in the $AuCu_3$ structure.

[b]These are averages of the values obtained in Buschow *et al.* [5]; T_N is taken from Tsuchida and Wallace [7].

The majority of the RIn_3 compounds order antiferromagnetically at low temperatures. $PrIn_3$ does not order, but instead develops Van Vleck paramagnetism. Buschow *et al.* [5] have analyzed the susceptibility behavior of $PrIn_3$ to obtain information about its overall crystal field splitting, E_c. They have estimated E_c for various values of the x parameter, which establishes the relative importance of the sixth- to fourth-order interaction. Their estimates for E_c (difference between the Γ_1 ground state and the highest energy Γ_5 state) range from 180 to 390°K depending upon the value chosen for x.

$YbIn_3$ presents a number of interesting features. Its lattice parameter suggests that in it Yb is dipositive, and hence nonmagnetic. As expected, its susceptibility measured at room temperature is rather small

TABLE 5-2

Magnetic Characteristics of R_3In Compounds[a]

	Paramagnetic state			Ordered state			
R	μ_{eff} (μ_B/R)	θ (°K)	Type	μ_{or} (μ_B/f.u.)	T_C (°K)	Type	Ref.
Ce	2.68	−33	C–W,??				[3]
Pr	3.48	9	C–W	0.7	62	Fi	[6]
Nd	3.4	10	C–W	1.1	114	Fi	[6]
Gd	8.8	196	C–W	15.6	213	Fi	[6]
Dy	10.4	113	C–W	17.5	138	Fi	[6]
Er	9.6	31	C–W	15.5	51	Fi	[6]

[a]The Ce, Pr and Nd compounds are Cu_3Au. The structure of the others is not known. It is possible that they are not single phase materials.

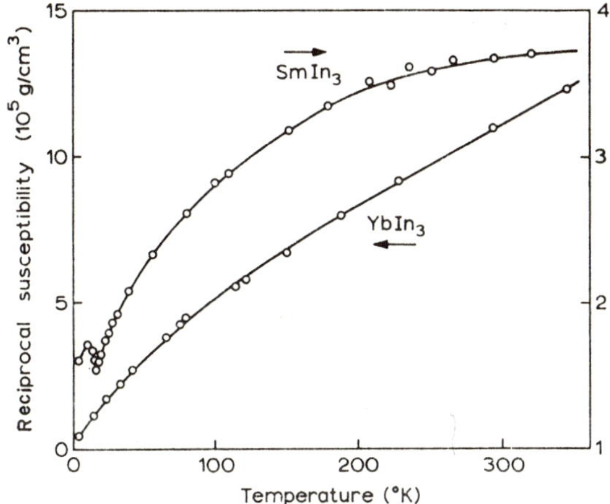

Fig. 5-1. Reciprocal susceptibility versus temperature for $SmIn_3$ and $YbIn_3$. Right-hand scale refers to $SmIn_3$ [5].

—two orders of magnitude below χ for the "normal" RIn_3 compounds. If Yb is indeed divalent, we expect $YbIn_3$ to behave as a Pauli paramagnet. This is not observed. Buschow *et al.* [5] find χ increasing sharply (Fig. 5-1) as temperature is reduced, approaching at 4°K the value expected of a system of Yb^{3+} ions. They ascribe the anomalous susceptibility behavior of this compound to a changing valence with temperature.

Interactions and ordering temperatures are somewhat larger for the R_3In compounds (Table 5-2). The low ordered moments observed for these materials suggest that they must be ferrimagnetic materials.

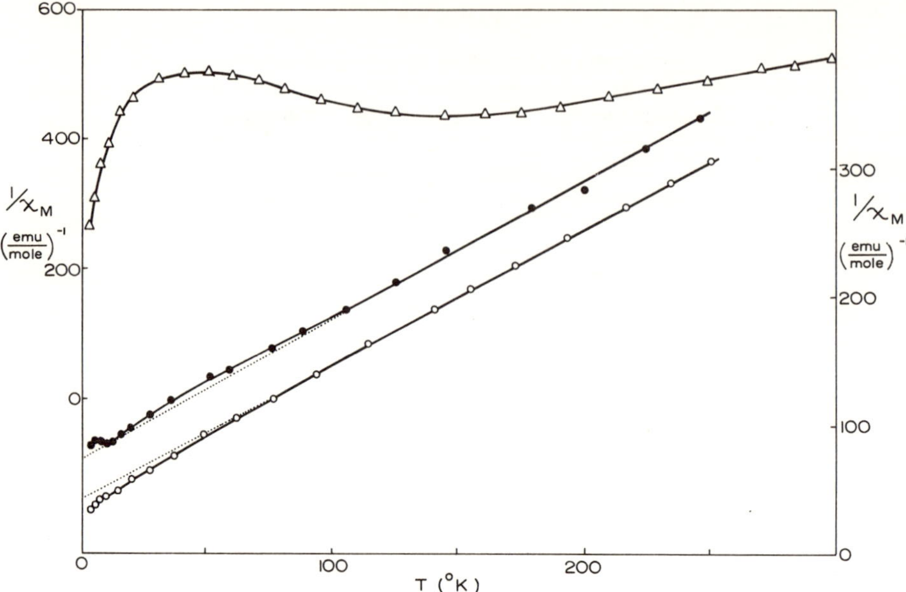

Fig. 5-2. Reciprocal susceptibilities versus temperature for $CeSn_3$ (Δ), $CeIn_3$ (●), and $CePb_3$ (O). Measurements were made at 19 kOe. The left-hand scale applies to $CeSn_3$, the right-hand scale to $CeIn_3$ and $CePb_3$ [7].

Neutron diffraction measurements would be helpful in clarifying the nature of their magnetic structures. $DyIn_3$ has been studied by this means, confirming [10] the ordering temperature and antiferromagnetic coupling inferred from bulk measurements.

2. Heat Capacity Studies

Heat capacities of RIn_3 with R = La, Ce, and Pr were determined over the ranges 7–300°K by Van Diepen *et al.* [8] and 1.5–4.2°K by Nasu *et al.* [11]. The data for $LaIn_3$ conformed to a Debye curve with $\theta_D =$ 170°K. These were used to represent the lattice and electronic heat capacity of $CeIn_3$ and $PrIn_3$. The difference between C_p of $PrIn_3$ and of $LaIn_3$, the crystal field contribution (C_{cf}), shown in Fig. 5-3, is Schottky type. Van Diepen *et al.* found a reasonable accounting for C_{cf} for $PrIn_3$ with an overall splitting of about 170°K, which lies at the lower limit of the value adduced by Buschow *et al.* from susceptibility measurements.

The difference between C_p for $CeIn_3$ and $LaIn_3$ is displayed in Figs. 5-4 and 5-5. The Schottky anomaly centered about 60°K is ascribed to excitation in the crystal field spectrum. Calculations were made assuming first the doublet Γ_7 state and then the quartet Γ_8 state to be lower lying. Experiment supports the former with the Γ_7 to Γ_8 spacing about 155°K. The λ-type anomaly at about 10°K is associated with the breakup of the antiferromagnetic structure. S for the process is found to be 6.0 mole^{-1} °K^{-1}, which is very near to $R \ln 2 = 5.76$. This supports the conclusion that the Γ_7 state is lower lying.

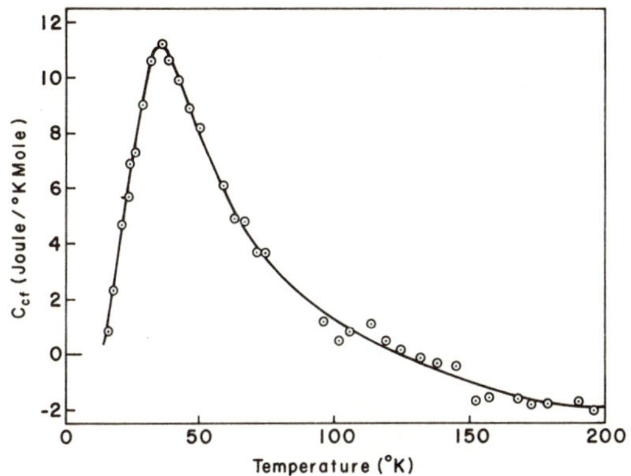

Fig. 5-3. Crystal field heat capacity of $PrIn_3$ [8].

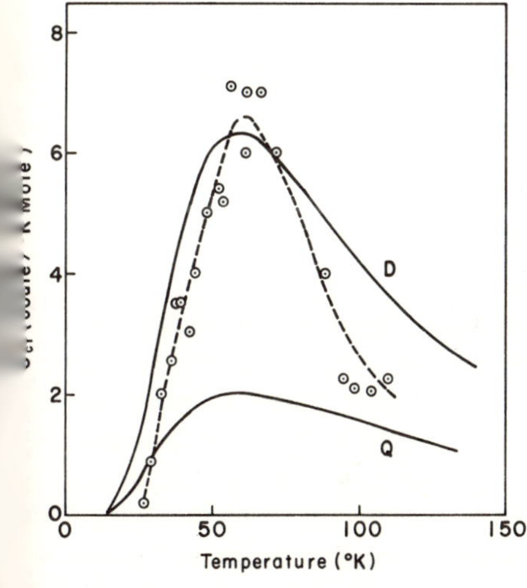

Fig. 5-4. Crystal field heat capacity of $CeIn_3$ ircles, dashed line), compared with theory r doublet lowest (curve D) and quartet west (curve Q) [8].

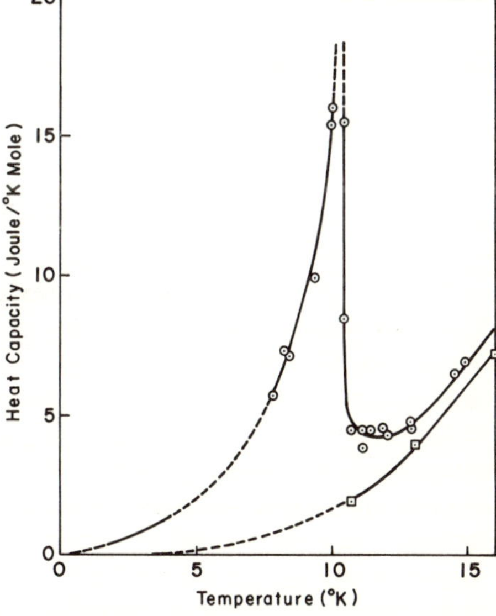

Fig. 5-5. Heat capacity of $CeIn_3$ and $LaIn_3$ (lower curve) [8].

Data for $LaIn_3$ and $PrIn_3$ in the 1.5–4.2°K range (Fig. 5-6) show the expected temperature dependence, i.e.,

$$C_p = \gamma T + \beta T^3. \qquad (5\text{-}1)$$

For C_p in mJ °K^{-1} mole^{-1} $\gamma = 6.28$ for $LaIn_3$ and 11.42 for $PrIn_3$. Corresponding values for β are 1.06 ($\theta_D = 104$°K) and 0.94 ($\theta_D = 202$°K).

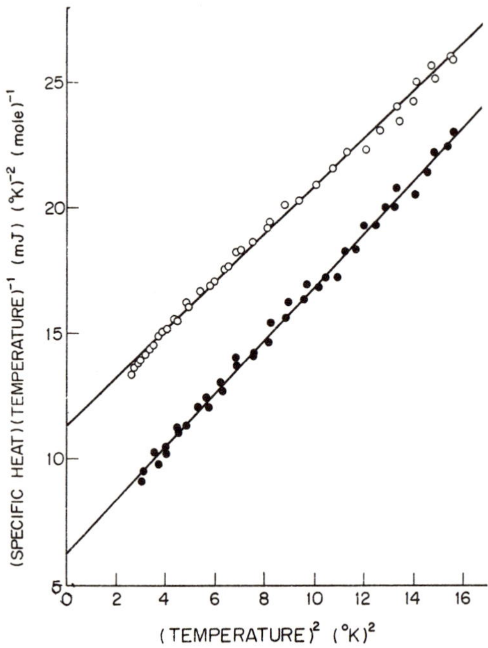

Fig. 5-6. Heat capacity of $LaIn_3$ (●) and $PrIn_3$ (○) [11].

Results for $CeIn_3$ do not conform to Eq. (5-1). They are, moreover, an order of magnitude larger than C_p for the other RIn_3 compounds. The source of the exceptionally large heat capacity is of some interest. We can readily exclude the possibility that the excess C_p is the tail of the Schottky anomaly in Fig. 5-3; calculations show this to be six orders of magnitude too small. (See Fig. 5-7.) The heat capacity of $CeIn_3$ is undoubtedly largely magnetic, the beginnings of the destruction of antiferromagnetism, but to establish this conclusively is not straightforward. In the simple case with $T << T_N$ and weak anisotropy the magnetic heat capacity of an antiferromagnet varies as T^3. However, $CeIn_3$ is not expected to exhibit this behavior for several reasons. First, the magnitude of the crystal field interaction indicates that it is not weakly anisotropic. Second, the demagnetization process occurs not within the entire J manifold but instead only within the Γ_7 doublet state. Finally, the two Γ_7 states have moments which vary with the degree of magnetic order. This is because the moments contain both permanent and induced components. The latter vary with magnetic order through the associated molecular field. These several complexities render uncertain the precise nature of the functional dependence of the magnetic heat capacity on temperature, but under the circumstances the cubic dependence on T seems extremely unlikely.

The excess heat capacity may be largely, but not exclusively, magnetic in origin. There may also be an enlarged density of states due to proximity of the 4f level to the Fermi energy. Spin fluctuation effects (see Chapter 12 for details involving $CePd_3$ and $LaPd_3$) may also be involved. Possibly some Kondo-type heat capacity is also included.

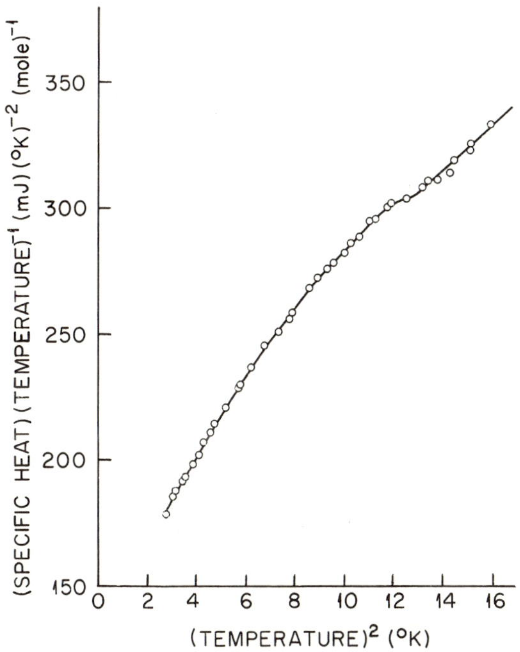

Fig. 5-7. Heat capacity behavior of $CeIn_3$ [11].

B. THE OTHER In, Ga, AND Tl COMPOUNDS

The several other compounds that have been examined are listed in Tables 5-3 and 5-4. They seem to merit little comment. The value of

μ_{eff} cited for $DyTl_3$ is obtained from data for $T < 200°K$. A break in the χ^{-1} versus T curve occurs at about 200°K with χ at higher temperatures falling below the "Curie–Weiss χ." Data for the Ce compounds are based upon data for $T > 73°K$. Hence, the low temperature behavior of these materials remains to be elucidated.

TABLE 5-3

Magnetic Characteristics of RGa_2 Compounds[a]

	Paramagnetic state			Ordered state		
R	μ_{eff} (μ_B/R)	θ (°K)	Type	T_N (°K)	Type	Ref.
Ce	2.76	−18	C–W,??			[3]
Pr	3.6	9	C–W,??	11	AF	[2]
Nd	3.7	10	C–W,??	14	AF	[2]
Tb	9.6	−20	C–W,??	18	AF	[2]
Dy	10.7	6	C–W,??	15	AF	[2]
Ho	10.7	2	C–W,??	10	AF,M	[2]

[a]These materials occur in the AlB_2 structure.

TABLE 5-4

Magnetic Properties of Miscellaneous In, Ga, and Tl Compounds

	Paramagnetic state				Ordered state				
	Structure	μ_{eff} (μ_B/R)	θ (°K)	Type	μ_{or} (μ_B/f.u.)	T_C (°K)	T_N (°K)	Type	Ref.
TbGa	CrB				9	155		F	[1]
TbIn	Complex Tetr.						190	AF	[1]
$DyTl_3$	$AuCu_3$	11.3	−20	C–W			11	AF	[4]
GdIn	?		−66				28	AF	[12]
Ce_3Ga	$AuCu_3$	2.64	−56	C–W,??[a]					[3]
Ce_3Ga_2	?	2.74	−40	C–W,??					[3]
CeGa	?	2.73	−21	C–W,??					[3]
Ce_2In	?	2.62	− 1	C–W,??					[3]
CeIn	?	2.63	+ 6	C–W,??					[3]
Ce_2In_3	?	2.73	− 8	C–W,??					[3]
Ce_2Tl	?	2.65	−33	C–W,??					[3]
CeTl	?	2.58	−22	C–W,??					[3]
$CeTl_3$	$AuCu_3$	2.40	− 4	C–W,??					[3]

[a]The compounds studied by Colombo and Olcese were not examined at temperatures below 75°K.

REFERENCES

1. J. W. Cable, W. C. Koehler, and E. O. Wollan, *Phys. Rev.* **136,** A240 (1964).
2. B. Barbara, C. Bécle, and E. Siaud, *J. Phys. C. Suppl.* **32,** 1–1126 (1971).
3. L. Colombo and G. L. Olcese, *Atti Lincei Rend. Sci. Fis. Mat. Nat.* **35,** 53 (1963))
4. C. E. Olsen, G. P. Arnold, and N. G. Nereson, *Proc. Rare Earth Conf.,* p. 63 (April 1970).
5. K. H. J. Buschow, H. W. de Wijn, and A. M. Van Diepen, *J. Chem. Phys.* **50,** 137 (1969).
6. R. D. Hutchens, W. E. Wallace, R. S. Craig, and V. U. S. Rao (to be published).
7. T. Tsuchida and W. E. Wallace, *J. Chem. Phys.* **43,** 3811 (1965).
8. A. M. Van Diepen, R. S. Craig, and W. E. Wallace, *J. Phys. Chem. Solids* **32,** 1867 (1971).
9. H. J. van Daal and K. H. J. Buschow, *Phys. Status Solidi* **3,** 853 (1970).
10. G. P. Arnold and N. G. Nereson, *J. Chem. Phys.* **51,** 1495 (1969).
11. S. Nasu, A. M. Van Diepen, H. H. Neumann, and R. S. Craig, *J. Phys. Chem. Solids* **32,** 2773 (1971).
12. K. Sekizawa and K. Yasukochi, *J. Phys. Soc. Japan* **21,** 684 (1966).

Chapter 6

COMPOUNDS WITH COPPER, SILVER, AND GOLD

The IB metals react with the rare earths to form two series of compounds that have been studied magnetically. These are represented by the formulas RM and RM_2, where M signifies copper, silver, or gold. The RM compounds are CsCl in structure [1], with the exception of CeCu, PrCu, and SmCu which occur in the FeB structure [2]. The RCu_2 series occurs in a distorted AlB_2 structure [3] whereas the RAg_2 and RAu_2 systems form in the $MoSi_2$ structure [4,5]. In addition, copper forms a series of compounds with stoichiometry given by the formula RCu_5; these are [6] of the cubic $AuBe_5$ structure. As regards magnetic behavior the 1:1 compounds have received the greatest attention to date.

A. THE RM COMPOUNDS

1. Binary Systems

Systematic studies of the RAg series were made by Walline and Wallace [7] and by Pierre and Pauthenet [8]. Results are accumulated in Table 6-1. SmAg, as expected, fails to exhibit Curie–Weiss paramagnetic behavior undoubtedly because of the multiplet structure of Sm^{3+}. The others show Curie–Weiss behavior at high temperatures but deviate at low temperatures. The behavior of CeAg (Fig. 6-1) could be ascribed to crystal field effects (Chapter 4) or spin compensation (Kondo phenomenon). The discrepancy between the measured χ and that computed from Curie's law is never more than a factor of 4 so that the reduction in χ could reasonably be ascribed solely to the influence of the crystal field (see discussion of $CeAl_2$ in Chapter 4). However, the Weiss constant for CeAg is −20°K [9] (or estimating from the data in Fig. 6-1 we get more nearly −50°K), which is quite a large negative quantity, so large in fact as to suggest a more complicated basis for the nonlinearity

TABLE 6-1

Magnetic Characteristics of RAg, RCu, and RAu Compounds[a]

	Paramagnetic state			Ordered state				
R	μ_{eff} (μ_B/R)	θ (°K)	Type	μ_{or} (μ_B/f.u.)	T_C (°K)	T_N (°K)	Type	Ref.
				Ag systems				
Ce	2.51	−20	C–W,?	0.95	9		F	[7–9]
Pr	3.44	2	C–W	1.55	14		F	[7,8]
Nd	3.59[b]	− 3	C–W,CF			22	AF	[7,8]
Sm			N,m					[7]
Gd	8.40[b]	−77[b]	C–W			150	AF	[7,8]
Tb	9.75[b]	−23[b]	C–W			106	AF	[7,8]
Dy	10.51[b]	−23	C–W			55	AF	[7,8]
Ho	10.09[b]	−12[b]	C–W			32	AF	[7,8]
Er	9.22	− 9.5	C–W,CF			18	AF	[7,8]
Tm	7.34[b]	− 5.5[b]	C–W			10	AF	[7,8]
				Cu Systems				
Ce	2.15	0	C–W			2.7(?)	AF(?)	[10]
Pr	3.66	−11	C–W,VV	Probably does not order				[10]
Nd	3.69	− 2	C–W					[10]
Sm			N,m			40(?)	AF(?)	[10]
Gd	8.46	−26	C–W			41(?)	AF	[10]
Tb	9.68	−23	C–W			115[c]	AF	[10,11]
Dy	10.70[b]	−22[b]	C–W			61	AF	[8,10]
Ho	10.73	−13	C–W			28	AF	[10]
Er			N,?			17	AF	[10]
Tm	7.56	− 8				11	AF	[10]
				Au systems				
Gd	7.92	29	C–W			37	AF	[12]
Tb	9.54	23	C–W			43	AF	[12]
Dy	10.22	7	C–W,?			14	AF(?)	[12,13]
Ho	10.50	0	C–W			10	AF	[12,14]
Er	9.42	− 4	C–W					[12,14]
Tm	7.32	− 5	C–W					[12,14]
Yb			N,?					[12]

[a]All these compounds occur in the CsCl structure except CeCu, PrCu, and SmCu, which have the FeB structure.

[b]These are averages of values found in Walline and Wallace [7] and Pierre and Pauthenet [8].

[c]TbCu shows a second susceptibility maximum [7] at 48°K (see text).

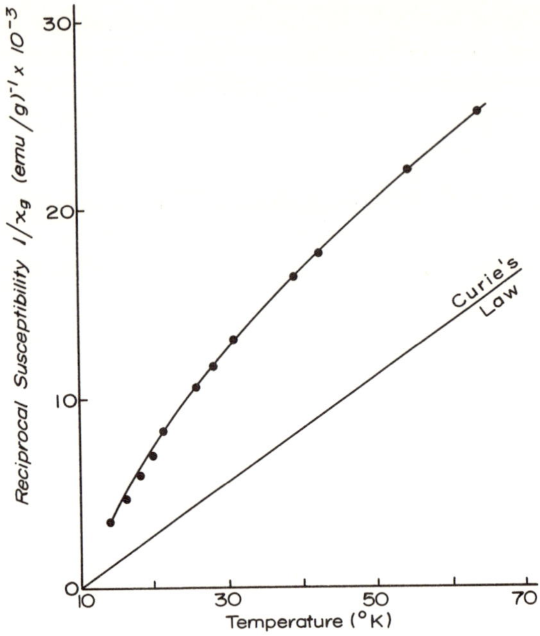

Fig. 6-1. Reciprocal susceptibility-temperature curve for CeAg at 6.48 kOe [7]. Curie law behavior for Ce^{3+} is also shown.

of χ^{-1} versus T. It seems likely that CeAg is a Kondo system, but information about its resistivity at low temperatures which could confirm this is lacking. The existence of an ordered AF structure in DyAg has

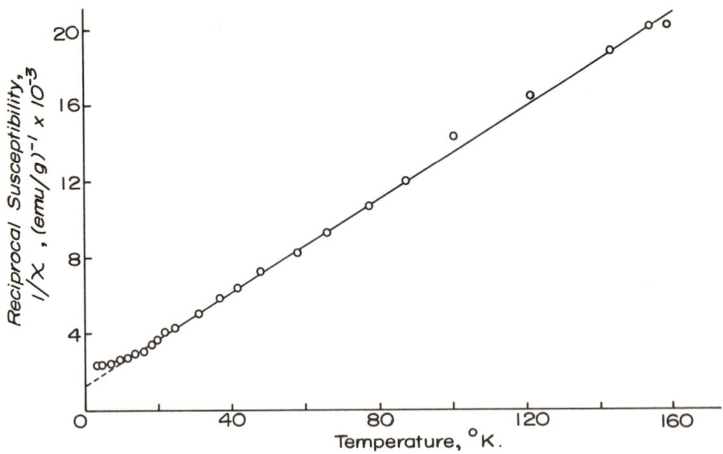

Fig. 6-2. Reciprocal susceptibility-temperature curve for PrCu at 6.47 kOe [10].

been confirmed by Arnold *et al.* [15] who established T_N to be 51°K.

Magnetization of the RCu compounds was investigated by Walline and Wallace [10] and the Dy and Tb compounds by Pierre and Pauthenet [8,11]. The status of the compounds with R = Ce through Sm is not satisfactorily clarified. Except for SmCu these compounds exhibit Curie–Weiss behavior with reasonable effective moments. There is no unambiguous indication of ordering for any of the four compounds, although SmCu and CeCu exhibit weak susceptibility peaks and hence are listed as antiferromagnets in Table 6-1. PrCu (Fig. 6-2) seems to remain paramagnetic and exhibits field independent susceptibility behavior at 2°K. There is some indication with it of the onset of Van Vleck paramagnetism. Representative results for the heavy rare earth compounds are shown in Figs. 6-3 and 6-4. Curie–Weiss behavior is

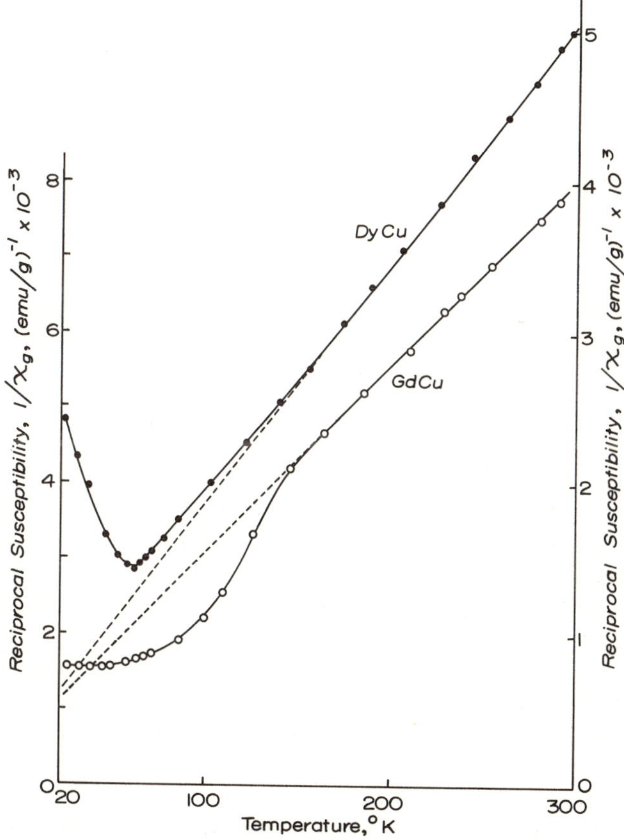

Fig. 6-3. Reciprocal susceptibility-temperature curve for GdCu (left scale) and DyCu (right scale) at 6.47 kOe [10].

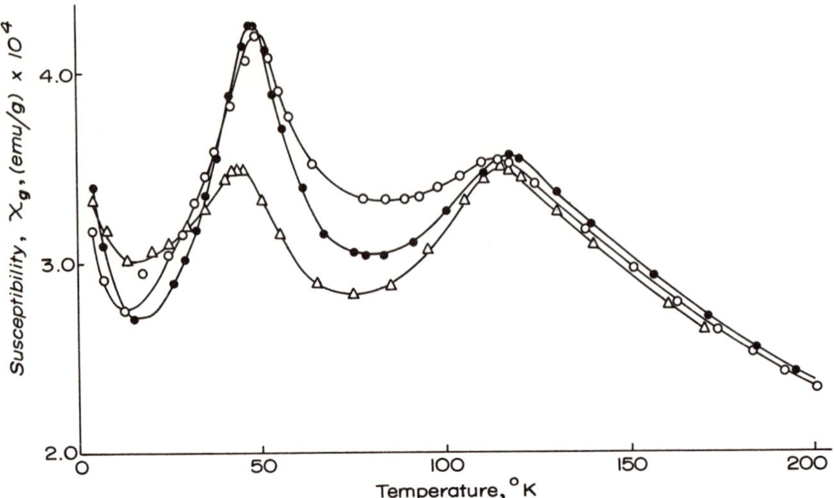

Fig. 6-4. Susceptibility-temperature curves for TbCu (2 samples) [10]. Sample 1, $H=6.50$ kOe (O), $H=19.1$ kOe (Δ); sample 2, $H=6.50$ kOe (●).

observed, but for GdCu the moment is somewhat higher than expected for an assemblage of Gd^{3+} ions. This, which is also true for GdAg, is ascribed to the contribution of the polarized conduction electrons. GdCu shows a weak susceptibility maximum at 41°K which may be due to antiferromagnetism. There are two susceptibility maxima for TbCu and also for HoCu [16]. The higher temperature maxima is the Néel temperature. Cable *et al.* [17] have confirmed by neutron diffraction studies than an antiferromagnetic structure develops at 115°K and from similar work Pierre has inferred [11] a change in direction of magnetization at about 45°K, which seemingly produces the lower temperature susceptibility maximum.

The heavy rare earth–gold compounds were studied by Kissell and Wallace [12] and DyAu was investigated by Kaneko [13] at fields up to 80 kOe. Except for YbAu, the paramagnetic behavior is straightforward. Weiss constants and ordering temperatures are low, implying rather weak interactions. There is no indication of ordering at 2°K for the Er, Tm, and Yb compounds. This is not unreasonable since on the basis of T_N for GdAu (37°K) and the relative de Gennes function, ordering temperatures lower than 4°K are expected for these compounds. Magnetization versus field data for ErAu are shown in Fig. 6-5 compared with the Brillouin function for $J = 15/2$. This suggests a paramagnetic material at 4°K but with appreciable antiferromagnetic exchange. YbAu represents an interesting situation (Fig. 6-6). The linear dependence of reciprocal susceptibility on temperature above about 60°K appears to be

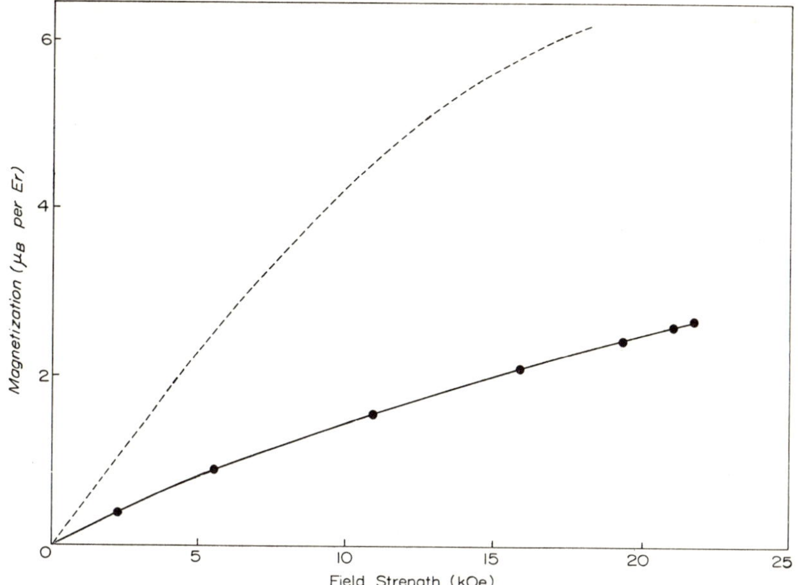

Fig. 6-5. Magnetization–field strength curve for ErAu at 4.2°K [12]. The dashed curve gives the Brillouin function for $J = 15/2$.

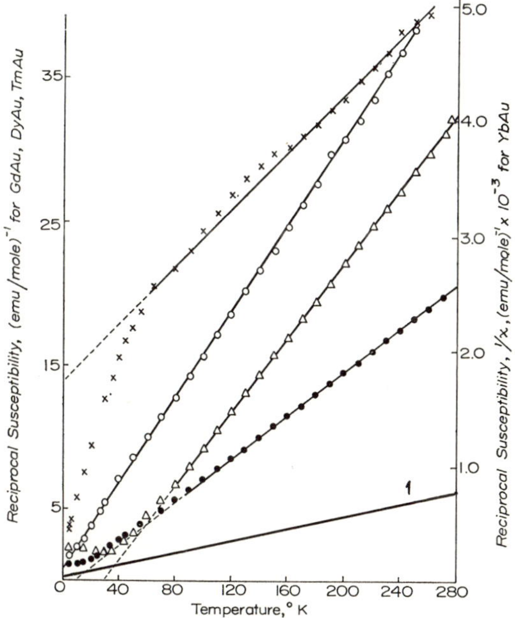

Fig. 6-6. Reciprocal susceptibility–temperature behavior for GdAu (Δ), DyAu (●), TmAu (O), YbAu (×), measured at 19.3 kOe [12]. Curie's law for Yb^{3+} is also shown, 1.

fortuituous since the effective moment, 0.81 μB, obtained does not correlate with that expected for any reasonable electronic configuration of ytterbium. Its χ is a factor of 6 or more smaller than the Curie law susceptibility for tripositive Yb. The suggestion was put forward earlier [12] that the small susceptibility was due to a mixture of Yb^{2+} and Yb^{3+} in YbAu, the former being nonmagnetic. There is, however, a problem in this concept in that it requires (because of the rapid increase in slope below 60°K) that the proportion of the divalent (i.e., the larger) species be increasing as temperature is reduced. We expect from Le Chatelier's principle an increase in concentration of Yb^{3+} because of the shrinkage of the lattice. Perhaps the behavior of YbAu is a crystal field effect or a consequence of spin compensation, or a combination of the two.

Chao has examined [14] the resistivity behavior of many of the silver, copper, and gold compounds. The decrease of ρ at the ordering temperature because of the disappearance of the spin-disorder contribution is very evident in his measurements. The ordering temperatures inferred by him from his ρ versus T determinations are in good agreement with those listed in Table 6-1.

2. Ternary Systems

The first work on RM ternaries appears to have been that of Sekizawa and Yasukochi [18] on the $RAg_{1-x}In_x$ system. They found, in a celebrated paper, that antiferromagnetic GdAg can be converted into a ferromagnetic material by replacement of Ag with In (Fig. 6-7). This is ascribed to the changing electron concentration and the effect on the RKKY sum in Eq. (2-3). The effect here is like that in the $Eu_{1-x}La_xAl_2$ ternaries described in Chapter 4, but we note that the work of Sekizawa and Yosukochi was carried out earlier. Similar work has been performed by Alfieri *et al.* [19] on $GdAg_{1-x}Zn_x$ and $GdCu_{1-x}Zn_x$ ternaries and by Pierre [11] on $TbCu_{1-x}Zn_x$ ternaries. In each case, the replacement of Cu or Ag by Zn produces a change in coupling from antiferromagnetic to ferromagnetic. Pierre [20] has examined the change in ordering quantitatively in terms of Eq. (2-3), using the reasoning similar to that of Swift and Wallace which was described for the $Eu_{1-x}La_zAl_2$ ternaries in Chapter 4. He finds that the behavior can be accounted for under the assumption that $k_F \approx 0.85\ k_F^0$. We recall that when a similar procedure was employed with the Al ternaries, k_F was obtained to be approximately $0.94\ k_F^0$.

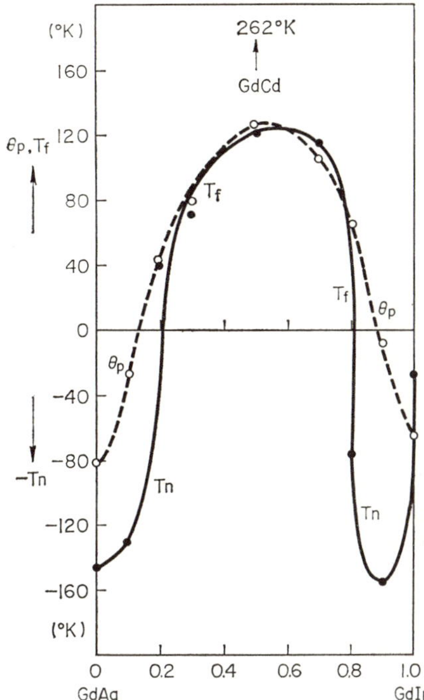

Fig. 6-7. The Néel temperature T_N, the ferromagnetic Curie temperature T_f, and the paramagnetic Curie temperature θ_p, of the compounds in Gd(Ag,In) system, plotted against the proportion of GdIn [18].

B. THE RM_2 COMPOUNDS

The RCu_2 series was investigated by Sherwood *et al.* [21] by conventional bulk magnetic methods. Atoji has examined a number of the Ag and Au compounds by means of neutron diffraction [22]. Miura *et al.* [23] have determined the magnetization of $DyAg_2$ and $DyAu_2$ at fields up to 100 kOe. A summary of the results obtained is given in Tables 6-2 and 6-3. No information is provided concerning the ordering temperature of the Ce through Sm compounds, probably because these compounds failed to order at 4.2°K. Yb is divalent, as is Eu, in the RCu_2 compounds and hence the Yb and Lu compounds are Pauli paramagnets. The heavy rare earth compounds show well-defined susceptibility peaks indicating antiferromagnetism. Magnetization measurements to 80 kOe show a breakup of the low field structure at fields ranging upwards from 5 kOe, so that these are metamagnetic materials. The critical field for $EuCu_2$ seems to be quite high, $>$50 kOe.

TABLE 6-2

Magnetic Characteristics of RCu_2 Compounds

R	Paramagnetic state		Ordered state			Ref.
	μ_{eff} (μ_B/R)	Type	μ_{or} (μ_B/f.u.)	T_N (°K)	Type	
Ce		N,?	0.8[a]			[21]
Pr	3.51	C–W	2.3[a]			[21]
Nd	3.56	C–W	1.9[a]			[21]
Sm		N,m	0.1			[21]
Eu	7.4		5.8	14	AF	[21,24]
Gd	8.4	C–W	6.0	41	AF	[21]
Tb[b]	9.8	C–W	7.4[a]	54	AF(M)	[21]
Dy	10.8	C–W	8.7[a]	24	AF(M)	[21]
Ho	10.5	C–W	9.2[a]	9	AF(M)	[21]
Er	9.4	C–W	5.6[a]	11		[21]
Tm	7.49	C–W	4.2			[21]
Yb		P				[21]
Lu		P				[21]

[a]Moments were measured at H up to 80 kOe and 4.2°K and extrapolated to infinite field. μ_{or} for the other compounds is the value at 80 kOe and 4.2°K.

[b]This compound shows a pronounced change in slope for the χ^{-1} versus T curve at about 150°K.

TABLE 6-3

Magnetic Characteristics of RAg_2 and RAu_2 Compounds

R	RAg_2 compounds				RAu_2 compounds				Ref.
	μ_{eff} (μ_B/R)	θ (°K)	T_N (°K)	Type	μ_{eff} (μ_B/R)	θ (°K)	T_N (°K)	Type	
Tb			35	AF			55	AF[a]	[22]
Dy	10.5	−25	15.9[b]	AF[a]			33.8,31[b]	AF[a]	[22,23]
Ho			5.7	AF[a]			9.2	AF	[22]
Er			5.15	AF			6.8	AF	[22]
Tm			<1.5	AF			3	AF	[22]
Yb							<1.5	AF	[22]

[a]There is a second antiferromagnetically ordered form existing below T'. T' is as follows in °K: $TbAu_2$–42.5, $DyAu_2$–25.5, $DyAg_2$–9.5, $HoAg_2$–3.8.

[b]These Néel temperatures are μ_{eff} and θ are from Miura *et al.* [23]. The rest of the tabulated data are from Atoji [22].

C. THE RCu_5 COMPOUNDS

Bulk magnetic characteristics of these compounds were measured by Buschow *et al.* [25]. Results obtained are summarized in Table 6-4. They also made ^{63}Cu NMR measurements on these compounds and analyzed their results in the manner similar to that used in work on the RAl_2 compounds (see Chapter 4). The Weiss constant and Knight shifts (which were negative) can be interpreted on the basis of $k_F/k_F^0 = 0.78$ and $J \approx 0.1$ eV [see Eq. (4-1)].

TABLE 6-4

Magnetic Characteristics of RCu_5 Compounds

	Paramagnetic state			Ordered state				
R	μ_{eff} (μ_B/f.u.)	θ (°K)	Type	μ_{or} (μ_B/f.u.)	T_N (°K)	T_C (°K)	Type	Ref.
Tb	9.6	2	C–W		15		AF	[25]
Dy	10.9	2	C–W	6.5	7		AF(M)	[25]
Ho	10.8	~0	C	7.6			F	[25]
Er	9.7	~0	C	7.3		~7	F	[25]
Tm	7.4	~0	C	4.9			F	[25]

REFERENCES

1. C. C. Chao, H. L. Luo, and P. Duwez, *J. Appl. Phys.* **34,** 1971 (1963).
2. A. C. Larson and D. T. Cromer, *Acta Crystallogr.* **14,** 946 (1961).
3. A. C. Larson and D. T. Cromer, *Acta Crystallogr.* **14,** 73 (1961).
4. A. R. Storm and K. E. Benson, *Acta Crystallogr.* **16,** 701 (1963).
5. A. E. Dwight, J. W. Downey, and R. A. Conner, *Acta Crystallogr.* **22,** 745 (1967).
6. K. H. J. Buschow, A. S. Van der Goot, and F. Birkan, *J. Less Common Metals* **19,** 433 (1969).
7. R. E. Walline and W. E. Wallace, *J. Chem. Phys.* **41,** 3285 (1964).
8. J. Pierre and R. Pauthenet, *C. R. Acad. Sci. Paris* **260,** 2739 (1965).
9. G. L. Olcese, *Atti Accad. Naz. Lincei Rend. Cl. Sci. Fis. Mat. Natur.* **34,** 642 (1963).
10. R. E. Walline and W. E. Wallace, *J. Chem. Phys.* **42,** 604 (1965).
11. J. Pierre, *C. R. Acad. Sci. Paris* **265B,** 1169 (1967); and also in *Les Éléments des Terres Rares Coll. Int. C.N.R.S.* No. 180, Tome II, p. 55 (1970).
12. F. Kissell and W. E. Wallace, *J. Less Common Metals* **11,** 417 (1966).
13. T. Kaneko, *J. Phys. Soc. Japan* **25,** 905 (1968).
14. C. C. Chao, *J. Appl. Phys.* **37,** 2081 (1966).
15. G. Arnold, N. Nereson, and C. Olsen, *J. Chem. Phys.* **46,** 4041 (1967).
16. R. E. Walline, Ph.D. Dissertation, Univ. of Pittsburgh (1964).

17. J. W. Cable, W. C. Koehler, and E. O. Wollan, *Bull. Amer. Phys. Soc.* **9,** 213 (1964).
18. K. Sekizawa and K. Yasukochi, *J. Phys. Soc. Japan* **21,** 684 (1966).
19. G. T. Alfieri, E. Banks, and K. Kanematsu, *J. Appl. Phys.* **37,** 1254 (1966).
20. J. Pierre, *Solid State Commun.* **7,** 165 (1969).
21. R. C. Sherwood, H. J. Williams, and J. H. Wernick, *J. Appl. Phys.* **35,** 1049 (1964).
22. M. Atoji, *J. Chem. Phys.* **48,** 560, 3380 (1968). He has also made magnetic structure determinations on RAu_2 and RAg_2 with Dy, Ho, Er, Tm and Yb and communicated the results of this extensive work to the author privately.
23. S. Miura, T. Kaneko, M. Ohashi, and K. Kamigaki, *J. Phys. Suppl.* **32,** C 1–1124 (1971).
24. H. H. Wickman, J. H. Wernick, R. C. Sherwood, and C. F. Wagner, *J. Phys. Chem. Solids* **29,** 181 (1968).
25. K. H. J. Buschow, A. M. Van Diepen, and H. W. de Wijn, *J. Appl. Phys.* **41,** 4609 (1970).

Chapter 7

COMPOUNDS WITH Si, Ge, Sn, AND Pb

The rare earths form many compounds with the group IVA elements. Although these elements lie well to the right in the Periodic Table, approaching the typically ionic halides and rather ionic chalcogenides, the structures involved and the electrical transport properties indicate that the compounds with the silicon group elements are indeed intermetallics. For example, we find the equiatomic germanium and silicon compounds occurring (Table 7-1) in the CrB and FeB structures, in which we also find such typical intermetallics as the compounds involving Cu and Ag. The RSn_3 and RPb_3 series occur in the characteristic metallic structure $AuCu_3$. Buschow and Fast [1] indicate the conductivity of the RGe compounds to be comparable with the pure rare earths.

Compounds between the rare earths and the IVA elements occur with a variety of stoichiometries and while their magnetic properties have received a moderate amount of attention, it is clear that they have not been exhaustively investigated. Results of work to date are summarized in Tables 7-1 through 7-6.

A. THE EQUIATOMIC COMPOUNDS

These have been studied by Buschow and Fast [1], Ruggiero and Olcese [2], and Nhung *et al.* [3]. The latter investigators studied only TbSi but examined it by bulk magnetic measurements and neutron diffraction. More recently the work has been extended [4] to include the Ho, Er, and Tm compounds and $TbGe_{0.2}Si_{0.8}$. Results in the main are straightforward. The Sm compound fails to exhibit Curie–Weiss behavior because of excitation out of the ground state multiplet. The other

TABLE 7-1

Magnetic and Structural Characteristics of 1:1 Ge and Si Compounds

	Paramagnetic state				Ordered state				
	Structure	μ_{eff} (μ_B/R)	θ (°K)	Type	μ_{or} (μ_B/R)	T_C (°K)	T_N (°K)	Type	Ref.
			Ge compounds						
La	FeB			P					[1]
Ce	FeB	2.50	0	C			10	AF	[1,2]
Pr	CrB	3.69	30	C–W	1.73	39		F	[1]
Nd	CrB	3.74	18	C–W	1.64	28		F	[1]
Sm	CrB			N,m					[1]
Gd	CrB	8.21	−13	C–W			62	AF	[1]
Tb	CrB	9.76	− 5	C–W			48	AF	[1]
Dy	CrB	10.60	0	C			36	AF	[1]
Ho	CrB	10.77	− 5	C–W	6.5		18	AF(M)	[1]
Er	CrB	10.0	−15	C–W			7	AF	[1]
			Silicon compounds						
Ce	FeB	2.50	− 7	C–W					[2]
Tb	FeB	10.1	− 8	C–W			57	AF	[2,3]
Ho	CrB	10.6	2	C–W			25	AF	[4]
Er	CrB	9.4	− 5	C–W			10	AF	[4]
Tm	CrB	7.5	−10	C–W			10	AF	[4]
			Ternary phase						
$TbGe_{0.2}Si_{0.8}$	CrB	9.9	17	C–W			56	AF	[4]

systems show paramagnetic moments in accordance with expectation for assemblages of free tripositive ions. The GdGe moment exceeds $g[J(J+1)]^{1/2}$ for Gd, but this is not unique. Similar behavior was noted for Gd compounds with Cu and Ag (Tables 6-1 and 6-2) and with In (Tables 5-1 and 5-2). The excess moment is regarded as a consequence of the conduction electron polarization.

High field magnetization measurements on HoGe show metamagnetic behavior for this compound; two critical fields were noted by Buschow and Fast with a final moment of 6.5 μ_B/f.u. achieved in fields in excess of 80 kOe. This, and the moments for PrGe and NdGe, are small compared to gJ and are interpreted as being reduced because of the crystal field interaction.

Nhung *et al.* [3] find a collinear structure for TbSi.

B. THE 5:3 AND 5:4 COMPOUNDS

1. Binary Systems

The R_5Ge_3 compounds were studied by Buschow and Fast [5] and the Ce compound by Ruggiero and Olcese [2] with the results summarized in Table 7-2. The large paramagnetic moment for the Gd compound is also apparent here; the other compounds show μ_{eff} well in excess of $g[J(J+1)]^{1/2}$ indicating a significant contribution from the polarized conduction electrons for all the R_5Ge_3 compounds.

TABLE 7-2

Magnetic Characteristics of R_5Ge_3 Compounds[a]

	Paramagnetic state			Ordered state				
	μ_{eff} (μ_B/R)	θ (°K)	Type	μ_{or} (μ_B/R)	T_C (°K)	T_N (°K)	Type	Ref.
La			P					[5]
Ce	2.66	−50	C–W		<4.2		F(?)	[2,5]
Pr	3.80	10	C–W,VV					[5]
Nd	3.73	20	C–W	1.92	45		F or F_i	[5]
Gd	8.42	70	C–W			48	AF	[5]
Tb	9.98	93	C–W			85	AF	[5]
Dy	10.47	45	C–W			40	AF	[5]
Ho	11.10	10	C–W			10	AF	[5]
Er	9.48	35	C–W			31	AF(M)	[5]

[a]These occur in the Mn_5Si_3 structure ($D8_8$).

The heavy rare earth compounds order antiferromagnetically. Metamagnetism is observed [5] for Er_5Ge_3 with a critical field of about 15 kOe. Buschow and Fast indicate qualitatively similar behavior for the corresponding Dy and Tb compounds but were unable to establish the critical field. Ce_5Ge_3 and Nd_5Ge_3 are taken to be ferromagnets, with moment in the latter case partially quenched by the crystal field, whereas the magnetization temperature curve for Pr_5Ge_3, which shows a small maximum at 12°K, is interpreted as implying antiferromagnetism. The Pr compound thus is not in line with its neighbors in the sequence and it seems likely that the interpretation of the magnetization maximum is in error. The maximum in this case is probably a special manifestation of Van Vleck paramagnetism which is briefly discussed for $PrNi_5$ in Chapter 9. Appropriate calculations have not been made for Pr^{3+} in the symmetry which prevails in Pr_5Ge_3 to permit a similar analysis; hence the present suggestion as regards its low temperature behavior must be regarded as merely a plausible postulate.

The 5:4 compounds, studied by Holtzberg *et al.* [6] (Table 7-3), show one particularly striking characteristic—T_C for Gd_5Si_4 is substantially increased (58°) over the Curie temperature of elemental Gd. There are only two cases in which T_C for Gd is increased when it is chemically united with a nonmagnetic partner. Gd_5Pd_2 (see Chapter 12) with T_C = 335°K is the other intermetallic for which an increased T_C is observed. Despite the same stoichiometry the Ge and Si compounds are not isostructural. The differences are reflected in the contrasting magnetic structures (F for Gd_5Si_4 and AF for Gd_5Ge_4) and in the large difference in ordering temperatures. In regard to the magnetic interactions, they are in striking conformity with Eq. (2-3) in the excellent linearity of the Weiss constants (Fig. 7-1) with the de Gennes function $(g-1)^2J(J+1)$.

TABLE 7-3

Magnetic Characteristics of 5:4 Ge and Si Compounds[a]

	Paramagnetic state			Ordered state				
	μ_{eff} (μ_B/R)	θ (°K)	Type	μ_{or} (μ_B/f.u.)	T_C (°K)	T_N (°K)	Type	Ref.
				R_5Ge_4 series				
Gd	8.10	94	C–W			15	AF	[6]
Tb	9.61	80	C–W,??			30	AF	[6]
Dy	10.5	43	C–W,??			40	AF	[6]
Ho	10.7	16	C–W,??			21	AF	[6]
Er	9.75	10	C–W,??			7	AF	[6]
				R_5Si_4 series				
Gd	8.10	349	C–W	36.2	336		F	[6]
Tb	9.31	216	C–W,??	32.4	225		F	[6]
Dy	10.3	133	C–W,??	35.4	140		F	[6]
Ho	11.1	69	C–W,??	37.2	76		F	[6]
Er	9.9	20	C–W,??	30.6	25		F	[6]

[a] These are all orthorhombic materials but according to Holtzberg *et al.* [6] the Ge and Si compounds are not isostructural.

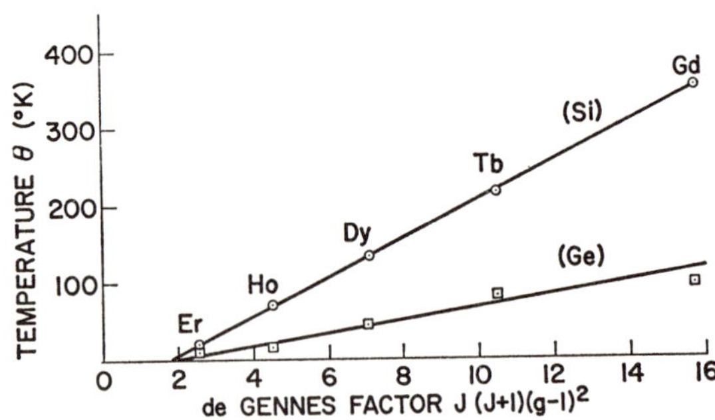

Fig. 7-1. The dependence of θ values of the RE_5Si_4 and RE_5Ge_4 compounds on the deGennes factor [6].

The R_5Ge_4 compounds have large positive Weiss constants but surprisingly they order antiferromagnetically.

Except for Gd_5Si_4, ordered moments are well below $5gJ$, the value expected for free tripositive ions. The discrepancy is ascribed to incomplete saturation or partial quenching of the moment or both by the crystal field.

2. Ternary 5:4 Systems

Holtzberg *et al.* [6] also examined the pseudobinary system Gd_5Si_4–Gd_5Ge_4. Salient features of this work are set forth in Figs. 7-2 and 7-3. They find that considerable Si in Gd_5Si_4 can be replaced with Ge with little effect on the magnetic properties. However, the converse is not true in that small amounts of Si incorporated into Gd_5Ge_4 produce profound magnetic changes. The coupling is changed to ferromagnetic by replacing one Ge atom out of 32 by silicon.* The large positive values of

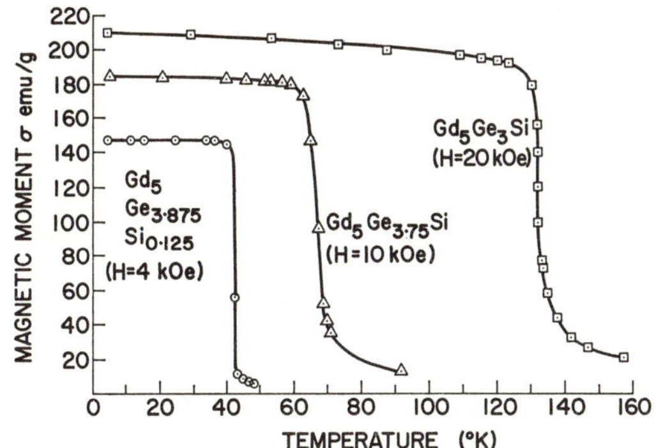

Fig. 7-2. Magnetic moment versus temperature behavior representative of $Gd_5Si_{4-x}Ge_x$ solid solutions in the range $x = 3$ to 4 [6].

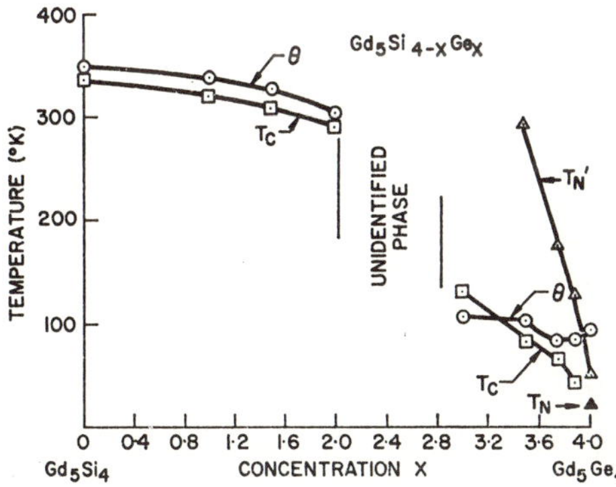

Fig. 7-3. The variation of Curie temperature T_C, Néel temperature T_N and $T_{N'}$, and θ on concentration (x) in the system $Gd_5Si_{4-x}Ge_x$ [6].

*The Néel temperature of Gd_5Ge_4 is taken to be 15°K (see Table 7-3). However, a second weak susceptibility is seen at about 58°K, which Holtzberg *et al.* call T_N'. This indicates that Gd_5Ge_4 may order in two stages.

θ suggest that ferromagnetic interactions are strong in the R_5Ge_4 series. Evidently replacing a small amount of Ge with Si allows them to become dominant. The anomalous inversion of ordering temperatures between Gd_5Ge_4 and Tb_5Ge_4 is further indication that the compound is in a critical condition in regard to magnetic ordering.

C. THE RSi_2 AND RGe_2 COMPOUNDS

These systems have been examined by Sekizawa and Yasukochi [7,8] and the cerium compounds by Ruggiero and Olcese [2] (Table 7-4). All except $PrGe_2$ and $NdGe_2$ are antiferromagnetic. These two compounds are ferromagnetic but with an inverted order of curie temperatures, at least as viewed in terms of the de Gennes treatment of exchange.

TABLE 7-4

Magnetic and Structural Characteristics of RGe_2 and RSi_2 Compounds

		Paramagnetic state			Ordered state				
	Structure[a]	μ_{eff} (μ_B/R)	θ (°K)	Type	μ_{or} (μ_B/f.u.)	T_C (°K)	T_N (°K)	Type	Ref.
				RGe_2 series					
Ce	C_c	2.46	5	C−W,??					[2]
Pr	C_c	3.6	22	C−W,??	2.18	19		F	[7]
Nd	C_c	3.7	7	C−W,??		3.6		F	[7]
Sm	C_c			N,m					[7]
Gd	OR	8.1	−54	C−W,??			28	AF	[7]
Tb	OR	9.9	−48	C−W			42	AF	[7]
Dy	OR	10.9	−26	C−W,??			28	AF	[7]
Ho	?	10.7	−4	C−W,??			11	AF	[7]
Er	?	9.4	+20	C−W,??			< 4		[7]
				RSi_2 series					
Ce		2.94	−299	C−W,??					[2]
Gd	dC_c	2.94	−58	C−W			27	AF	[8]
Tb	dC_c	9.9	−18	C−W			17	AF	[8]
Dy	dC_c	10.4	−8	C−W			17	AF	[8]
Ho	dC_c	10.4	−5	C−W,??			18	AF	[8]
Er	C32	9.3	0	C,??			< 4	?	[8]

[a] C_c is the $\alpha ThSi_2$ structure and dC_c is a distorted version of this structure. OR denotes an unsolved orthorhombic structure. It is to be noted that Lundin [9] regards the true stoichiometry of the RSi_2 compounds to be R_5Si_3 for R = Gd through Er.

Sekizawa [7] has drawn attention to this, to the fact that the ordering temperature in the RGe_2 series is maximal not at Gd but at Tb, and to the change of the sign of the interactions in the sequence. She postulates a non-RKKY contribution, perhaps due to small direct exchange, which increases in importance with decreasing atomic number of R and which is competitive with the normal conduction electron polarization exchange. The rising importance of this accounts for the reduced Néel temperature of $GdGe_2$ compared to $TbGe_2$ and the change to weak and then stronger ferromagnetism in $NdGe_2$ and $PrGe_2$, respectively. The trend in these two compounds is evident not only in T_C but in θ as well.

The mode of ordering in $SmGe_2$ is unknown. It remains paramagnetic to 4°K.

Except for $CeSi_2$, the paramagnetic behavior of the 1:2 compounds is in keeping with expectations. $SmGe_2$ does not exhibit Curie–Weiss behavior but the other members of both series do, with reasonable effective moments. The very large negative θ for $CeSi_2$ is puzzling. We would expect to see antiferromagnetic order in excess of liquid nitrogen temperatures if the Weiss constant provided a measure of the magnetic interactions in this compound. The measurements in this case extended [2] only to 75°K and we do not know the low temperature behavior of this system. The linearity is probably a consequence of a fortuitous cancellation of effects and the moment and intercept, particularly the latter, do not lend themselves to any simple interpretation.

D. RSn_3 AND RPb_3 COMPOUNDS

The plumbides exhibit behavior which may now be considered normal for rare earth intermetallics. $PrPb_3$ is an excellent example of Van Vleck paramagnetism (Figs. 7-4 and 7-5). Analysis of these data by the Penney–Schlapp method (Chapter 3) gives an overall crystal field splitting of 80°K.

The RSn_3 compounds, except for the Ce compound, also seem straightforward. Data [11] for $CeSn_3$ (together with results for $CeIn_3$ and $CePb_3$) are shown in Fig. 5-1. The temperature dependence of χ is confirmed by independent measurements of Shenoy *et al.* [12] and Ruggiero and Olcese [2] although, because of their restricted temperature range, the latter authors were able to detect only the first beginnings of the anomalous low temperature behavior. Tsuchida and Wallace drew attention to the aberrant behavior and suggested that the behavior might be due to a valence change such as has been postulated [13,14] for elemental Ce (Table 7-5). Shenoy *et al.* have examined $CeSn_3$ by ^{119}Sn Mössbauer spectroscopy over the range of temperature from 1.6 to 293°K. They find no change in the isomer shift or the electric field gradient over the range of temperature in which the valence change was

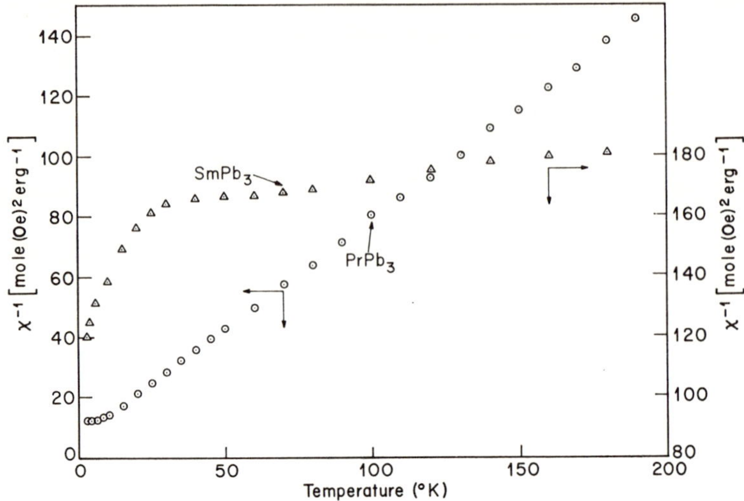

Fig. 7-4. Inverse susceptibility versus temperature for $SmPb_3$ and $PrPb_3$ [10]. H_{appl} = 19 kOe.

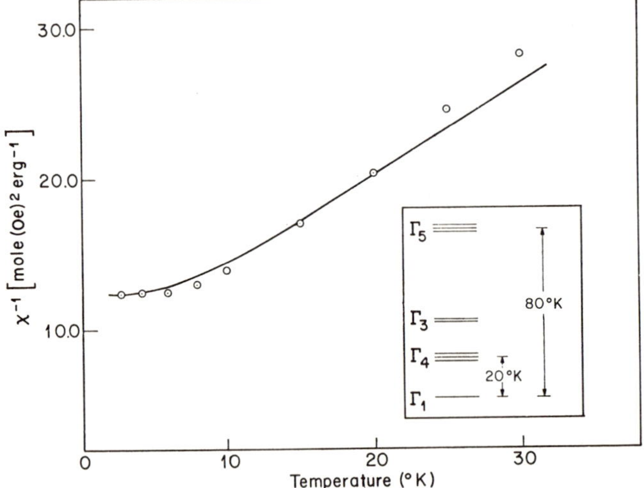

Fig. 7-5. Inverse susceptibility versus temperature for $PrPb_3$. The solid line is calculated using the crystal field splittings shown in the inset. Points are experimental [10].

presumed to be occurring. Thus the effects operating in $CeSn_3$ remain a mystery. They probably originate somehow in the interaction between the local moment and the conduction electrons.

Cooper *et al.* [16] measured the heat capacity of $CeSn_3$ over the range 1.5–11°K. C_p/T versus T^2 is linear and γ, the electronic coefficient, was found to be 53 mJ/mole °K^2.

TABLE 7-5

Magnetic Characteristics of RSn_3 and RPb_3 Compounds[a]

	Paramagnetic state			Ordered state		
	μ_{eff} (μ_B/R)	θ (°K)	Type	T_N (°K)	Type	Ref.
			RSn_3 series			
Ce	2.8		C–W,?			[2,11,12]
Pr	3.42	− 8	C–W	8.6	AF	[11,12,15]
Nd	3.6	−22	C–W	4.7	AF	[2,11,12]
Sm			N,m	12	AF	[11,13]
Gd	8.0	−73	C–W,?			[11]
			RPb_3 series			
Ce	2.74	−30	C–W,CF			[2,11,10]
Pr	3.7	−10	C–W,VV			[10]
Nd	3.5	−25	C–W			[10]
Sm			N,m			[10]
Eu	7.5	−47	C–W	22	AF	[10]
Gd	7.6	−68	C–W	17	AF	[10]

[a] All compounds occur in the $AuCu_3$ structure.

E. MISCELLANEOUS IVA COMPOUNDS

Several other intermetallics involving Si, Ge, Sn, and Pb have been studied by Ruggiero and Olcese [2], Hutchens and Wallace [10], and Percheron *et al.* [17]. Results obtained are accumulated in Table 7-6. The Ce compounds with Ge and Si show rather large negative values of θ,

TABLE 7-6

Magnetic Characteristics of Ge, Si, Sn, and Pb Compounds[a]

	Paramagnetic state			Ordered state		
	μ_{eff} (μ_B/R)	θ (°K)	Type	T_N (°K)	Type	Ref.
Ce_2Pb	2.44	− 1	C–W			[2]
Eu_2Pb		− 7	C–W	32,42	AF	[10]
CePb	2.49	1	C–W			[2]
Ce_2Sn	2.57	−13	C–W			[2]
Ce_2Sn_3	2.55	− 8	C–W			[2]
Ce_3Ge	2.57	−41	C–W			[2]
Ce_3Si	2.58	−60	C–W			[2]
Ce_2Si	2.55	−46	C–W			[2]
Sm_2Sn_3			N,m	11	AF	[17]
Sm_5Sn_3			N,m		?	[17]

[a] Structures are: $Eu_2Pb-PbCl_2$; $Sm_5Sn_3-Mn_5Si_3$. The structures of the other materials were not given and are presumably unknown.

suggesting the operation of the factor or factors responsible for the peculiar susceptibility behavior of $CeSn_3$, but to a reduced extent.

Eu_2Pb is characterized by two peaks in the susceptibility temperature curve—at 32 and 42°K. This may mean two polymorphic ordered magnetic forms stable in different temperature regions, both antiferromagnetic; or alternatively it may connote the disordering of the two crystallographically different types of Eu which occur in this material.

REFERENCES

1. K. H. J. Buschow and J. F. Fast, *Phys. Status Solidi* **16,** 467 (1966).
2. A. F. Ruggiero and G. L. Olcese, *Atti Acad. Lincei—Rend. Sci. Fis. Mat. Nat.* **37,** 169 (1964).
3. N. Van Nhung, J. Laforest, and J. Sivardiere, *Solid State Commun.* **8,** 23 (1970).
4. N. Van Nhung, A. Bertlett, and J. Laforest, *J. Phys. C Suppl.* **32,** 1–1133 (1971).
5. K. H. J. Buschow and J. F. Fast, *Phys. Status Solidi* **21,** 593 (1967).
6. F. Holtzberg, R. J. Gambino, and T. R. McGuire, *J. Phys. Chem. Solids* **28,** 2283 (1967).
7. K. Sekizawa, *J. Phys. Soc. Japan* **21,** 1137 (1966).
8. K. Sekizawa and K. Yasukochi, *J. Phys. Soc. Japan* **21,** 274 (1966).
9. C. E. Lundin, *in* "Rare Earth Research" (E. V. Kleber, ed.), p. 306. Macmillan, New York, 1961.
10. R. D. Hutchens and W. E. Wallace, *J. Solid State Chem.* **3,** 564 (1971).
11. T. Tsuchida and W. E. Wallace, *J. Chem. Phys.* **43,** 3811 (1965).
12. G. K. Shenoy, B. D. Dunlap, G. M. Kalvius, A. M. Toxen, and R. J. Gambino, *J. Appl. Phys.* **41,** 1317 (1970).
13. J. M. Lock, *Proc. Phys. Soc. ,London.* **B70,** 566 (1957).
14. K. A. Gschneidner, Jr. and R. Smoluchowski, *J. Less Common Metals* **5,** 374 (1963).
15. A. Czopnik, Z. Kletowski, and B. Stalinski, *Phys. Status Solidi* **3,** K263 (1970).
16. J. R. Cooper, C. Rizzuto, and G. L. Olcese, *J. Phys. C Suppl.* **32,** 1–1136 (1971).
17. A. Percheron, P. Lethuillier, J. L. Feron, and O. Gorochov, *Proc. Rare Earth Conf., 8th, Reno* p. 322 (April 1970).

Chapter 8

COMPOUNDS WITH Be, Mg, Zn, Cd, AND Hg

These compounds have not been investigated with any real degree of thoroughness. The only systematic investigations of an entire series covering temperatures down into the helium range are those of DebRay *et al.* on the RZn_2 compounds [1] and Alfieri *et al.* [2] in the RCd compounds. Iandelli [3] and Olcese [4] have studied a number of compounds, but unfortunately have not made measurements below liquid nitrogen temperatures. Results of these and other studies are summarized in Tables 8-1 to 8-4.

The systems involving Zn, Cd, and Hg are noteworthy in that the Eu compound usually forms, whereas in most of the series of compounds treated in this monograph the Eu (and Yb) members are missing because they exist in a different valence state than the other rare earths.

A. THE RM COMPOUNDS

1. Binary Compounds

These compounds, which occur in the CsCl structure, have been studied by Alfieri *et al.* [2], Iandelli and Palenzona [3], and Olcese [4]; Pierre [5] has investigated TbZn. The RCd compounds, except for GdCd, have been studied, both in the ordered and paramagnetic states. Apart from SmCd, all exhibit Curie–Weiss behavior; slopes are appropriate except for YbCd. The paramagnetic moment for this material, 2.1 μ_B, does not correlate with any electronic state of Yb. The lattice parameter suggests divalent ytterbium, which is nonmagnetic. It thus appears that the linear χ^{-1} versus T in this instance is an artifact—either a consequence of the limited temperature range covered or a fortuitous combination of contributions to χ. SmCd is interesting in that below about 110°K it is diamagnetic; it only becomes paramagnetic at higher temperatures. The weak paramagnetism of the ground ($^6H_{5/2}$) state of Sm^{3+} is apparently outweighed by the diamagnetism of the closed shells and that of the conduction electrons for $T < 110$°K, but at higher temperatures the population of the more strongly magnetic $^6H_{7/2}$ state becomes appreciable and the paramagnetism increases to predominance.

PrCd gives no indication of ordering. The other compounds order ferromagnetically but with moments well below gJ, the value expected for R^{3+} ions. This appears to be primarily due to the low fields used in making the measurements (4 kOe) and the consequent lack of saturation.

The paramagnetic moment of EuCd (7.89 μ_B) clearly indicates Eu^{2+}. This is also evident from its lattice parameter (7.96 Å) which is 0.20 Å higher than the value expected by interpolation between the a spacings for SmCd and GdCd. Divalent Eu is also indicated by the paramagnetic moments for the zinc and mercury compounds (Table 8-2).

TABLE 8-1

Magnetic Characteristics of RCd Compounds[a]

	Paramagnetic state			Ordered state				
R	μ_{eff} (μ_B/R)	θ (°K)	Type	μ_{or} (μ_B/R)	T_C (°K)	T_N (°K)	Type	Ref.
Ce	2.2	12	C–W		~12		F(?)	[2]
Pr	3.2	49	C–W,?					[2]
Nd	3.3	83	C–W,CF	3.2	105		F	[2]
Sm			N,m		265		F	[2]
Eu	7.89	−3	C–W					[3]
Gd				5.9	185		F	[2]
Tb	9.2	177	C–W	4	105		F	[2]
Dy	9.5	80	C–W	6	30		F	[2]
Ho	10.0	36	C–W		14		F	[2]
Er	9.2	10	C–W					[2]
Yb	2.1		C–W					[2]

[a] All the compounds contained excess rare earth. The value of x in the formula R_xCd was 1.2 to 1.3. Synthesis of the compounds from mixtures containing equal atomic ratios of the constituents led to X-ray patterns showing the presence of the RCd_2 phases. The extra phase was eliminated by incorporating a stoichiometric excess of R.

The cerium compounds seem to exhibit normal (i.e., Curie–Weiss) behavior although we note that measurements on CeZn and CeHg did not extend below liquid nitrogen temperatures. Possibly some of the

TABLE 8-2

Paramagnetic Characteristics of the RZn, RHg, RZn_2, RCd_2, and RHg_2 Compounds[a]

	CeZn	CeHg	EuZn	TbZn[c]	EuHg	TbHg[c]	Ref.
μ_{eff} (μ_B/R)	2.48	2.73	7.82	9.13	7.63		[3–5]
θ (°K)	−5	−32	14	205	11		
	$CeZn_2$	$CeCd_2$	$CeHg_2$	$EuZn_2$[c]	$EuCd_2$	$EuHg_2$	Ref.
μ_{eff} (μ_B/R)	2.72	2.18	2.73	8.23	7.95	8.00	[3,4]
θ (°K)	−38	−75	−33	45	50	29	

[a] See also Table 8-4.

[b] All compounds obey the Curie–Weiss law over the temperature range studied, which was 4–300°K for TbZn and −200–+200°C for the others. The 1:1 compounds are all CsCl structure. $CeZn_2$ has the KHg_2 structure. $CeCd_2$, $CeHg_2$, and $EuHg_2$ occur in the AlB_2 structure and $EuZn_2$ in the $CaZn_2$ structure. The structure of $EuCd_2$ is unknown.

[c] TbZn becomes ferromagnetic at 207°K [5]. EuZn becomes antiferromagnetic at 23°K [9]. The ordered moment of TbZn is 8.36 μ_B/Tb^{3+}. Cable *et al.* find by neutron diffraction measurements antiferromagnetic structures for TbZn and TbHg with Néel temperatures of 160 and 80°K, respectively [7].

unusual magnetic features of compounds involving cerium, to which attention is repeatedly drawn throughout this monograph, would have been evident had measurements been made at lower temperatures. However, measurements by Alfieri *et al.* [2] on CeCd did extend to 4°K and they report no aberrant behavior.

2. Ternary Systems

Ternaries based on GdZn and on TbZn have been studied by Sekizawa and Yasukochi [8] and Alfieri *et al.* [9], and Pierre [5], respectively. In each case, the objective of the work was to establish whether a modification of the magnetic structure could be brought about by changing electron concentration. In the main, the expected modifications were observed. These studies are briefly summarized in Chapter 6.

B. THE RM_2 COMPOUNDS

The complete RZn_2 series has been studied [1,6] (Table 8-3), and also the compounds of Ce and Eu with Cd and Hg (Table 8-2). The former studies involved temperatures to 4°K, whereas the other investigations did not extend below liquid nitrogen temperatures.

Salient features of these studies are: (1) $LaZn_2$ exhibits Pauli para-

TABLE 8-3

Magnetic Characteristics of RZn_2 Compounds[a]

	Paramagnetic state			Ordered state				
R	μ_{eff} (μ_B/R)	θ (°K)	Type	μ_{or} (μ_B/f.u.)	T_N (°K)	T_C (°K)	Type	Ref.
La			P					[1]
Ce	2.25	24	C–W,?		8		AF	[1]
Pr	3.56	14	C–W,VV					[1]
Nd	3.40	28	C–W,CF	1.82		24	F	[1]
Sm			N,m		45		AF	[1]
Eu			N,m		20		AF	[1]
Gd	8.48	70	C–W	6.91	68		AF(M)	[1]
Tb	9.62	58	C–W		55		AF	[1]
Ho	9.71	26	C–W		12		AF	[1]
Yb			N,?					[1]

[a] KHg_2 structures.

magnetism; (2) the cerium compounds exhibit essentially normal paramagnetic behavior (Fig. 8-1), but we recall that $CeCd_2$ and $CeHg_2$ are yet to be examined below 75°K; (3) $PrZn_2$ seems to be developing the Van Vleck paramagnetism at low temperatures, but this conclusion is only a tentative one (see Fig. 8-2); (4) $SmZn_2$ exhibits the usual nonlinear χ^{-1} versus T behavior; (5) $EuZn_2$ behaves anomalously, i.e., χ^{-1} versus T is nonlinear, according to DebRay *et al.* [1] (Fig. 8-3), but exhibits Curie–Weiss behavior according to Iandelli and Palenzona [3]; (6) $YbZn_2$ appears to involve Yb^{3+}, at least in part, since χ is strongly dependent on temperature (Fig. 8-1), but the slope of χ^{-1} versus T is not constant; (7) the other compounds, except $NdZn_2$, order antiferromagnetically; (8) $NdZn_2$ orders ferromagnetically.

In regard to the $NdZn_2$ behavior we note that $GdZn_2$ is actually metamagnetic with a critical field <5 kOe [1]. It is possible that $NdZn_2$ is similar, with an even lower critical field so that the antiferromagnetic state has not been detected.

The non-Curie–Weiss behavior is expected in view of the multiplet structure of the Eu^{3+} ion and the possibility that the dipositive and tripositive europium ions coexist in the zinc compound. χ is far too large to be ascribed to an assemblage of Eu^{3+} ions, which exhibits only Van Vleck paramagnetism. The nonlinearity of χ^{-1} versus T suggests that not all Eu is in the divalent state and the proportion which is changes with temperature. The magnetic properties of this material are thus complicated by the narrow spacing of the multiplet structure for Eu^{3+} and the possible dependence of the relative proportions of Eu^{3+} and Eu^{2+} on temperature.

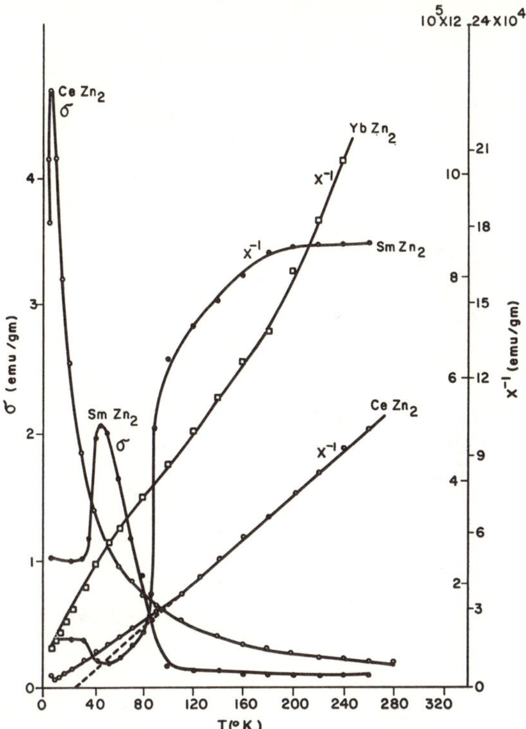

Fig. 8-1. Magnetization, σ, and reciprocal susceptibility, χ^{-1}, versus T °K plots for $CeZn_2$, $SmZn_2$, and $YbZn_2$. On the χ^{-1} ordinate axis the left-hand scale applies to $YbZn_2$, the right-hand scale to the others [1].

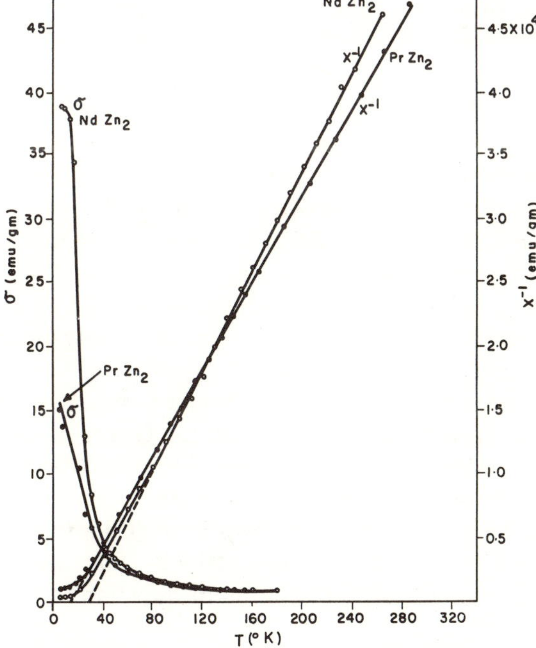

Fig. 8-2. Magnetization, σ, and reciprocal susceptibility, χ^{-1}, versus T plots for $PrZn_2$ and $NdZn_2$ [1].

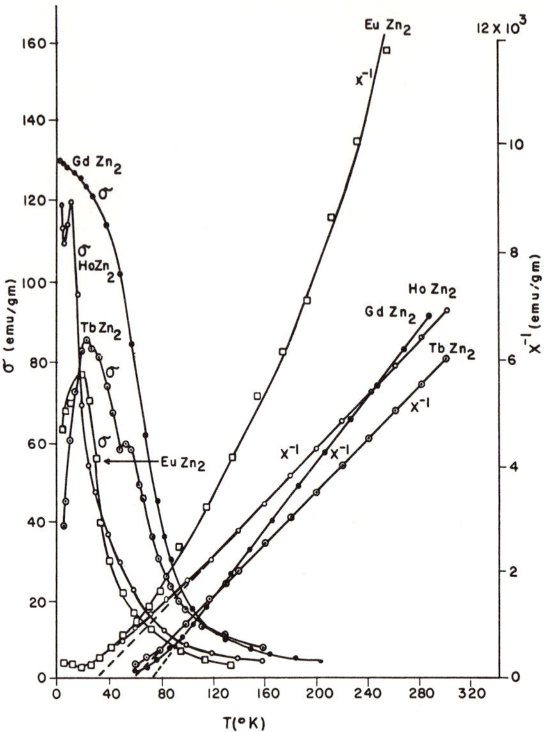

Fig. 8-3. Magnetization, σ, and reciprocal susceptibility, χ^{-1}, versus T plots for $EuZn_2$, $GdZn_2$, $TbZn_2$, and $HoZn_2$ [1].

The magnetic behavior of $YbZn_2$ also seems to imply mixed valencies with Yb^{3+} predominating at low temperatures.

Weiss constants of the RZn_2 series have been analyzed in terms of the de Gennes Equation (2-3). The RKKY sum for the KHg_2 structure is shown in Fig. 8-4. Since θ is positive for the RZn_2 compounds, this sum must be negative, indicating that k_F must lie between N and M in Fig. 8-4. For $GdZn_2$, $k_F{}^0 = 1.51$ Å^{-1}, which is outside this region. The effective k_F must then be less than $k_F{}^0$. Detailed analysis, and comparison with the results for $GdCu_2$, led DebRay *et al.* to the conclusion that k_F must lie close to the point M in Fig. 8-4 or k_F must be approximately 80% of $k_F{}^0$. The $k_F/k_F{}^0$ value established for $GdZn_2$ is in accord with earlier work on other rare earth systems. The reduction of the effective k_F from $k_F{}^0$ was noted earlier in R–Al systems by Van Diepen *et al.* (Chapter 4) and in R–Cu systems by Pierre (Chapter 6), who found $k_F/k_F{}^0$ values of 0.94 and 0.85, respectively, that were required to account for the behavior of the systems in question.

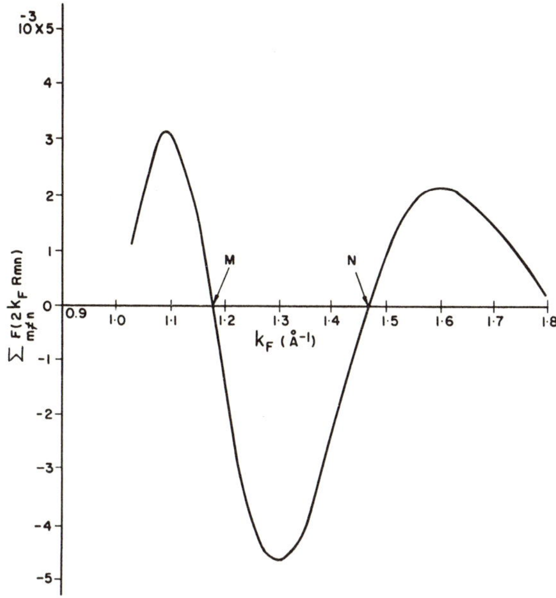

Fig. 8-4. Dependence of Σ $F(2K_FR_{mn})$ on conduction electron wave vector, K_F, in the KHg_2-type structure, after DebRay *et al.* [1]. R_{mn} is the distance between two rare earth ions. Interatomic distances in $GdZn_2$ were used.

Using $k_F = 0.8\ k_F^0$ we calculate $|\Gamma|$ [Eq. (2-3)] to be 0.15 eV which compares favorably with 0.09 eV obtained from esr measurements [10].

The ordered moment of $NdZn_2$ is well below gJ ($=3.3\ \mu_B$) indicating substantial crystal field quenching. DebRay *et al.* [11] have examined $CeZn_2$ and $TbZn_2$ by neutron diffraction techniques. These measurements confirm the antiferromagnetism of these two compounds and give moments of 1.5 and 8.9 μ_B for Ce and Tb, respectively. The quenching effect is evident in the behavior of Ce but not of Tb.

C. OTHER COMPOUNDS

The data in Tables 8-4 and 8-5 indicate that Eu is divalent when combined with Mg, Zn, Cd, and Hg. These electropositive metals, in contrast with the situation in less strongly electropositive metals such as Pd (Chapter 12), encourage Eu into the dipositive state. This is true for the 1:2, 2:17, and 1:3 compounds. However, for $EuBe_{13}$ the susceptibility agrees with that expected for Eu^{3+}. Evidently when combined with the small Be^{2+} ion, Eu is compressed to the extent that it adopts the lower volume characteristic of the trivalent state. μ_{eff} for $YbBe_{13}$ also indicates trivalence; the mechanism operating is probably similar.

TABLE 8-4

Paramagnetic Characteristics of 1:5 Compounds of the Rare Earths with Zn, Cd, and Hg[a]

	$CeZn_5$	$CeHg_5$	$EuZn_5$	$EuCd_5$	$EuHg_5$	Ref.
μ_{eff} (μ_B/R)	2.54	2.70	8.06	7.80	7.79	[3,4]
θ	−44	−8	−18	6	10	

[a] Structures of $CeHg_5$ and $EuHg_5$ have not been solved. The other compounds occur in the $CaCu_5$ structure.

TABLE 8-5

Paramagnetic Characteristics of Miscellaneous Be, Mg, Zn, Cd, and Hg Compounds

	$EuMg_2$	Eu_2Mg_{17}	$EuBe_{13}$[a]	$YbBe_{13}$	$EuZn_{13}$	$CeCd_3$	$CeHg_3$	$EuHg_3$
μ_{eff} (μ_B/R)	7.9	8.0		4.0	7.50	2.44	2.93	7.63
θ (°K)					18	−29	−53	−69
Structure	C14	Th_2Ni_{17}	$NaZn_{13}$	$NaZn_{13}$	$NaZn_{13}$	$BiLi_3$	$MgCd_3$	$MgCd_3$
Ref.	[12]	[12]	[12]	[12]	[3]	[4]	[4]	[3]

	$CeHg_4$	$CeZn_9$	$CeZn_{11}$	$CeCd_{11}$	$PrCd_{11}$	$EuCd_{11}$
μ_{eff} (μ_B/R)	3.03	2.64	2.54	2.47		7.56
θ (°K)	−55	0	26	32		18
Structure	bcc	?	$BaCd_{11}$	$BaCd_{11}$	$BaCd_{11}$	$BaCd_{11}$
Ref.	[4]	[4]	[4]	[4]	[13]	[3]

[a] χ versus T is indicative of Eu^{3+}.

Olcese [14] has studied $CeBe_{13}$ and reported rather anomalous magnetic results for it. Above 150°K the χ versus T curve has a small negative slope, corresponding to an effective moment of 3.04 μ_B and a Weiss constant of −240°K. χ varies only about 20% between 150 and 450°K. With decreasing temperature χ passes through a weak maximum and decreases slowly until about 50°K, at which point the curve takes a precipitous drop. χ becomes negative for $T < 10$°K. Olcese also observes diamagnetism, or at least $\chi < 0$, in the same temperature range for other cerium compounds, e.g., $CeSi_2$ and $CeSn_3$. His work in the case of $CeSn_3$ is in contradiction with other studies [15]. It appears that until substantiated independently the results of Olcese on $CeBe_{13}$ should be regarded with caution, at least those for temperatures below 50°K.

The other cerium compounds in Tables 8-1 and 8-5 show normal behavior. However, we recall that measurements on them below 75°K have not been made.

D. HEAT CAPACITIES

The only heat capacity work on the compounds discussed in this chapter appears to be that of Cooper *et al.* [16], who made measurements on $CeBe_{13}$ between 2.5 and 14°K. They find linearity of C/T with T^2 over this temperature range. The derived values for the electronic specific heat coefficient (γ) is 155 mJ/mole °K^2. This γ is quite large implying a large density of states or a substantial electron–phonon enhancement factor or both.

REFERENCES

1. D. K. DebRay, W. E. Wallace, and E. Ryba, *J. Less Common Metals* **22,** 19 (1970).
2. G. T. Alfieri, E. Banks, K. Kanematsu, and T. Ohoyama, *J. Phys. Soc. Japan* **23,** 507 (1967).
3. A. Iandelli and A. Palenzana, *Atti. Accad. Naz. Lincei Rend. Sci. Fis. Mat. Nat. XXXVII,* 165 (1964).
4. G. L. Olcese, *Atti Accad. Naz. Lincei Rend. Sci. Fis. Mat. Nat.* **37,** 48 (1963).
5. J. Pierre, *C. R. Acad. Sci. Paris* **265B,** 1169 (1967).
6. H. H. Wickman, J. H. Wernick, R. C. Sherwood, and C. F. Wagner, *J. Phys. Chem. Solids* **29,** 181 (1968).
7. J. W. Cable, W. C. Koehler, and E. O. Wollan, *Phys. Rev.* **136,** A240 (1964).
8. K. Sekizawa and K. Yasukochi, *J. Phys. Soc. Japan* **21,** 684 (1966).
9. G. T. Alfieri, E. Banks, and K. Kanematsu, *J. Appl. Phys.* **37,** 1254 (1966).
10. D. K. DebRay and E. Ryba, *J. Phys. Suppl.* **32,** C 1–1130 (1971).
11. D. K. DebRay, M. Sougi, and P. Meriel, paper presented at the *Conf. Rare Earths Actinides, Durham, England* July (1971).
12. W. Klemm and W. Mühlpfordt, *Rev. Chim. Minérale* **5,** 561 (1968).
13. M. A. Aoyagi and W. E. Wallace, unpublished measurements.
14. G. L. Olcese, *Les Éléments des Terres Rares, Coll. Int. du C.N.R.S.* Tome 1,14 (1970); *Atti Accad. Naz. Lincei Rend. Sci. Fis. Mat. Nat.* **XL,** 1052 (1966).
15. T. Tsuchida and W. E. Wallace, *J. Chem. Phys.* **43,** 3811 (1965).
16. J. R. Cooper, C. Rizzuto, and G. Olcese, *J. Phys. Suppl.* **32,** C 1-1136 (1971).

Part III

RARE EARTH–TRANSITION METAL SYSTEMS

Chapter 9

NICKEL COMPOUNDS

$CeNi_2$, along with $CeCo_2$, was the first rare earth intermetallic to be structurally characterized; it was investigated in 1942 by Fülling *et al.* [1]. Since then a very large number of rare earth intermetallics with varied stoichiometries have been isolated. Rare earth–nickel systems are particularly rich in respect to compound formation. Buschow has asserted [2] that there are eleven stable stoichiometries in the Er–Ni system—$ErNi_4$, Er_4Ni_{17}, and Er_5Ni_{22} in addition to those commonly observed (see Table 9-1). The eight series for which more or less complete structural and magnetic data have been reported are listed in Table 9-1. In most of the R–Ni compounds nickel is nonmagnetic, as will be brought out later, electrons apparently being transferred from R to nickel until the nickel d-band is filled. The RNi_5 compounds are the structural counterparts of the RCo_5 compounds, which are discussed in detail in the following chapter, but are magnetically simpler since the d-transition metal sublattice is nonmagnetic. For this reason the RNi_5 compounds are of particular interest, since they constitute a stepping stone toward a better understanding of the industrially significant, but more complex, RCo_5 compounds. The RNi_5 compounds are of interest for another reason. Lemaire and Paccard [21] have drawn attention to the fundamental role of RNi_5 in establishing the structure of all the R–Ni intermetallics. The 2:17 structure, for example, can be generated by the ordered replacement of one-third of the rare earth atoms by pairs of nickel atoms, the Laves phases by replacement of R by Ni, etc.

TABLE 9-1

Structures of Rare Earth–Nickel Compounds

R	R_3Ni	R_7Ni_3	RNi	RNi_2	R_2Ni_7	RNi_3	RNi_5	R_2Ni_{17}
La	Fe_3C	Th_7Ni_3		C15	Ce_2Ni_7		$CaCu_5$	
Ce		Th_7Ni_3	CrB	C15	Ce_2Ni_7	hex	$CaCu_5$	
Pr	Fe_3C	Th_7Ni_3	CrB	C15	*a*	$PuNi_3$	$CaCu_5$	
Nd	Fe_3C		CrB	C15	*a*	$PuNi_3$	$CaCu_5$	Th_2Ni_{17}
Sm	Fe_3C		CrB	C15	*a*	$PuNi_3$	$CaCu_5$	Th_2Ni_{17}
Eu							$CaCu_5$	
Gd	Fe_3C		CrB	C15	*a*	$PuNi_3$	$CaCu_5$	Th_2Ni_{17}
Tb	Fe_3C		*b*	C15	*a*	$PuNi_3$	$CaCu_5$	Th_2Ni_{17}
Dy	Fe_3C		FeB	C15	*a*	$PuNi_3$	$CaCu_5$	Th_2Ni_{17}
Ho	Fe_3C		FeB	C15	Gd_2Co_7	$PuNi_3$	$CaCu_5$	Th_2Ni_{17}
Er	Fe_3C		FeB	C15	Gd_2Co_7	$PuNi_3$	$CaCu_5$	Th_2Ni_{17}
Tm	Fe_3C		FeB	C15		$PuNi_3$	$CaCu_5$	Th_2Ni_{17}
Yb						$PuNi_3$		
Lu				C15			$CaCu_5$	
Ref.	[3]	[4]	[5–8]	[9–11]	[12–14]	[15,16]	[10,11,17]	[18–20]

[a] Coexistence of hexagonal Ce_2Ni_7-type and rhombohedral Gd_2Co_7-type.
[b] Coexistence of CrB and FeB.

A. RNi_2 COMPOUNDS

1. Binary Systems

a. General Magnetic Features

These cubic Laves phase materials have been very extensively studied in a number of laboratories. In many ways they resemble the RAl_2 compounds; the magnetic interactions involve only the R sublattice, but are transmitted in the case of the RNi_2 compounds through a sublattice which consists of a recently filled d-transition metal. In the course of the present discussion we shall show that the R–R′ coupling systematics (R and R′ are two different rare earths) are the same in RNi_2 systems as in RAl_2 systems, despite the different nature of the chemical species (nontransition metal versus recently filled transition metal) on the non-rare-earth sublattice.

Bulk magnetic characteristics of the RNi_2 compounds were reported by Wallace *et al.* [22,23] and by Ross and Crangle [24]. Farrell's saturation magnetizations are higher and are accepted as being more nearly representative of the true moment of the RNi_2 compounds. Results are summarized in Table 9-2, based primarily on Farrell's measurements.

Perhaps the most striking feature of the observations represented by Table 9-2 is that nickel in these compounds appears to be nonmagnetic, a point which was anticipated in the preliminary remarks above. $LaNi_2$ and $LuNi_2$, in which nickel is in chemical union with a nonmagnetic partner, are Pauli paramagnets; they display only the paramagnetism associated with the conduction electrons. The effective moments measured for the others accord well with $g[J(J+1)]^{1/2}$, that expected for the free tripositive rare earth ion. Thus the paramagnetic susceptibility can be accounted for solely in terms of the rare earth component. Of course, there is a small contribution due to the conduction electrons. Except for its share in the Pauli contribution nickel appears as a nonmagnetic entity as regards the paramagnetic behavior of the RNi_2 compounds.

TABLE 9-2

Magnetic Characteristics or RNi_2 Compounds

	Paramagnetic state				Ordered state				
R	θ (°K)	μ_{eff} (μ_B/R)	$g[J(J+1)]^{1/2}$	Type	T_C (°K)	μ_{or} (μ_B/f.u.)	gJ	Type	Ref.
La				P					[25]
Ce				N,k					[23,26]
Pr	4	3.57	3.58	C–W,VV		0.86[b]	3.20		[23,27]
Nd	10	3.74	3.62	C–W	16	1.80	3.27	F	[23]
Sm				N,m	21	0.25	0.71	F	[23]
Gd	78	7.82	7.95	C–W	85	7.1	7.00	F	[23]
Tb	35	9.82	9.72	C–W	45	7.8	9.00	F	[23]
Dy	23	10.4	10.60	C–W	30	9.2	10.00	F	[23]
Ho	12	10.5	10.60	C–W	22	8.4	10.00	F	[23]
Er	11	9.37	9.60	C–W	21	6.8	9.00	F	[23]
Tm	0	7.28	7.60	C–W,VV		3.20[b]	7.00		[23]
Lu				P					[22]

[a]The Curie temperature was measured in a field of 20 kOe. T_C at zero field would be lower. For example, T_C for $NdNi_2$ obtained from heat capacity measurements [28] was 10°K.

[b]These are not ordered moments but are instead moments measured at 4.2°K in an applied field of 20 kOe (see text).

The ordered moments measured also support the conclusion that nickel in these compounds is nonmagnetic. We note that most of the values of μ_{or} in Table 9-2 are well below gJ, that expected for an assemblage of free R^{3+} ions. We ascribe this to the quenching effect of the crystal field, to which we shall presently return. Gd^{3+}, being in an S state, is largely free of this interaction and is expected to display the full moment characteristics of its $4f^7$ configuration, i.e., 7 μ_B. The measured moment is within 0.1 μ_B of this, the difference being reasonably ascribed to polarization of the conduction electrons. Thus the ordered moment in $GdNi_2$ originates exclusively with the rare earth component. It appears safe to assume that the situation is the same in the other RNi_2 compounds, although, as noted, μ_{sat} is not equal to gJ because of the influence of the crystal field.

Elemental nickel, as is well known, has approximately 0.6 vacancies per atom in its d band. Alloying it with an electron-rich metal such as copper diminishes its moment [29] progressively as the copper content increases. This has been traditionally ascribed to the filling of the nickel d band by the extra electrons supplied by Cu. Similar effects have been noted for intermetallics, e.g., $MgNi_2$. Wollam and Wallace observed [30] that $MgNi_2$ exhibited only Pauli paramagnetism, indicating electron transfer from the strongly electropositive Mg component to less strongly electropositive Ni until the d band was filled. Evidently a similar effect occurs in the RNi_2 series, and, for that matter, in all of the R–Ni compounds discussed in this chapter with the exception of the R_2Ni_{17} series. In these latter compounds the R concentration is too low to provide sufficient electrons to enable the nickel d band to be filled. However, as we shall point out below, we infer, on the basis of the observed nickel moment in the R_2Ni_{17} series, that essentially complete transfer of the three valence electrons ($6s^25d$) from R to Ni has taken place in the 2:17 compounds.

b. $PrNi_2$ and $TmNi_2$ as Van Vleck Paramagnets

$PrNi_2$ and $TmNi_2$ show a tendency toward the Van Vleck temperature-independent paramagnetism below 20°K. $PrNi_2$ has been studied in detail [27] to ascertain whether it should more properly be regarded as a Van Vleck paramagnet with susceptibility appreciably enhanced by exchange or a weak ferromagnetic, i.e., a ferromagnetic material with a moment substantially quenched by the action of the crystal field. The former possibility is permitted for $PrNi_2$ (and for $TmNi_2$ as well) since the ground state can be a singlet (see Fig. 9-1). To answer the question as to whether $PrNi_2$ is paramagnetic or ferromagnetic, the system $Pr_{1-x}Y_xNi_2$ has been studied [27]. Dilution of Pr with Y suppresses exchange but leaves the crystal field interaction unmodified. Results obtained are shown in Table 9-3. All the Pr–Y–Ni ternaries exist in the C15 ($MgCu_2$) structure; they exhibit Curie–Weiss behavior at elevated temperatures but showed deviations at low temperatures in a matter to suggest the onset of Van Vleck paramagnetism. Since ferromagnetism is observed for most of the RNi_2 compounds, ferromagnetic exchange is expected to predominate in $PrNi_2$. Thus, on the basis of the magnetic interactions alone a positive θ is anticipated. Actually θ as given in Table 9-3 is negative and becomes increasingly

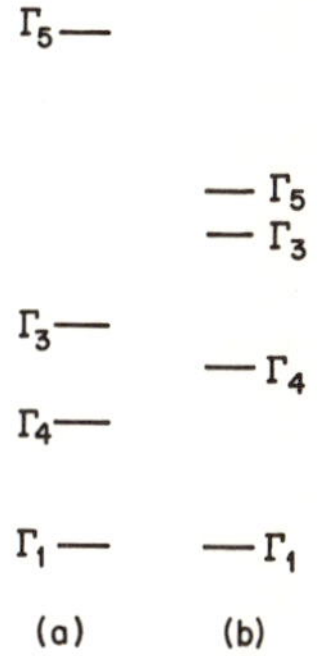

Γ_5 —
— Γ_5
— Γ_3
Γ_3 —
— Γ_4
Γ_4 —
Γ_1 — — Γ_1
(a) (b)

Fig. 9-1. Crystal field spectrum for Pr^{3+} in $PrNi_2$ as deduced from magnetic measurements [27]. Case a—pure fourth-order interactions; case b—four- and sixth-order interactions involved with $x = 0.80$. (x is the parameter introduced by Lea *et al.* [36] to indicate the relative importance of fourth- and sixth-order interactions.)

TABLE 9-3

Magnetic Characteristics of $Pr_{1-x}Y_xNi$ Ternaries[a]

	χ_0 (emu/mole)[a]	μ_{eff} (μ_B/R)	θ (°K)
$PrNi_2$	0.222	3.60	−1
$Pr_{0.5}Y_{0.5}Ni_2$	0.222	3.72	−7
$Pr_{0.2}Y_{0.8}Ni_2$	0.125	3.67	−12
$Pr_{0.1}Y_{0.9}Ni_2$	0.087	3.59	−12
YNi_2		Pauli paramagnet	

[a]This is the susceptibility per mole of Pr^{3+}, extrapolated to 0°K.

negative as Pr^{3+} is diluted. This is not unanticipated. Penney and Schlapp [31] pointed out in their classic paper in 1932 that the crystal field interaction for an assemblage of Pr^{3+} ions introduces a positive contribution to θ. Thus,

$$\theta_{obs} = \theta_{ex} + \theta_{cf}. \tag{9-1}$$

From the fact that θ_{obs} decreases with dilution we infer a positive value of θ_{ex} in accordance with expectations. The value of θ_{ex} decreases with dilution and should vanish when the Pr concentration is sufficiently low. On the other hand, θ_{cf} should be essentially unchanged by dilution since the concentration change entails merely the replacement of one tripositive ion by another. We thus accept the limiting value of θ_{obs}

to be θ_{cf} and hence from Eq. (9-1) compute θ_{ex} for $PrNi_2$ to 11°K, corresponding to a molecular field constant γ of 7 Oe^2 mol/erg.*

We can now discuss the behavior of $PrNi_2$ in terms of Eq. (4-12) or, more simply, Eq. (4-10). We recall that T_C is obtained with sufficient reliability by making the condition that the denominator, $(Q - \alpha B \lambda)$, in Eq. (4-10) vanish. Thus, to find T_C we merely evaluate $Q - \alpha B \lambda$ at several temperatures and establish the temperature at which $Q = \alpha B \lambda$. Mader and Wallace [27] showed that $\alpha B \lambda$ is always less than unity, indicating that ordering will not occur at any temperature, since the minimal value of Q is unity at 0°K. Thus, exchange is below the threshold value for the bootstrap process to operate $PrNi_2$ and it remains paramagnetic to the lowest observed temperatures.

The Penney–Schlapp formalism leads to a relationship between χ_0 in Table 9-3 and E_c, the overall crystal field splitting. For a pure fourth-order interaction [31]

$$E_c = 12.34/\chi_0. \tag{9-2}$$

Segal [32] has made calculations, corresponding to the Penney–Schlapp, or perhaps more nearly to the Schumacher–Hollingsworth [33] calculations, for combinations of fourth- and sixth-order interactions. Assuming the sixth-order contribution to be the same for $PrNi_2$ as in elemental Pr and in $PrAl_2$ [34,35],

$$E_c = 8.86/\chi_0. \tag{9-3}$$

In Eqs. (9-2) and (9-3), E_c is in °K when χ_0 is in erg Oe^{-2} mole $^{-1}$. E_c = 136 and 94°K from Eqs. (9-2) and (9-3), respectively. The crystal field spectrum corresponding to these two overall splittings is shown in Fig. 9-1. χ versus T, obtained from Eq. (3-28) using the crystal field splittings shown in Fig. 9-1, is given in Fig. 9-2.

From the foregoing discussion it seems clear that $PrNi_2$ is a Van Vleck paramagnet and probably this is also the case for $TmNi_2$. There is, however, one feature of $PrNi_2$ which upon superficial examination seems to be in contradiction with this viewpoint—its susceptibility at 4.2°K is not field-independent; rather, its magnetization shows signs of saturation. If the Pr^{3+} ions are in the Γ_1 state with moments linearly dependent on field [Eq. (3-27)], it is not clear, at least at the outset, why its magnetization should show saturation effects.

*$\lambda = \theta_{ex}/C$, where C is the Curie constant. $C = 1.62$ for Pr^{3+}.

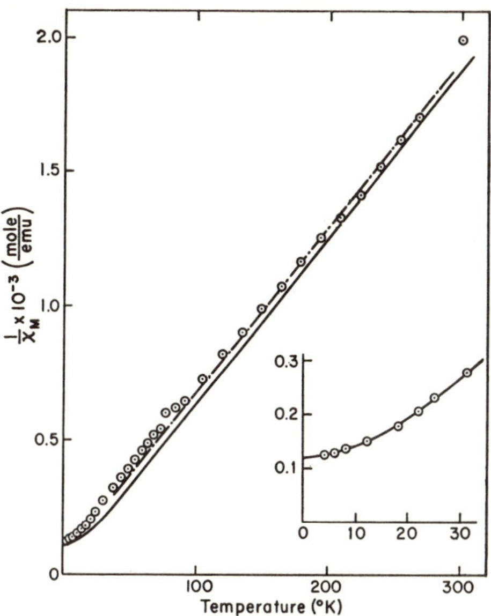

Fig. 9-2. Plot of $1/\chi$ versus T for $Pr_{0.1}Y_{0.9}Ni_2$ [27]. Points give the experimental results. The line is computed regarding the crystal field interaction as pure fourth-order. The dashed line is the corresponding calculation with sixth-order contribution corresponding to $x = 0.80$. The overall crystal field splitting E_c, is taken to be 134°K if the interaction is pure fourth order or 94°K if the sixth-order contribution is included. Calculated results for low temperatures (inset) are insignificantly different for the two cases. Reprinted from *Inorg. Chem.* 7, 1627 (1968). Copyright 1968 by the American Chemical Society. Reprinted by permission of the copyright owner.

The molar magnetization M is given by the expression

$$M = \alpha(A + \mathrm{B}\ \mathrm{H}_0)/(\theta - \alpha B\lambda). \tag{4-10}$$

M can not be solved for directly from Eq. (4-10) since A, Q, and B all involve M, through the effective H, $H_{\mathrm{eff}} = H_0 + \lambda M$. To evaluate the magnetization M is expressed in terms of H_{eff} using Eq. (4-10) and the value of H_0 to produce this effective field is obtained from the relationship

$$H_0 = H_{\mathrm{eff}} - \lambda M. \tag{9-4}$$

The magnetization versus H_0 is shown in Fig. 9-3, which brings out the interesting and rarely appreciated point that the magnetization of a Van Vleck paramagnet does *not* vary linearly with the strength of the applied field. Instead it tends toward saturation as the applied field is increased if exchange is appreciable.

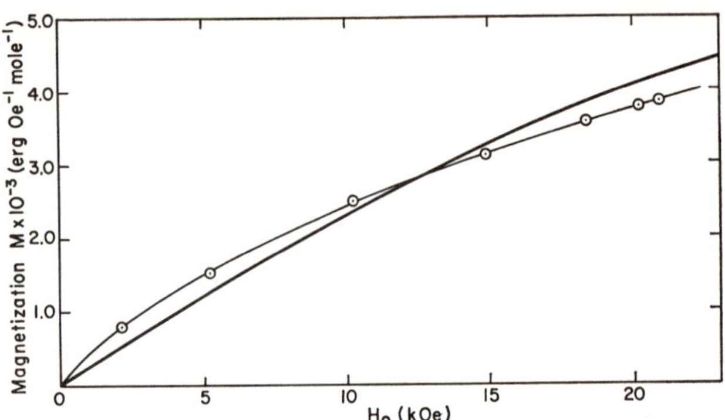

Fig. 9-3. Magnetization versus applied field for $PrNi_2$ at 4.2°K. Points are experimental [27]. The heavy line is computed (see text). Results calculated with and without the sixth-order contribution cited in Fig. 9-1 are insignificantly different. Reprinted from *Inorg. Chem.* **7**, 1627 (1968). Copyright 1968 by the American Chemical Society. Reprinted by permission of the copyright owner.

The deficiency of ordered moment in the RNi_2 compounds as compared to gJ has been alluded to earlier. The possibility that this might result from antiferromagnetic R–Ni coupling is excluded by the observation that the paramagnetic behavior of the several compounds and the ferromagnetic behavior of $GdNi_2$ can be accounted for using a model in which Gd is the only magnetic species present. As noted earlier, we ascribe the diminished moment to the quenching effect of the crystal field and make use of the case of $NdNi_2$, already discussed in Chapter 3, to illustrate the point. As noted there, ordering takes place in the Γ_6 ground state, having a permanent moment 1.33 μ_B/Nd^{3+}. The exchange field, estimated from the ordering temperature by the molecular field theory, is 113 kOe, from which we compute an induced moment of 0.27 μ_B/Nd^{3+} and an overall moment of 1.60 μ_B, in reasonable agreement with experiment.

Similar calculations are not possible for the remaining RNi_2 compounds because E_c for them has not been determined. However, as noted above, there is every reason to believe that in those cases the reduction in moment originates in a similar fashion. As regards $DyNi_2$, we find from bulk measurements $\mu_{or} = 9.2\ \mu_B$/f.u. whereas Ofer and Segal [37] find from ^{161}Dy Mössbauer spectroscopy that the Dy moment is 9.8 μ_B. This suggests that incomplete saturation in the bulk magnetic measurements may also be a factor, albeit a minor one, in the low observed moment.

c. *Curie Temperatures in Relation to de Gennes' Theory*

The trend of the Curie temperatures in Table 9-2 is interesting to examine in the light of de Gennes' treatment of the interactions, cited in Chapter 2. Equation (2-3) indicates that θ varies as the de Gennes function $F_{dG}=(g-1)^2J(J+1)$. Since in the molecular field approximation T_C and θ are identical, we expect T_C to vary with F_{dG}. Data shown in Fig. 9-4 are in impressive conformity with the de Gennes treatment.

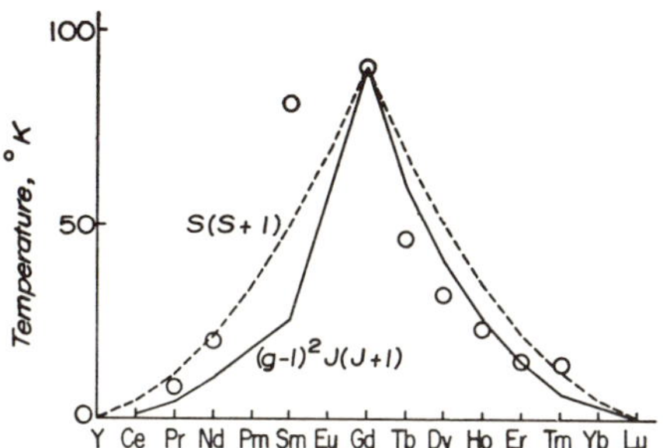

Fig. 9-4. Comparison of observed Curie temperatures with the deGennes functions for the RNi_2 series [22].

d. Heat Capacities

Heat capacities of RNi_2 with R = La, Ce, Pr, and Nd have been measured in the range 6–300°K [28,38] and 1.7–4.2°K [28,38–40]. Representative results are shown in Figs. 9-5 to 9-9. The excess heat capacity for $PrNi_2$, shown in Fig. 9-5, is ascribed to excitation within the crystal field spectrum. There is no λ-type thermal anomaly indicating the development of magnetic order. This is, of course, consistent with the concept of Pr^{3+} in $PrNi_2$ derived from analysis of the magnetic results.

$NdNi_2$ exhibits a λ-type thermal anomaly peaking at about 10°K and excess heat capacity over nonmagnetic $CeNi_2$ at higher temperatures. The ground state of the Nd^{3+} ion in a cubic environment splits [33,36] into a Γ_6 state and two Γ_8 states. If the singlet Γ_1 state is the ground state for Pr^{3+} in $PrNi_2$ we expect the Γ_6 state to be lowest lying for Nd^{3+} in $NdNi_2$. The λ-type thermal anomaly is taken (see Chapter 3) to be a consequence of the destruction of magnetic order within the Γ_6 Kramers doublet, the components of which have equal but opposite magnetic moments [31,33]. Thus, when they are equally populated, the spontaneous magnetization is lost. We expect an entropy change of $R \ln 2$ associated with the destruction of magnetic order. Experiment shows an entropy change of $1.06\, R \ln 2$, confirming Γ_6 as the ground state. The excess heat capacity at temperatures between 10° and 100°K is ascribed to excitation of Nd^{3+} ions from the Γ_6 to the Γ_8 states.

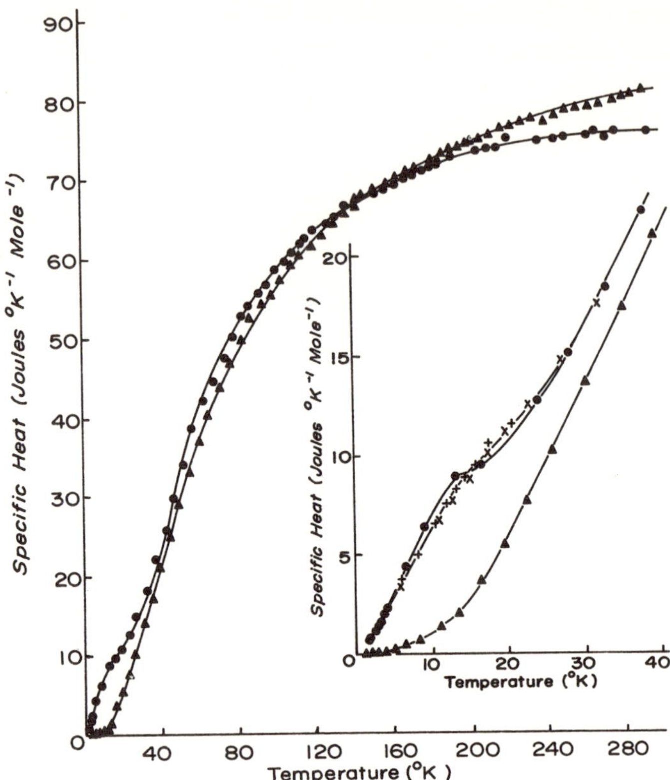

Fig. 9-5. Molar heat capacities of $CeNi_2$ and $PrNi_2$ [28]. $PrNi_2$ (●), $CeNi_2$ (▲). Insert: $CeNi_2$ (▲); $PrNi_2$ (●), after cycling to 300°K (+), after cycling to 70°K (×).

Of the four compounds studied below 4°K only $LaNi_2$ exhibits behavior that can be considered "normal" or "as expected." Results for $CeNi_2$ will be discussed in the next section. The $LaNi_2$ and $NdNi_2$ data (Fig. 9-8) are accurately represented by an expression of the form

$$C_p = \gamma T + \beta T^3. \tag{9-5}$$

For $LaNi_2$, $\gamma = 12.52$ and $\beta = 0.413$. For $NdNi_2$, $\gamma = 101.1$ and $\beta = 10.8$. The units are such that C_p in (9-5) is in millijoules per mole (°K). The value of γ for $LaNi_2$ is reasonable; it is comparable with that of $LaAl_2$ obtained by Hungsberg (see Chapter 4). However, γ for $NdNi_2$ is an order of magnitude larger, too large to be reasonably ascribed to excitation of the conduction electrons. The source of the large heat capacity of this material is of considerable interest. As noted above, the cooperative magnetic phase in $NdNi_2$ is destroyed at 10°K and it seems

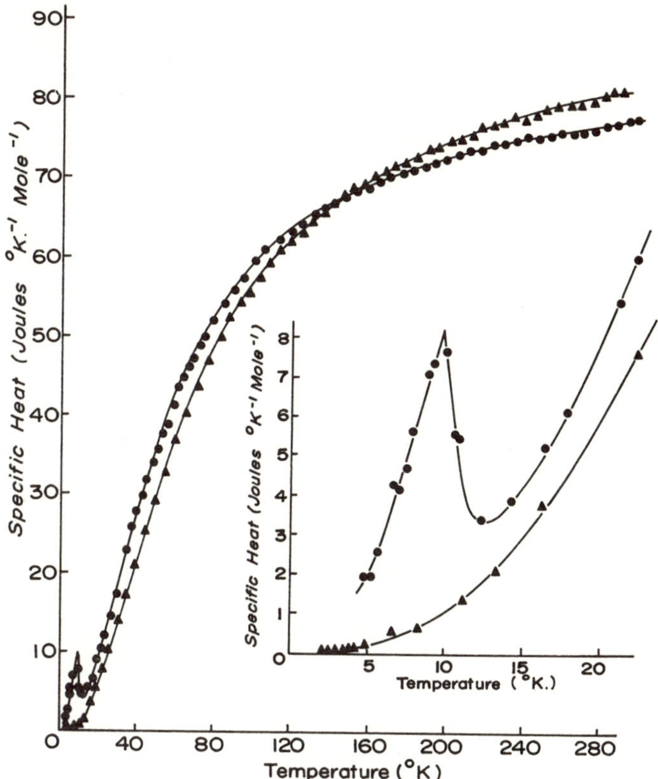

Fig. 9-6. Molar heat capacities of $CeNi_2$ and $NdNi_2$ [28]. $NdNi_2$ (●), $CeNi_2$ (▲).

likely that there is a significant magnetic contribution to its heat capacity between 1.5 and 4°K. Moreover, there is, as noted above, a crystal field contribution to C_p in the paramagnetic region associated with excitation into the Γ_8 levels. Thus C_p for $NdNi_2$ is the sum of four terms:

$$C_p = C_e + C_l + C_m + C_{cf}. \tag{9-6}$$

With an overall splitting of 431°K [28], $C_{cf} \leqq 0.3$ mJ mole^{-1} deg^{-1} in the liquid helium range and can hence be ignored. Simple spin wave theory gives [41] for C_m, the magnetic heat capacity

$$C_m = AT^{3/2} + BT^{5/2} + CT^{7/2}. \tag{9-7}$$

At $T \ll T_C$ only the first term in Eq. (9-7) is important. If results for

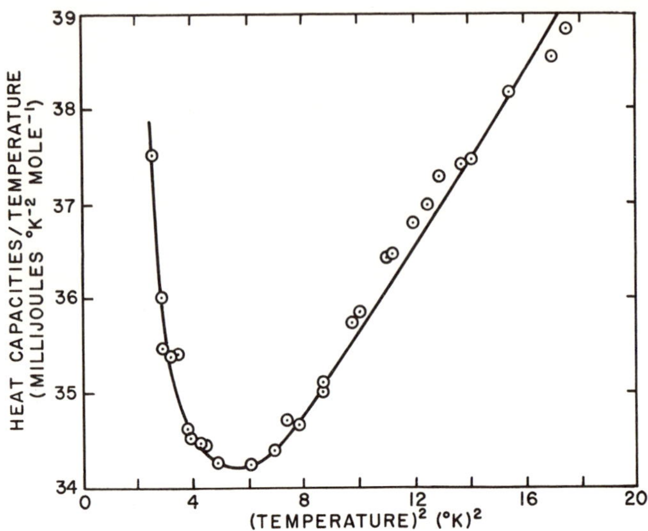

Fig. 9-7. Heat capacity behavior of $CeNi_2$ between 1.5 and 5°K [28].

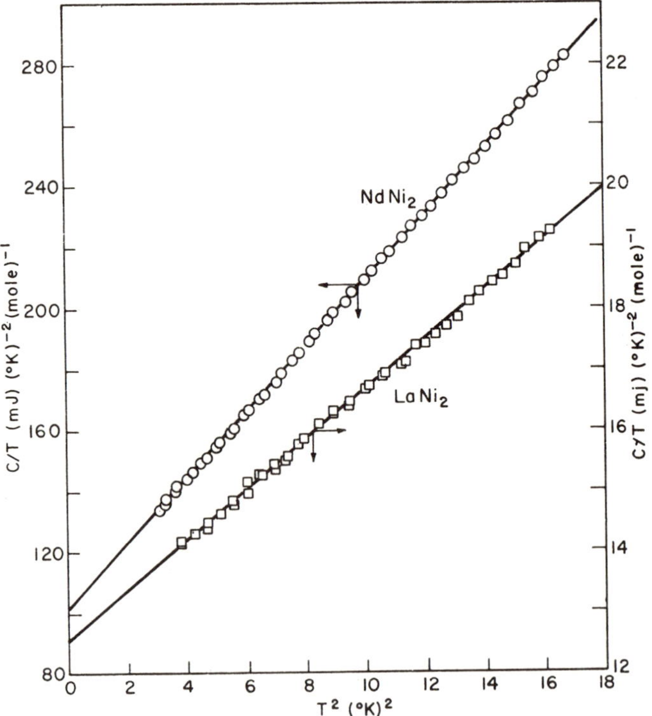

Fig. 9-8. Heat capacity behavior of $NdNi_2$ and $LaNi_2$ [40].

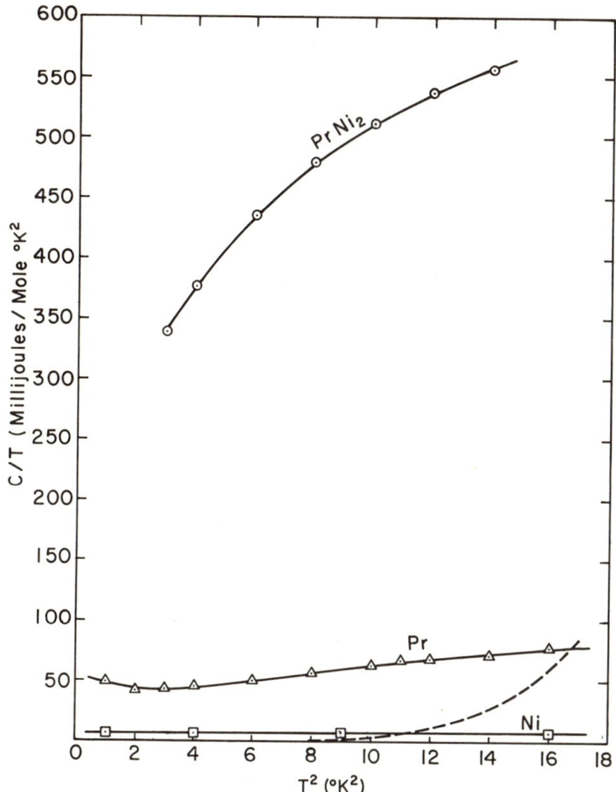

Fig. 9-9. Heat capacity of $PrNi_2$ together with that of the constituent elements. The crystal field contribution is shown as the dashed curve. This is computed assuming spectrum a in Fig. 9-1.

$LaNi_2$ are accepted as representing $C_e + C_l$ for $NdNi_2$, then its magnetic heat capacity is

$$C_m = 88.6T + 10.4T^3. \tag{9-8}$$

The variance of form of Eqs. (9-7) and (9-8) is obvious. Of course, the functional dependence of C_m on T is controlled by the nature of the spin wave spectrum. The form of Eq. (9-8) suggests an unusual spin wave spectrum for $NdNi_2$. To date measurements to provide information on this interesting point have not been made.

C_p data for $PrNi_2$ are shown in Fig. 9-9 along with similar data for its constituent elements. The very large heat capacity of the compound is quite evident. The crystal field heat capacity of $PrNi_2$ is also shown*

*This is calculated assuming a pure fourth interaction and an overall splitting of 136°K, the value derived from analysis of the susceptibility of $PrNi_2$.

in Fig. 9-9. The observed heat capacity of this compound exceeds that expected on the basis of lattice, electronic, and crystal field contributions by roughly an order of magnitude. A proper explanation of the large heat capacity of this compound and its trend with temperature remains to be provided.

e. $CeNi_2$

This compound exhibits a number of unusual features. In the first magnetic studies of $CeNi_2$ it was regarded [22,23] as a Pauli paramagnet since its susceptibility seemed to be temperature independent. More recent studies show [26] (Fig. 9-10) a significant variation with temperature, but a susceptibility rather small in comparison with expectation based on Curie's law—about 100-fold smaller at 4.2°K, for example. The Pauli paramagnetism was taken to imply loss of the 4f electron to the conduction band and hence a valence state of +4 for Ce in $CeNi_2$. However, the rise in χ below 100°K and the heat capacity behavior below 4°K (Fig. 9-7) are not consistent with the simple view of quadipositive Ce. C_p values for $CeNi_2$ and $LaNi_2$ are of the same order of magnitude but data for the former compound to not show the simple, expected functional dependence on temperature [Eq. (9-5)]. The mechanism responsible for the anomalous behavior in C_p, particularly the

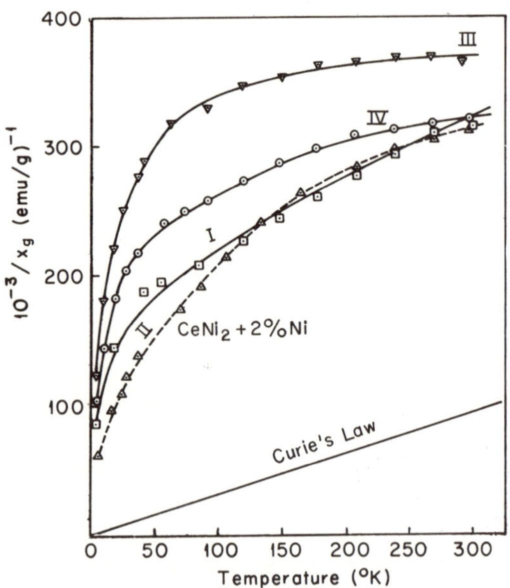

Fig. 9-10. Reciprocal susceptibility–temperature data (measured in an applied field of 20 kOe) for three $CeNi_2$ samples (I, III, and IV) and one $CeNi_2$ samples deliberately contaminated with 2 wt % excess Ni (II) [26,42].

upturn for $T < 2.5°K$, is not known. It can not be ascribed to a hyperfine interaction since the various stable cerium isotopes have neither a magnetic nor a quadrupole moment. Moreover, if Ce is indeed quadripositive, the upturn can not originate with the onset of magnetic ordering, unless a portion of Ce is tripositive, as has been sometimes suggested for Ce systems. Mixture of valencies in the compound is a simple idea which has its attractive features but is probably untenable. It is difficult to conceive of different valence states in an assemblage of structurally equivalent Ce ions.

We also note anomalous heat capacity behavior for $CeNi_2$ at high temperatures (Fig. 9-5); C_p exceeds the duLong–Petit limit for $T > 200°K$ and it continues to increase with rising temperature in an approximately linear fashion near room temperature.

Although the evidence is by no means conclusive, we are inclined to ascribe to various observed features of $CeNi_2$ to strong antiferromagnetic interaction between the conduction electrons and the localized moment of Ce^{3+}. We consider the low χ caused by spin compensation, i.e., the Kondo phenomenon, and hence we regard $CeNi_2$ as a concentrated Kondo system. For a dilute Kondo system, χ^{-1} should vary as $T^{1/2}$ at low temperatures. The properties of concentrated Kondo systems are imperfectly understood and it is not clear that they will display the quadratic dependence of inverse susceptibility on temperature, the resistance minimum, etc. It is of interest to note that the measurements of Volkmann, Craig, and Wallace on $CeNi_2$ confirm [26] the temperature dependence of susceptibility expected for dilute Kondo systems

TABLE 9-4
Heat Capacities and Derived Thermodynamic Properties of RNi_2 Compounds[a]

	$LaNi_2$	$CeNi_2$	$PrNi_2$	$NdNi_2$
C_p	77.06	81.59	77.62	77.67
$(H-H_0^0)/T$	56.35	57.80	57.78	58.03
S	121.95	124.32	136.58	134.97
$-(F-H_0^0)/T$	65.60	66.52	78.81	76.94
ΔS_f	5.14	−5.21	3.66	1.84
γ	12.52	*b*	*b*	101.1[c]
θ (°K)	2.42	*b*	*b*	*c*

[a] In units of joules $mole^{-1}$ $°K^{-1}$ for the quantities in the first five rows. γ is in millijoules $mole^{-1}$ $°K^{-1}$. Data are for 298.15°K; they have been computed from the C_p results given in Wallace *et al.* [28] and Marzouk *et al.* [38].

[b] Data could not be obtained since C_p/T was not linear with T^2.

[c] The large value of γ for $NdNi_2$ indicates the importance of other contributions besides electronic excitation. See text for further details.

(actually $\chi^{-1} \propto T^{0.48}$). However, resistivity measurements fail [43] to reveal a minimum in the range 2–400°K. Despite the lack of a resistance minimum we feel that the Kondo interaction is the dominant influence in determining the magnetic and heat capacity behavior of $CeNi_2$. We recall the case of $CeAl_2$ (Chapter 4) in which there was evidence in heat capacity measurements of the local moment–conduction electron interaction at rather high temperatures. We regard the situation in $CeNi_2$ as similar and ascribe the excess heat capacity above 200°K to the progressive weakening of this interaction. It is possible that the upturn in C_p/T below 2.5°K is also produced as a consequence of this interaction. Thermodynamic properties of the RNi_2 compounds are summarized in Table 9-4.

2. Ternary Systems

Four investigations of ternary systems based on the RNi_2 compounds have been carried out: (1) $RNi_{2-x}Al_x$; (2) $Pr_{1-x}Y_xNi_2$; (3) $GdNi_{2-x}Cu_x$; and (4) $R_{1-x}R_x'Ni_2$ systems. The $RNi_{2-x}Al_x$ system was discussed in Chapter 4 and the $Pr_{1-x}Y_xNi_2$ system above in this chapter.

$GdNi_2$ was found [26] to be able to incorporate Cu in the lattice up to the composition $GdNi_{1.65}Cu_{0.35}$ while maintaining the C15 structure. The $GdCu_2$ structure exists over the range GdNiCu to $GdCu_2$. A two-phase region separates the two regions of primary solubility. Magnetic measurements on the C15 phase revealed with increasing copper content a progressive rise of the strength of the interactions as indicated by the Curie temperature and Weiss constants, which for $GdNi_{1.65}Cu_{0.35}$ were 105 and 92°K, respectively. Paramagnetic and ferromagnetic moments were substantially independent of composition.

The study [44] involving the $R_{1-x}R_x'Ni_2$ system was the exact analogue of earlier work (Chapter 4) involving $R_{1-x}R_x'Al_2$ ternaries and was undertaken to ascertain whether the Ni and Al systems showed the same coupling systematics. We find in the studies of four systems (R,R′ = Gd,Dy; Gd,Ho; Gd,Nd; Ho,Nd) that the spins always couple parallel. This gives ferromagnetic coupling for the first two systems (Fig. 9-11) and antiferromagnetic coupling in the others (Fig. 9-12), as was observed for the corresponding aluminum compounds. This indicates that the sign of the coupling is the same irrespective of whether the rare earth partner is a nontransition metal or a transition metal with a recently filled d-band. The variation of T_C and θ is shown in Figs. 9-11 and 9-12. T_C is linear with composition for all ternaries studied. θ is also linear with composition for the heavy–heavy combination but not for the combination of one light and one neavy rare earth. In the latter case θ more nearly parallels the magnetization, rather as expected if the molecular field treatment applies.

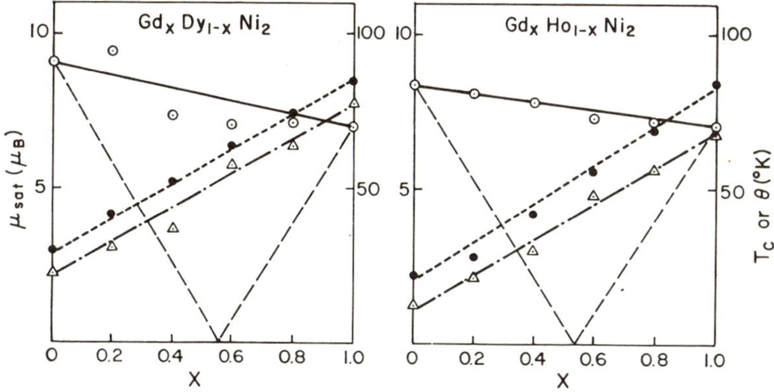

Fig. 9-11. Magnetic data for two ternary Laves phase systems containing two heavy lanthanides combined with nickel. The variation of saturation moment (⊙), Weiss constant (Δ) and Curie temperature (●) with composition are shown. The full and dashed lines show the expected μ_{sat} values for ferromagnetic and antiferromagnetic coupling, respectively [44].

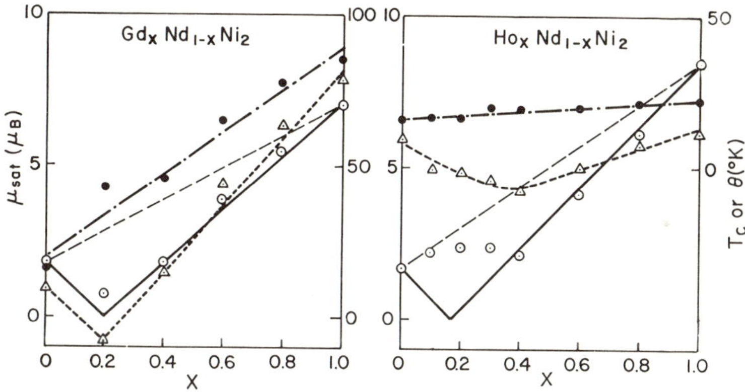

Fig. 9-12. Magnetic data for two ternary Laves phase systems containing one heavy and one light lanthanide combined with nickel. The legend is the same as for Fig. 9-11 except that the full and dashed lines refer to antiferromagnetic and ferromagnetic coupling, respectively [44].

All of the $R_{1-x}R_x'Ni_2$ ternaries exhibit Curie–Weiss behavior in the paramagnetic region with effective moments in close agreement with those expected for an assemblage of free tripositive rare earth ions.

Leon and Wallace studied the systems $Pr_{1-x}Dy_xNi_2$ and $Pr_{1-x}Ho_xNi_2$ ternaries [45]. As noted above, Pr^{3+} in $PrNi_2$ is in the nonmagnetic Γ_1 state at low temperatures and the compound exhibits Van Vleck paramagnetism. This is also true in the ternaries dilute in Dy or Ho. However, as the Dy or Ho content is increased Pr suddenly develops a

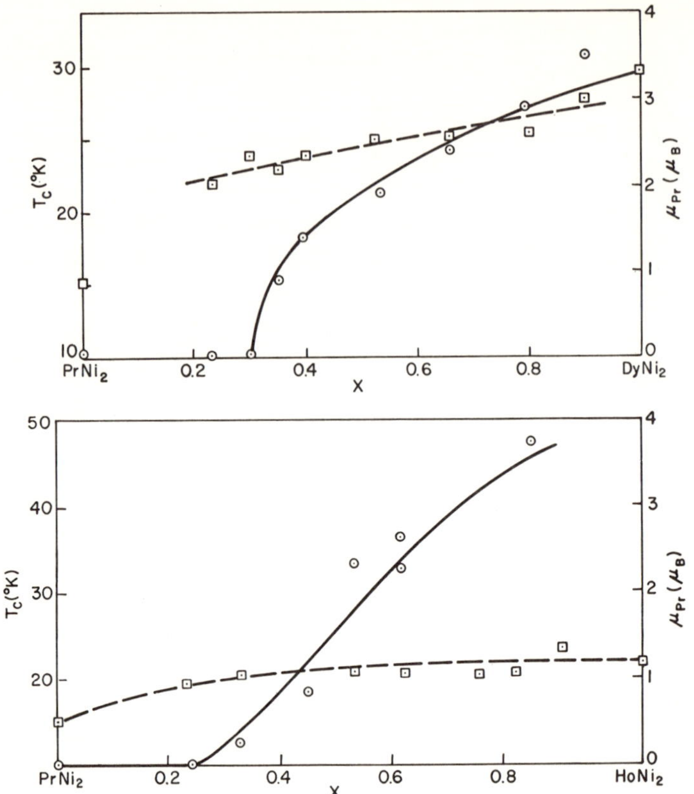

Fig. 9-13. Composition dependence of Curie temperature T_C (□) and praseodymium moment μ_{Pr} (O) in Pr–Ni ternaries. μ_{Pr} is evaluated under the assumptions that Ni is nonmagnetic and Pr and Dy are aligned antiparallel [45].

moment (Fig. 9-13) which then couples antiferromagnetically with Dy or Ho. Seemingly the strong field around the Dy^{3+} or Ho^{3+} ion induces and sustains a moment in otherwise nonmagnetic Pr^{3+}. Exchange is too weak for $PrNi_2$ to form an ordered structure at low temperatures but with the help of the field generated by the heavy rare earth ion the magnetic state is the stable form of Pr^{3+}. Judging from T_C and θ the internal field in $DyNi_2$ is higher than in $HoNi_2$. We should then expect the magnetic form of Pr^{3+} to become stable at a lower concentration of Dy^{3+} than Ho^{3+}. This is borne out by experiment.

B. RNi_5 COMPOUNDS

1. General Magnetic Properties

The first investigation of the magnetic characteristics of these compounds was carried out by Nesbitt *et al.* [46]. They examined eight compounds of the formula RNi_5 and established that with R = Pr, Sm, and Yb the compounds remained paramagnetic to the lowest temperatures studied whereas with R = Gd to Er ferromagnetic ordering occurred at temperatures ranging from roughly 10 to 25°K. They first drew attention to the weakly or nonmagnetic nature of nickel in these systems by noting that the magnetic moment of the compound was very close to that expected for the rare earth component alone. Thus the behavior of the RNi_5 compounds in this respect seems to parallel that of the RNi_2 series.

Wallace and Aoyagi [47] extended the work of Nesbitt *et al.* to include the several RNi_5 compounds not studied in the original work and to provide information concerning these materials in the paramagnetic state. Results obtained to date are summarized in Table 9-5.

The Pauli paramagnetism observed for $LaNi_5$ and $LuNi_5$ supports the notion that nickel is nonmagnetic or at most only weakly magnetic. It appears that nickel is nonmagnetic in the heavy rare earth compounds but perhaps just by the narrowest of margins. Assuming initially nickel with 0.6 vacancy per atom we have for the RNi_2 series 1.2 vacancies/f.u. whereas with RNi_5 we have 3.0 vacancies/f.u. In each case the vacancies are to be filled by the three valence electrons supplied by the rare earth. Thus the situation in regard to electron transfer is somewhat more marginal in th Haucke phase compounds (RNi_5) than in the Laves phase materials (RNi_2).

TABLE 9-5

Magnetic Characteristics of RNi_5 Compounds[a]

	Paramagnetic state			Ordered state			
R	μ_{eff} (μ_B/R)	θ (°K)	Type	μ_{or} (μ_B/R)	T_C (°K)	Type	Ref.
La			P				[48]
Ce	2.65	−90	C–W,VC(?)				[47]
Pr			N,x				[46,47,49]
Nd			N,x	2.27	13	F(?)	[47]
Sm				0.7	25	F	[7,46]
Gd	7.87	30	C–W	6.8	33	F	[46,47]
Tb				7.0	27	F	[7,46]
Dy				9.9	15	F	[7,46]
Ho	10.5	8	C–W	7.8	23	F	[46,47]
Er	9.85	12		7.7	13	F	[7,46,50]
Tm	7.88	60	C–W,VV	6.7	22	F	[47]
Lu			P				[51]

[a] $YbNi_5$ has also been investigated [46]. Nesbitt *et al.* say it remains paramagnetic to 1.4°K, but may become ferromagnetic for $T < 1.4$°K. No analysis of its paramagnetic behavior was made.

The evidence now to be presented suggests that the transfer to fill the nickel d band does occur with R = Gd to Lu (excluding $YbNi_5$ possibly) but filling does not occur in the light rare earth compounds, Pr and Nd, and in fact dilution of nickel with cobalt in the heavy rare earth compounds may confer a moment on nickel. Let us consider the second point first, using the system $DyNi_{5-x}Co_x$. Saturation magnetizations of these ternaries were determined by Wallace *et al.* [52]. ^{161}Dy Mössbauer measurements by Nowik and Wernick [53] showed that within a few percent the Dy moment in $DyNi_5$ and $DyCo_5$ correspond to that expected for the free ion. It seems reasonable that the Dy moment is the same in the Dy–Ni–Co ternaries. Thus assuming the Dy–Ni and Dy–Co coupling to be antiferromagnetic, we can calculate the combined Ni and Co moment. The results showed that if nickel is assumed nonmagnetic in these ternaries, cobalt has a higher moment in the ternaries than in pure $DyCo_5$. This does not seem reasonable. The results can be interpreted alternatively as indicating that nickel carries a small moment in the ternaries, rising to about 1/2 μ_B (Table 9-6). The trend of nickel moment with composition does not seem unreasonable. In the ternaries Co and Ni compete for the three electrons that can be supplied by Dy, hence it appears that that which was only marginally possible in $DyNi_5$ now becomes impossible in say $DyNi_2Co_3$—nickel than has an incomplete shell and acquires a moment.

TABLE 9-6

Nickel Moments in $DyNi_{5-x}Co_x$[a]

x	Nickel moment (μ_B/Ni)
0.0	0
1.0	0.16
2.0	0.33
3.0	0.49
4.0	0.47

[a] Wallace *et al.* [52].

Susceptibility data are shown [47] for $NdNi_5$ in Fig. 9-14. Linearity seems to be developing at higher temperature, although this may be an artifact of the restricted range of temperature covered. If the slope is used to establish an effective moment, we compute a moment well in excess of the free Nd^{3+} ion moment. Efforts* to account for the tempera-

*See detailed analysis of $PrNi_5$ results below as an indication of the kind of efforts involved.

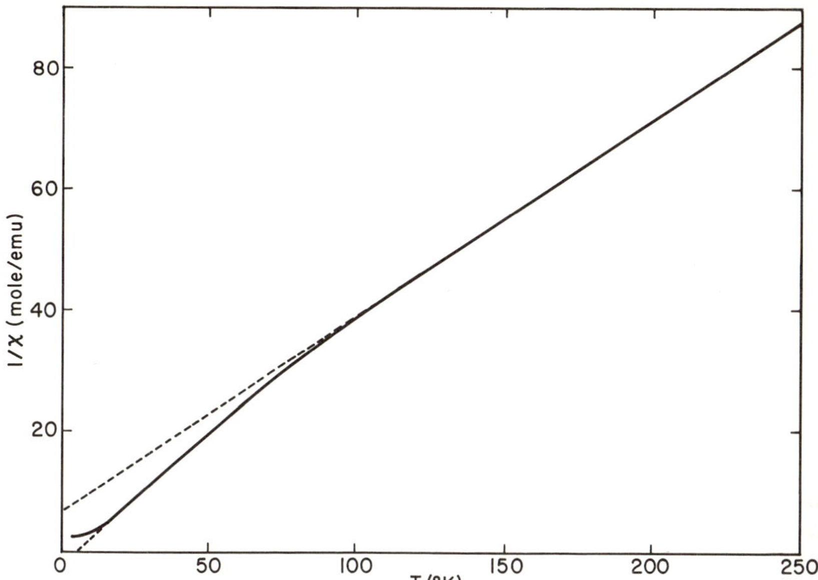

Fig. 9-14. Susceptibility data for $NdNi_5$ (taken from Wallace and Aoyagi [47]) showing the nonlinearity of χ^{-1} versus T.

ture dependence of susceptibility, except by assuming nickel to be a moment-carrying species, have been unsuccessful. Thus while we can safely regard nickel as nonmagnetic in the heavy eare earth RNi_5 compounds, we can not assume the same for the earlier members of the series. Nor, for that matter, can we regard nickel as nonmagnetic in ternary systems in the absence of explicit supporting evidence. This conclusion, in regard to binary systems, is buttressed by neutron diffraction work carried out on the RNi_5 compounds [54,55]. Nickel was found to be nonmagnetic in $ErNi_5$ and $TbNi_5$. The moments found were 7.1 μ_B/Tb^{3+} and 7.7 μ_B/Er^{3+}, in good agreement with results obtained in bulk magnetic measurements (Table 9-5).

2. $PrNi_5$ and $TmNi_5$

The magnetic and heat capacity behavior of $PrNi_5$ presents some very interesting features. Its magnetization passes through a maximum (Fig. 9-15) at about 13°K. Initially this was taken to mean that $PrNi_5$ was antiferromagnetic below 13°K. However, C_p measurements failed [49] to show a λ-type thermal anomaly near 13°K indicating that the susceptibility maximum does not correspond to a Néel point. The difference between the heat capacity of $PrNi_5$ and that of its nonmagnetic counterpart, $LaNi_5$ (Fig. 9-16), shows two Schottky-type thermal anomalies. These maxima and the magnetization maximum seem to be consequences of the crystal field interaction in $PrNi_5$. Its properties have been examined with the aid of the Hamiltonian

$$\mathcal{H} = g\mu_B(\mathbf{J}\cdot\mathbf{H}) + B_2^0 O_2^0 + B_4^0 O_4^0 + B_6^0 O_6^0 + B_6^6 O_6^6 \qquad (9\text{-}9)$$

or putting Eq. (9-9) in a form more convenient for calculation

$$\mathcal{H} = g\mu_B(\mathbf{J}\cdot\mathbf{H}) + W\left\{(1-|y|)\frac{O_2}{F_2} + y\left[x\frac{O_4}{F_4} + (1-|x|)\frac{O_6}{F_6}\right]\right\}, \qquad (9\text{-}10)$$

where O_6 represents $O_6^0 + 77/8\, O_6^6$, the operator obtained using the point charge value for the ratio $B_6^0/B_6^6 = 8/77$. Details of the calculation of the energy levels and moments are given by Segal and Wallace [56]. The new symbols appearing in (9-10) are defined by the relationships

$$B_4F_4 = Wx, \qquad B_6F_6 = W(1-|x|), \qquad \text{and} \qquad B_2F_2 = W(1-|y|). \qquad (9\text{-}11)$$

The appearance of the second-order term is a consequence of the non-ideal axial ration in $PrNi_5$.

In a hexagonal field the nine-fold degenerate multiplet state for Pr^{3+} splits into three singlet states (Γ_1, Γ_3, and Γ_4) and three doublet Γ_6 and two Γ_5's. The parameters in (9-10) were optimized for best fit with experiment, both as regards χ and C_p. Pertinent data for the Γ_i states in $PrNi_5$ are given in Table 9-7 and the computed χ and C_{cf} are shown in Figs. 9-16 and 9-17.

The procedure just described is able to account in a general way for the most conspicuous features of $PrNi_5$ but is insufficiently powerful to account for experiment in quantitative detail. This lack of quantitative agreement is ascribed to various limitations in both the experimental and calculated results. For further details the reader is referred to the original work [49]. We do direct attention here to one fact of the discrepancy between theory and experiment—the widening discrepancy above 100°K between the computed and observed susceptibilities. No variation of the parameters reduces this discrepancy. We are forced to ascribe the enhanced experimental χ to a nickel contribution and thus conclude that nickel in $PrNi_5$ is a magnetic species as it is in the $DyNi_{5-x}Co_x$ ternaries. The susceptibility behavior of $NdNi_5$ at higher temperatures is similar to that of $PrNi_5$, indicating that nickel may also be magnetic in the neodymium compound.

$TmNi_5$ shows a clear tendency toward Van Vleck paramagnetism at reduced temperatures. It fails to develop, however, because of the intervention of magnetic ordering.

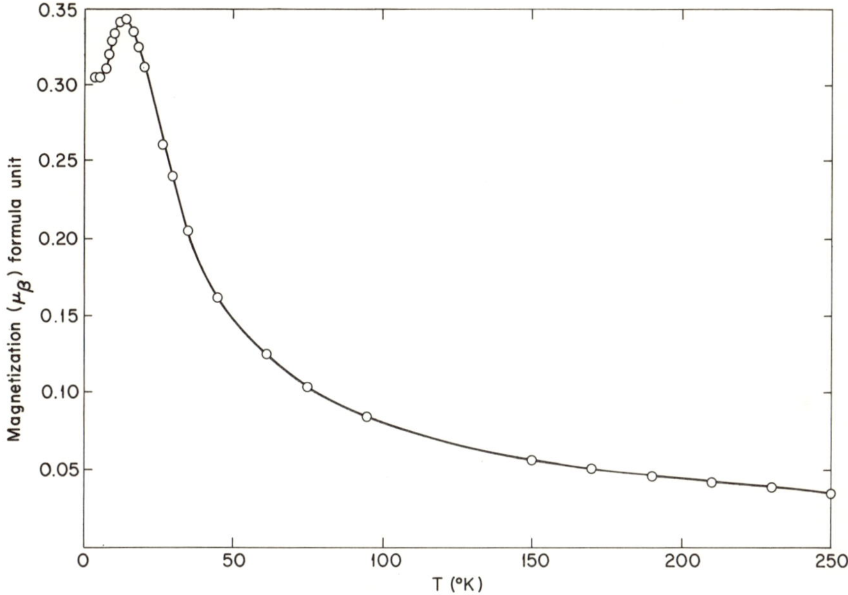

Fig. 9-15. Temperature dependence of the magnetization of $PrNi_5$ [49].

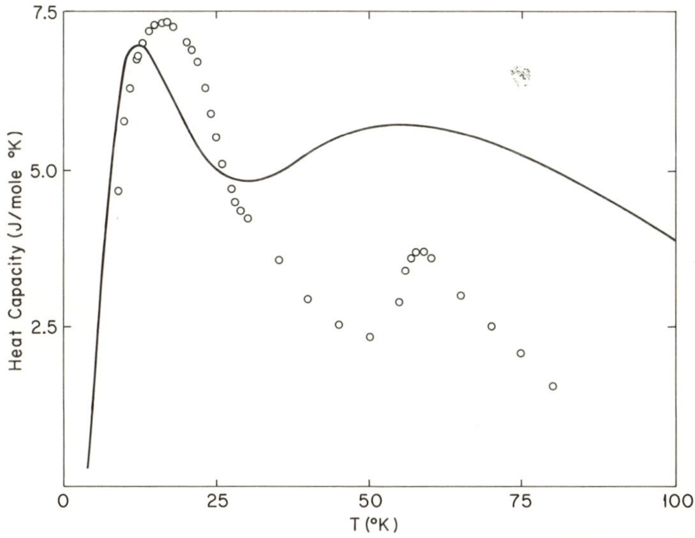

Fig. 9.16. Points are the difference in heat capacity of $PrNi_5$ and $LaNi_5$. Curve is the calculated crystal field heat capacity [49].

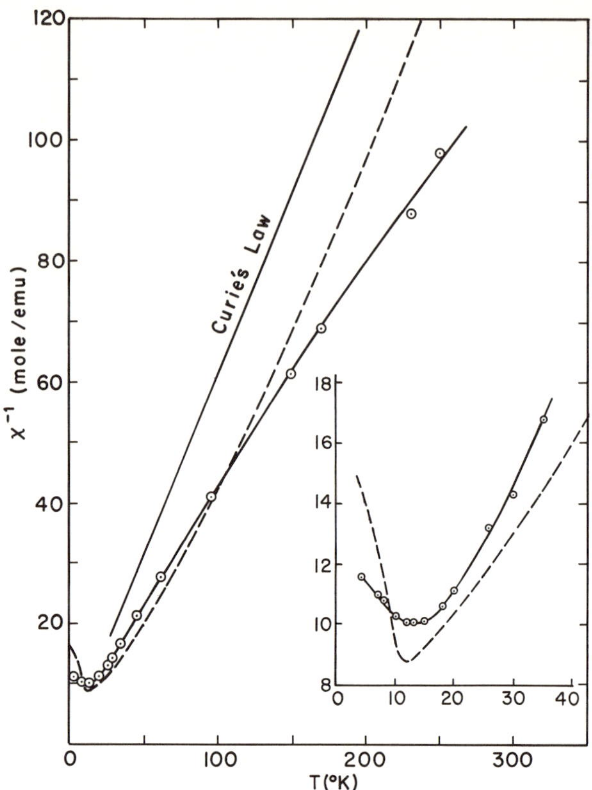

Fig. 9-17. Inverse susceptibility versus temperature for $PrNi_5$, measured (O) and calculated (---). For comparison, Curie's law for Pr^{3+} is also shown [49].

TABLE 9-7

Characteristics of the Crystal Field States of Pr^{3+} in $PrNi_5$

State	Energy (°K)[a]	Eigenfunction[a]	Magnetic moment[b]
Γ_6	99.5	$\|\pm\rangle$	±0.250
Γ_1	80.5	$\|0\rangle$	0.000
Γ_5	79.5	$0.180\|\mp4\rangle - 0.987\|\pm2\rangle$	$+0.445, -0.454$
Γ_3	-61.5	$0.707[\|3\rangle + \|-3\rangle]$	-0.073
Γ_5	-112.2	$0.987\|\mp4\rangle - 0.180\|\pm2\rangle$	$+0.955, -0.949$
Γ_4	-150.0	$0.707[\|3\rangle - \|-3\rangle]$	0.073

[a] In zero magnetic field.

[b] In a magnetic field of 20 kOe. The quantities tabulated are given as a fraction of gJ, the free ion moment in Bohr magnetons.

3. Heat Capacities

Measurements have been made [28,49,57,58] on the five compounds listed in Table 9-8 over the temperature range extending from 1.6 to

300°K. Results for $PrNi_5$ for the 6–300°K range have been discussed in the preceding section. Its heat capacity and derived thermodynamic properties for 25°C are listed in Table 9-8 along with similar data for the other RNi_5 compounds. The C_p data for $LaNi_5$, $CeNi_5$, and $NdNi_5$ are straightforward, the first two compounds showing behavior in keeping with that expected for a normal solid. $NdNi_5$ behaves in a like manner except for the λ-type transition peaking at 6.4°K* associated with the destruction of ferromagnetic order.

TABLE 9-8

Heat Capacities and Derived Thermodynamic Properties of RNi_5 Compounds[a]

	$LaNi_5$	$CeNi_5$	$PrNi_5$	$NdNi_5$	$GdNi_5$
C_p	155.00	154.90	161.20	155.40	150.40
$(H-H_0^\circ)/T$	106.12	106.13	109.98	107.09	103.76
S	213.29	213.84	230.14	231.35	221.37
$-(F-H_0^\circ)/T$	107.17	107.71	120.16	124.26	117.61
ΔS_f	6.87	−5.30	7.61	8.61	6.04
γ	34.3	40.0	37.0	*b*	*b*
θ (°K)	341	347	334	*b*	*b*

[a] Units for the quantities in the first five rows are joules $mole^{-1}$ $°K^{-1}$. Units for γ are millijoules $mole^{-1}$ $°K^{-2}$. Data are for 298.15°K.

[b] Quantities could not be determined. See text.

In a field of hexagonal symmetry the tenfold degenerate ground state of Nd^{3+} is split into five doublets, one or more of which are obviously lowest lying. We therefore expect the entropy associated with the λ point to be $R \ln 2$ if one is lowest lying. Experimentally it is 11.0 J/mole-°K, i.e., 0.96 $R \ln 4$. This indicates that two doublets are involved in the formation of the cooperative magnetic state. Excess C_p (over $LaNi_5$) at temperatures above T_C is observed in consequence of the repopulation of the excited doublets. This excess peaks at about 50°K. As yet calculations for $J=9/2$ have not been made using the Hamiltonian in Eq. (9-8). Hence analysis of the crystal field heat capacity, such as was done for $PrNi_5$, to give E_c for $NdNi_5$ is as yet not possible.

Results obtained for $GdNi_5$ (Fig. 9-18) are surprising in that two closely spaced λ-type thermal anomalies are observed at 29.8 and 30.6°K. We see that in this compound magnetic order is broken down in two stages. While it is possible that this is an impurity effect, it seems un-

*This temperature is in contrast with the Curie point listed for $NdNi_5$ in Table 9-5, 13°K. The latter was obtained from magnetic measurements made at 20 kOe whereas the heat capacity data were obtained in zero field.

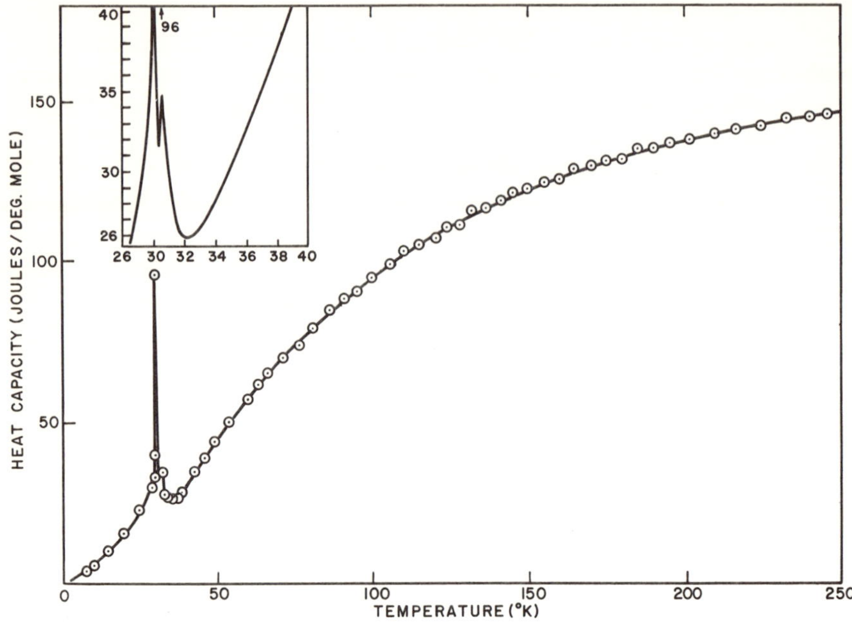

Fig. 9-18. Molar heat capacity of $GdNi_5$ [28].

likely since no other Gd–Ni intermediate phases have ordering temperatures near 30°K.

Excess entropies of $PrNi_5$, $NdNi_5$, and $GdNi_5$ over $LaNi_5$ are (Table 9-8) 16.9, 18.1, and 14.8 J/mole °K, respectively. These are 92, 95, and 86% of $R \ln(2J + 1)$ in the several cases. Agreement is excellent for the Pr and Nd compounds but somewhat less than completely satisfying for $GdNi_5$, for reasons which are as yet not clear.

The lack of a λ-type transition for $CeNi_5$ is striking, as is the negative ΔS_f for it compared with a positive value of ΔS_f for the other RNi_5 compounds. The absence of a magnetic anomaly is formally accounted for under the rubric that Ce in $CeNi_5$ is quadripositive. However, the ΔS_f values make it appear as if its entropy is low by 12 to 13 J/mole^{-1} °K^{-1}, which is 80 to 85% of $R \ln(2J + 1)$ for Ce^{3+}. These considerations suggest that $CeNi_5$ might have a transition and an associated thermal anomaly at a temperature lower than those covered in the experiments to date. Estimating the probable Curie point of $CeNi_5$ from the de Gennes factor and T_C for $NdNi_5$ we obtain 0.6°K. Perhaps $CeNi_5$ orders below 1°K.

In Table 9-5 we very tentatively ascribe the non-Curie–Weiss behavior of $CeNi_5$ to variable valence, primarily because this has been the fashion in regard to cerium intermetallics in the recent past. This may not be correct. Spin compensation (Kondo effect) or the crystal field interaction or both may be operating to reduce the susceptibility. As regards the former possibility we note that no resistance minimum for $CeNi_5$ has been found [57]. Clearly additional work on $CeNi_5$ is necessary to elucidate its strange behavior.

Data below 4°K show conformity with Eq. (9-5) for the La, Ce, and Pr compounds, but not for $NdNi_5$ and $GdNi_5$. The γ values are high (Fig. 9-19). Even so they are less than the sum of γ for the constituents, reflecting the effect of the filling of the nickel d-band. The γ values are, as expected, of the same order as that observed for transition metal systems. There is a large magnetic contribution in C_p for $NdNi_5$ in the 1.5–4°K range due to proximity of T_C (6.5°K) and hence the C_p data can not be analyzed to obtain γ and θ. The C_p values obtained for $GdNi_5$ are high and are regarded [58] as impurity dominated.

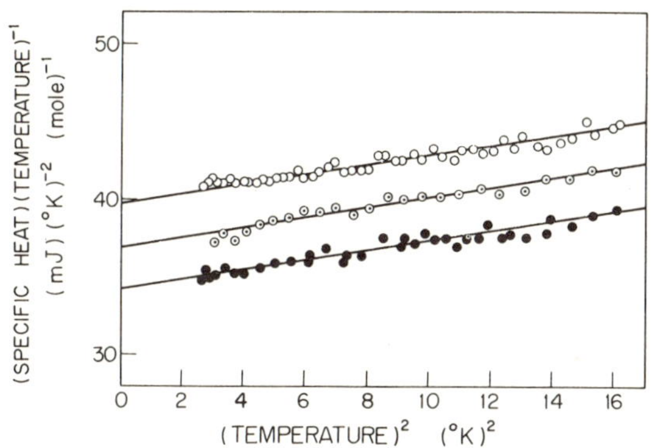

Fig. 9-19. Heat capacity behavior of $LaNi_5$ (●), $CeNi_5$ (○), and $PrNi_5$ (⊙) [58].

4. Ternary Systems

The work on $R_{1-x}R_x'Ni_2$ ternaries by Hopkins, Lehman, and Wallace has been referred to above. Similar measurements were made [44] on $R_{1-x}R_x'Ni_5$ ternaries with R,R′ = Gd,Dy; Gd,Ho; Gd,Nd; and Ho,Nd. The coupling systematics observed in the Laves phase systems are also observed in the $(R,R')Ni_5$ ternaries: Gd couples ferromagnetically with Dy or Ho and Nd couples antiferromagnetically with Gd or Ho. Results are shown in Figs. 9-20 and 9-21.

Curie–Weiss behavior is observed for Gd–Dy–Ni and Gd–Ho–Ni ternaries with effective moments in close agreement with that expected for an assemblage of free tripositive lanthanide ions. Curie–Weiss behavior is not observed for any of the alloys in the $Gd_xNd_{1-x}Ni_2$

system but is observed for the $Ho_xNd_{1-x}Ni_5$ system with $x > 0.2$. The effective moments are somewhat in excess of that expected from the rare earth component, as if nickel were contributing to μ_{eff}. These observations are compatible with the viewpoint advanced above to indicate that even when nickel is nonmagnetic in RNi_5 systems it is close to the threshold for developing a moment and can be pushed over that threshold by a suitable change of environment.

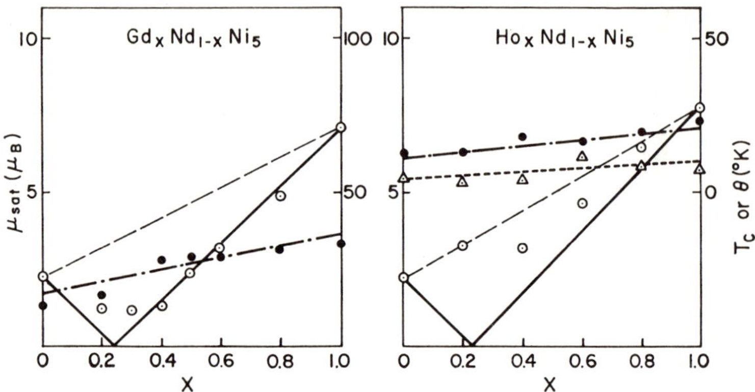

Fig. 9-20. Magnetic data for two Haucke phase systems containing two heavy lanthanides combined with nickel. The legend is the same as for Fig. 9-12 [44].

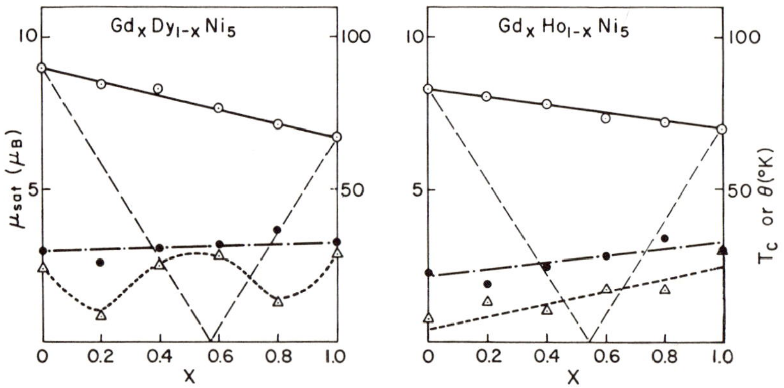

Fig. 9-21. Magnetic data for two Haucke phase systems containing one heavy and one light lanthanide combined with nickel. The legend is the same as for Fig. 9-11 [44].

C. THE RNi AND R_2Ni_{17} COMPOUNDS

Bulk magnetic characteristics observed [6,7] for these compounds are set forth in Tables 9-9 and 9-10. Neutron diffraction work has been carried out on RNi, HoNi, and ErNi by Lemaire and Paccard [59]. This work confirms the nonmagnetic nature of nickel, which had been inferred earlier from bulk magnetic measurements—the Pauli paramagnetism of LaNi and LuNi and μ_{or} for GdNi and μ_{eff} for all the RNi compounds. The rare earth moments obtained by the two techniques are in excellent agreement for NdNi and HoNi. Surprisingly, considering the difficulty of saturating these compounds, the bulk moment for ErNi exceeds that obtained by the diffraction method. The latter method confirms that NdNi is ferromagnetic. The work of Lemaire and his associates shows the magnetic structures of HoNi and ErNi to be rather complex. Moments lie in planes parallel to the 010 planes. Coupling is essentially ferromagnetic but the spins are not collinear so that there is a significant antiferromagnetic component.

TABLE 9-9

Magnetic Characteristics of RNi Compounds

	Paramagnetic state			Ordered states			
R	μ_{eff} (μ_B/R)	θ (°K)	Type	μ_{or} (μ_B/R)	T_C (°K)	Type	Ref.
La			P				[7]
Ce			P				[6,7]
Pr	3.9	23	C–W	2.26	22	F	[6,7]
Nd	3.7	24	C–W	2.74 (2.7)[a]	28	F	[6,7,59]
Sm				0.23	45	F	[6,7]
Gd	8.1	77	C–W	7.40	71	F	[6,7]
Tb	9.7	40	C–W	7.74	52	AF(M)	[6,7,60]
Dy	10.7	64	C–W	8.75	62	F	[6,7]
Ho	10.7	36	C–W	8.65 (8.6)	37	F[b]	[6,7,59]
Er	9.8	13	C–W	8.10 (7.0)	13	F[b]	[6,7,59]
Tm			N,VV	5.15	8	F	[6,7]
Lu			P				[7]

[a]Numbers in parentheses are rare earth moments obtained from neutron diffraction measurements [59].

[b]There is also an AF component (see text).

The ordered moments of all the RNi compounds are less than gJ indicating significant quenching of the orbital contribution to the moment by the crystal field. The moment of GdNi exceeds gJ by about the same amount as in elemental Gd [63]. These additional moments are ascribed to the polarized conduction electrons [64] (see Chapters 6 and 7).

In the paramagnetic region Curie–Weiss behavior is observed for all the RNi compounds containing magnetic rare earths except TmNi. χ^{-1} versus T in this case is nonlinear, exhibiting a tendency toward

TABLE 9-10

Magnetic Characteristics of R_2Ni_{17} Compounds

	Saturation magnetization (μ_B/f.u.)			T_C (°K)		
R	From [61]	From [62]	Calc.[a]	From [61]	From [62]	Type
Sm	4.5	5.25	6.09	186	641	F
Gd	8.8	9.36	9.33	205	623	F_i
Tb	8.5	12.2	13.4	178	615	F_i
Dy	8.05	14.7	15.4	168	604	F_i
Ho	12.2	13.8	15.4	162	611	F_i
Er	9.1	11.0	13.4	166	602	F_i
Tm		7.31	9.33	152	603	F_i
Lu		5.0			601	F

[a] Assuming the free ion moment for R (i.e., gJ) and 0.27 μ_B/Ni atom; R and Ni spins are taken to couple antiparallel so that Sm_2Fe_{17} is ferromagnetic and the others, excluding Lu_2Ni_{17}, are ferrimagnetic.

Van Vleck paramagnetism, which is possible in this case since Tm^{3+} could have a singlet ground state.

The bulk magnetic properties of 2:17 compounds (R_2Ni_{17}) were studied by Laforest *et al.* [61] and Carfagna and Wallace [62] with somewhat disparate results emerging. In both studies it was concluded that unlike the situation in the RNi, RNi_2, and RNi_5 series nickel in the R_2Ni_{17} is a magnetic species. Because of the low concentration of R in these materials electron transfer is insufficient to fill the nickel d-band. Thus Y_2Ni_{17} and Lu_2Ni_{17} are ferromagnetic with moments per Ni atom of 0.27 and 0.29, respectively. These moments can be accounted for on the following simple model: The two R atoms contribute six electrons to 17 nickel atoms each lacking 0.6 electron to fill its d shell. This electron transfer corresponds to 0.35 electron/Ni atom leaving 0.25 hole/atom, which is in good accord with the atomic moment observed in the Y and Lu compounds.

The principal difference between the results of Laforest *et al.* and Carfagna and Wallace is in the reported Curie temperature. The former workers reported T_C ranging from 152°K (for Tm_2Ni_{17}) to 205°K (for Gd_2Ni_{17}), whereas in the latter study Curie temperatures were found to lie in a narrow range, 601–623°K, which is close to T_C for Ni in the elemental form, 631°K. In the Carfagna work magnetization for Y_2Ni_{17} and Lu_2Ni_{17} was lost in two stages, the temperature of the first corresponding roughly to the T_C reported by Laforest *et al.* and the latter being close to T_C for Ni. This raised the possibility of a substantial nickel contaminant in the samples studied by Carfagna, the possibility being heightened by the fact that Laforest's samples had a nickel impurity of ~2%, for which a correction was applied. Rather exhaustive examination on the R_2Ni_{17} samples and on the closely related $Y_2Ni_{17-x}Cu_x$ ternaries [65] failed to show evidence of any significant nickel contaminant. Carfagna and Wallace advanced the suggestion that Laforest *et al.*, in correcting for their "Ni impurity," may have subtracted out a contribution which was indeed characteristic of R_2Ni_{17}. Carfagna *et al.* [65] have presented detailed arguments based on the rather complex behavior of $Y_2Ni_{17-x}Cu_x$ ternaries to the effect that the low temperature drop in magnetization is a result of an electronic change in nickel and that the drop for $T > 600$°K is a consequence of the breakup in order in R_2Ni_{17} and hence is the true T_C. Carfagna and Wallace summarized their position by stating it to be their belief that their data are "more nearly characteristic of the Ln_2Ni_{17} compounds, due to better control and assessment of impurities." Savitskii *et al.* [66] have studied Eu_2Ni_{17} and established T_C for it to be greater than 600°K reinforcing the conclusion that the higher T_C's are correct.

Both investigations lead to the conclusion that the R–Ni coupling system is the same as in the closely related R_2Co_{17} and R_2Fe_{17} compounds discussed in subsequent chapters: ferromagnetic when R is light and antiferromagnetic when R is heavy. Results, summarized in Table 9-10, show rather good agreement between the calculated values and the data of Carfagna and Wallace but substantial discrepancies with the data of Laforest *et al.*, adding to the notion that the latter authors, in correcting for nickel, may have subtracted part of the moment for R_2Ni_{17}. The low moments of Laforest *et al.* and the high T_C values, and especially the close proximity to T_C for nickel, arouse suspicion that we may not yet be seeing the true characteristics of R_2Ni_{17}. Additional work is needed to clarify these matters.

D. THE R_3Ni, R_7Ni_3, RNi_3, AND R_2Ni_7 AND Er_5Ni_3 COMPOUNDS

Er_3Ni was studied by Buschow [69] and the entire series listed in Table 9-11 by Feron *et al.* [67]. Neutron diffraction studies by Bécle *et al.* [60] showed it to be a complex noncollinear antiferromagnet. Kissell *et al.* [68] studied Ce_7Ni_3 and Pr_7Ni_3. Neither of these compounds showed signs of ordering at 4°K. Low temperature deviations from linearity for χ^{-1} versus T are ascribed to the crystal field interaction. The RNi_3 series has been investigated by Paccard and Pauthenet [70]. Buschow [69] studied $ErNi_3$ and Er_2Ni. Magnetic behavior of the entire R_2Ni_7 series has been investigated by Lemaire *et al.* [71].

Nickel is nonmagnetic in R_3Ni, R_7Ni_3, and Er_5Ni_3. Pauthenet and his colleagues believe [70,71] that nickel carries a small moment, 0.05 or

TABLE 9-11

Magnetic Characteristics of Miscellaneous Rare Earth–Nickel Compounds

	Paramagnetic state			Ordered state				
	μ_{eff} (μ_B/R)	θ (°K)	Type	μ_{or} (μ_B/R)	T_C (°K)	T_N (°K)	Type	Ref.
Pr_3Ni	3.7	−24	C–W			2	AF	[67]
Nd_3Ni	3.6	0	C–W			15	AF	[67]
Gd_3Ni	8.1	60	C–W			100	AF	[67]
Tb_3Ni	10.0	−5	C–W			62	AF	[67]
Dy_3Ni	10.6	29	C–W			35	AF	[67]
Ho_3Ni	11.1	−6	C–W			20	AF	[67]
Er_3Ni	9.8	−5	C–W			7	AF(M)	[60,67,69]
Tm_3Ni	7.4	0	C–W	3.7	12		F	[67]
Ce_7Ni_3	2.41	−38	C–W,CF					[68]
Pr_7Ni_3	3.60	−2	C–W,VV					[68]
Er_5Ni_3	9.77	31	C–W	8.1	13		F	[69]
$PrNi_3$				1.57	20		F	[70]
$NdNi_3$		30	C–W	1.88	27		F	[70]
$SmNi_3$				0.33	85		F	[70]
$GdNi_3$		115	C–W	6.55	116		F_i	[70]
$TbNi_3$				6.84	98		F_i	[70]
$DyNi_3$		70	C–W	7.0		69	F_i	[70]
$HoNi_3$				7.84		66	F_i	[70]
$ErNi_3$	9.75	15		5.77		64	F_i	[60,70]
$TmNi_3$				3.86		43	F_i	[70]
$YbNi_3$				2.0		<20	F_i	[70]
Ce_2Ni_7			N,IEL	0.26[a]	48		F	[71]
Pr_2Ni_7			N,IEL	4.36[a]	85		F	[71]
Nd_2Ni_7			C–W,CF	4.14[a]	87		F	[71]
Gd_2Ni_7			N,IEL	12.65[a]	118		F_i	[71]
Tb_2Ni_7			N,IEL	10.72[a]	101		F_i	[71]
Dy_2Ni_7			N,IEL	13.30[a]	81		F_i	[71]
Ho_2Ni_7			N,IEL	12.57[a]	70		F_i	[71]
Er_2Ni_7	9.56	22	C–W	12.28[a]	70		F_i	[69,71]

[a] These are per formula unit of R_2Ni_7.

0.06 μ_B, in the RNi_3 and R_2Ni_7 compounds when R is nonmagnetic. Probably the moment is larger in the presence of a magnetic rare earth. We find from the measured magnetizations that the R–Ni coupling follows the same systematics as in the R_2Ni_{17} phases, i.e., ferromagnetic when R is light and antiferromagnetic when R is heavy. The ferrimagnetic coupling accounts for some of the deficiency in moment, as compared with that expected from the R component, in the RNi_3 and R_2Ni_7 series, but not all. Probably these are not fully saturated and some reduction of the moment of the rare earth by the crystal field interaction is also likely to be taking place.

For most of the R_2Ni_7 series, χ^{-1} fails to be linear with T. The existence of a nickel moment suggests that the nonlinearity originates with the nickel component and is due to the fact that the d electrons for nickel either are delocalized or at least are incompletely localized.

REFERENCES

1. W. Fülling, K. Moeller, and R. Vogel, *Z. Metallk.* **34,** 254 (1942).
2. K. H. J. Buschow, *J. Less Common Metals* **16,** 45 (1968).
3. R. Lemaire and D. Paccard, *Bull. Soc. Fr. Mineral Cristallogr.* **XC,** 311 (1967).
4. F. Kissell, T. Tsuchida, and W. E. Wallace, *J. Chem. Phys.* **44,** 4651 (1966).
5. J. J. Finney and A. Rosenzweig, *Acta Crystallogr.* **14,** 69 (1961).
6. R. E. Walline and W. E. Wallace, *J. Chem. Phys.* **41,** 1587 (1964).
7. S. C. Abrahams, J. L. Bernstein, R. C. Sherwood, J. H. Wernick, and H. J. Williams, *J. Phys. Chem. Solids* **25,** 1069 (1964).
8. A. E. Dwight, R. A. Conner, and J. W. Downey, Jr., *Acta Crystallogr.* **18,** 835 (1965).
9. R. Vogel and W. Fülling, *Z. Metallk.* **32,** 254 (1947).
10. K. Nassau, L. V. Cherry, and W. E. Wallace, *J. Phys. Chem. Solids* **16,** 123 (1960).
11. F. Endter and W. Klemm, *Z. Anorg. Chem.* **252,** 64 (1943).
12. D. T. Cromer and A. C. Larson, *Acta Crystallogr.* **12,** 855 (1959).
13. R. Lemaire, D. Paccard, and R. Pauthenet, *C. R. Acad. Sci. Paris* **265,** 1280 (1967).
14. K. H. J. Buschow and A. S. Van der Goot, *J. Less Common Metals* **22,** 419 (1970).
15. D. T. Cromer and C. E. Olsen, *Acta Crystallogr.* **12,** 689 (1959).
16. D. Paccard and R. Pauthenet, *C. R. Acad. Sci. Paris* **264,** 1056 (1967).
17. H. Nowotny, *Z. Metallk.* **34,** 247 (1942).
18. K. H. J. Buschow, *J. Less Common Metals* **11,** 204 (1966).
19. J. Laforest, R. Lemaire, D. Paccard, and R. Pauthenet, *C. R. Acad. Sci. Paris* **264,** 676 (1967).
20. P. D. Carfagna and W. E. Wallace, *J. Appl. Phys.* **11,** 5259 (1968).
21. R. Lemaire and D. Paccard, *Bull. Soc. Fr. Mineral. Cristallogr.* **92,** 9 (1969).
22. E. A. Skrabek and W. E. Wallace, *J. Appl. Phys.* **34,** 1356 (1963).
23. J. Farrell and W. E. Wallace, *Inorg. Chem.* **5,** 105 (1966).
24. J. W. Ross and J. Crangle, *Phys. Rev.* **133,** A509 (1964).
25. N. Marzouk, W. E. Wallace, and R. S. Craig, unpublished measurements.
26. W. E. Wallace, T. V. Volkmann, and R. S. Craig, *J. Phys. Chem. Solids* **31,** 2185 (1970).
27. K. H. Mader and W. E. Wallace, *Inorg. Chem.* **7,** 1627 (1968).
28. W. E. Wallace, R. S. Craig, A. Thompson, C. Deenadas, M. Dixon, M. Aoyagi, and N. Marzouk, *Les Éléments des Terres Rares, Coll. Int. C.N.R.S.* No. 180, 427 (1970).
29. See, for example, E. W. Pugh, B. R. Coles, A. Arrott, and J. E. Goldman, *Phys. Rev.* **105,** 814 (1956).
30. J. S. Wollam and W. E. Wallace, *J. Phys. Chem. Solids* **13,** 212 (1960).
31. W. G. Penney and R. Schlapp, *Phys. Rev.* **41,** 194 (1932).
32. E. Segal, private communication (1970).
33. D. P. Schumacher and C. A. Hollingsworth, *J. Phys. Chem. Solids* **27,** 749 (1966).
34. B. Bleaney, *Proc. Roy. Soc., London.* **A276,** 39 (1963).
35. K. Mader, E. Segal, and W. E. Wallace, *J. Phys. Chem. Solids,* **30,** 1 (1969); and W. E. Wallace, F. Kissell, E. Segal, and R. S. Craig, *ibid.* **30,** 13 (1969).

36. K. R. Lea, M. J. M. Leask and W. P. Wolf, *J. Phys. Chem. Solids* **23,** 1381 (1962).
37. S. Ofer and E. Segal, *Phys. Rev.* **141,** 844 (1966).
38. N. Marzouk, R. S. Craig, and W. E. Wallace, measurements on $LaNi_2$ (to be published).
39. W. E. Wallace, *Progr. Rare Earth Sci. Technol.* **3,** 1 (1968).
40. H. H. Neumann, S. Nasu, R. S. Craig, N. Marzouk and W. E. Wallace, *J. Phys. Chem. Solids* **32,** 2788 (1971).
41. J. Van Kranendonk and J. H. Van Vleck, *Rev. Mod. Phys.* **30,** 1 (1958).
42. W. E. Wallace, *Progr. Solid State Chem.* **6,** 155 (1971).
43. V. U. S. Rao, R. S. Craig, and W. E. Wallace, unpublished measurements.
44. W. E. Wallace, H. P. Hopkins, Jr., and K. Lehman, *J. Solid State Chem.* **1,** 39 (1969).
45. B. Leon and W. E. Wallace, *J. Less Common Metals* **22,** 1 (1970).
46. E. A. Nesbitt, H. J. Williams, J. H. Wernick, and R. C. Sherwood, *J. Appl. Phys.* **33,** 1674 (1962).
47. W. E. Wallace and M. Aoyagi, *Chem. Monatsh.* **102,** 1455 (1971).
48. N. Marzouk, R. S. Craig, and W. E. Wallace, unpublished measurements.
49. R. S. Craig, S. G. Sankar, N. Marzouk, V. U. S. Rao, W. E. Wallace, and E. Segal, *J. Phys. Chem. Solids* **33,** 2267 (1972).
50. M. Aoyagi and W. E. Wallace, unpublished measurements.
51. J. J. Farrell and W. E. Wallace, unpublished measurements.
52. W. E. Wallace, T. V. Volkmann, and H. P. Hopkins, Jr., *J. Solid State Chem.* **3,** 510 (1971).
53. I. Nowik, and J. H. Wernick, *Phys. Rev.* **140,** A131 (1965).
54. R. Lemaire and D. Paccard, *C. R. Acad. Sci. Paris* **270,** 1131 (1970).
55. L. M. Corliss and J. M. Hastings, *La diffraction et la diffusion des neutrons. Coll. Int. C.N.R.S., Grenoble* **126,** 133 (1963).
56. E. Segal and W. E. Wallace, *J. Solid State Chem.* **2,** 347 (1970).
57. N. Marzouk, R. S. Craig, and W. E. Wallace, *J. Phys. Chem. Solids* **34,** 15 (1973).
58. S. Nasu, H. H. Neumann, N. Marzouk, R. S. Craig, and W. E. Wallace, *J. Phys. Chem. Solids* **32,** 2779 (1971).
59. R. Lemaire and D. Paccard, *Les Éléments des Terres Rares, Coll. Int. C.N.R.S.* No. 180, Tome 2, 231 (1970).
60. C. Bécle, R. Lemaire, and D. Paccard, *J. Appl. Phys.* **41,** 855 (1970).
61. J. Laforest, R. Lemaire, D. Paccard, and R. Pauthenet, *C. R. Acad. Sci. Paris* **264,** 676 (1967).
62. P. D. Carfagna and W. E. Wallace, *J. Appl. Phys.* **39,** 5259 (1968).
63. W. C. Thoburn, S. Legvold, and F. H. Spedding, *Phys. Rev.* **110,** 1298 (1958).
64. S. H. Liu, *Phys. Rev.* **123,** 470 (1961).
65. P. D. Carfagna, W. E. Wallace, and R. S. Craig, *J. Solid State Chem.* **2,** 1 (1970).
66. E. M. Savitskii, V. Terekhova and R. S. Torchinova, *Metalloved. Term. Obrab. Mtal.* **2,** 25 (1967).
67. J. L. Feron, R. Lemaire, D. Paccard, and R. Pauthenet, *C. R. Acad. Sci. Paris* **267B,** 371 (1968).
68. F. Kissell, T. Tsuchida, and W. E. Wallace, *J. Chem. Phys.* **44,** 4651 (1966).
69. K. H. J. Buschow, *J. Less Common Metals* **16,** 45 (1968).
70. D. Paccard and R. Pauthenet, *C. R. Acad. Sci. Paris* **264B,** 1056 (1967).
71. R. Lemaire, D. Paccard, and R. Pauthenet, *C. R. Acad. Sci. Paris* **265B,** 1280 (1967).

Chapter 10

COBALT COMPOUNDS

The cobalt compounds played a pivotal role in kindling interest in rare earth intermetallics. Prior to 1959 there had been but a few scattered attempts [1–3] to study the magnetic behavior of systems containing the rare earths in chemical union with other metals. The results of most of these early studies proved upon re-examination to have been untrustworthy. Then in rapid succession in 1959 and 1960 three highly significant papers were published. Nesbitt *et al.* [4] presented magnetic results for a series of Gd–Co alloys, which included the intermetallics $GdCo_2$, $GdCo_3$, and $GdCo_5$ and showed that the Gd and Co sublattices couple antiparallel in each of these phases, i.e., that they are ferrimagnetic materials. Nassau *et al.* [5] published results for the commercially significant compound $SmCo_5$ and for the series RCo_5 with R = Ce, Gd, Dy, Ho, and Y, and confirmed the ferrimagnetism of $GdCo_5$. $DyCo_5$ and $HoCo_5$ were observed to behave similarly, whereas the coupling between the light rare earths, Ce and Sm, was observed to be ferromagnetic. Hubbard *et al.* [6] studied Gd–Mn and Gd–Co alloys, including $GdCo_2$ and $GdCo_5$, and observed for $GdCo_5$ a large coercive force which they ascribed to a large magnetocrystalline anisotropy. This was the first observation that the RCo_5 compounds have a feature, namely, a large coercive force, which could make them of interest from a technological viewpoint. These early and significant papers kindled strong interest in the rare earth intermetallics in general and were responsible in large measure for the proliferation of activity in the field.

Intermetallics which occur in rare earth–cobalt systems are summarized in Table 10-1 together with a very brief structural résume. Experiment shows that single phase materials are frequently difficult, if not impossible, to obtain. Examples are the R_2Co_{17} phases, which are often mixtures of the Th_2Zn_{17} (rhombohedral) and Th_2Ni_{17} (hexagonal)

TABLE 10-1

Structures of Rare Earth–Cobalt Compounds[a]

	R_3Co	R_4Co_3	RCo_2	RCo_3	R_2Co_7	RCo_5	R_2Co_{17}
La	Fe_3C				*b*	$CaCu_5$	
Ce			C15	$PuNi_3$	Ce_2Ni_7	$CaCu_5$	*d*
Pr	Fe_3C		C15	$PuNi_3$	Ce_2Ni_7	$CaCu_5$	Th_2Zn_{17}
Nd	Fe_3C		C15	$PuNi_3$	Ce_2Ni_7	$CaCu_5$	Th_2Zn_{17}
Sm	Sm_3Co		C15	$PuNi_3$	*b*	$CaCu_5$	*d*
Gd	Fe_3C	Ho_4Co_3	C15	$PuNi_3$	Gd_2Co_7 *c*	$CaCu_5$	*d*
Tb	Fe_3C		C15	$PuNi_3$	Gd_2Co_7	$CaCu_5$	*d*
Dy	Fe_3C	Ho_4Co_3	C15	$PuNi_3$	Gd_2Co_7	$CaCu_5$	*d*
Ho	Fe_3C	hex (Ho_4Co_3)	C15	$PuNi_3$	Gd_2Co_7	$CaCu_5$	Th_2Ni_{17}
Er	Fe_3C	Ho_4Co_3	C15	$PuNi_3$	Gd_2Co_7	$CaCu_5$	Th_2Ni_{17}
Tm	Fe_3C		C15	$PuNi_3$	Gd_2Co_7	$CaCu_5$	Th_2Ni_{17}
Lu	Fe_3C	Ho_4Co_3	C15	$PuNi_3$			Th_2Ni_{17}
Ref.	[7–11]	[11–13]	[14–16]	[11,17–19]	[10,17,18]	[15,20–22]	[11,23,27]

[a] In addition to the above compounds $LaCo_{13}$ forms in the $NaZn_{13}$ structure [25] and La_2Co, $R_{7/4}Co$ (R = Gd through Er), Sm_9Co_4, and La_2Co_3 are said to exist [26]. Since no magnetic data have been provided for the latter four compounds, we have not included them in the structural compilation.

[b] Ce_2Ni_7 (hexagonal [24]) and Gd_2Co_7 (rhombohedral) structures coexist.

[c] Gd_2Co_7 is a new rhombohedral structure with 54 atoms per unit all in the nonprimitive hexagonal description, which is usually used. Buschow and Van der Goot say [18] Gd_2Co_7 also always contains the rhombohedral and hexagonal forms.

[d] These exist in either the Th_2Zn_{17} (rhombohedral) or the Th_2Ni_{17} (hexagonal) structures depending on thermal history. See references cited for further details.

structures [27] and the R_2Co_7 phases, which are mixtures of the Ce_2Ni_7 (hexagonal) and Gd_2Co_7 (rhombohedral) structures [26]. Evidence also suggests [28,29] that at least in one of the series, RCo_5, perhaps more [10], the stoichiometry deviates substantially from that represented in Table 10-1. The so-called RCo_5 compounds become progressively richer in cobalt as the atomic number increases beyond that of Gd until the erbium compound is more nearly represented by the formula $ErCo_6$. The extra cobalt is accommodated by substitution at the Er sites.

A. THE RCo_5 SERIES

1. Binary Compounds

a. Magnetic Characteristics

As noted above, the essential magnetic features of the RCo_5 compounds were presented in the three pioneering papers dealing with these systems. Nesbitt *et al.* [4] established that the Curie temperature is high (~800°C) and noted the ferromagnetism of $GdCo_5$. The coupling systematics, which now appear to apply to all rare earth intermetallics involving not only cobalt but nickel and iron as well, and for all compounds regardless of stoichiometry, were established in a series of papers published [4,5,30,32,35,36] between 1959 and 1964, namely, ferromagnetic and antiferromagnetic for Co–light rare earths and Co–heavy rare earth combinations, respectively. The high coercive force which makes the light rare earth–Co compounds useful as permanent magnets was first noted by Hubbard *et al.* [6] for $GdCo_5$. The early measurements of moments, which were crude, were soon superseded by improved determinations [30]. Pertinent magnetic data for the Haucke phases (RCo_5), and also for the Laves phases (RCo_2), are accumulated in Table 10-2. In regard to the tabulated ordered moments

TABLE 10-2

Magnetic Characteristics of RCo_5 and RCo_2 Compounds in the Ordered State

	RCo_5				RCo_2[a]			
	μ_{or} (μ_B/f.u.)	T_C (°K)	Type	Ref.	μ_{or} (μ_B/f.u.)	T_C (°K)	Type	Ref.
La	7.1	840	F	[28]				
Ce	7.4	737	F	[5,30,31]	0		P	[35–37]
Pr	9.9	912	F	[5,30–32]	3.2	50	F	[35–37]
Nd	11.7	910	F	[5,30–32]	3.8	120	F	[35–37,38]
Sm	7.8	1020	F	[5,30–32]	2.0	240	F_i	[35–37]
Gd	1.6	1008	F_i	[5,30,31]	4.9	408	F_i	[35–37,39]
Tb	0.7	980	F_i	[5,30,31,33]	6.7	250	F_i	[35–37]
Dy	1.6	966	F_i	[5,30,31]	7.6	155	F_i	[35–37]
Ho	1.9	1000	F_i	[5,30,31,34]	7.8	95 (T_N)	AF,M	[31,35–37]
Er	1.4[b]	986[b]	F_i	[5,30,31]	7.0	40	F_i	[35–37]
Tm	1.6	1020	F_i	[5,30,31]	4.7	25	F_i	[35–37]
Lu							P	[35]

[a] In the paramagnetic region Curie–Weiss behavior was observed only for $ErCo_2$, $NdCo_2$, and $DyCo_2$. However, in these cases the Curie constant bore no rational relationship to the electronic characteristics of the constituent atoms. Probably the linearity of χ^{-1} with temperature is fortuitous. The aberrant paramagnetic behavior of the RCo_2 is ascribed to incomplete localization of the cobalt d electrons.

[b] Buschow *et al.* [29] indicate that this phase is not $ErCo_5$ but is instead $ErCo_6$ with an ordered moment of 2.1 μ_B/f.u. and an ordering temperature of 1050°K.

(μ_{or}), when independent investigations are in disagreement, we have chosen the higher moment as being more nearly the saturation value. We have only deviated from this practice if there appeared to be some overt reason for doing so—if, for example, the sample studied was demonstrably unsatisfactory (multiphasic, impure). Variations of 10 to 20° are noted in the reported values of T_C obtained by different investigators due to different applied fields or different methods or both. Values tabulated are averages unless there is good reason to regard one particular value as more trustworthy.

The ferrimagnetism in the RCo_5 systems with R = Gd to Er which was initially reported by Nassau *et al.* [4] on the basis of the measured moments was confirmed soon after by Nesbitt *et al.* [30] by determinations of magnetization (*M*) versus temperature. They observed a minimum in *M*, called the compensation point, for $GdCo_5$, $DyCo_5$, etc. at temperatures between 50 and 125°K, which was ascribed to internal cancellation of *M* due to antiferromagnetic coupling of the two sublattices. They observed, moreover, a reversal in *M* as the sample was warmed through the region in which *M* is minimal. Data in Fig. 10-1, taken from work published by Cherry and Wallace [40], exemplify the compensation points and also indicate some special magnetic effects in $PrCo_5$ and $NdCo_5$. Peaks in *M* are also observed for $DyCo_5$ (Fig. 10-2) and $TbCo_5$, the origin of which was at the time obscure (*vide infra*).

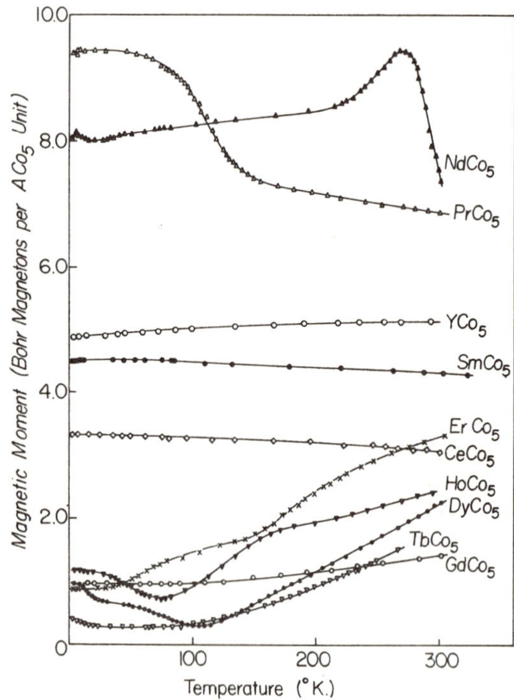

Fig. 10-1. Variation of magnetic moment with temperature for RCo_5 compounds in an applied field of 7200 Oe [40].

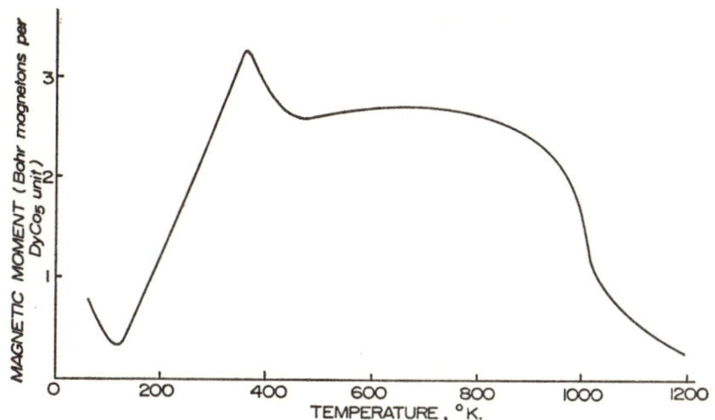

Fig. 10-2. Magnetization–temperature curve for $DyCo_5$ at 7100 Oe [41].

As regards $PrCo_5$ the original supposition, which has never been adequately confirmed [40] or refuted, was that at $T \approx 125°K$ the Pr moments randomize while the Co moments remain ordered to T_C, ~ 900°K. The factors responsible for the peak in magnetization of $NdCo_5$ have been clarified through the neutron diffraction studies of Bartholin *et al.* [42]. They found for $T > 300°K$ collinear coupling of the Nd and Co moments with moments lying along the *c* axis. However, on cooling the arrangement changed so that for $T < 220°K$ moments lie in the basal plane. The peak in M is associated with this rearrangement of moments. Lemaire and Schweizer found [33] for $TbCo_5$, the change in direction of magnetization taking place over the range of temperature 370 to 440°K. The alteration in magnetic structure observed for $NdCo_5$ and $TbCo_5$ is identical with that noted earlier by James *et al.* [34] for $HoCo_5$. On the basis of neutron diffraction they found that its preferred axis of magnetization lies in the basal plane at 4.2°K but along the *c* axis at room temperature. Velge and Buschow suggest [28] that $DyCo_5$ behaves like $TbCo_5$. We thus see that several of the RCo_5 systems are uniaxial only at high temperatures and the possibility exists that this may be true of the entire series. If so, it is fortunate from the point of view of applications (*vide infra*) that the temperature dependence of the aniostropy is as it is, rather than the reverse, so that uniaxiality occurs at high temperatures.

$CeCo_5$, as is the case with most cerium intermetallics, is exceptional. Its unit cell size is anomalously small, 84.5 Å^3 compared, for example, to 87.4 Å^3 for $PrCo_5$, suggesting that Ce is ionized beyond the tripositive state in this material. Lemaire and Schweizer [33] by neutron diffraction measurements put an upper limit of 0.3 μ_B to the moment of Ce in $CeCo_5$. Comparison of moments of $CeCo_5$ and $LaCo_5$ suggests (Table 10-2) that Ce may be carrying a moment close to the maximum suggested by Lemaire and Schweizer. However, to interpret the low moment and small size as connoting Ce^{4+} may be oversimplistic reasoning. Spin compensation could reduce the moment and could also alter the band structure so as to give Ce an apparent small size, a suggestion advanced by Edelstein [43] in his considerations of elemental cerium.

By ^{161}Dy Mössbauer spectroscopy Nowik and Wernick [44] established the hyperfine field of Dy in $DyCo_5$ to be 4% higher than the free ion hyperfine field, suggesting an effective Dy moment of 10.4 μ_B. Co in the heavy rare earth RCo_5 compounds has the same moment [29,33, 39] as in the elemental state, namely 1.72 μ_B. Hence for ferrimagnetic $DyCo_5$ we expect 1.8 μ_B/f.u., which compares rather favorably with the value obtained by conventional bulk methods (Table 10-2).

b. Heat Capacities

The only heat capacity work on the RCo_5 compounds appears to be that of Saba and Wallace [41] on $DyCo_5$ (Fig. 10-3). An estimate of the functional dependence of the C_p of hypothetical nonmagnetic $DyCo_5$ on temperature was made (dashed curves) as follows: C_v of hypothetical nonmagnetic Dy was estimated from its Debye temperature (158°K) and the difference between C_p and C_v was evaluated using the known compressibility and expansivity of Dy. Then the Kopp–Neumann rule of additivity of heat capacities was employed to compute a lattice and electronic C_p of $DyCo_5$ from the calculated C_p of nonmagnetic Dy and the measured C_p of Co. Although this estimate is somewhat crude, it seems sufficiently reliable to indicate that magnetic entropy is being introduced at the lowest temperatures. This is consistent with the M versus T data (Fig. 10-2) which show very substantial changes in magnetization throughout the temperature range covered in the heat capacity studies.

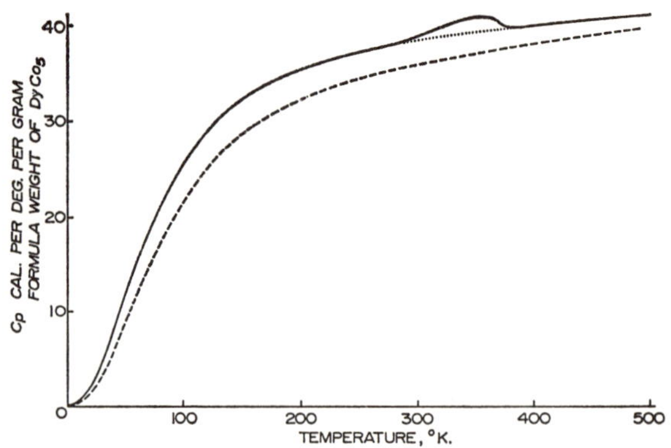

Fig. 10-3. Heat capacity of $DyCo_5$ as a function of temperature. Measured results —; estimate of results without magnetic contribution - - - [41].

The hump in the heat capacity curve around 350°K is undoubtedly associated with the change in the direction of easy magnetization in $DyCo_5$. The energy and entropy are 351 J $mole^{-1}$ and 1.04 J $°K^{-1}$ $mole^{-1}$, respectively. These data show that the structure in which the moments lie along the *c* axis has substantially higher entropy than the arrangement with moments in the basal plane; it is the increase in entropy which is responsible for the transformation in magnetic structure at elevated temperatures for $DyCo_5$ and probably for $NdCo_5$, $TbCo_5$, and $HoCo_5$, as well.

2. Ternary Systems

Ternary systems based on the RCo_5 systems have consistently attracted attention throughout the past decade. Introduction of a third component is of significance for a wide variety of reasons. Alloying is useful in attempts to achieve the mechanical properties needed for permanent magnet fabrication (*vide infra*), to convert the peritectically formed binary into a congruent melting material, etc. There is, of course, no *a priori* reason that the binary should have optimal magnetic properties; improvement in magnetic properties might be achieved by forming an appropriate ternary. Discussion of ternaries in this section will be confined to the effect of alloying on their fundamental magnetic properties. Nassau *et al.* studied [5] $Gd_{1-x}Y_xCo_5$ ternaries in 1960. Nesbitt *et al.* [45] examined the $RCo_{5-x}Cu_x$ with R = Tb and Dy, and $Gd_{0.3}Nd_{0.7}Co_5$ in 1961. Hubbard and Adams [46] studied $GdCo_{5-x}Fe_x$ ternaries in 1962. However, this latter work seems to be of limited significance since it was analyzed in terms of $GdFe_5$, which is now known not to exist. The studies of Gd–Y–Co ternaries showed a rise in moment as Y replaces Gd, providing additional support that the Gd–Co coupling is antiferromagnetic. The Gd–Nd coupling in $Gd_{0.3}Nd_{0.7}Co_5$ was found to be antiferromagnetic, indicating that the coupling systematics observed for $R_{1-x}R_x'Al_2$ and $R_{1-x}R_x'Ni_2$ ternaries (Chapters 4 and 9) also applied here. Examination of the Cu-containing ternaries was largely confined to studying the effect of Cu on the compensation temperature.

A systematic study of $GdCo_{5-x}Cu_x$ and of $GdCo_{5-x}Al_x$ was made by Shidlovsky and Wallace [39], the results of which are largely summarized in Table 10-3 and in Figs. 10-4 and 10-5. This work was undertaken in the spirit of the investigations carried out by Mader and Wallace [47] (Chapter 4) on $Eu_{1-x}La_xAl_2$ ternaries and by Sekizawa and Yasukochi [48] on $GdAg_{1-x}In_x$ ternaries. In these studies the coupling was reversed by altering the electron concentration. It was expected that replacement of Co in $GdCo_5$ with Cu or Al would raise the electron concentration and bring about a change in the Gd–Co coupling from antiferromagnetic to ferromagnetic. This does not happen probably because, contrary to expectation, the electron concentration is not increased by replacing cobalt with Cu or Al. Co seems to absorb the extra electrons supplied by Cu or Al into the vacancies in its d-band [as evidenced by the fall in Co moment (Table 10-3) as the Cu or Al content increases] so that the electron concentration is actually not changed.

TABLE 10-3

Structural and Magnetic Characteristics of $GdCo_5$-based Ternaries[a]

	$GdCo_{5-x}Cu_x$				$GdCo_{5-x}Al_x$			
x	a_0 (Å)	c_0 (Å)	μ_{sat} [b]	μ_{Co} [c]	a_0 (Å)	c_0 (Å)	μ_{sat} [b]	μ_{Co} [c]
0	4.98	3.98	1.55	1.71	4.98	3.98	1.55	1.71
0.25					5.00	3.99	1.10	1.71
0.50	4.99	3.99	0.65	1.70				
1.00	4.98	4.00	0.46	1.64	5.01	4.04	2.16	1.21
1.50					5.04	4.06	3.99	0.86
1.75					5.05	4.07	4.85	0.66
2.00	5.00	4.02	2.48	1.51				
3.00	5.01	4.04	4.40	1.30				
4.00	5.02	4.07	6.28	0.72				
4.20	5.03	4.08	6.43	0.71				

[a] Shidlovsky and Wallace [39].

[b] Saturation magnetization at 4.2°K in μ_B per formula unit.

[c] Value of the cobalt moment in μ_B per atom calculated under the assumptions that the Gd moment is 7.00 μ_B and the Gd–Co coupling is antiferromagnetic.

The early work of Nesbitt *et al.* [45] on $Gd_{0.3}Nd_{0.7}Co_5$ has been followed by a more systematic study by Wallace *et al.* [38]. The ternaries $R_{1-x}R_x'Co_5$ were studied with the following R,R′ combinations: Gd, Nd; Dy, Pr; Dy, Nd; Gd, Dy; and Gd, Ho. Most of the results are summarized in Table 10-4. Curves of M versus T show compensation points for many of these ternaries. For details concerning this aspect of the work the reader should consult the original work. Measured saturation moments for the $R_{1-x}R_x'Co_5$ ternaries are given in column 2 of Table 10-4. In column 3 are presented moments calculated on the basis of the coupling scheme described above for RCo_5 compounds and for ternaries involving the rare earths with Ni (Chapter 9) and Al (Chapter 4) namely:

1. the heavy lanthanides (Gd, Dy, Ho) couple antiparallel to cobalt;
2. the light lanthanides (Pr, Nd) couple parallel to cobalt;
3. the heavy–heavy and heavy–light coupling is parallel and antiparallel, respectively.

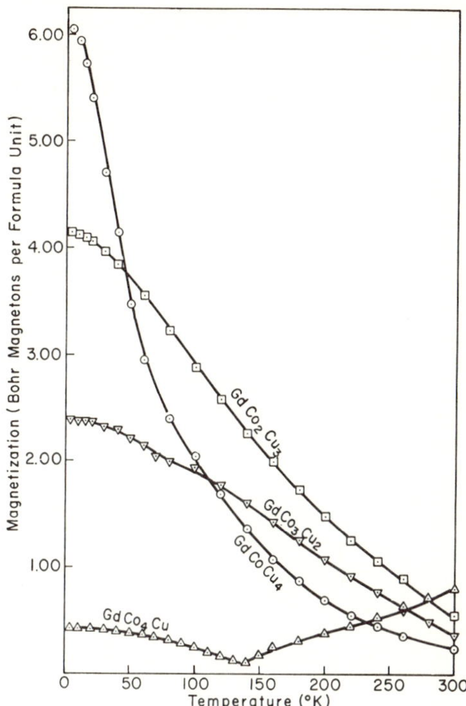

Fig. 10-4. Magnetization versus temperature for $GdCo_{5-x}Cu_x$ alloys [39].

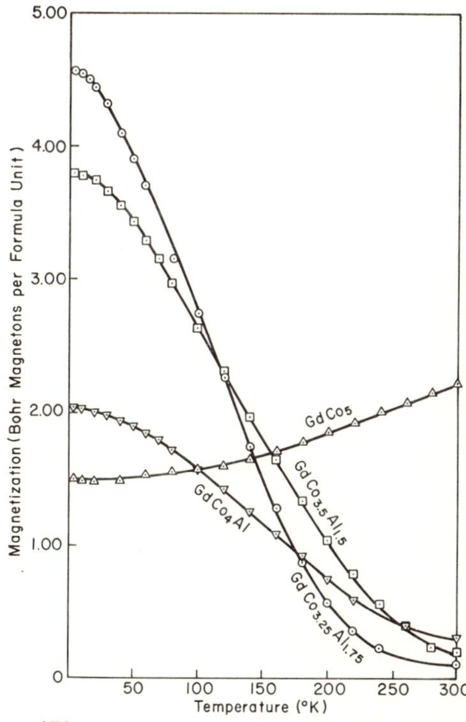

Fig. 10.5. Magnetization versus temperature for $GdCo_{5-x}Al_x$ alloys [39].

TABLE 10-4

Magnetic Characteristics of $R_{1-x}R_x'Co_5$ Ternaries [38]

x	Magnetic moment (μ_B/f.u.) Meas.	Calc.[a]	T_C (°K)
	$Gd_xNd_{1-x}Co_5$		
0.0 ($NdCo_5$)	11.7	11.8	910
0.2	8.5	9.7	950
0.4	7.0	7.7	965
0.6	5.6	5.6 (14.0)	1000
0.8	3.2	3.2	1000
1.0 ($GdCo_5$)	1.3	1.5	1008
	$Dy_xPr_{1-x}Co_5$		
0.0 ($PrCo_5$)	10.0	11.7	912
0.2	8.0	9.1	955
0.4	5.7	6.4	955
0.6	3.3	3.8 (14.0)	960
0.8	1.36	1.14	975
1.0 ($DyCo_5$)	1.6	1.5	966
	$Dy_xNd_{1-x}Co_5$		
0.2	8.6	9.2	920
0.4	7.8	6.5	930
0.6	7.3	3.9 (15.8)	940
0.8	1.6	1.2	950
0.85	0.8	1.0	
0.9	0.5	0.2	
	$Gd_xDy_{1-x}Co_5$		
0.2	0.89	0.90	1000
0.4	0.54	0.30	1000
0.6	0.47	0.30 (16.7)	1000
0.8	1.12	0.90	1005
	$Gd_xHo_{1-x}Co_5$		
0.0 ($HoCo_5$)	1.9	1.5	1000
0.2	1.02	0.90	1010
0.4	1.25	0.30	1010
0.6	1.39	0.30 (16.7)	1015
0.8	1.57	0.90	1010

[a]Calculated assuming the coupling scheme described in the text. Moments used (μ_B/atom): Co – 1.7; Pr – 3.2; Nd – 3.27; Gd – 7.0; Dy and Ho – 10.0. The numbers in parenthesis at the $x = 0.6$ composition give the magnetic moment for ferromagnetic coupling of all three species.

For simplicity this is termed the ferrimagnetic coupling scheme. The moment calculated for all three species coupled parallel is shown for $x = 0.6$ in each case. It is clear that the first, i.e., the ferrimagnetic coupling scheme agrees much more closely with experiment than one involving ferromagnetic coupling. Even so we find substantial deviations between the calculated and measured values for ternaries containing light lanthanides. It appears that the deviations largely originate with the variable moment of cobalt and the light lanthanide component.

Bleaney [49] appears to have been the first to draw attention to the dependence of the cobalt moment on the nature of the rare earth element with which it was united; the exchange field of the lanthanide was presumed to produce a substantial induced component to the cobalt moment. The postulate of a variable cobalt moment was later confirmed [31,33] by neutron diffraction studies (*vide infra*). More recently Leon and Wallace [50] showed that the Pr^{3+} ion behaves similarly. Thus the assumption of fixed ionic moments for calculating the saturation magnetizations in Table 10-4 is valid only as a rough approximation.

While variation in ionic moments seems to be the main influence in the deviation of computed and observed moments of most of the $R_{1-x}R_x'Co_5$ ternaries studied, it does not appear to be the sole effect. The deviation noted for $Dy_{0.6}Nd_{0.4}Co_5$ seems to be too large to be reasonably ascribed to variation in the individual ion moments. It seems in this case as if the magnetic structure (i.e., the coupling scheme) differs from that postulated. It is of interest to note that similar aberrations were observed [51] in the earlier studies of $R_xR'_{1-x}Ni_5$ ternaries with R = Nd and R' = Ho.

The situation in regard to the $R_{1-x}R_x'Co_5$ ternaries can thus be summarized as follows: Despite the fact that agreement between computed and observed results is in a few cases only fair, there is little reason to doubt the essential correctness of the postulated ferrimagnetic coupling scheme except possibly for $Dy_{0.6}Nd_{0.4}Co_5$. The possibility that all species are ferromagnetically coupled can be excluded. These conclusions are supported not only by the magnitude of magnetic moments but also by the shape of the magnetization–temperature curves.

3. Utility in Permanent Magnets

Recent work has shown that some of the RCo_5 compounds have characteristics which make them attractive possibilities as materials for use in the fabrication of permanent magnets. Hoffer and Strnat [52] in the course of measuring anisotropy constants appear to be the first to have drawn attention to their potential in this respect. In this section we provide a brief account of this facet of the RCo_5 compounds. More details are to be found in the excellent summaries by Becker [53]. His article in *The Scientific American* is particularly helpful and informative to individuals unfamiliar with permanent magnet technology.

In the ensuing discussion we first review the salient features of the hysteresis loop and make use of it to stipulate the figures of merit normally used to assess the effectiveness of a permanent magnet material. We next show the fundamental quantities which are involved in setting upper limits for the figures of merit and then present the desiderata for a permanent magnet material. Finally we present expressions for the limiting values of the figures of merit and assess the RCo_5 compounds and other permanent magnet materials in terms of these limiting values.

a. Desiderata for a Permanent Magnet Material

Before specifying the properties desired in a material to be used in permanent magnet fabrication we shall find it convenient to make use of the hysteresis loop in Fig. 10-6 in defining the remanence (B_r), the coercivity (${}_BH_c$), and the maximum energy product $(BH)_{max}$. The meaning of B_r and ${}_BH_c$ is obvious from the diagram. $(BH)_{max}$ is the maximum value of the product of B and H in the second quadrant.*

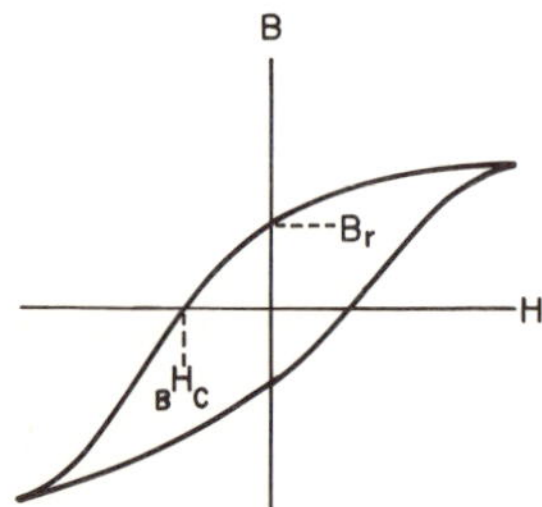

Fig. 10-6. Schematic B–H curve for a ferromagnetic or ferrimagnetic material. The meaning of B_r and ${}_BH_c$ is obvious from the diagram. $(BH)_{max}$ is the maximum value.

For a material to serve as a superior permanent magnet it must have the ability to produce a large field and to continue to produce this field for an extended period of time under a variety of perturbing influences. If, for example, the magnet is to be used in an electromechanical device such as a motor, it is necessary that it be able to sustain its magnetized condition against the influence of the reverse magnetic field to which it

*There is, of course, a generally similar diagram when the magnetization (M) is plotted against H. The value of the reverse field to bring M to zero is designated as ${}_MH_c$. ${}_BH_c$ and ${}_MH_c$ are termed the coercivity and intrinsic coercivity, respectively. When H_c is used in in the discussion in this section, it is to be understood that the coercivity (${}_BH_c$) is being referred to.

will be exposed. The generation of a large field requires a large energy product* and the sustaining of the field against reverse magnetic fields requires a high coercivity. Thus two properties which a material must possess are a large $(BH)_{max}$ and a high coercivity. These two quantities are the figures of merit most often used in assessing permanent magnet materials. To these must be added the requirement of a high Curie temperature (T_C) if the magnet is to be useful over a significant temperature span. There are, of course, other factors such as cost, chemical stability, etc.; but the essential magnetic features are high T_C and large values of $(BH)_{max}$ and ${}_BH_c$.

We now inquire further into the nature of the energy product and the coercivity. Experience shows that these are not state functions in a thermodynamic sense, depending on the temperature, pressure, and composition of the sample. Rather they are markedly dependent on the state of subdivision of the material once it has been increased beyond a threshold value. The reasons why these quantities depend on particle size have been clarified as a result of the activity of numerous workers in the field during the past two or three decades. It is due to the influence of magnetocrystalline anisotropy and the existence of particles consisting of a single domain. We must now consider how the anisotropy and the existence of single domain particles exert an effect on the hysteresis loop.

In Chapter 3 we have discussed in reasonable detail the effect of the crystalline electric field on the magnetic properties of a system of rare earth ions. One of the crystal field effects, which is given little emphasis there, is the generation of a force that acts on the atomic moments in such a way as to cause them to lie in certain directions with respect to the crystallographic axes. These forces may be very powerful so that the moments are very strongly constrained to the so-called easy axis of magnetization. This preference for direction is the magnetocrystalline anisotropy.

If we have a material with a large magnetocrystalline anisotropy, we have the potential for a permanent magnet material of superior quality (other factors, of course, also being favorable). To take advantage of this property, however, it is necessary to arrange matters so that in responding to a reversed magnetic field the magnetization is forced to rotate against the anisotropy forces. In the absence of this restriction, the magnetization will rotate at lower fields by the process of domain wall motion giving rise to a thin hysteresis loop and the low coercivity

*Superficially one might expect B_r to be the criterion of the field which the magnet can generate. Actually it is $(BH)_{max}$ which provides a measure of the maximum field that can be produced (by a given volume of material and in a fixed configuration). There is, of course, a correlation between the maximum energy product and the remanence, but they are not proportional.

shown schematically in Fig. 10-7a. If now the particle is subdivided into small spherical particles comparable to or smaller than the domain wall thickness (less than about 10^{-5} cm), it becomes energetically unfavorable for it to contain a domain wall. Under these circumstances each

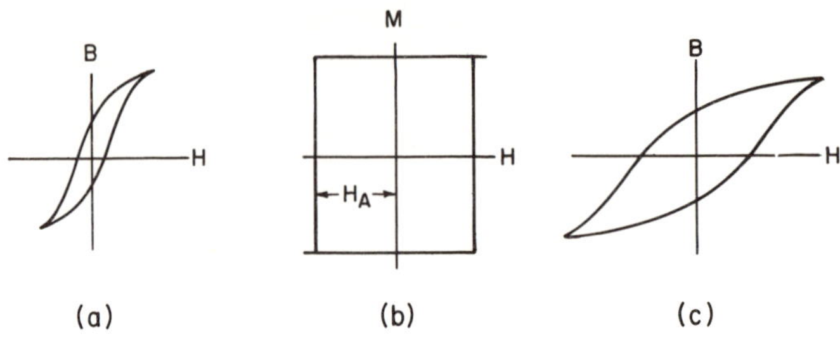

Fig. 10-7. Schematic hysteresis loops for a low coercivity material (a), spherical single domain particles (b), and a high coercive material (c). The intrinsic coercive force in (b) equals the anisotropy field. In (a), reversal of magnetization occurs by domain wall motion. In (c), the magnetocrystalline anisotropy is involved (see text).

particle becomes a single domain and the reversal of magnetization takes place against the constraint of the crystal anisotropy. The characteristics of an assemblage of single domain particles have been treated by a number of individuals. Stoner and Wohlfarth [54] in a classic paper showed that if the magnetization is coherent, that is all moments in the particle rotate coherently, the original arrangement will persist until the reverse field exceeds some critical value, H_a, called the anisotropy field, at which point the magnetization will suddenly reverse. Similar behavior will occur with increasing positive field and a hysteresis loop such as shown in Fig. 10-7b results. It can be shown that [54,55]

$$H_a = 2K/M_s, \tag{10-1}$$

where K is the magnetocrystalline anisotropy constant* and M_s is the saturation magnetization. Anisotropy fields of the order of 10^2 kOe are found for rare earth intermetallics (*vide infra,* Table 10-5), indicating that in the ideal case (i.e., single domain particles experiencing coherent rotation against the crystal anisotropy) we would have coercive forces of the order of 10^2 kOe and, since $B_{sat} \approx 10^4$ G, $(BH)_{max}$ of the order of 10^8 G·Oe. These quantities, which are far larger than those exhibited by currently available permanent magnet materials, are of significance

*If the particles are ellipsoidal, K in Eq. (10-1) becomes $K_0 + M_s^2/2\,(N_b - N_a)$. The second term is a shape factor and is usually quite small compared to K_0, the crystal anisotropy. N_b and N_a are the demagnetization factors along the major and minor axes of the ellipse.

only as upper limits. From these considerations, however, the fundamental characteristics desired of a permanent magnet material now become clear. We need the following:

1. a high Curie temperature (T_C)
2. a large saturation magnetization (M_s) or induction (B_s)
3. strong anisotropy (large K or H_a).

b. Limiting Values of the Energy Product and Coercivity

We can now set forth the limiting values of coercivity and energy product for a material having a high anisotropy field, i.e., $H_a \gg B_s$. We find that

$$H_c = B_s = 4\pi M_s \tag{10-2}$$

and

$$(BH)_{\max} = 4\pi^2 M_s^2 = B_s^2/4. \tag{10-3}$$

The origin of these expressions is made clear by reference to Fig. 10-8. Since B is the vector sum of H and $4\pi M$, it is possible to reduce B to zero, by applying a field opposed to the magnetization, without altering the magnetization. This possibility exists since H_a greatly exceeds the needed field. Thus B in the second quadrant of Fig. 10-8 decreases to

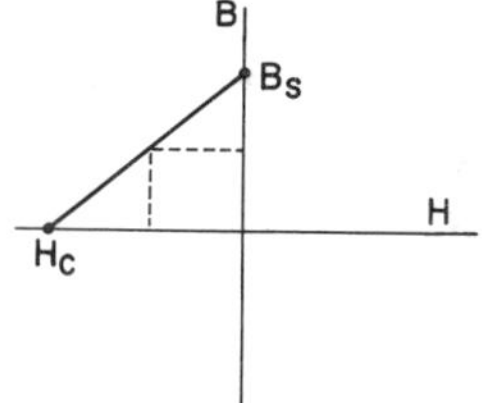

Fig. 10-8. B versus H in the second quadrant of the hysteresis loop for the case in which the anisotropy field is much larger than B_s. $(BH)_{\max} = 4\pi^2 M_s^2$ in this case.

zero at $H = B_s$, which is the condition expressed in (10-2). The maximum energy product is obviously $[\tfrac{1}{2}(4\pi M_s)]^2$ or $4\pi M_s^2$. If $B_r < B_s$ and $H_a \geq B_r$, similar considerations hold, indicating a limiting value of the maximum energy product given by Eq. (10-4):

$$(BH)_{\max} = B_r^2/4. \tag{10-4}$$

As noted earlier we observe for present day permanent magnet materials figures of merit substantially below those given by Eqs. (10-2) to (10-4). It is not possible to achieve the idealized conditions required, i.e., coherent rotation against the magnetocrystalline anisotropy. The hysteresis loop observed for a good permanent magnet material is shown schematically in Fig. 10-7c. It is considerably wider than that shown in Fig. 10-7a because some advantage is taken of the crystal anisotropy but the width is typically only 10 to 20% of the limiting value [Eq. (10-2) and Fig. 10-7b].

c. Work Involving the RCo_5 Compounds

The work of Hoffer and Strnat [52] has been referred to above. They measured the anisotropy constant K for a single crystal of YCo_5 and found it to be 5.7×10^7 ergs/cm^3, which is one or two orders of magnitude higher than that of permanent magnet materials currently in use. This indicated a very large anisotropy field and the possibility of a large coercivity if the material could be reduced to fine particle condition. The anisotropy field can be obtained from Eq. (10-1) (after first establishing K) or more directly by measuring M versus H in the magnetically easy and hard direction. Hoffer and Strnat used both procedures for YCo_5. Velge and Buschow [28] used the second procedure to obtain H_a for the entire series of RCo_5 compounds, employing oriented powders. In this procedure one finds, usually by extrapolation, the field needed to achieve saturation in the hard direction (see Fig. 10-9). The H_a values are shown in Table 10-5.

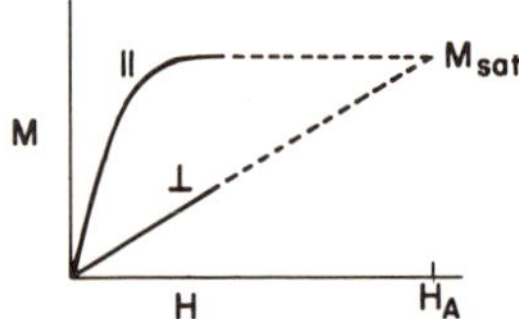

Fig. 10-9. Magnetization (M) for field applied parallel and perpendicular to the easy axis. The anisotropy field (H_a) is approximated by extrapolating the linear curve to M_{sat}.

TABLE 10-5

Approximate Value of the Anisotropy Field in kOe for RCo_5 Compounds[a]

YCo_5	130	$GdCo_5$	270
$LaCo_5$	175	$TbCo_5$	*b*
$CeCo_5$	180	$DyCo_{5.2}$	25
$PrCo_5$	145	$HoCo_{5.5}$	135
$NdCo_5$	~30	$ErCo_6$	100
$SmCo_5$	210		

[a] From Velge and Buschow [28] and Hoffer and Strnat [52].

[b] $TbCo_5$ has an easy plane at room temperature.

As regards T_C and H_a, all the RCo_5 compounds listed in Table 10-5 are exceedingly attractive candidates for permanent magnet materials. However, when B_s is included, the compounds with R = Gd through Er must be removed from consideration. We recall that coupling in these materials is ferrimagnetic and hence B_s is too small for them to be useful. The RCo_5 compounds with R = Y, La, Ce, Pr, and Sm seem to offer excellent possibilities, as does $MMCo_5$. (MM represents misch metal.) $MMCo_5$ is attractive because of cost considerations. Of the several RCo_5 possibilities $SmCo_5$, for reasons which are as yet not clear, is the most attractive in that its coercivity and energy product most closely approach the theoretical values. For $SmCo_5$, $B_s = 10.5$ kG, and hence the limiting $(BH)_{max}$ and H_c are 24×10^6 G·Oe and 10.5 kOe, respectively. To appreciate the exceptional size of these quantities we note that $(BH)_{max}$ and H_c for ordinary steels are 10^4 G·Oe and 0.2 kOe, respectively. As early as 1969, $SmCo_5$ magnets were produced [56,57] with coercive forces in the range 10 kOe and energy products approximately 20×10^6 G·Oe, which is roughly 75% of the limiting value. Energy products and coercivities of a number of the RCo_5 compounds and other common magnetic materials are listed in Table 10-6, which is taken from the compilations of Becker [53] and Strnat [58].

TABLE 10-6

Figures of Merit of Various Permanent Magnet Materials[a]

	$(BH)_{max} \times 10^{-6}$(G · Oe)		${}_MH_C$ Expt.[b]
	Expt.	Theory	
$CeCo_5$	8	16	2.8
$MMCo_5$	2	21	
$LaCo_5$		22	3.6
$SmCo_5$	20	24	30
YCo_5	1	28	2.6
$PrCo_5$	8	37	5.7
Ba Ferrite	4		3
MnBi	5.5		
PtCo	9.2		5.5
Alnico 5	6		
Alnico 9	10		2

[a]From Becker [53] and Strnat [58].

[b]If the magnetic moments rotated coherently, the intrinsic coercive force would correspond with the anisotropy field listed in Table 10-5. The experimental value is only a fraction of H_a indicating the possibility of greatly increased values of ${}_MH_C$.

We should appreciate that the coercive force is as important, if not more so, than the energy product. It is, for example, the dominant consideration if the permanent magnet is to be used in the construction of a motor—one of the developments currently being pursued. The usual magnetic materials are demagnetized by fields associated with the armature, whereas the coercivity of the RCo_5 materials is sufficiently high that this does not occur.

Experiment shows (Table 10-6) the properties of the RCo_5 compounds to fall far short of the theoretical limits set forth above; as cast materials fall very far short of these limits. This is undoubtedly because the demagnetization in the first and second quadrants of the hysteresis loop does not take place by coherent rotation of single domains. The process is more complex and depends upon the microscopic structural details of the magnetic material, its lattice imperfections, grain boundary structure, etc. Best results have been achieved on RCo_5 compounds of small particle size, which more nearly approach the idealized model described above.

Fine particle characteristics have been brought out not only by grinding the compounds as cast materials but also by metallurgical procedures [59,60] in which the Co in $SmCo_5$ is partially replaced by Cu or Cu plus Fe. The constitution of these ternary or quaternary systems has not as yet been established except to indicate that in many, perhaps all, cases they are two-phase materials. Their favorable properties undoubtedly originate with their two-phase character, RCo_5 being precipitated out in an RCu_5 matrix, or the converse, so that in effect the magnetic species is in fine particle form. These have some advantages over the ground materials since the latter appear to be susceptible to atmospheric attack with concomitant degradation of their magnetic properties. The mechanism involved in this degradation is yet to be clarified. Probably the oxidation which occurs during atmospheric attack alters the surface in a way to facilitate the motion of the domain boundaries with consequent reduction of B_r, H_c, and $(BH)_{max}$. The possibility also exists that the samples become contaminated with hydrogen during the oxidation process. Zijlstra and Westendorf [61] have shown that $SmCo_5$ readily absorbs hydrogen up to 2.5 moles of H_2 per mole of intermetallic (room temperature and 20 atm). They note a severe degradation of coercive force with added hydrogen.

One final aspect of the use of the RCo_5 as permanent magnet materials deserves comment. It is of interest to note that in the compounds which are being found useful the rare earth makes little contribution to B_s (or B_r). The magnetization originates primarily with the cobalt sublattice. The role of the rare earth is to define the crystallography, producing a uniaxial crystal with a favorable anisotropy. The ferrimagnetism of the heavy rare earth RCo_5 compounds alluded to above prevents us from making use of these compounds and of the potentially substantial contribution which these rare earths could make to B_s and hence the energy product.

Clearly the RCo_5 have very significant possibilities as regards applications. It is equally clear that this important chapter in the development of rare earth intermetallics has just begun. Even so, the progression to date—from fundamental inquiries to the threshold of applications—is very satisfying. This is the kind of progression—coupling science and technology—that can be so invigorating to pure science and so essential to applications technology.

B. THE RCo_2 SERIES

1. Binary Systems

a. Magnetic Properties

These have been studied by Wallace *et al.* [35,37], by Crangle and Ross [36,62], and by Lemaire [31]. Results are given in Table 10-2. Several noteworthy features are evident:

1. Ce behaves similarly in $CeCo_2$ and in $CeCo_5$, appearing formally as if it is in the quadripositive state. However, as noted above the true situation may be somewhat more complex.
2. The loss of magnetization occurs in two stages for $NdCo_2$, $SmCo_2$, and $HoCo_2$ (Fig. 10-10).
3. The ferromagnetic-paramagnetic transitions in $ErCo_2$ [37] and $HoCo_2$ [31] are observed to be extremely sharp. Lemaire asserts that the $HoCo_2$ transition is first order and Petrich and Mössbauer [63] reach a similar conclusion about $ErCo_2$; the latter investigators find a latent heat at the ordering temperature. Lemaire further finds $HoCo_2$ to be metamagnetic in the temperature interval 75–95°K.

The coupling systematics are in accord with those noted above for the R_2Ni_{17} and RCo_5 compounds; spins of R and Co always couple antiparallel so that compounds containing light rare earths are ferromagnetic and those containing heavy rare earths are antiferromagnetic.

Several of the RCo_2 compounds have been examined by neutron diffraction techniques with the results shown in Table 10-7. These studies confirm the coupling pattern inferred earlier from bulk magnetic measurements. They also show variation in cobalt moment to which attention will be directed below.

Ofer and Segal [66] have established the Dy moment in $DyCo_2$ by ^{161}Dy Mössbauer work to be 10.4 μ_B. This, with $\mu_{Co} = 1.0\ \mu_B$, leads to a

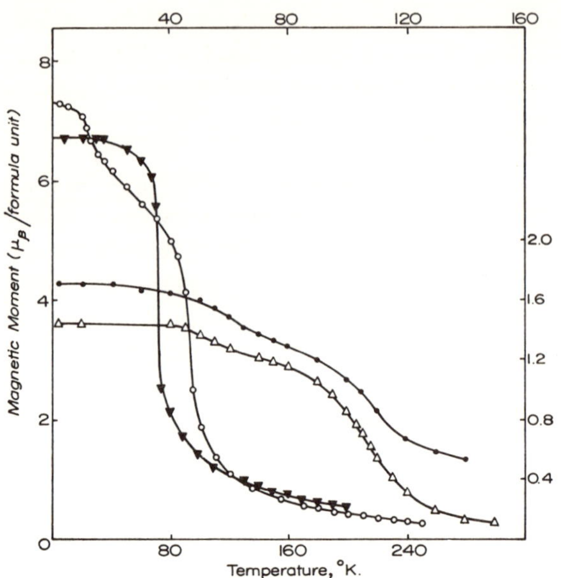

Fig. 10-10. Magnetic moments versus temperature for $NdCo_2$ (Δ), $SmCo_2$ (●), $HoCo_2$ (O) and $ErCo_2$ (▼) [37]. The upper scale applies to $NdCo_2$ and $ErCo_2$; the lower scale is for the Sm and Ho compounds. The right-hand scale is for $SmCo_2$, and the left-hand one is for the other three compounds. Reprinted from *Inorg. Chem.*, **5**, 105 (1966). Copyright 1966 by the American Chemical Society. Reprinted by permission of the copyright owner.

TABLE 10-7

Moments in RCo_2 Compounds from Neutron Diffraction Studies

	Moment			Moment (μ_B/f.u.)	
	R (μ_B/atom)	Co (μ_B/atom)	RCo_2 (μ_B/f.u.)	Bulk measurements	Ref.
$PrCo_2$	2.7	0.5	3.7	3.2	[37,64]
$NdCo_2$	2.6	0.8	4.2	3.8	[37,65]
$TbCo_2$	8.8	1.0	6.8	6.7	[37,65]
$HoCo_2$	9.5	1.0	7.5	7.8	[37,65]
$ErCo_2$	8.9	1.0	6.9	7.0	[37,65]

moment of 8.4 μ_B/f.u. for $DyCo_2$, which is 10% higher than the value obtained from bulk measurements. This suggests that the data given in Table 10-2 probably do not correspond to saturation for $DyCo_2$ and perhaps for some of the other compounds as well.

Bloch *et al.* [67] have confirmed the nonlinearity of χ^{-1} with T observed earlier by Farrell and Wallace and in addition established the dependence of T_C on pressure for a number of RCo_2 compounds. T_C is decreased at the rate of roughly 1°K per kilobar over the range 0 to 4.5 kilobars, whereas T_C for isomorphous $GdNi_2$ is essentially unchanged by a pressure increase in this amount.

b. Heat Capacities and Entropies

Deenadas *et al.* [68] made measurements on $CeCo_2$, $PrCo_2$, and $NdCo_2$ over the range 6–300°K. This appears to be the only study of the thermal properties of the RCo_2 compounds which has been made. $PrCo_2$ shows a doublet thermal anomaly (Fig. 10-11) in the vicinity of its Curie temperature suggesting that its magnetic order is lost in two stages. This is also true of $NdCo_2$ (Fig. 10-12) although in this case the thermal anomalies are well separated. As noted above magnetization in $NdCo_2$ appears to be lost in two stages (see Fig. 10-10). The thermal anomalies coincide with the temperatures at which the decline in magnetization occurs, indicating a common cause. It is not clear at this time whether the lower anomaly is a result of a magnetic structure change or merely a shift in the easy axis of magnetization.

Thermodynamic properties are accumulated in Table 10-8.

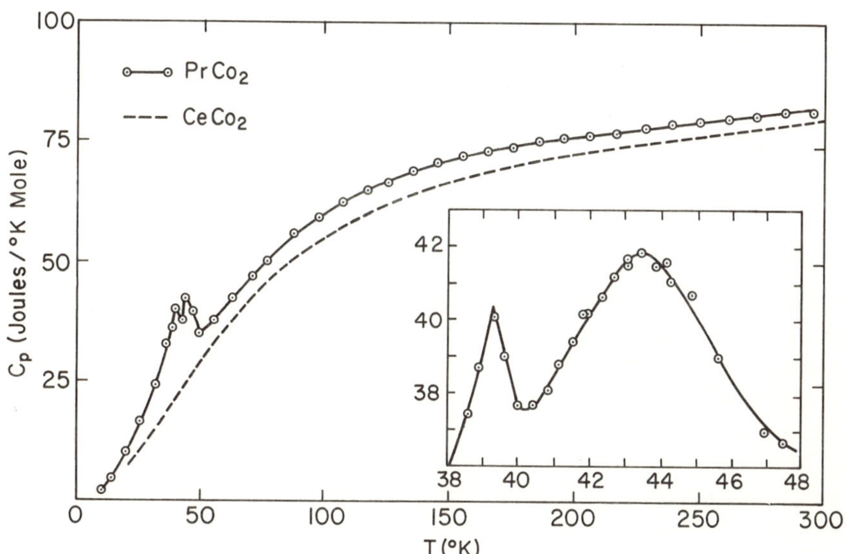

Fig. 10-11. Heat capacities for $PrCo_2$ and $CeCo_2$ versus temperature [68].

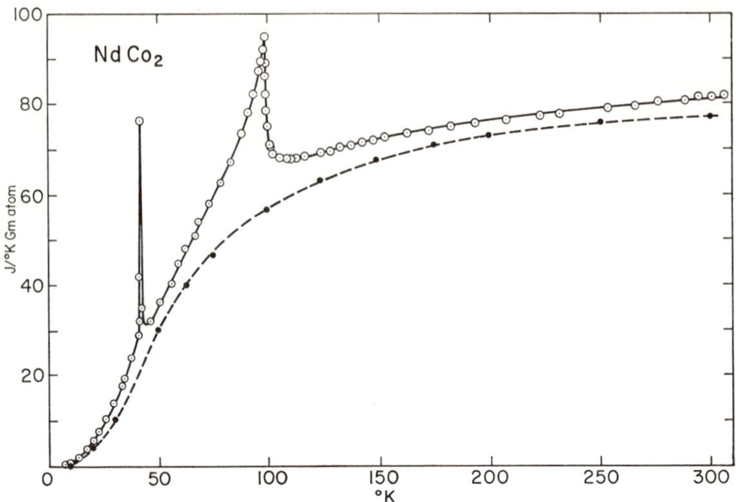

Fig. 10-12. Heat capacities of $NdCo_2$ and $CeCo_2$ versus temperature [68].

TABLE 10-8

Thermodynamic Characteristics of RCo_2 Compounds[a]

	C_p	$(H - H_0^0)/T$	S	$-(F - H_0^0)/T$	$-\Delta S$ (formation)
$CeCo_2$	80.38	56.63	121.99	65.37	7.88
$PrCo_2$	82.20	60.92	139.83	78.91	−6.57
$NdCo_2$	81.40	62.65	140.76	78.11	−7.29

[a] Units are joules $mole^{-1}$ $°K^{-1}$. Data are from Deenadas *et al.* [68].

2. Ternary Systems

a. Rare Earth Substitution

The systems $Pr_{1-x}Dy_xCo_2$ and $Pr_{1-x}Ho_xCo_2$ were studied by Leon and Wallace [69]. Saturation magnetizations (Table 10-9) are consistent with a coupling scheme of Pr and Co moments parallel and Dy or Ho moment antiparallel to the Pr and Co moments, if one allows for variation in the Co moment (0.5–1.0 μ_B as x varies from 0 to 1) and in the Pr moment (2.6–3.2 μ_B). The Pr moment increases as x increases because of the increasing exchange field when Dy or Ho is inserted in the lattice. The situation here is analogous to $Pr_{1-x}M_xNi_2$ (M = Dy or Ho) discussed in Chapter 9 except that Pr has [64] a moment of about 2.7 μ_B for $x = 0$. The shape of the magnetization–temperature curve (see Leon and Wallace [69] provides additional evidence for the ferrimagnetic character of these ternaries.

TABLE 10-9

Saturation Magnetization[a] *of $Pr_{1-x}Dy_xCo_2$ and $Pr_{1-x}Ho_xCo_2$ at 4.2°K*

$Pr_{1-x}Dy_xCo_2$		$Pr_{1-x}Ho_xCo_2$	
x	μ_{sat} (μ_B/f.u.)	x	μ_{sat} (μ_B/f.u.)
0.10	2.39	0.10	2.80
0.34	0.19	0.20	1.48
0.44	0.44	0.25	0.91
0.53	1.73	0.31	0.46
0.60	2.58	0.46	1.85
0.69	3.56	0.61	3.29
0.81	5.10	0.72	4.33
0.91	6.67	0.85	5.86

[a]From Leon and Wallace [69].

$Y_{1-x}Gd_xCo_2$ ternaries have been studied by Lemaire [31] and Schweizer [70] and by Taylor *et al.* [71]. Neutron diffraction measurements [70] show YCo_2 to be nonmagnetic whereas the cobalt moment in $TbCo_2$ or $ErCo_2$ is [65], as noted above, 1.0 μ_B. The cobalt moment in the $Y_{1-x}Gd_xCo_2$ ternaries is [31] given in Table 10-10 together with other magnetic characteristics of the system. The cobalt moment is obtained by assuming 7.0 μ_B for Gd and antiferromagnetic collinear Gd–Co coupling. Taylor *et al.* obtained similar results and interpreted their data in terms of noncollinear Gd–Co coupling. Their analysis, put forth before the cobalt moment variability was fully appreciated, seems to be superceded by that of Lemaire [31].

TABLE 10-10

Magnetic Characteristics of $Y_{1-x}Gd_xCo_2$ Compounds[a]

x	Magnetization (μ_B/f.u.)	T_C (°K)	Co moment (μ_B/Co)
0	~0	<2	~0
0.1	~0	<2	~0
0.2	0.48	60	0.46
0.3	0.80	146	0.65
0.4	1.34	205	0.73
0.6	2.48	280	0.86
0.8	3.60	350	1.00
1.0	4.90	404	1.05

[a]From Lemaire [31].

b. Cobalt Substitution

Oesterreicher and Wallace [72,73] examined the pseudobinary systems RAl_2–RCo_2 with R = Pr, Gd, and Er. $PrAl_2$ dissolves 21 mole % $PrCo_2$ and $PrCo_2$ dissolves 8 mole % $PrAl_2$. $PrAl_2$ is ferromagnetic material with $T_C = 34°K$. With substituted Co its magnetization temperature characteristics resemble those of $PrNi_5$ (Chapter 9) and $PrAl_{2-x}Ni_x$ (Chapter 4). $PrAl_2$ is a bootstrap ferromagnet. It appears that when Co or Ni is inserted in partial replacement of Al, exchange is weakened to the point that ordering does not occur. Instead it exhibits Van Vleck paramagnetism, its magnetization passing through a peak for reasons* which have been presented in the discussion of $PrNi_5$. The $PrAl_{1.55}Co_{0.45}$ is ferromagnetic with $T_C = 24°K$. Replacement of Co with Al in $PrCo_2$ to the phase boundary leaves the material ferromagnetic but with a diminished T_C and μ_{sat}, as might be expected [72].

Unlike the Pr–Co–Al system, the $GdAl_2$–$GdCo_2$ and $ErAl_2$–$ErCo_2$ pseudobinary systems show [73] a C14 ($MgZn_2$) structure at the composition RCoAl. The terminal phases in each case extend to about 25 mole % of the added component. Magnetic characteristics are summarized in Table 10-11. Similar studies have been made on $GdFe_2$–$GdAl_2$ pseudobinaries and are summarized in the next chapter.

If we assume antiferromagnetic coupling between R and Co (or R and Fe), we can compute the Co or Fe moment in the ternaries from the observed saturation moment. Results obtained are shown in Fig. 10-13.

TABLE 10-11

Magnetic Characteristics of RCo_2–RAl_2 Ternaries[a]

	Gd system			Er system		
Mole% RAl_2	μ_{or} (μ_B/f.u.)	μ_{eff} (μ_B/R)	T_C (°K)	μ_{or} (μ_B/f.u.)	μ_{eff} (μ_B/R)	T_C (°K)
0	4.94		420	7.00		40
10	5.56		291	7.84	*b*	86
17.5	6.09					
25				6.60		30
50[c]	7.00	7.71	116	8.31	10.4	25
77.5	7.53					
85				5.76		24
92.5	7.53	9.75	20	8.41	10.0	25
100	6.95	7.96	180	7.1		22

[a] From Oesterreicher and Wallace [73].
[b] Studied in the paramagnetic region. Did not follow Curie-Weiss Law.
[c] C14 structure. All the others are C15 structure.

*Replacing Al in $PrAl_2$ with 15% Fe, Cu, and Ag produces a similar effect. However, 15% replacement by Mn leaves the material ferromagnetic but with a reduced T_C [72].

We note that the Co or Fe moments are reduced by 0.3 and 0.7 μ_B, respectively, when combined with Gd and a further reduction occurs upon replacement of the transition metal by Al.

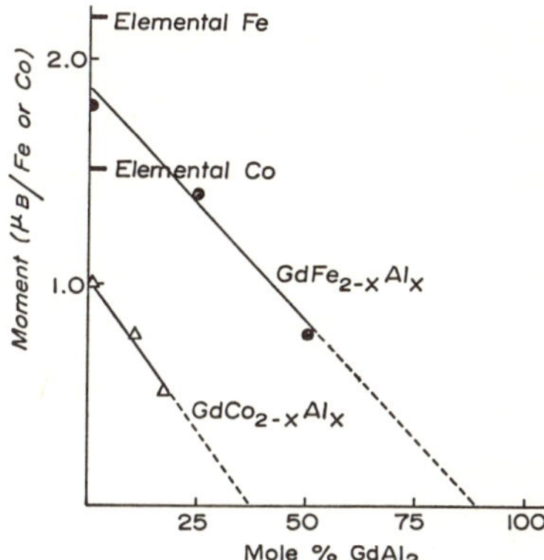

Fig. 10-13. Fe and Co moments in Gd–Co–Al and Gd–Fe–Al alloys [73].

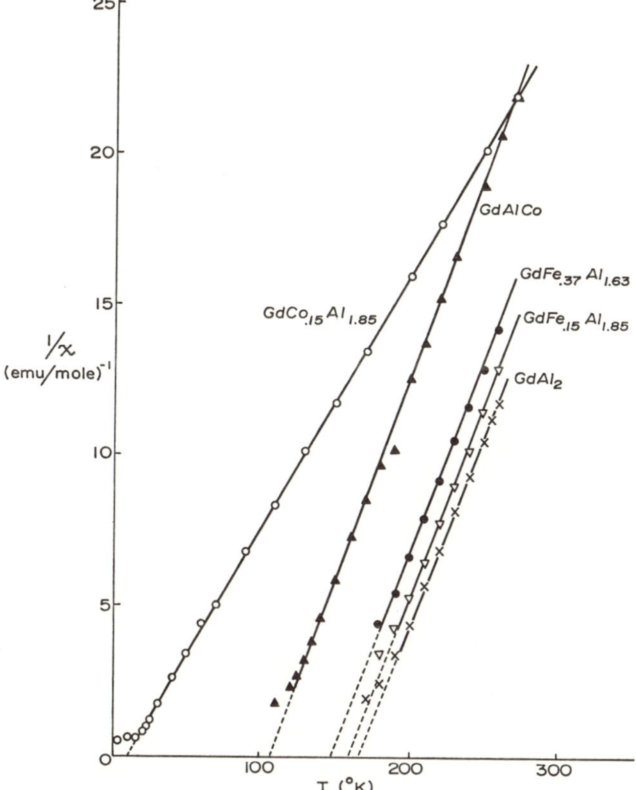

Fig. 10-14. Inverse susceptibility–temperature data for $GdAl_2$ and Gd–Co–Al and Gd–Fe–Al ternaries [73].

Paramagnetic behavior of the Gd–Co–Al and Gd–Fe–Al ternaries is shown in Fig. 10-14. The decline in Weiss constant with concentration of Co is striking, i.e., a decrease of 100° by replacement of only 7.5% of the Al by Co. We note a rise in ordered moment (Table 10-11) as Al in RAl_2 is replaced by Co (or Fe—see next chapter). It seems unlikely that this is due to ferromagnetic R–Co or R–Fe coupling since not only is this contrary to the coupling systematics noted for these pairs of species but also because unreasonably large Fe or Co moments would be required—7.1 and 3.9 μ_B/atom, respectively. It seems instead that the extra moment is a consequence of the contribution from rather striking changes in the conduction electron polarization. This, i.e., the striking change, is also suggested by the sharp drop in T_C and the Weiss constant.

The formation of the C14 structure as Al replaces Co in RCo_2 is ascribed to Brillouin zone filling effects similar to those observed by Laves and Witte [74] in their classic studies of the prototype Laves phase structures. For details on this interesting facet of Laves phase material, which lies outside the scope of this monograph, the reader is referred to the original work and related studies [75].

C. THE R_2Co_{17} COMPOUNDS

Although the existence of compounds of the formula R_2Co_{17} was known and structures had been worked out as early as 1962, no studies of their magnetic properties had been made until 1966. In that year the results of two independent investigations appeared [76–79]. Magnetic and structural data are summarized in Table 10-12. Most of the compounds exist in two closely related structural types—Th_2Ni_{17} or Th_2Zn_{17}—depending on their thermal history and it has proved difficult to obtain materials in one form free of the other. This fact should be borne in mind in considering the magnetic results. However, in view of the closeness of the structural types and their similar coupling systematics it is unlikely that the existence of a second allotropic form as an impurity phase renders significant uncertainty in the tabulated values of μ_{or} or T_C.

The coupling mode can be inferred by comparing μ_{or} for a particular R_2Co_{17} compound with that of Lu_2Co_{17} (27.4 μ_B/f.u.) or Y_2Co_{17} (27.6 μ_B/f.u.). Pr_2Co_{17} and Nd_2Co_{17} have moments in excess of the moment of the cobalt sublattice, from which we infer a ferromagnetic coupling scheme. The deficiency in moment for R = Gd through Tm, compared to the cobalt sublattice, indicates these compounds to be ferrimagnetic. Ferromagnetic coupling for Sm_2Co_{17} is expected but the moment is below 27.5 μ_B/f.u., the cobalt sublattice contribution. Strnat *et al.* ad-

TABLE 10-12

Magnetic and Structural Characteristics of R_2Co_{17} Compounds

	Structure[a]	Ref.	Characteristics in the ordered state			Ref.
			μ_{or} (μ_B/f.u.)	T_C (°K)	Type	
Ce	R,H(q)	[76,79,80]	26.1	1080	F	[76–79]
Pr	R,R(q)	[76,79]	32.9	1170	F	[76–79]
Nd	R,R(q)	[76,79]	32.1	1155	F	[76–79]
Sm	R,H(q)	[76,79]	23.6	1190	F_i(?)	[76–79]
Gd	R,H(q)	[76,79,81,82]	14.4	1210	F_i	[76–79]
Tb	R,H(q)	[76,79]	10.8	1185	F_i	[76–79]
Dy	R,H(q)	[76,79]	8.3	1165	F_i	[76–79]
Ho	H,H(q)	[76,79]	7.8	1180	F_i	[76–79]
Er	H,H(q)	[76,79]	10.6	1180	F_i	[29,76–79]
Tm	H,H(q)	[76,79]	13.6	1185	F_i	[76–79]
Lu	H	[76]	27.4	1210	F	[76,78]

[a] H = the hexagonal Th_2Ni_{17} structure; R = the rhombohedral Th_2Zn_{17} structure. Some compounds are found in different structures depending upon whether quenched or annealed. The former are designated (q). Unless otherwise stated, structures are for annealed materials.

vance the possibility that the Sm compound is ferrimagnetic [76]. This seems unlikely, however, in view of the Sm moment required, ~2 μ_B/Sm. Another possibility is that the sample used was unsatisfactory; difficulties are generally experienced in achieving satisfactory Sm samples because of the high volatility of Sm.

Results for the Ce compound as usual are anomalous. Lattice parameters of Ce_2Co_{17} and to a lesser extent those of Pr_2Co_{17} are anomalously small, suggesting to Ostertag *et al.* [78] that Ce and Pr are ionized beyond the tripositive state to Ce^{4+} and $Pr^{3.3+}$. On this basis the Ce sublattice is nonmagnetic and the cobalt sublattice in Ce_2Co_{17} carries a magnetization of 1.54 μ_B/Co, compared with 1.62 and 1.67 μ_B/Co in Lu_2Co_{17} and Gd_2Co_{17}, respectively. The extra 4f electron is presumed to be contributed to the Co d-band, reducing the moment by about 0.11 μ_B per Co atom to 1.51 or 1.56 μ_B/Co (depending on whether we use Lu_2Co_{17} or Gd_2Co_{17} figure for the cobalt moment) in rather striking conformity with the observed Co moment in Ce_2Co_{17}. Thus Strnat *et al.* [76] were able to produce a consistent picture of the structural and magnetic features of Ce_2Co_{17} in terms of quadrivalent Ce, but, as noted earlier, the true situation may be somewhat more complex.

The large saturation moments of R_2Co_{17} compounds with R=Ce through Sm together with their high Curie temperatures make them attractive candidates for permanent magnet fabrication along with the

RCo_5 systems mentioned above. Hoffer and Strnat [52] examined single crystals of Y_2Co_{17} and established that its anisotropy is unfavorable since the moments lie in the basal plane. Thus, there is, instead of an easy axis, an easy plane, disqualifying Y_2Co_{17} as a permanent magnet material. Becker [53] quotes Hoffer to the effect that the Ce, Pr, and Nd compounds also have an easy plane, but states that Sm_2Co_{17} has an easy axis and a high anisotropy field, ~100 kOe. Studies of its capability in permanent magnet fabrication are currently being made.

D. THE RCo_3, R_2Co_7, R_4Co_3, AND R_3Co COMPOUNDS

Magnetic studies by Lemaire *et al.* [79,83] have led to the results given in Table 10-13. We note that the R_2Co_7 series has not been extensively investigated. The magnitude of the moments and the magnetization–temperature behavior of the R_2Co_7 systems establishes that they are ferrimagnetic. The same is true of RCo_3 compounds with R = Gd through Tm. Neutron diffraction studies in $PrCo_3$ and $NdCo_3$ show [83] R moments of 2.7 μ_B for Pr^{3+} and 2.6 μ_B for Nd^{3+}. The cobalt moment estimated from these and μ_{or} is 0.44 and 1.0 in $PrCo_3$ and $NdCo_3$, respectively. We thus see in the RCo_3 series a steady weakening of the cobalt moment in the progression Gd to Ce, from 1.6 to 0.07 μ_B (if Ce is regarded as nonmagnetic).

TABLE 10-13

Magnetic Characteristics of RCo_3 and R_2Co_7 Compounds

	RCo_3 series				R_2Co_7 series			
	μ_{or} (μ_B/f.u.)	T_C (°K)	Type	Ref.	μ_{or} (μ_B/f.u.)	T_C (°K)	Type	Ref.
Ce	0.2	78	F	[79,83]				
Pr	3.8	349	F	[79,83]				
Nd	5.6	395	F	[79,83]				
Gd	2.2	612	F_i	[79,83]	2.4	775	F_i	[79]
Tb	3.4	506	F_i	[79,83]	5.3	717	F_i	[79]
Dy	4.3	450	F_i	[79,83]				
Ho	5.6	418	F_i	[79,83]	6.0	670	F_i	[79]
Er	3.9	401	F_i	[79,83]				
Tm	3.0	370	F_i	[79,83]				

The R_3Co systems have been studied by Gignoux *et al.* [8] and Feron *et al.* [7] and the R_4Co_3 compounds by Berthet-Colimnas *et al.* [84] (Table 10-14). The latter compounds exhibit Curie–Weiss behavior with effective moments close to those of the rare earth constituent only. This implies a rather small cobalt moment, which is confirmed by the observed μ_{or} for Gd_4Co_3. The R_3Co compounds exhibit Curie–Weiss behavior at temperatures in excess of their ordering temperatures. Because of the low Co moment and concentration it is not possible to establish the coupling pattern for these materials from their measured moments. However, magnetization-field measurements are concave upwards at low fields for Gd_3Co, Dy_3Co, and Ho_3Co, making it likely that these involve antiferromagnetic coupling. It seems reasonable to believe that these materials obey the same coupling systematics as the other rare earth–cobalt compounds. However, owing to lack of complete details the several compounds in Table 10-14 are designated as metamagnetic on the basis of information given by Feron *et al.* [7]. Neutron diffraction work on Er_3Co gives [8] an Er moment of 6.0 μ_B, which is in excellent agreement with results obtained from bulk magnetic measurements. Er_3Co has a complex noncollinear structure.

TABLE 10-14

Magnetic Characteristics of R_3Co and R_4Co_3 Compounds

	μ_{or} (μ_B/f.u.)	μ_{eff} (μ_B/R)	θ (°K)	T_C (°K)	T_N (°K)	Type	Ref.
			R_3Co series				
Pr	3.3		18	7		F	[7]
Nd			35		14	M	[7]
Gd			159		127	M	[7,85]
Tb			85		82	M	[7]
Dy			65		45	M	[7]
Ho			44		24	M	[7]
Er	18		20	7		F	[7,8]
Tm	10.5		0	5		F	[7]
			R_4Co_3 series				
Gd	26.3	7.9	240	230		F_i	[84]
Dy	20.9	10.6	62	55		F_i(?),M	[84]
Ho	29.6	10.6	50	44		F_i(?),M	[84]
Er	23.1	9.6	28	25		F_i(?),M	[84]

E. CONCLUDING REMARKS

The rare earth–cobalt compounds present many interesting features, several of which seem sufficiently significant to warrant repetition here for purposes of emphasis—the high Curie temperatures, the strong anisotropy fields (H_A), the consistent coupling scheme, and the variable cobalt moment.

As noted above, attention was first drawn to the latter point by Bleaney [49] who reasoned on the basis of work by Skrabek and Wallace [35] that the cobalt moment was proportional to the exchange field produced by the rare earth partners. He set this exchange field equal to $(g-1)\mu'_{\rm or}/g$ where $\mu'_{\rm or}$ represents the ordered moment of the rare earth sublattice. Since then, many others have also drawn attention to the variable Co moment that must be assumed to account for the properties of the cobalt compounds. Lemaire *et al.* have discussed this feature in some detail [79,86]. In Fig. 10-15 taken from the publication of Lemaire *et al.* [86] the cobalt moment ($M_{\rm Co}$) is plotted against the number of R atoms in chemical union with one cobalt atom ($N_{\rm R}$) for the $\rm GdCo_x$ and $\rm YCo_x$ series. Two effects are evident—that (1) $M_{\rm Co}$ decreases with increasing $N_{\rm R}$ and (2) $M_{\rm Co}$ is larger when R is magnetic, the effect to which Bleaney drew attention. The latter is an induction effect; the former is ascribed to electron transfer from R to cobalt, decreasing $M_{\rm Co}$ as the Co band is filled. The electron transfer is more extensive the more R atoms there are per cobalt atom. If Co is considered to be in a $3d^{8.3}$ configuration in the elemental state with one half band filled, the decline in $M_{\rm Co}$ due to electron transfer has the slope shown in Fig. 10-15. This is approximated by the $\rm YCo_x$ systems.

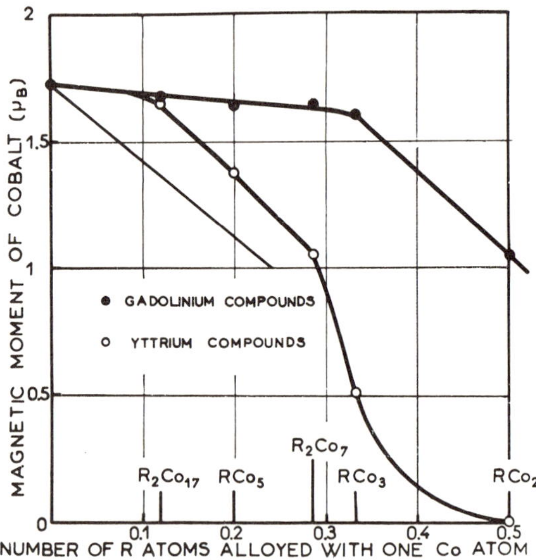

Fig. 10-15. Variations of the magnetic moment of cobalt in its intermetallic compounds with yttrium or gadolinium [86].

The high Curie temperatures imply importance, usually dominance, of the Co–Co interactions over the R–R interactions. Curie temperatures versus atomic number for the several series taken from data in Tables 10-2, 10-13, and 10-14 are shown in Fig. 10-16. The progressive increase in importance of the R–R interactions as the proportion of R increases is apparent. Despite this, the transition metal interaction remains the dominant interaction even in the RCo_2 series. T_C for $GdNi_2$, where only the R–R interaction is involved, is 85°K as compared with 408°K for $GdCo_2$. The higher T_C for the cobalt compound is a clear indication of the importance of the Co–Co interaction.

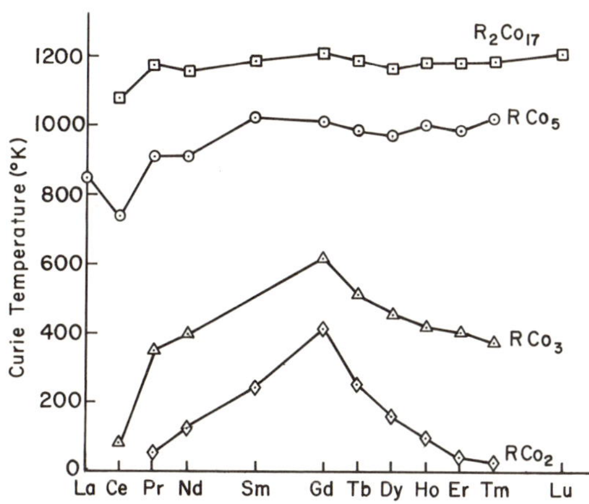

Fig. 10-16. Curie temperatures for several rare earth cobalt series (Tables 10-2, 10-12, and 10-13).

The origin of the high anisotropy fields in the RCo_5 compounds is not clear. If we try to ascribe them exclusively to the cobalt sublattice, we encounter the problem (Table 10-5) that H_a varies with the nature of R. If the high H_a is allocated to the R sublattice, we also meet with difficulties. Consider, for example, $GdCo_5$. It has a high anisotropy field, which is in striking contrast with expectation if the anisotropy is regarded as stemming from the Gd sublattice. Rare earth anisotropy is governed by the strength of the crystal field interaction. Thus H_a for Gd^{3+} should be weak since it is in an S state. Hence we must conclude that both sublattices are involved.

Attention has been drawn earlier to the fact that the RCo_5 compounds with R = Nd, Tb, Ho, and probably Dy are uniaxial only at high temperatures. At reduced temperatures the direction of easy magnetization lies in the basal plane. It is observed that $CeCo_5$ [33] and YCo_5 [34] are uniaxial at all temperatures; in these materials Ce and Y are nonmagnetic. It thus appears that the uniaxial behavior of the RCo_5 originates with the cobalt sublattices. It is reasonable to ascribe the development of an easy basal plane at low temperatures to the influence of the rare earth. The behavior indicates that in regard to magnetic anisotropy the cobalt sublattice is dominant at high temperatures whereas at low temperatures the rare earth sublattices is the dominant influence.

The constant coupling scheme (R and Co spins always couple antiparallel) is also an enigma if the RKKY interaction is dominant. Since the RKKY sums are structure sensitive, we might reasonably expect with the diversity of structures involved some variation in coupling would ensue. We can set forth three possible reasons why this variation does not occur: (1) The diversity in structure is more apparent than real, the several structures being similar particularly in their nearest neighbor relationships, which are important in establishing the sign and magnitude of the RKKY sums; (2) the interaction is via the conduction electrons but the RKKY formalism is too crude to relate properly to the phenomena under discussion; (3) the interaction does not involve the polarization of the conduction electrons at least as portrayed by the RKKY description. Which of these reasons, if any, is the pertinent one is not clear as yet.

As a final comment we note that since the coupling systematics are the same in $R_{1-x}R_x'M_z$ ($z = 2$ or 5) ternaries with M = Co, Al, or Ni, the R–R interactions appear to be the same whether M is magnetic or not. This point was alluded to earlier in Chapter 2 in the preliminary discussion of the R–R interactions in rare earth intermetallics.

REFERENCES

1. R. Vogel, *Z. Anorg. Chem.* **99,** 25 (1917).
2. W. Klemm, "Naturforschung und Medizin in Deutschland 1939–46," Vol. 25, part III, p. 255. Dietrich'sche Verlagshandlung, Wiesbaden, 1948.
3. J. Wucher, *J. Phys. Radium* **13,** 278 (1952).
4. E. A. Nesbitt, J. H. Wernick, and E. Corenzwit, *J. Appl. Phys.* **30,** 365 (1959).
5. K. Nassau, L. V. Cherry, and W. E. Wallace, *J. Phys. Chem. Solids* **16,** 131 (1960).
6. W. M. Hubbard, E. Adams, and J. V. Gilfrich, *J. Appl. Phys.* **31,** 368S (1960).
7. J. L. Feron, D. Gignoux, R. Lemaire, and D. Paccard, *Les Eléménts des Terres Rares, Coll. Int. CNRS* No. 180, Tome 2, 75 (1970).
8. D. Gignoux, R. Lemaire, and D. Paccard, *Solid State Commun.* **8,** 391 (1970).
9. D. T. Cromer and A. C. Larson, *Acta Crystallog.* **14,** 1226 (1961).
10. K. H. J. Buschow and A. S. Van der Goot, *J. Less Common Metals* **14,** 323 (1968); **18,** 309 (1969); **19,** 153 (1969).
11. F. Givord and R. Lemaire, *Solid State Commun.* **9,** 341 (1971).
12. R. Lemaire and J. Schweizer, *C. R. Acad. Sci. Paris* **264,** 642 (1967).
13. C. Berthet-Colimnas, J. Laforest, R. Lemaire, R. Pauthenet, and J. Schweizer, *Cobalt* **39,** 97 (1968).
14. W. Fülling, K. Moeller, and R. Vogel, *Z. Metallk.* **34,** 254 (1942).
15. K. Nassau, L. V. Cherry, and W. E. Wallace, *J. Phys. Chem. Solids* **16,** 123 (1960).

16. J. H. Wernick and S. Geller, *Trans. AIME* **218,** 806 (1960).
17. E. F. Bertaut, R. Lemaire, and J. Schweizer, *Bull. Soc. Fr. Mineral. Cristallogr.* **LXXX-VIII,** 580 (1965).
18. K. H. J. Buschow and A. S. Van der Goot, *J. Less Common Metals* **17,** 249 (1969).
19. J. H. Van Vucht, *J. Less Common Metals* **10,** 146 (1965).
20. T. Heuman, *Nachr. Ges. Wiss. Göttingen* **1,** 21 (1948).
21. J. H. Wernick and S. Geller, *Acta Crystallogr.* **12,** 662 (1959).
22. S. E. Haszko, *Trans. AIME* **218,** 763 (1960).
23. G. Bouchet, J. Laforest, R. Lemaire, and J. Schweizer, *C. R. Acad. Sci. Paris* **262,** 1227 (1966).
24. D. T. Cromer and A. C. Larson, *Acta Crystallogr.* **12,** 855 (1959).
25. K. H. J. Buschow and W. A. J. J. Velge, *J. Less Common Metals* **13,** 11 (1967).
26. K. H. J. Buschow, *Les Éléments des Terres Rares, Coll. Int. CNRS* No. 180, Tomes 1, 101 (1970).
27. K. H. J. Buschow, *J. Less Common Metals* **11,** 204 (1966).
28. W. A. J. J. Velge and K. H. J. Buschow, *J. Appl. Phys.* **39,** 1717 (1968). See also *Z. Angew. Phys.* **26,** 157 (1969).
29. K. H. J. Buschow, J. F. Fast, and A. S. Van der Goot, *Phys. Status Solidi* **29,** 825 (1968).
30. E. A. Nesbitt, H. J. Williams, J. H. Wernick, and R. C. Sherwood, *J. Appl. Phys.* **32,** 342S (1961).
31. R. Lemaire, *Cobalt* **32,** 132 (1966).
32. L. V. Cherry and W. E. Wallace, *J. Appl. Phys.* **33,** 1515 (1962).
33. R. Lemaire and J. Schweizer, *J. Phys.* **28,** 216 (1967).
34. W. James, R. Lemaire, and E. F. Bertaut, *C. R. Acad. Sci. Paris* **255,** 896 (1962).
35. W. E. Wallace and E. A. Skrabek, *in Rare Earth Res., Proc. Rare Earth Conf., 3rd* (K. S. Vorres, ed.), p. 431. Gordon and Breach, New York, 1964.
36. J. W. Ross and J. Crangle, *Phys. Rev.* **133,** A509 (1964).
37. J. Farrell and W. E. Wallace, *Inorg. Chem.* **5,** 105 (1966).
38. W. E. Wallace, T. V. Volkmann, and H. Hopkins, Jr., *J. Solid State Chem.* **3,** 510 (1971). (1971).
39. I. Shidlovsky and W. E. Wallace, *J. Solid State Chem.* **2,** 193 (1970).
40. L. V. Cherry and W. E. Wallace, *J. Appl. Phys.* **33,** 1515 (1962).
41. W. G. Saba and W. E. Wallace, *J. Chem. Phys.* **35,** 689 (1961).
42. H. Bartholin, B. Van Laar, R. Lemaire, and J. Schweizer, *J. Phys. Chem. Solids* **27,** 1287 (1966).
43. A. S. Edelstein, *Phys. Rev. Lett.* **20,** 1348 (1968).
44. I. Nowik and J. H. Wernick, *Phys. Rev.* **140,** A131 (1965).
45. E. A. Nesbitt, H. J. Williams, J. H. Wernick, and R. C. Sherwood, *J. Appl. Phys.* **33,** 1674 (1962).
46. W. M. Hubbard and E. Adams, *J. Phys. Soc. Japan Suppl. B-1* **17,** 143 (1962).
47. K. H. Mader and W. E. Wallace, *J. Chem. Phys.* **49,** 1521 (1968).
48. K. Sekizawa and K. Yasukochi, *J. Phys. Soc. Japan* **21,** 684 (1966).
49. B. Bleaney, *in Rare Earth Res., Proc. Rare Earth Conf., 3rd* (K. S. Vorres, ed.), Vol. 2, p. 499. Gordon and Breach, New York, 1964.
50. B. Leon and W. E. Wallace, *J. Less Common Metals* **22,** 1 (1970).
51. W. E. Wallace, H. P. Hopkins, Jr., and K. Lehman, *J. Solid State Chem.* **1,** 39 (1968).
52. G. Hoffer and K. Strnat, *IEEE Trans. Magnet.* **2,** 487 (1966).
53. J. J. Becker, *J. Appl. Phys.* **41,** 1055 (1970); *Sci. Amer.* **223,** 92 (December 1970).
54. E. C. Stoner and E. P. Wohlforth, *Phil. Trans. Roy. Soc. London* **A240,** 599 (1948). See also A. H. Morrish, "The Physical Principles of Magnetism," p. 345. Wiley, New York, 1966.

55. A. E. Berkowitz and E. Kneller, "Magnetism and Metallurgy," Vol. 1, p. 372. Academic Press, New York, 1969.
56. K. H. J. Buschow, P. A. Naastepad, and F. F. Westendorf, *J. Appl. Phys.* **40,** 4029 (1969).
57. F. F. Westendorf and K. H. Buschow, *Solid State Commun.* **7,** 639 (1969).
58. K. Strnat, *Proc. Rare Earth Conf. 7th, Coronado, Colorado,* p. 17. (1968)
59. E. A. Nesbitt, R. H. Willens, R. C. Sherwood, E. Buehler, and J. H. Wernick, *Appl. Phys. Lett.* **12,** 361 (1968).
60. E. A. Nesbitt, *J. Appl. Phys.* **40,** 1259 (1969).
61. H. Zijlstra and F. F. Westendorf, *Solid State Commun.* **7,** 857 (1969).
62. J. Crangle and J. W. Ross, *Proc. Int. Conf. Magnet. Nottingham,* p. 240. Inst. of Phys. and the Phys. Soc., London (1964).
63. G. Petrich and R. L. Mössbauer, *Phys. Lett.* **16A,** 403 (1968).
64. J. Schweizer, *Phys. Lett.* **24A,** 739 (1967).
65. R. M. Moon, W. C. Koehler, and J. Farrell, *J. Appl. Phys.* **36,** 978 (1965).
66. S. Ofer and E. Segal, *Phys. Rev.* **141,** 448 (1966).
67. D. Bloch, F. Chaissé, F. Givord, and E. Burzo, *J. Phys. Suppl.* **32,** C1–659 (1971).
68. C. Deenadas, R. S. Craig, N. Marzouk, and W. E. Wallace, *J. Solid State Chem.* **4,** 1 (1972). $CeCo_2$ data are to be published.
69. B. Leon and W. E. Wallace, *J. Less Common Metals* **22,** 1 (1970).
70. R. Lemaire and J. Schweizer, *Phys. Lett.* **21,** 366 (1966).
71. K. N. R. Taylor, H. D. Ellis, and M. I. Darby, *Phys. Lett.* **20,** 327 (1966).
72. H. Oesterreicher and W. E. Wallace, *J. Less Common Metals* **13,** 91 (1967).
73. H. Oesterreicher and W. E. Wallace, *J. Less Common Metals* **13,** 475 (1967).
74. F. Laves and H. Witte, *Metallwirtschaft* **14,** 645 (1935); **15,** 840 (1936).
75. W. E. Wallace and R. S. Craig, *in* "Phase Stability in Metals and Alloys" (P. S. Rudman, J. Stringer, and R. I. Jaffee, eds.), p. 255. McGraw-Hill, New York, 1967.
76. K. Strnat, G. Hoffer, W. Ostertag, and J. C. Olson, *J. Appl. Phys.* **37,** 1252 (1966).
77. J. Laforest, R. Lemaire, R. Pauthenet, and J. Schweizer, *C. R. Acad. Sci. Paris* **262B,** 1260 (1966).
78. W. Ostertag, K. Strnat, and G. I. Hoffer, Tech Rep. AFML-TR-66-420, Air Force Mater. Lab., Wright-Patterson Air Force Base, December 1966. (The report was released in December 1966; however, the cover carries the date February 1967).
79. R. Lemaire, *Cobalt* **33,** 201 (1966).
80. O. S. Zarechnyuk and P. I. Kripyakevich, *Sov. Phys.-Crystallogr.* **7,** 436 (1963).
81. E. M. Savitskii, V. Terekhova, and I. Burov, *Zh. Neorg. Khim.* **7,** 2572 (1962).
82. P. I. Kripyakevich, V. Terekhova, O. Zarechnyuk, and I. V. Burov, *Kristallographiya* **8,** 268 (1963).
83. R. Lemaire, R. Pauthenet, J. Schweizer, and I. S. Silvera, *J. Phys. Chem. Solids* **28,** 2471 (1967).
84. C. Berthet-Colominas, J. Laforest, R. Lemaire, R. Pauthenet, and J. Schweizer, *Cobalt* **39,** 97 (1968).
85. O. A. W. Strydom and L. Alberts, *J. Less Common Metals* **22,** 503 (1970).
86. R. Lemaire, D. Paccard, R. Pauthenet, and J. Schweizer, *J. Appl. Phys.* **39,** 1092 (1968).

Chapter 11

IRON AND MANGANESE COMPOUNDS

The rare earths form fewer compounds with iron and manganese than with nickel or cobalt. The several stoichiometries observed are given in Table 11-1. The earliest reported iron and manganese–rare earth intermetallics were $GdFe_2$ and $GdMn_2$ by Endter and Klemm [1] in 1943. The Ce–Fe phase diagram was established [18] by Jepson and Duwez some years later. They interpreted their results to indicate the existence of $CeFe_2$ and $CeFe_5$. Nassau *et al.* [2] and almost simultaneously Wernick and Geller [3] prepared a number of RFe_2 and RMn_2 compounds and showed them to have the C15 ($MgCu_2$) structure. Nassau *et al.* [2] attempted unsuccessfully to prepare $CeFe_5$ and other RFe_5 compounds. Subsequent work [7,12] has shown that neither the RFe_5 phases nor the RFe_4 phases, that were presumably studied in some of the early work, are stable in the R–Fe system.

The high concentration phases have subsequently been shown to be R_6Fe_{23}, formerly thought to be RFe_4, and R_2Fe_{17}, which seemingly was taken to be RFe_5.* Both of these high concentration Fe phases are difficult to prepare free of extraneous phases. The rhombohedral and hexagonal forms are frequently found together as was the case with the R_2Co_{17} compounds. Givord and Lemaire [14] have made a thorough study of Gd_2Fe_{17} and find three allotropic modifications—a hexagonal modification (Th_2Ni_{17} structure) if quenched from just below the peritectic temperature, a rhombohedral form (Th_2Zn_{17} structure) if annealed at 1050°C, and a hexagonal variety isomorphous with $CaCu_5$ if annealed at 900°C. We term these h_1, r, and h_2 forms, respectively. As

*Also a RFe_7 phase with R = Ce through Nd reported in earlier studies is now known to be R_2Fe_{17} [13].

TABLE 11-1

Structures of the Iron and Manganese Compounds

R	RFe_2	RMn_2	RFe_3	R_6Fe_{23}	R_6Mn_{23}	R_2Fe_{17}	RMn_{12}
Ce	C15					R and H[a]	
Pr						R	
Nd		C15		Th_6Mn_{23}	Th_6Mn_{23}	R	$ThMn_{12}$
Sm	C15		$PuNi_3$		Th_6Mn_{23}	R	
Gd	C15	C15	$PuNi_3$	Th_6Mn_{23}	Th_6Mn_{23}	*b*	$ThMn_{12}$
Tb	C15	C15	$PuNi_3$	Th_6Mn_{23}	Th_6Mn_{23}	R and H	
Dy	C15	C15	$PuNi_3$	Th_6Mn_{23}	Th_6Mn_{23}	H	$ThMn_{12}$
Ho	C15	C15	$PuNi_3$	Th_6Mn_{23}	Th_6Mn_{23}	H	
Er	C15	C14	$PuNi_3$	Th_6Mn_{23}	Th_6Mn_{23}	H	$ThMn_{12}$
Tm	C15	C14	$PuNi_3$	Th_6Mn_{23}	Th_6Mn_{23}	H	
Lu	C15		$PuNi_3$	Th_6Mn_{23}	Th_6Mn_{23}	H	
Ref.	[1–4]	[2,3,5,6]	[4,7–9]	[4,10,11]	[11]	[4,7,12–15]	[16,17]

[a] R and H denote the rhombohedral Th_2Zn_{17} and hexagonal Th_2Ni_{17} structures, respectively.

[b] This compound has been shown [14] to exist in three allotropic modifications (see text).

noted in Chapter 9, the 2:17 compounds can be viewed as $(R_{2/3}P_{1/3})Fe_5$ in the h_2 form and P's are randomly distributed over the rare earth sites (P represents a pair of iron atoms). In the h_1 and r forms the substitution is ordered so that adjacent layers containing R atoms are different. The stacking sequence in the h_1 form is *ABAB* . . . ; that in the r form is *ABC,ABC*, ... We can readily appreciate that h_1 and r frequently coexist in cast samples since they differ only in stacking sequence. The R_2T_{17} (T=Fe, Co, or Ni) can be viewed as the culmination of the process observed by Buschow in the RCo_5 phases (see Chapter 10) in which a stoichiometric excess of Co is observed when R=Tb through Er, the excess being in the forms of pairs of cobalt atoms substituting for R. When the concentration of this excess grows to 1 pair of T atoms for each 2 R atoms, the R_2T_{17} structure forms.

Similar to the situation in the rare earth–iron systems, it was once thought [19] that a Mn-rich phase of composition, RMn_4, existed. This is now known to be the R_6Mn_{23} phase.

Structural information for the R–Fe and R–Mn intermetallics is accumulated in Table 11-1.

A. R–Fe COMPOUNDS

1. RFe_2 Series

a. Binary Systems

Nesbitt *et al.* [20] in 1959 studied the bulk magnetic characteristics of a series of Gd–Fe alloys, including $GdFe_2$, and concluded that they are ferrimagnetic materials. More complete studies of the entire RFe_2 series came later, in 1964 [21–24] (Table 11-2). These and later studies have confirmed that the R–Fe coupling is antiferromagnetic when R is a heavy rare earth and in this respect the R–Fe systems are analogous to the R–Co, R–Ni systems. However, there is no counterpart to the observed ferromagnetic coupling in light rare earth–cobalt or –nickel systems since light rare earth–iron compounds either do not form or, if they do form, the purity achieved to date has not warranted determination of their magnetic behavior. The Ce compounds do form, but in these materials Ce behaves as if it is nonmagnetic and hence is described as quadripositive. $SmFe_2$ also forms and seem to be ferri-

TABLE 11-2

Magnetic Characteristics of RFe_2 and RFe_3 Compounds

R	μ_{or} (μ_B/f.u.)	T_C (°K)	Type	Ref.
		RFe_2		
Ce	2.48	235	F	[15,23,24]
Sm	2.5	700	F_i(?)	[21,24]
Gd	3.35, 3.80	782	F_i	[22,24,25]
Tb	3.68, 4.72	705	F_i	[21,24]
Dy	4.91, 5.50	638	F_i	[4,22,24]
Ho	6.1	614	F_i	[15,24,26]
Er	4.75, 5.02	596	F_i	[7,21,24,27]
Tm	2.52, 2.92	610	F_i	[21,27]
Lu	2.97	610	F	[21]
		RFe_3		
Sm		651		[28]
Gd	1.6, 1.51	728	F_i	[9,25,28]
Tb	3.6	648	F_i	[9,28,29]
Dy	4.6	600	F_i	[4,9,28,29]
Ho	4.6	567	F_i	[9,28]
Er	4.8	553	F_i	[7,9,28]
Tm	1.6	539	F_i	[9,28]
Lu		529		[9]

magnetic. However, results obtained in the single study of this compound [21] need to be confirmed before antiferromagnetic Sm–Fe coupling is accepted.* As mentioned previously, samarium compounds are notoriously difficult to prepare in pure form due to the volatility of this rare earth; it is not certain that the sample studied by Skrabek and Wallace was sufficiently pure to warrant unambiguous conclusions

*Crangle and Ross have also studied $SmFe_2$ but did not obtain a saturation moment. They only measured magnetization versus temperature at an unspecified field.

about its magnetic structure. The shape of the magnetization–temperature curve confirms the ferrimagnetism of $ErFe_2$ and $TmFe_2$, compensation points being observed [21] at 480 and 225°K, respectively. Antiferromagnetic coupling between R and Fe has been established by neutron diffraction studies by Moreau *et al.* [26] for $HoFe_2$, and in $ErFe_2$ by Bargouth and Will [30].

The iron moment shows some variations throughout the series. This is taken up in Section 3.

b. Ternary Systems

Shidlovsky [31], Oesterreicher [32], and Wallace have studied $R_{1-x}Zr_xFe_2$ with R = Gd, Dy, and Ho, and $GdFe_{2-x}Al_x$ ternaries. The purpose of incorporating Zr in the lattice was part of attempts to induce ferromagnetic coupling in heavy rare earth–iron intermetallics by varying electron concentration [Eq. (2-3)] similar to the work on $Eu_{1-x}La_xAl_2$ and $GdCo_{5-x}Al_x$ ternaries described in Chapters 4 and 10, respectively. The C15 structure exists for all values of x in $R_{1-x}Zr_xFe_2$ ternaries for R = Dy or Ho, but oddly there is a miscibility gap extending from $x = 0$ to $x = 0.8$ in the $Gd_{1-x}Zr_xFe_2$ system. Magnetization versus composition shown in Fig. 11-1 makes it clear that the antiferromagnetic coupling in $GdFe_2$ persists when Gd is replaced by Zr, a result which was unexpected. We might have expected that replacing tripositive Gd with quadripositive Zr (if indeed it is quadrivalent)

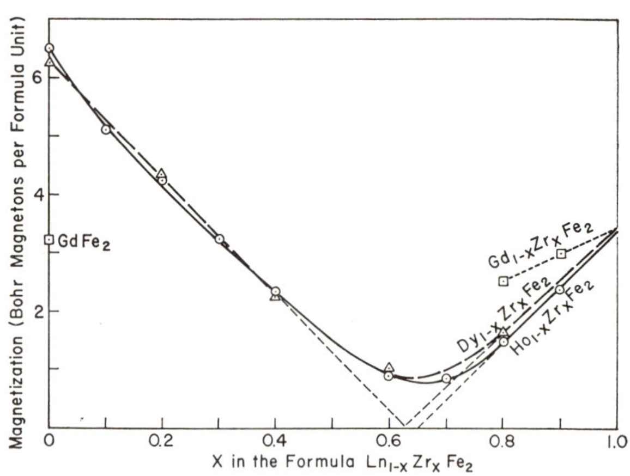

Fig. 11-1. Saturation moments at 4.2°K for $R_{1-x}Zr_xFe_2$ ternaries [31].

would have converted the ferrimagnetic material into a ferromagnetic substance. Noting that the Fe moment in $GdFe_2$ is 1.9 μ_B and in $ZrFe_2$ it is 1.6 μ_B, we conclude that electron transfer from Zr to Fe is more extensive than from Gd to Fe because of the larger number of valence electrons supplied by Zr. This suggests that as Zr replaces Gd the extra electron is absorbed by Fe maintaining a nearly constant electron concentration (e.c.). The argument is thus the same as that given earlier in Chapter 10 in the discussion of $GdCo_{5-x}M_x$ ternaries with M = Cu or Al.

Shidlovsky and Wallace reasoned [31] that Fermi surface—Brillouin zone boundary interactions are of importance in RT_2 systems (T = Fe, Co, or Ni) and in ternaries based on RT_2. Earlier work on $GdCo_{2-x}Al_x$, $ErCo_{2-x}Al_x$, and $RNi_{2-x}Al_x$ (R=all the rare earths except Ce, Eu, and Yb) ternaries showed that in each case the C15 structure becomes unstable when the e.c. reaches about 1.15. This corresponds rather well to the e.c. required for impingement of the Fermi surface on the 220 Brillouin zone boundaries. The estimate [31] of e.c. in $GdFe_2$ and $ZrFe_2$ and the $R_{1-x}Zr_xFe_2$ ternaries indicates it to be constant at about 1.1, in close agreement with the 1.15 figure for Ni and Co compounds. This suggests that as R and Fe or Zr and Fe are brought together to form the cubic Laves phase structure, Fe absorbs electrons into its d-shell to a point such that the Fermi surface is reduced in size until it fits into the zone formed by the 220 planes. As composition is varied, the Fe absorbs or releases electrons to preserve this configuration between the Fermi surface and the zone boundary, which is an energetically favorable condition. This constancy of e.c. ensures a concomitant constancy in coupling in the ternary systems.

The work on the $GdFe_{2-x}Al_x$ system is very similar to that on $GdCo_{2-x}Al_x$ reported in the preceding chapter. $GdFe_2$ dissolves 26 mole % $GdAl_2$ whereas $GdAl_2$ is able to incorporate 33 mole % $GdFe_2$, the structure in all cases being C15. The C14 (hexagonal) structure is the stable form between 27 and 52 atom % $GdAl_2$.* Magnetic characteristics are summarized in Table 11-3. Attention has already been drawn in Chapter 10 to the shrinking Fe moment at the Al content of the material increases. This is undoubtedly due to electron transfer from R and Al to Fe. By extrapolating (Fig. 10-13) we reach nonmagnetic Fe at $GdAl_{1.8}Fe_{0.2}$. The rise in ordered moment for $GdAl_2$ as Fe is introduced is similar to that observed in the corresponding Co system. Here, as in the Gd–Al–Co ternaries, the extra moment is ascribed to conduction electron polarization effects.

The $R_{1-x}Y_xFe_2$ ternaries for R = Gd, Tb, Dy, and Er have been studied by Buschow and Van Stapele [29] and by Piercy and Taylor [33]. The Curie temperature of these ternaries is observed to decrease

*DebRay and Wallace [34] have found generally similar behavior in the $ErFe_{1-x}Al_x$ system.

TABLE 11-3

Magnetic Characteristics of R_6Fe_{23} and R_2Fe_{17} Compounds

R	μ_{or} (μ_B/f.u.)	T_C (°K)	Type	Ref.
		R_6Fe_{23}		
Nd		492		[28]
Gd	14.8	468		[35]
Tb		574		[28]
Dy	15^a	524		[4,28]
Ho	15^a	501		[28]
Er	5.4^a, 7.5	493		[7,28]
Tm	18^a	475		[28]
Lu		471		[25,28]
		R_2Fe_{17}		
Ce	30.6	70	F^b	[15]
Pr		282		[36]
Nd		327		[28]
Sm		395		[28]
Gd	21.2, 22.8	472	F_i	[25,28,35,37]
Tb	16.8	408	F_i	[28,37]
Dy	15.6	363	F_i	[5,28,37]
Ho	15.0	325	F_i	[28,37]
Er	16.4, 18.0	310	F_i	[8,28,37,38]
Tm		271	F_i	[28,37]
Lu	34.7	100	F	[25]

[a] These were obtained for alloys of formula R_6Fe_{24} (i.e., RFe_4) [30].

[b] This compound in the rhombohedral form is antiferromagnetic between 70°K (T_C) and 270°K (T_N) according to Buschow and Van Wieringen [15]. Lu_2Fe_{17} behaves similarly [25] with T_N = 270°K, and perhaps other R_2Fe_{17} compounds as well.

linearly or almost so with increasing x. The former investigators report an iron moment of about 1.5 μ_B which is independent of x for systems rich in Y; they claim to find the iron moment increasing to 2.2 μ_B for small x. Clearly, calculation of the moment of Fe (from μ_{or} and the assumed moment for R) can be made only for Fe in the Gd system, since crystal field quenching renders the R moment uncertain in the other systems. The case for a large Fe moment in Gd-rich Gd–Y–Fe ternaries rests upon the observation of a rather small value of μ_{or} for these materials. For $GdFe_2$ Buschow and Stapele report 2.80 μ_B/f.u., leading to an Fe moment of 2.1 μ_B. Other studies (Table 11-2) give $\mu_{or} = 3.35$ or 3.80 μ_B/f.u. and an Fe moment of 1.82 or 1.60 μ_B. Thus, the clain of an iron moment close to that of elemental iron is not substantiated by independent investigations, which give higher and presumably more reliable values of μ_{or}.

2. RFe_3, R_6Fe_{23}, and R_2Fe_{17} Compounds

The RFe_3 and R_6Fe_{23} series have not been extensively investigated to date. Results obtained, given in Tables 11-2 and 11-3 seem to indicate the usual coupling systematics, not only on the basis of the magnitude of the moment, but also in the variation of magnetization with temperature. However, we note some difficulties with Gd_6Fe_{23}, assuming the usual coupling scheme. The ordered moment (Table 11-3) indicates an Fe moment of 1.18 or 1.47 μ_B (depending on whether the Fe sublattice magnetization is smaller or larger than that of the Gd sublattice), neither of which accords well with Fe moments observed in other R–Fe intermetallics. This could mean that the simple concept of each sublattice coupled ferromagnetically and antiferromagnetic intersublattice coupling in the R_6Fe_{23} compounds requires revision. However, in a number of cases compensation points [25,39] have been observed, lending support to the concept of antiferromagnetic coupling between the heavy rare earth and iron, and Moreau *et al.* [26] have confirmed the antiferromagnetic coupling in Ho_6Fe_{23} by neutron diffraction techniques. Ferromagnetic coupling between the light rare earths and iron is presumed to occur, but as yet corroborative evidence is lacking, since satisfactory measurements of moments of 1:3 or 6:23 light rare earth–iron compounds have not been reported.

TABLE 11-4

Magnetic Characteristics of $GdFe_{2-x}Al_x$ Ternaries[a]

x	μ_{or} (μ_B/f.u.)	μ_{eff} (μ_B/f.u.)	T_C (°K)
0.00	3.35		782
0.50	4.85		>300
1.00	6.17		264
1.63	7.18	8.00	132
1.85	8.03	7.96	138
2.00	6.95	7.96	180

[a] From Oesterreicher and Wallace [32].

Somewhat more information is available (Table 11-3) for the R_2Fe_{17} compounds due to greater ease of preparation. Strnat *et al.* [37] and Salmans [28] observed evidence of two magnetic transitions in some of the R_2Fe_{17} compounds. Recently this has been confirmed by Buschow and Van Wieringer [15] by Mössbauer spectroscopy for Ce_2Fe_{17} and by Givord *et al.* [28] for Lu_2Fe_{17} by neutron diffraction. The lutetium compound is ferromagnetic below 100°K. Between 100 and 270°K it exists in a helical structure. Results indicate [28,37] that Gd_2Fe_{17} and Tm_2Fe_{17}

may also have an antiferromagnetic (in contrast with a ferrimagnetic) form. In an attempt to clarify this and other questions, ^{57}Fe Mössbauer work has been carried out [15,36] on a number of R_2Fe_{17} compounds. Since there are four crystallographically distinct types of Fe present, the patterns are complex and, as yet, have not been successfully analyzed so as to give, for example, the hyperfine field for the different kinds of Fe. This work does, however, seem to support the concept of two magnetic transitions in the 2:17 iron compounds involving Pr, Gd, Tm, and Lu.

3. Variation in the Iron Moment

The variable cobalt moment in the rare earth intermetallics was brought out in the preceding chapter. We find a similar, but not quite so striking, phenomena in the iron compounds. In the 2:17 cobalt compounds the cobalt moment is about 1.7 μ_B, the same as in the element, and it drops off to zero in YCo_2. In Gd_2Fe_{17} the iron moment (M_{Fe}) is 2.1 μ_B; it decreases to about 1.6 μ_B in $GdFe_2$. Even in $LuFe_2$, M_{Fe} is 1.45 μ_B. We see from the difference between M_{Fe} in $GdFe_2$ and in $LuFe_2$ that M_{Fe} has an induced component. In this respect it behaves somewhat like cobalt. However, M_{Fe} contains [21] a permanent component so that $LuFe_2$, YFe_2, and $ZrFe_2$ are ferromagnetic materials whereas the corresponding Co compounds are not [40]. The difference between the behavior of Fe and Co is probably a consequence of the fact that for the materials containing cobalt one of the d half bands is filled and electron transfer from R causes a sharp decrease in uncompensated spins and hence the moment. In Fe probably both half bands are unfilled [31] but of course contain different numbers of electrons, so that electron transfer occurs into both half bands and the decrease in the number of unpaired spins is considerably reduced [25].

We note that M_{Fe} in a cerium compound is less than that of Fe in other members of the series. For example, in Ce_2Fe_{17}, M_{Fe} is 1.8 μ_B as contrasted with 2.1 μ_B in Gd_2Fe_{17}. Similarly, M_{Fe} is 1.2 μ_B in $CeFe_2$ and 1.6 in $GdFe_2$. The difference in M_{Fe} in the Laves phase also shows in the hyperfine field measured by ^{57}Fe Mössbauer spectroscopy [41]. The reduced Fe moment in the Ce compounds is ascribed to the quadrivalence of Ce and the consequent enhanced electron transfer to Fe.

The $GdFe_{2-x}Al_x$ ternaries represent the most advanced situation for electron transfer to the Fe d band (Fig. 10-13 and Table 11-4). As Al enters the Fe sublattice, the extra electrons brought in are absorbed by Fe until its moment falls to about 0.85 μ_B.

Moreau *et al.* [26,42] report a variation in iron moment between the various crystallographically distinguishable sites in Ho_6Fe_{23}. They state that 68 of the Fe atoms in the unit cell have moments of 2.4 μ_B and that the other 24 carry a moment of 1.9 μ_B.

4. Potentiality for Use in Permanent Magnets

On the basis of experience to date the rare earth–iron intermetallics appear to offer little in the way of opportunity for commercial use either in permanent magnet development or otherwise. We might *a priori* have reasoned to the contrary, considering the generally similar magnetic behavior of alloy systems containing iron and cobalt and the exciting possibilities which the rare earth–cobalt compounds represent (Chapter 10). The reasons for the difference are twofold: (1) Curie temperatures of the iron compounds are substantially lower than those of the corresponding cobalt compounds (Tables 10-2, 11-2, and 11-3), and (2) there is no iron counterpart of the technologically significant RCo_5 systems. Moreover, the light rare earth systems, in which ferromagnetic coupling is expected, either do not form or can not be formed free of extraneous phases; generally only the antiferromagnetically coupled heavy rare earth systems occur, and these have magnetizations which are insufficiently large to be of interest. Thus the outlook for applications of rare earth–iron systems seems to be quite unpromising, except possibly through ternaries involving iron and cobalt with the rare earths.

B. R–Mn COMPOUNDS

1. Binary Systems

The decrease in number of rare earth intermetallics, noted on passing from Co to Fe, continues in the progression from Fe to Mn. There are only three rare earth–Mn compounds: RMn_2, R_6Mn_{23}, and RMn_{12}. If we continue to the left in the 3d transition series we reach the elements that do not enter into chemical union with the rare earths—Sc, Ti, and V.

The RMn_2 compounds, with R = Gd through Ho, were identified in structural work by Nassau *et al.* [2] and Wernick and Geller [3] in 1960; they occur in the C15 structure. As with the R–Fe systems considerable confusion existed concerning the compositions of the Mn-rich phases for a number of years. Initially it was thought [2,43] that they corresponded to the formula RMn_5. Later Cherry and Wallace concluded [44] that the phase was more nearly RMn_4. In time Wang *et al.* established [16] the true stoichiometry of the socalled RMn_4 phase to be R_6Mn_{23}, and Kirchmayer [17], using the amalgam preparation process, identified another Mn-rich phase, RMn_{12}.

Magnetic properties of the manganese compounds are given in Table 11-5. The magnetic characteristics of these materials present many puzzling features. The Laves phases with Gd through Ho appear to become antiferromagnetic at or below liquid nitrogen temperatures, since magnetization (Fig. 11-2) passes through a maximum [47]. However, magnetizations below the maxima are dependent upon the history of the sample. If cooled in a field, the decline in magnetization at lower temperatures does not occur. These materials in this respect resemble the behavior of Tb–Y solid solutions [50] observed a number of years ago, and the mechanism postulated for these binary alloys probably holds for the RMn_2 compounds. The stable form of the RMn_2 at 4.2°K is probably an antiferromagnetic state and this develops if the

TABLE 11-5

Magnetic Characteristics of Rare Earth–Manganese Compounds

	Paramagnetic state			Ordered state				
	μ_{eff} (μ_B)	θ (°K)	Type	μ_{or} (μ_B/f.u.)	T_C (°K)	T_N (°K)	Type	Ref.
$NdMn_2$	4.9	−90						[45]
$GdMn_2$	8.86	45	C–W			86	AF	[46,47]
$TbMn_2$						40	AF	[47,48]
$DyMn_2$	10.9	−15	C–W				AF	[46,47]
$HoMn_2$							AF	[47]
$ErMn_2$	10.1	−90	C–W				F	[46,48]
$TmMn_2$							F	[48]
Nd_6Mn_{23}				9.6	437		F_i	[49]
Sm_6Mn_{23}	5.04	85	C–W,?	3.0	439		F_i	[49]
Gd_6Mn_{23}	10.04	−90	C–W,?	50.2	468		F(?)	[44,46,49]
Tb_6Mn_{23}				45.0			F_i	[49]
Dy_6Mn_{23}	11.47	25	C–W,?	51.6	443		F_i	[46,49]
Ho_6Mn_{23}	11.83	−5		55.3	434		F_i	[44,46,49]
Er_6Mn_{23}	10.56	30	C–W,?	46.2	415		F_i	[46,49]
Tm_6Mn_{23}				31.8			F_i	[49]
Lu_6Mn_{23}				9.2			F	[49]
$NdMn_{12}$	8.6	−196			135			[45]
$GdMn_{12}$	1.2	−25	C–W					[46]
$DyMn_{12}$	16.3	−398	C–W		110			[46]
$ErMn_{12}$	13.2	−65	C–W					[46]

[a] μ_{eff} is for a quantity of compound which contains one rare earth ion.

substance is cooled in the absence of a field. The antiferromagnetic structure is sufficiently stable at 4.2°K to persist when a field is applied.

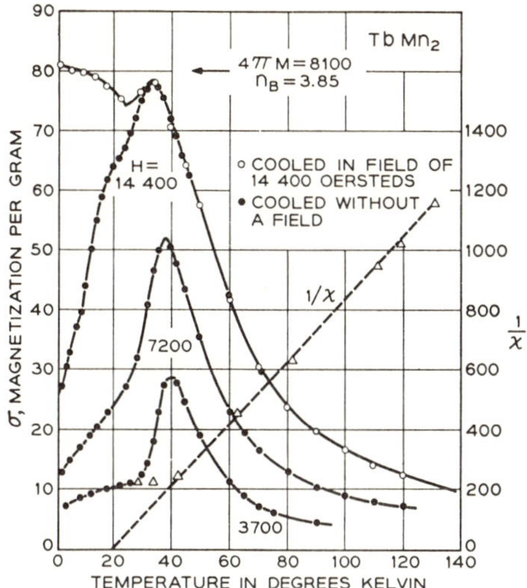

Fig. 11-2. Magnetic behavior of $TbMn_2$ [47].

However, if a field is applied before the material is cooled through its Néel temperature, it orders in a ferromagnetic state which is the stable form in an applied field. The RMn_2 compounds have not been extensively studied, so that many questions about them remain to be answered—such as, does the ferromagnetic form persist at 4.2°K if the field is taken off, etc.

Neutron diffraction work on $ErMn_2$ and $TmMn_2$ by Felcher *et al.* [51] showed these to exist in the C14 (hexagonal Laves phase) structure and to be ferromagnetic materials. The Er and Tm moments were found to be 7.72 and 4.95 μ_B, respectively, which is lower than gJ for the free ion, presumably because of the quenching effect of the crystal field. No moment was detected at the Mn site. This is consistent with the work of Marei *et al.* [52], who found no evidence of a local moment on Mn in YMn_2 by bulk magnetic measurements.

The R_6Mn_{23} compounds have received somewhat more attention than the Laves phases. Cherry and Wallace measured [44] the ordered moments (as RMn_4) of the Gd and Ho compounds in 1962, and DeSavage *et al.* [49] studied the entire series in 1965. A short time later their paramagnetic behavior was established by Kirchmayr [46].

Paramagnetic results shown for the Gd compound (Fig. 11-3) are typical. Curie-Weiss behavior is observed at temperatures well above T_C but the susceptibility increases more rapidly than expected by the C–W law as temperature is decreased. In addition there is a wide discrepancy between T_C and θ.* Curiously, Sm_6Mn_{23} exhibits C–W behavior. This indicates that the paramagnetism originates primarily with its Mn content. We can see that this is the case for, were the Sm influence dominant, C–W behavior would not occur because of the narrowness of the Sm multiplet spacing. Effective moments of the 6:23 compounds can be accounted for within about 0.5 μ_B by ascribing a moment of 2.8 μ_B to Mn. This accords with two unpaired spins and a formal valence state of -1 for manganese in these compounds.† The rapid increase in χ below 100°K may connote ionization of this anion into the Mn^0 or Mn^{+1} state brought on by the thermal contraction of the

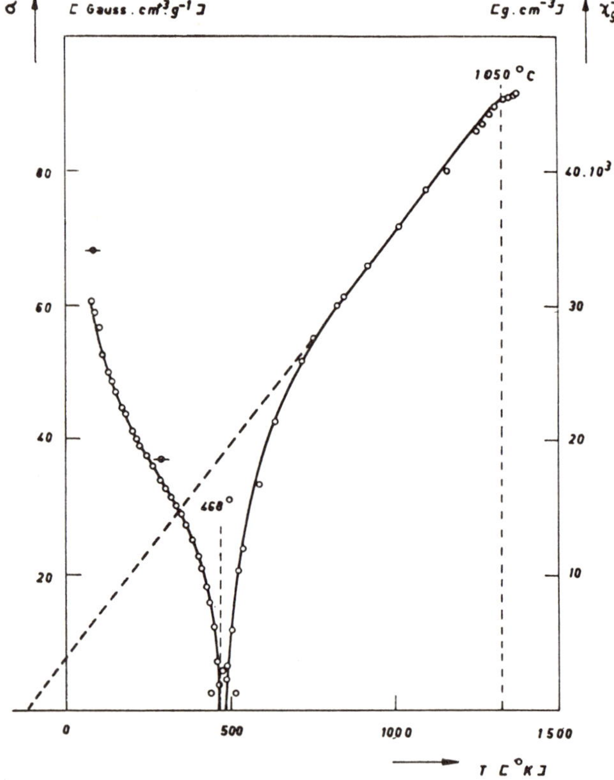

Fig. 11-3. Magnetization, σ, and reciprocal mass susceptibility, $1/\chi_g$, of Gd_6Mn_{23} versus temperature. O: saturation magnetization [46].

*Y_6Mn_{23} behaves differently. It follows the C–W law down to T_C and, moreover, for it, T_C and θ are approximately equal.

† Linearity of χ^{-1} versus T for Y_6Mn_{23} and Sm_6Mn_{23} implies a local moment description for Mn in these phases, at least for higher temperatures.

lattice. If so, ordered Mn moments in excess of 2 μ_B are expected—3 μ_B for Mn^0 and 4 μ_B for Mn^{+1}.

In their study of the 6:23 compounds in the ordered state DeSavage *et al.* reached the surprising conclusion [49] that the R–Mn coupling is antiferromagnetic for all compounds including light and heavy rare earths with the exception of Gd_6Mn_{23}. They regard the coupling to be ferromagnetic for this compound. These findings, if substantiated, would indicate that the R_6Mn_{23} compounds do not follow the coupling systematics observed in all the compounds involving Fe, Co, and Ni. The conclusion of DeSavage *et al.* could be correct but we incline toward another possibility—that the Mn moment and valence state vary throughout the series* and it is this variation which makes the coupling in these compounds seem atypical. Consider Gd_6Mn_{23} as an example. We can rationalize its observed moment, 50.2 μ_B/f.u., without postulating ferromagnetic coupling if we assume Mn^{1+} and a Mn moment of 4 μ_B. Assuming antiferromagnetic coupling of Gd^{3+} and Mn^{1+} we obtain 50 μ_B/f.u., in very close agreement with experiment.

The low moment for the Lu compound is noteworthy. It is not clear whether this signifies ferrimagnetism, a noncollinear structure, or a substantially altered valence condition for Mn. Clarification of the observed magnetic characteristics of this and the other 6:23 compounds calls for additional work—neutron diffraction, etc.

The paramagnetic moment of $ErMn_{12}$ accords well with that expected if Mn contributes 2.8 μ_B, i.e., manganese present as the anion Mn^{1-}. Those of Gd and Dy require Mn moments of 4.4 μ_B and 4.1 μ_B, respectively, corresponding approximately to Mn^{1+}. Although it is interesting to speculate as to the electronic configuration of Mn in the 1:12 compounds, it may not be very instructive to do so considering the experimental limitations. As Kirchmayr [46] has pointed out, the 1:12 compounds are difficult to obtain free of the strongly ferrimagnetic 6:23 phases. In addition, the range of temperature covered was restricted and the reliability of the paramagnetic moments are reduced accordingly.

We have drawn attention above to variable magnetic moments for Fe and Co. Manganese is a celebrated example [53] of variable valency; it is well recognized that in the metallic state several ionization states of Mn exist with comparable energy. Thus we expect even greater variation of moment for Mn than for Fe or Co. Hence, the need to invoke different moments in different series of compounds or in the paramagnetic state compared to the ferrimagnetic state does not seem to be a particularly troublesome concept in this instance.

*The structural variation in the RMn_2 series (Table 11-1) implies an analogous variation in valence state of Mn in the Laves phases.

2. Ternary Systems

The only satisfactory work on rare earth–manganese based ternary systems appears to be that due to Kirchmayr *et al.* [35,38,54] on $R_6(Mn_{1-x}Fe_x)_{23}$ systems, with R = Gd, Er, and Y. Williams *et al.* made magnetic measurements on a series of alloys of composition RFe_3Mn with R = Dy, Ho, Er, and Tm [55]. Since we now know that 1:4 stoichiometry is incorrect, their samples were probably a mixture of predominantly $R_6(Mn_{0.78}Fe_{0.18})_{23}$ and a small amount of the R_2Fe_{17} phase. Results obtained with these mixtures are of limited significance but they did show effects similar to those observed for RMn_2 compounds (R = Gd through Ho) and Tb–Y solid solutions (see above): namely, a magnetization at low temperatures ($T < 40°K$) which varied appreciably with the history of the sample, depending upon whether or not it was cooled in a field.

The work of Kirchmayr and associates has revealed some truly extraordinary features. Consider the Gd_6Mn_{23}–Gd_6Fe_{23} ternaries. Within experimental error these two compounds have identical Curie temperatures, 468°K. We might reasonably expect the ordering temperature to be unmodified for the ternaries. However, experiment shows that incorporation of either of these compounds into the other leads to a precipitous drop in the Curie point (Fig. 11-4). The Er system behaves

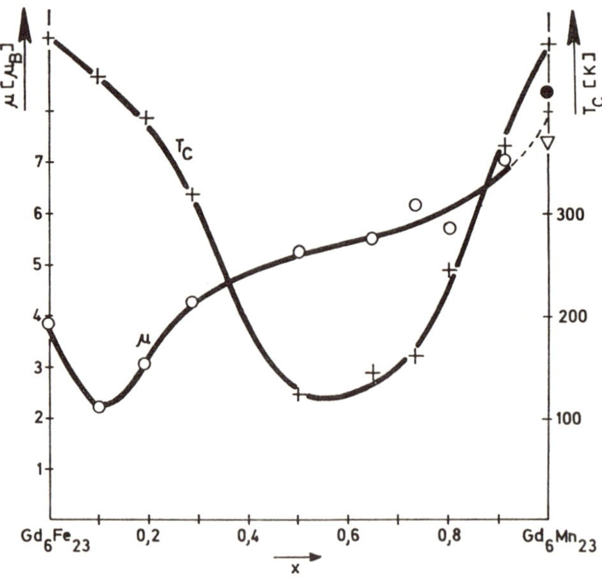

Fig. 11-4. Variation of ordered moment and T_C for Gd–Fe–Mn with composition [35]. Used with permission of Masson et Cie, Paris.

[38] similarly (Fig. 11-5). Even more striking, although generally similar, behavior is noted for $Y_6(Fe_{1-x}Mn_x)_{23}$ ternaries (Fig. 11-6). They give no evidence of ordering in alloys ranging from 48 to 68 mole % Y_6Mn_{23}. It is clear that these ternaries are burdened with many complexities and interpretation of the observation is not at all straightforward. As noted above, there appear to be two iron species in Ho_6Fe_{23} from a magnetic point of view, as well as four crystallographically distinguishable kinds of iron. This opens the possibility of very considerable diversity of coupling modes between the several sublattices and the three magnetic species present.

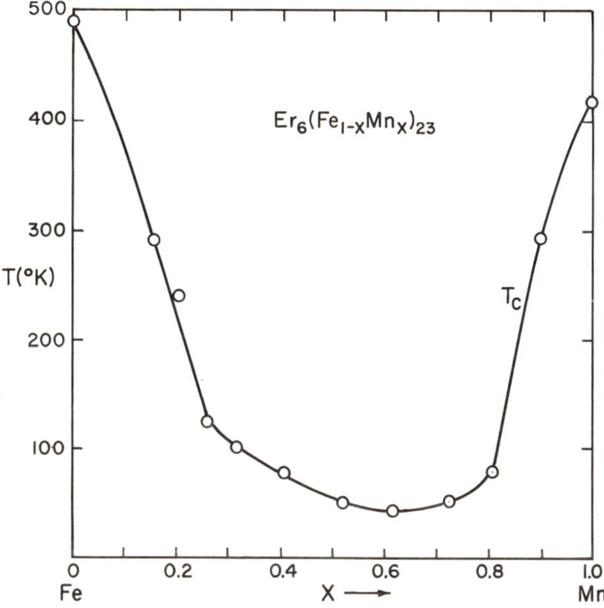

Fig. 11-5. Curie temperatures of Er–Fe–Mn ternaries [38].

It is noteworthy that Curie–Weiss behavior seems to obtain for the binary 6:23 systems but not for the ternaries in which Fe and Mn are present in nearly equal proportions. This may imply a significantly different band structure for the Fe and Mn compounds. If so, the rapid fall in T_C with introduction of a third component could be taken to imply the disruption of the band structure by introduction of the foreign atom—that is the introduction of Mn in Gd_6Fe_{23} or of Fe in Gd_6Mn_{23}.

Kirchmayr [45] finds a much weaker dependence of the magnetic characteristics on composition for Gd_6Mn_{23} when Y replaces Gd than

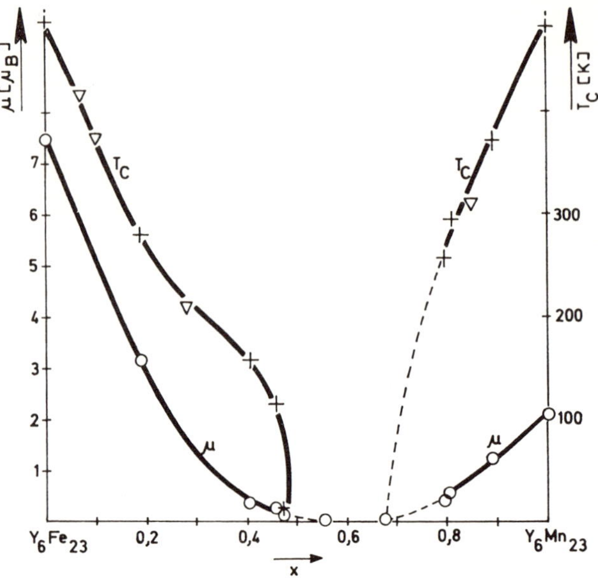

Fig. 11-6. Magnetization of Y–Fe–Mn ternaries [35]. Used with permission of Masson et Cie, Paris.

when Fe replaces Mn (Table 11-6). This result, like the others obtained with those ternaries, is quite unexpected. Obviously additional information will be required to elucidate the 6:23 ternaries. Neutron diffraction studies and density of states determinations should prove particularly valuable in resolving the interesting questions raised by the studies of Kirchmayr and his associates.

TABLE 11-6

Magnetic Characteristics of Gd_6Mn_{23}–Y_6Mn_{23} Ternaries[a]

	θ (°K)	μ_{eff} (μ_B/f.u.)[a]	T_C (°K)
Gd_6Mn_{23}	−90	10.0	468
$(Gd_{0.48}Y_{0.52})_6Mn_{23}$	30	8.9	468
Y_6Mn_{23}	475	2.35	486

[a] From Kirchmayr [45].
[b] The formula unit in this case is $RMn_{3.83}$ or $Gd_{0.48}Y_{0.52}Mn_{3.83}$.

REFERENCES

1. F. Endter and W. Klemm, *Z. Anorg. Chem.* **252,** 64 (1943).
2. K. Nassau, L. V. Cherry, and W. E. Wallace, *J. Phys. Chem. Solids* **16,** 123 (1960).
3. J. H. Wernick and S. Geller, *Trans. AIME* **218,** 866 (1960).
4. A. S. Van der Goot and K. H. J. Buschow, *J. Less Common Metals* **21,** 151 (1970).
5. H. R. Kirchmayr, *Z. Metallk.* **60,** 699 (1969).
6. J. H. Wernick and S. E. Haszko, *J. Phys. Chem. Solids* **18,** 352 (1961).
7. K. H. J. Buschow and A. S. Van der Goot, *Phys. Status Solidi* **35,** 515 (99).
8. J. H. Van Vucht, *J. Less Common Metals* **10,** 146 (1965).
9. G. I. Hoffer and L. R. Salmans, *Proc. Rare Earth Conf., Coronado, California* p. 371 (October, 1968).
10. P. I. Kripyakevich and D. P. Frankevich, *Kristallographia* **10,** 560 (1965).
11. P. I. Kripyakevich, D. P. Frankevich, and Y. V. Voroshilov, *Porosch. Met.* **55,** (1965).
12. K. H. J. Buschow, *J. Less Common Metals* **11,** 204 (1966).
13. Q. Johnson, D. H. Wood, G. S. Smith, and A. E. Ray, *Acta Crystallogr.* **B24,** 274 (1968).
14. F. Givord and R. Lemaire, *J. Less Common Metals* **21,** 463 (1970).
15. K. H. J. Buschow and J. S. Van Wieringen, *Phys. Status Solidi* **42,** 23 (1970).
16. F. E. Wang, J. V. Gilfrich, D. W. Ernst, and W. M. Hubbard, *Acta Crystallogr.* **17,** 931 (1964).
17. H. R. Kirchmayr, *Z. Kristallogr.* **124,** 152 (1967).
18. J. O. Jepson and P. Duwez, *Trans. Amer. Soc. Metals* **47,** 543 (1955).
19. L. V. Cherry and W. E. Wallace, *J. Appl. Phys. Suppl.* **32,** 340S (1961); *J. Appl. Phys.* **33,** 1619 (1962).
20. E. A. Nesbitt, J. H. Wernick, and E. Corenzwit, *J. Appl. Phys.* **30,** 365 (1959).
21. W. E. Wallace and E. A. Skrabek, *in* "Rare Earth Research," (K. S. Vorres, ed.), Vol. 2, p. 431. Gordon and Breach, New York, ,1964.
22. M. Mansmann and W. E. Wallace, *J. Chem. Phys.* **40,** 1167 (1964).
23. J. Farrell and W. E. Wallace, *J. Chem. Phys.* **41,** 1524 (1964).
24. J. Crangle and J. W. Ross, *Proc. Int. Conf. Nottingham* p. 240. *Inst. of Physics and the Phys. Soc., London* (1964).
25. D. Givord, F. Givord, and R. Lemaire, *J. Phys. Suppl.* **32,** C 1–668 (1971).
26. J. M. Moreau, C. Michel, M. Simmons, T. O'Keefe, and W. J. James, *Proc. Rare Earth Conf. 8th, Reno, Nevada* p. 34 (April 1970); see also *J. Phys. Suppl.* **32,** C 1–670.
27. H. J. Williams and E. A. Nesbitt, quoted in G. K. Wertheim and J. H. Wernick, *Phys. Rev.* **125,** 1937 (1962).
28. L. R. Salmans, Tech. Rep. from Wright-Patterson Air Force Base, AFML-TR-68-159 (1968).
29. K. H. J. Buschow and R. P. Van Stapele, *J. Appl. Phys.* **41,** 4066 (1970).
30. M. O. Bargouth and G. Will, *J. Phys. Suppl.* **32,** C 1–675 (1971).
31. I. Shidlovsky and W. E. Wallace, *J. Solid State Chem.* **2,** 193 (1970).
32. H. Oesterreicher and W. E. Wallace, *J. Less Common Metals* **13,** 91 (1967).
33. A. R. Piercy and K. N. R. Taylor, *J. Appl. Phys.* **39,** 1096 (1968).
34. D. DebRay and W. E. Wallace, unpublished measurements.
35. H. R. Kirchmayr and W. Steiner, *J. Phys. Suppl.* **32,** C 1–665 (1971).
36. L. M. Levinson, E. Rosenberg, A. Shanlov, S. Shtrikman, and K. Strnat, *J. Appl. Phys.* **41,** 910 (1970).
37. K. Strnat, G. Hoffer, and A. E. Ray, *IEEE* **Mag-2,** 489 (1966).

38. G. Hilscher, H. R. Kirchmayr, and W. Steiner (in press).
39. E. A. Nesbitt, H. J. Williams, J. H. Wernick, and R. C. Sherwood, *J. Appl. Phys.* **33,** 1674 (1962).
40. A. W. Abel and R. S. Craig, *J. Less Common Metals* **16,** 77 (1968).
41. W. E. Wallace, *J. Chem. Phys.* **41,** 3857 (1964).
42. J. M. Moreau, C. Michel, M. Simmons, T. J. O'Keefe, and W. J. James, *J. Phys. Suppl.* **32,** C 1–670 (1971).
43. H. R. Kirchmayr and K. H. Schindl, *Z. Angew. Phys.* **19,** 517 (1965).
44. L. V. Cherry and W. E. Wallace, *J. Appl. Phys.* **33,** 1619 (1962).
45. H. R. Kirchmayr, *Z. Angew. Phys.* **27,** 18 (1969).
46. H. R. Kirchmayr, *IEEE Trans. Magnet.* **Mag-2,** 493 (1966).
47. E. A. Nesbitt, H. J. Williams, J. H. Wernick, and R. C. Sherwood, *J. Appl. Phys.* **34,** 1347 (1963).
48. L. M. Corliss and J. M. Hastings, *La Diffraction et la diffusion des neutrons. Coll. Int. C.N.R.S., Grenoble* **126,** p. 133 (1963).
49. B. F. DeSavage, R. M. Bozorth, F. E. Wang, and E. R. Callen, *J. Appl. Phys.* **36,** 992 (1965).
50. S. Weinstein, R. S. Craig, and W. E. Wallace, *J. Chem. Phys.* **39,**1449 (1963).
51. G. P. Felcher, L. M. Corliss, and J. M. Hastings, *J. Appl. Phys.* **36,** 1001 (1965).
52. S. A. Marei, R. S. Craig, W. E. Wallace, and T. Tsuchida, *J. Less Common Metals* **13,** 391 (1967).
53. W. Hume-Rothery and G. V. Raynor, "The Structure of Metals and Alloys," pp. 44, 81, Inst. of Metals, London, 1956.
54. H. R. Kirchmayr, "Habilitations schrift." Technische Hochschule, Vienna, 1968.
55. H. J. Williams, R. C. Sherwood, J. H. Wernick, and E. A. Nesbitt, *J. Appl. Phys.* **35,** 1032 (1964).

Chapter 12

COMPOUNDS WITH 4d AND 5d TRANSITION METALS

Rare earths form fewer compounds with the 4d and 5d transition metals than with the transition metals in the first long period and the compounds which do form have been less extensively studied. The lack of attention appears to be partly a matter of the costliness of the Pt and Pd group metals and partly due to their nonmagnetic character. It is the magnetic nature of Mn, Fe, Co, and Ni which makes their intermetallics with the rare earths so intriguing both from a fundamental and a practical point of view. This feature is missing in phases involving 4d and 5d metals.

Stoichiometries observed are RT_2, RT_3, and RT_5, where T represents one of the 4d or 5d metals. The RT_2 compounds are [1] Laves phases, either the C14 or C15 type. The RT_3 phases have [2] the $AuCu_3$ structure and the RT_5 series form [3] in the $CaCu_5$ structure. Other compounds reported [4] are Gd_5Pd_2, GdPd, $EuPd_2$, GdPt, DyPt, LuRh, Dy_2Pt_3, Gd_2Pt_3, La_6Ru, and Ce_6Ru. However, magnetic information has been reported only for the first three of this latter rather miscellaneous group of compounds.

A. RT_2 COMPOUNDS

1. Magnetic Characteristics

Work by Bozorth *et al.* [5] on the RRu_2, ROs_2, and RIr_2 series in 1959 is of the same vintage as the early work on the rare earth–cobalt compounds dealt with in Chapter 10. This work was followed by the study of Crangle and Ross [6] of the RRh_2 and RPt_2 series and by the Vlasov and Wallace study [7] of the platinum compounds in the paramagnetic state. The latter represents the only instance in which examination has been made of the paramagnetic behavior of compounds involving the rare earths and 4d and 5d transition metals. Results are summarized in Tables 12-1–12-3.

Bozorth and his associates tacitly assumed that the transition metal is without magnetic moment and hence the magnetic characteristics of these systems—their ordered moments and ordering temperatures—are determined by the rare earth component. Subsequent work has confirmed this, although the interpretation of μ_{or} is ambiguous since it is generally less than gJ, the value expected for R^{3+}. The origin of the deficiency was for a time unclear—whether it was a consequence of crystal field quenching, as initially postulated by Bozorth *et al.*, or of ferrimagnetism. The ambiguity was not removed until considerably later when Felcher and Koehler [11] examined $TbIr_2$ and $HoIr_2$ by neutron diffraction means and established the reduced R moments and the lack of an Ir moment. Additional support was provided by the observation that the paramagnetic moment of the platinum com-

TABLE 12-1

Magnetic and Structural Characteristics of Rare Earth Compounds with Ru and Rh

R	Structure	T_C (°K)	μ_{or} (μ_B/f.u.)	Type	Ref.
			RRu_2		
Pr[a]	C15	40(?)	0.61	?	[1,5]
Nd	C15	35		F	[1,5]
Sm	C15				
Gd	C14	83	6.35	F	[1,5]
Tb					
Dy	C14				
Ho	C14				
Er	C14	13		F	[1,5]
			RRh_2		
Pr[a]	C15	8.6	2.24	F(?)	[1,6]
Nd	C15	~6	1.67	F	[1,6]
Sm	C15	~22	0.53	F	[6]
Gd	C15	~75	6.80	F	[1,6]
Tb	C15	~40	7.07	F	[6]
Dy	C15	~35	8.16	F	[6]
Ho	C15	~16	7.74	F	[6]
Er	C15	~7	7.26	F	[6]

[a] $PrRu_2$ and $PrRh_2$ may be Van Vleck paramagnets and hence their T_C and μ_{or} may be without significance (see text).

TABLE 12-2

Magnetic and Structural Characteristics of Rare Earth Compounds with Os and Ir

R	Structure	T_C (°K)	μ_{or} (μ_B/f.u.)	Type	Ref.
			ROs_2		
Ce	C15		0	P	[1,5]
Pr[a]	C14,C15	~28(?)	1.46	?	[1,5]
Nd	C14	23	1.34	F	[1,5]
Sm	C14	34		F	[1]
Gd	C14	66	6.71	F	[1,5]
Tb	C14	34	7.31	F	[5]
Dy	C14	15	6.73	F	[5]
Ho	C14	9	6.0	F	[5]
Er	C14	3	5.31	F	[5]
			RIr_2		
Ce	C15		0	P	[1,5]
Pr[a]	C15	18.5(?)	1.76	?	[1,5]
Nd	C15	12	1.47	F	[1,5]
Sm	C15	37	0.24	F	[5,9]
Gd	C15	89	6.83	F	[1,5,8]
Tb	C15	45	6.95	F	[5]
Dy	C15	23	7.65	F	[5,9]
Ho	C15	12	7.45	F	[5]
Er	C15	3	6.10	F	[5]
Tm	C15	1	2.92	F	[5]
Yb	C15	~0	1.70	F	[5]

[a] $PrOs_2$ and $PrIr_2$ may be Van Vleck paramagnets and hence their T_C and μ_{or} may be without significance (see text).

TABLE 12-3

Magnetic and Structural Characteristics of RPt_2 Compounds

	Paramagnetic state			Ordered state			
R	θ (°K)	μ_{eff} (μ_B/R)	Type	μ_{or} (μ_B/f.u.)	T_C (°K)	Type	Ref.
Ce	−4	2.33			1.7	AF	[13,14]
Pr[a]	0	3.43	VV(?)	1.67	13.5(?)	?	[1,6,7]
Nd	0	3.60	C–W	2.11	10	F	[1,6,7]
Eu		7.9			105	F	[10]
Gd	32	8.10	C–W	6.77	~50	F	[1,6,7]
Tb	17	9.61	C–W	7.08	26	F	[6,7]
Dy	7	10.6	C–W	7.34	25	F	[6,7]
Ho	2	10.6	C–W	8.88	19	F	[6,7]
Er	1	9.5	C–W	7.48	15	F	[6,7]

[a] $PrPt_2$ may be a Van Vleck paramagnet and hence its T_C and μ_{or} may be without significance (see text).

pounds is in close agreement with the effective moment of the rare earth component (Table 12-3). Thus the 4d and 5d transition metals behave in the Laves phases in a fashion similar to nickel.

The lack of transition metal moment is also suggested by the Curie temperatures which never exceed 85°K and hence are comparable with T_C for the corresponding nickel compounds. We note from the discussion of the Fe and Co compounds, in which both species are magnetic, that the ordering temperatures are substantially higher.

We recall that initially $PrNi_2$ was regarded as a ferromagnet with the Pr moment rather strongly quenched by the crystal field. Subsequently Mader and Wallace established [12] (Chapter 9) that at liquid helium temperatures this material is a Van Vleck paramagnet. It seems likely that the several PrT_2 compounds, with the possible exception of $PrRh_2$, are also Van Vleck paramagnets. Explicit attention has been drawn [7] to the questionable nature of T_C for $PrPt_2$: "Thus the use of the data to derive a Curie temperature, while formally indicating a transition temperature of reasonable magnitude, may actually be unwarranted." Magnetization versus field data for $PrPt_2$ shown in Fig. 12-1 are very similar to the corresponding data for $PrNi_2$ (Fig. 9-3) strongly suggesting that the platinum compound at 4.2°K is also a Van Vleck paramagnet. However, heat capacity data are not available to permit analysis similar to that made for the nickel compound. T_C obtained for $PrPt_2$ exceeds that for $NdPt_2$, which is surprising, since on the basis of the de Gennes function we expect T_C for the Nd com-

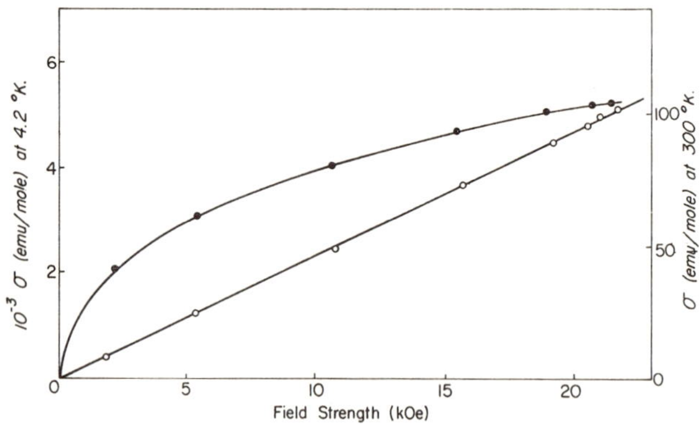

Fig. 12-1. Magnetization (σ) versus applied field for $PrPt_2$ at 4.2°K (●) and 300°K (O) [7]. Reprinted from *Inorg. Chem.* **7**, 2216 (1967). Copyright 1967 by the American Chemical Society. Reprinted by permission of the copyright owner.

pound to be higher by a factor of about 2.5. A similar inversion holds for the Ru, Rh, Os, and Ir compounds, to which attention was drawn by Bozorth *et al.* [5]: "The Curie points of the Pr compounds are consistently above those for Nd, though the theory as presently worked out gives no reason for this." These several features make it appear that the assumption of an ordering in the Pr compounds may be unwarranted. The T_C and μ_{or} values reported in the literature and listed in Tables 12-1–12-3 may be nothing more than artifacts of the method used in treatment of the data.

As noted above, $PrRh_2$ may be an exception to this generalization. Its moment is so large as to not easily be reconciled with Van Vleck paramagnetism. Perhaps in this case exchange is sufficiently strong to generate ferromagnetism by the "bootstrap" process.

We note that, except for $EuPt_2$, T_C is maximal in all cases for the Gd compound. The Curie temperatures scale roughly in accord with the de Gennes function. This observation, to which Bozorth *et al.* drew attention, appears to be the first instance in which the de Gennes treatment of exchange (see Chapter 2) was applied to rare earth intermetallics. The exceptional nature of $EuPt_2$ is undoubtedly a consequence of the divalent state of europium, which is in contrast with the trivalency of R in the other RT_2 compounds.

2. Heat Capacities

Joseph and Gschneidner [13] measured the heat capacities of $LaPt_2$, $CePt_2$, $LaRu_2$, and $CeRu_2$ from 1.4 to 8°K. For the La compounds C_p showed the normal temperature dependence,

$$C_p = \gamma T + \beta T^3 \tag{12-1}$$

with γ (mJ mole^{-1} °K^{-2}) equal to 1.17 for $LaPt_2$ and 13.6 for $LaRu_2$, and β (mJ mole^{-1} °K^{-4}) equal to 0.149 and 0.476. λ-type thermal anomalies were noted at 1.7°K for $CePt_2$ and at 5.9°K for $CeRu_2$. The latter is ascribed to the development of superconductivity whereas the $CePt_2$ anomaly is interpreted as a Néel point. The magnitude of the ordering temperature is close to that expected from the de Gennes function, namely 1°K, but the development of antiferromagnetic order is somewhat unexpected in view of the ferromagnetic nature of the other RPt_2 compounds. The situation thus resembles the Al Laves phase compounds (Chapter 4) in which $CeAl_2$ is also antiferromagnetic, whereas the others (except for $EuAl_2$, which is a special case) are ferromagnetic.

Joseph and Gschenidner estimate the entropy associated with the $CePt_2$ transition to be about $R \ln 1.75$, indicating that the ordering occurs in the doublet crystal field state. In this respect $CePt_2$ also resembles $CeAl_2$.

The value of γ for $CeRu_2$ is estimated to be 23.3 mJ mole^{-1} °K^{-2}. This is larger than γ for $LaNi_2$ (see Chapter 9) by about a factor of 2 whereas γ for $LaPt_2$ is an order of magnitude smaller. The low γ value for $LaPt_2$ is surprising; it is only 70% larger than the γ of Cu and on a per atom basis it is smaller by almost a factor of 2. This indicates a filled (or nearly filled) 5d band in $LaPt_2$, presumably by transfer of electrons from La. The large γ for $CeRu_2$ is in accordance with theoretical predictions by Coqblin [15] for Ce systems in which the f electron energies are close to the Fermi surface [14].

B. THE RPd_3 COMPOUNDS

1. Magnetic Characteristics

Magnetic results for these compounds have been published by Gardner *et al.* [16] and Hutchens *et al.* [17]. Exchange is weak in these materials and there is no indication of ordering for most of them. Gardner *et al.* report maxima in the susceptibility for $GdPd_3$ and $TbPd_3$ at 7.5 and 2.0°K, respectively, from which they infer antiferromagnetic ordering. Hutchens *et al.*, on the contrary, find evidence of ferromagnetic ordering in $GdPd_3$ with $T_C = 7$°K. The magnetization–temperature curve for $GdPd_3$ is typical of a ferromagnet (Fig. 12-2) and the magnetization–field curve at 4°K shows the saturation effects

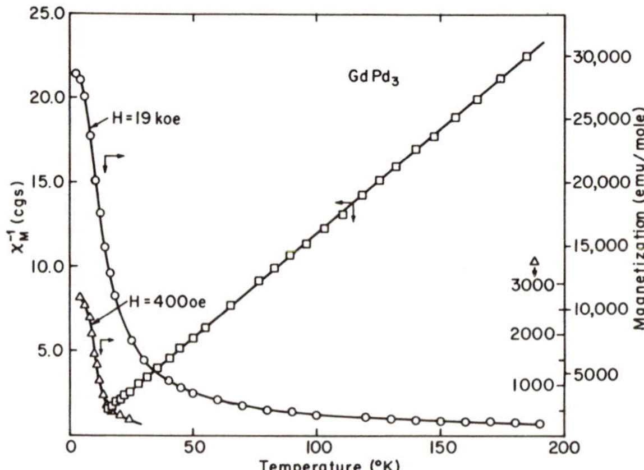

Fig. 12-2. Magnetization and inverse susceptibility data for $GdPd_3$ [17].

expected of a ferromagnet instead of being concave upward, which is expected if it were antiferromagnetic. The onset of ordering is readily apparent in the resistivity behavior of $GdPd_3$. There is a sharp decline in ρ at about 7°K associated with loss of the spin-disorder scattering when the ordered phase forms (Fig. 12-3).

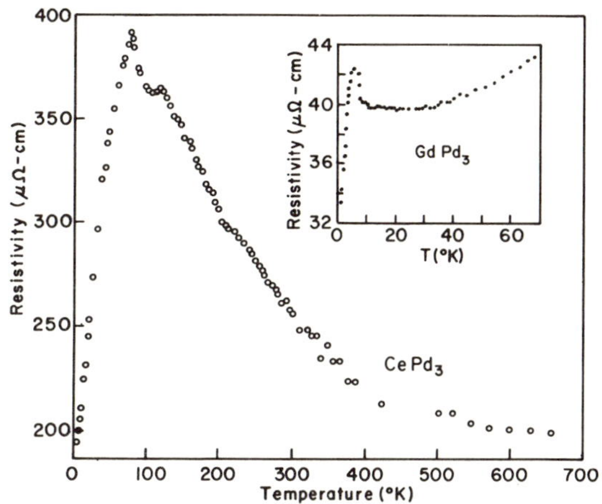

Fig. 12-3. Resistivity of $CePd_3$ and $GdPd_3$ [17].

Pertinent data for the RPd_3 phases are accumulated in Table 12-4, which is based primarily on the work of Hutchens *et al.* since Gardner *et al.* provided little in the way of quantitative information. $LaPd_3$ and $LuPd_3$ are Pauli paramagnets, confirming that Pd is nonmagnetic in these phases. Most of the remainder of the compounds exhibit Curie–Weiss behavior. Data for $YbPd_3$ (Fig. 12-4) are representative. Paramagnetic moments are in reasonable agreement with values expected for an assemblage of tripositive ions. Weiss constants are small, in agreement with the findings of Gardner *et al.* ($\theta < 5$°K), again implying weak magnetic interactions. The Ce, Sm, and Eu compounds do not exhibit Curie–Weiss behavior (Fig. 12-4). For the latter two materials this is because of the narrowness of the multiplet spacings. Gardner *et al.* have accounted quantitatively for the temperature dependence of χ for $SmPd_3$ in terms of the known Sm^{3+} multiplet structure. They took into account the decomposition of the sixfold degenerate ground state multiplet by the crystal field. Their analysis led to the conclusion that the Γ_8 (quartet) state is lowest lying, in accordance with expectations based on point charge calculations. An overall crystal field splitting of 100°K was indicated.

TABLE 12-4

Magnetic Characteristics of the RPd$_3$ Phases[a]

	Paramagnetic state			Ordered state			
	θ(°K)	μ_{eff} (μ_B/f.u.)	Type	μ_{or} (μ_B/f.u.)	T_C (°K)	Type	Ref.
La			P				[16]
Ce			N,K				[17]
Pr	−7	3.4	C–W				[17]
Nd	1	3.4	C–W				[17]
Eu			N,m	5.4			[16,17]
Gd	3	8.0	C–W		7	F	[17]
Sm			N,m				[16,17]
Tb	3	9.3	C–W		2.0[b]	AF	[16,17]
Dy	2	10.1	C–W				[17]
Ho	4	9.3	C–W				[17]
Er	0	9.5	C				[17]
Tm	−1	7.5	C–W				[17]
Yb	0	4.3	C				[17]
Lu			P				[16]

[a] All compounds occur in the AuCu$_3$ structure.
[b] Néel temperature.

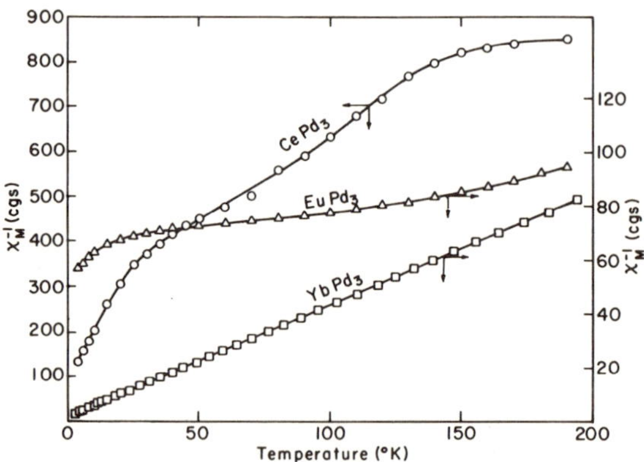

Fig. 12-4. Inverse susceptibility data for RPd$_3$ compounds [17].

Surprisingly Eu in EuPd$_3$ is trivalent. Usually Eu in intermetallics is divalent due to the special stability of the half-filled 4f shell, alluded to earlier in Chapter 4. Hutchens *et al.* ascribed the Eu trivalency in this case to strong tendency of Pd to fill its d band. Evidently in this case the special stability of the half-filled f shell is offset by the special stability of the filled d shell.* Since Eu^{3+} has $L=3$, $S=3$, and $J=0$, its susceptibility is exclusively of the Van Vleck type—the admixture of $^7F_1$, $^7F_2$, etc. into the ground $^7F_0$ state produced by the application of a field. Gardner *et al.* have accounted for the χ versus T behavior of $EuPd_3$ (except for the rise in χ below 50°K, which is regarded as an impurity effect) in terms of the known multiplet structure of Eu^{3+}. The $^7F_1$ and $^7F_2$ states are found to be 477 and about 1300°K, respectively, above the $^7F_0$ state.

$CePd_3$ represents a more complex situation. Its susceptibility behavior is strongly influenced by spin compensation (Kondo phenomenon). The value of χ is well below that expected from Curie's law for Ce^{3+}—by factors of about 4 and 20 at 300 and 4.2°K, respectively. The reduction is greater than can be ascribed to the crystal field interaction, which can reduce χ from the Curie law value at most by a factor of 5/21. (See, in Chapter 4, the discussion of crystal field effects on $CeAl_2$.) Resistivity measurements (Fig. 12-3) show a resistivity [17] which passes through a maximum (or possibly two maxima) and continues to decrease at temperatures well above room temperature. The Kondo minimum is above 650°K. The astonishingly high temperature is characteristic [18–20] of concentrated Kondo systems. The study of the resistivity behavior of $Ce_xLa_{1-x}Pd_3$ ternaries offers confirmation that $CePd_3$ is a Kondo system [20].

$YbPd_3$ exhibits behavior (Fig. 12-4) expected for an assemblage of tripositive ions.

The ordered moment for $GdPd_3$ is surprisingly low implying complexities may exist with this material which remain to be elucidated.

2. Heat Capacity Studies

Hutchens *et al.* [23] have published results of heat capacity measurements in the liquid helium range on $CePd_3$, $LaPd_3$, and solid solutions of these two compounds. The compounds and the solid solutions all form in the $AuCu_3$ structure. Although the lattice parameters of $LaPd_3$ and $CePd_3$ differ by less than 3%, the two substances are not miscible in all proportions. There is a two phase region separating the two regions of primary solid solubility and extending from about 0.3 to 0.5 mole fraction $CePd_3$.

Heat capacity data obtained on single phase materials are shown in Figs. 12-5 and 12-6. We see in most cases that the normal behavior, i.e., C/T linear with T^2, is not exhibited. Hutchens *et al.* ascribed the nonlinearity to spin fluctuation effects such as described by Doniach and

*Yb in $YbPd_3$ is also trivalent and for the same reasons. Wickman *et al.* [10] find by Mössbauer measurements that Eu is also trivalent in $EuRh_2$.

Englesberg [24]. They proposed that this would add a contribution to the specific heat proportional to $T^3 \ln T$, making the total expression

$$C/T = \gamma T + \beta T^2 + 2bT^2 \ln (T/T_s), \qquad (12\text{-}2)$$

where T_s is the characteristic spin fluctuation temperature and b is a constant. This can be put into the form

$$C/T = \gamma + aT^2 + 2bT^2 \ln T \qquad (12\text{-}3)$$

with $a = \beta - 2b \ln T_s$. Of course, a may be positive or negative, or it may vanish. Hutchens *et al.* fit their data to an equation of the form (12-3) with the results shown as the solid line in Figs. 12-5 and 12-6.

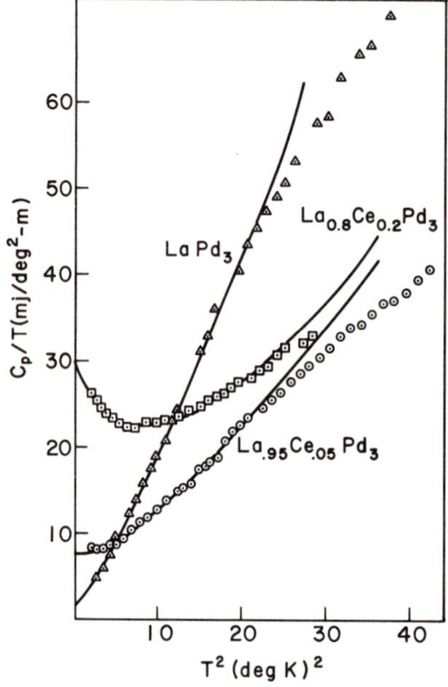

Fig. 12-5. C_p/T versus T^2 curves for $LaPd_3$ and for $La_{0.95}Ce_{0.05}Pd_3$ and $La_{0.8}Ce_{0.2}Pd_3$ [23].

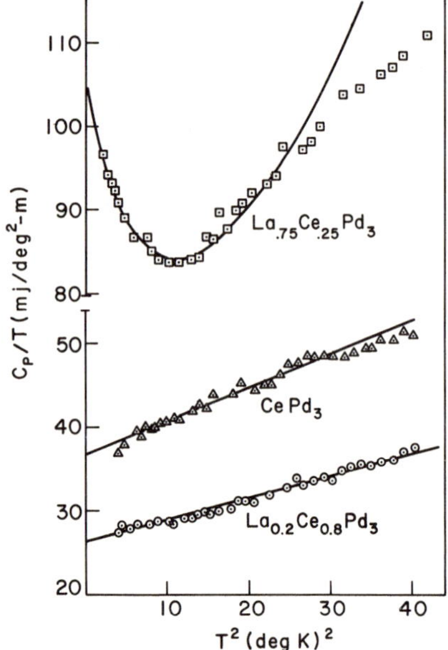

Fig. 12-6. C_p/T versus T^2 curves for $CePd_3$ and for $Ce_{0.8}La_{0.2}Pd_3$ and $La_{0.75}Ce_{0.25}Pd_3$ [23].

The γ values and a and b parameters obtained are shown in Table 12-5.

Hutchens *et al.* ascribe the several unusual aspects of this system, and their very odd transport properties, which lie outside the scope of this monograph, to the proximity of the 4f level to the Fermi energy of the

TABLE 12-5

Specific Heat Parameters for $Ce_{1-x}La_xPd_3$ Ternaries[a]

	γ (mJ/deg^2 m)	a (mJ/deg^4 m)	b (mJ/deg^4 m)
$LaPd_3$	1.674	0.808	0.412
$La_{0.95}Ce_{0.05}Pd_3$	7.96	−0.315	0.346
$La_{0.9}Ce_{0.1}Pd_3$	13.11	−1.590	0.637
$La_{0.8}Ce_{0.2}Pd_3$	30.32	−2.806	0.894
$La_{0.75}Ce_{0.25}Pd_3$	107.9	−7.052	2.063
$La_{0.2}Ce_{0.8}Pd_3$	28.26	0.084	0
$CePd_3$	36.64	0.404	0

[a] From Hutchens *et al.* [23].

system. They believe the 4f level to be first below the Fermi level in solutions dilute in Ce and to rise monotonically with increasing Ce content until the $CePd_3$ the 4f level is well above the Fermi limit. Thus Ce in $CePd_3$ behaves as a quadrivalent ion.

The small value of γ for $LaPd_3$ is noteworthy (cf. with that for $LaPt_2$ above). The rise in γ with Ce content is ascribed to enhancement effects and to the well-recognized increase in density of states when the 4f level is near the Fermi energy [15].

C. OTHER COMPOUNDS

$EuPd_2$ (C15 structure) has been studied by Wickman *et al.* [10] by bulk magnetic measurements and Mössbauer spectroscopy. In this compound Eu, unlike Eu in $EuPd_3$, is divalent. Its magnetic characteristics are listed in Table 12-6 along with those of GdPd and the compounds of

TABLE 12-6

Magnetic Characteristics of Miscellaneous Rare Earth–Palladium Compounds

	Paramagnetic state			Ordered state				
	θ (°K)	μ_{eff} (μ_B/R)	Type	μ_{or} (μ_B/f.u.)	T_C (°K)	T_N (°K)	Type	Ref.
$EuPd_2$	3	7.8			80		F	[10]
Gd_5Pd_2	333	7.6	C–W	34.5	335		F	[21]
$Tb_{5.10}Pd_{1.90}$	50	9.7	C–W		~30	62	F,AF	[21]
$Dy_{5.07}Pd_{1.93}$	45	10.7	C–W		~25	41	F,AF	[21]
$Ho_{5.04}Pd_{1.96}$	33	10.6	C–W		~10	27	F,AF	[21]
GdPd	40	8.31	C–W		39.5		F	[22]

formula R_5Pd_2. The 5:2 compounds, except for Gd_5Pd_2, show two magnetic transitions, the lower one corresponding to the onset of ferromagnetism. Above T_C probably an antiferromagnetic form exists for a range of temperature before magnetic order is destroyed. Gd_5Pd_2 is noteworthy in that its Curie temperature exceeds that of elemental Gd. Berkowitz *et al.* report [21] very high energy products (~25 million GOe) for the Tb and Dy compounds.

Pierre and Siaud have studied GdPd and established [22] that (1) its structure is FeB and (2) it becomes ferromagnetic at low temperatures. Pd is clearly nonmagnetic in the compounds listed in Table 12-5. Pierre and Siaud interpret their results on GdPd to indicate that in this compound the Pd 4d shell has been filled.

Vijayaraghavan *et al.* [25] have made NMR measurements on RPt_5 compounds ($CaCu_5$ structure). To interpret these data they also determined χ versus temperature for the La, Ce, Pr, and Nd compounds. They state that the slopes were as expected for tripositive rare earth ions and the θ values obtained, presumably for Ce, Pr, and Nd, are negative. However, they provide no quantitative information.

REFERENCES

1. V. B. Compton and B. T. Matthias, *Acta Crystallogr.* **12,** 651 (1959).
2. I. R. Harris and G. V. Raynor, *J. Less Common Metals* **9,** 263 (1965).
3. A. E. Dwight, *Trans. Amer. Soc. Met.* **53,** 479 (1961).
4. M. Hansen and K. Anderko, "Constitution of Binary Alloys." McGraw-Hill, New York, 1958. See also the supplement to this work by R. P. Elliott, McGraw-Hill, New York, 1965.
5. R. M. Bozorth, B. T. Matthias, H. Suhl, E. Corenzwit, and D. D. Davis, *Phys. Rev.* **115,** 1595 (1959).
6. J. Crangle and J. W. Ross, *Proc. Int. Conf. Magnet. Nottingham* p. 240. Inst. of Phys. and the Phys. Soc., London (1964).
7. W. E. Wallace and Y. G. Vlasov, *Inorg. Chem.* **6,** 2216 (1967).
8. S. Methfessel, *IEEE Trans. Magnet. Mag.* **1,** 144 (1965).
9. A. Heuberger, F. Pobelli, and P. Kienly, *Z. Phys.* **205,** 503 (1967).
10. H. H. Wickman, J. H. Wernick, R. C. Sherwood, and C. F. Wagner, *J. Phys. Chem. Solids* **29,** 181 (1968).
11. G. P. Felcher and W. C. Koehler, *Phys. Rev.* **131,** 1518 (1963).
12. W. E. Wallace and K. H. Mader, *Inorg. Chem.* **7,** 1627 (1968).
13. R. R. Joseph and K. A. Gschneidner, *Proc. Rare Earth Conf., 7th, Coronado, California* p. 7 (1968).
14. G. L. Olcese, *Estratto del Bollettino Sci. Della Facolta Chim. Ind. Bologna* **24,** 165 (1966).
15. B. Coqblin, *Proc. Rare Earth Conf., 7th, Coronado, California* p. 79 (1968).
16. W. E. Gardner, J. Penfold, and I. R. Harris, *J. Phys. C Suppl.* **32,** 1–1139 (1971).
17. R. D. Hutchens, V. U. S. Rao, J. E. Greedan, W. E. Wallace, and R. S. Craig, *J. Appl. Phys.* **42,** 1293 (1971).
18. F. E. Maranzana, *Phys. Rev. Lett.* **25,** 239 (1970).

19. H. J. van Daal, F. E. Maranzana, and K. H. J. Buschow, *J. Phys. C Suppl.* **32,** 1–424.
20. V. U. S. Rao, R. D. Hutchens, and J. E. Greedan, *J. Phys. Chem. Solids* **32,** 2755 (1971).
21. A. E. Berkowitz, F. Holtzberg, and S. Methfessel, *J. Appl. Phys.* **35,** 1030 (1964).
22. J. Pierre and E. Siaud, *C. R. Acad. Sci. Paris* **266B,** 1483 (1968).
23. R. D. Hutchens, V. U. S. Rao, J. E. Greedan, and R. S. Craig, *J. Phys. Soc. Japan* (in press).
24. S. Doniach and S. Engelsberg, *Phys. Rev. Lett.* **17,** 750 (1966).
25. R. Vijayaraghavan, S. K. Malik, and V. U. S. Rao, *Phys. Rev. Lett.* **20,** 102 (1968).

Appendix

I

ENERGIES, EIGENFUNCTIONS, AND MAGNETIC MOMENTS OF RARE EARTH IONS IN A HEXAGONAL FIELD*

$$\mathcal{H} = W\{(1-|y|)\frac{O_2}{F_2} + y[x\frac{O_4}{F_4} + (1-|x|)\frac{O_6}{F_6}]\}$$

The magnetic moment, μ, is given in units of $gJ\ \mu_B$; Γ_i is the eigenfunction; $\Gamma_i = \Sigma_{i=-J}^{J} a_i|M_i\rangle$; energy is given in terms of the parameter W; x and y describe the relative importance of the sixth- and second-order contribution, respectively, to the fourth-order term.

In the tables the first column gives the energy. The column headed "HEPAL" gives the component of μ along the hexagonal axis. "HEPER" indicates the component of μ perpendicular to the c axis. The remaining columns give a_i for the respective M_i values heading the column.

For the sake of brevity, results are given for only one integral and one half integral value of J, 8 and 15/2, respectively. Data for $|y|=1$ are not included since they have been tabulated by E. Segal and W. E. Wallace, *J. Solid State Chem.* **2**, 347 (1970). Results for the other values of J appearing in the rare earth series are available in pamphlet form and can be purchased from the senior author. The $J = 5/2$ case has been treated in detail in Chapter 3 as an illustrative example, the results appearing as the solution of the Hamiltonian given in Eq. (3-9).

*By W. E. Wallace and E. Segal.

RESULTS FOR $J = 8$

Y=-.80 X=-.80 OAS=4.1002E+02*W246 B4F4/B2F2= 3.20E+00 B4F4/B6F6=-4.00E+00 B2F2/B6F6=-1.25E+00 B4F4= 6.40E-01

ENERGY	HPAL		M=8	M=7	M=6	M=5	M=4	M=3	M=2	M=1	M=0
2.2440E+02	.9996	-M	0.000000	0.000000	0.000000	0.000000	.001174	0.000000	0.000000	0.000000	0.000000-M
		+M	.999721	0.000000	0.000000	0.000000	0.000000	0.000000	-.023591	0.000000	+M
2.2440E+02	-.9996	-M	.999721	0.000000	0.000000	0.000000	0.000000	0.000000	-.023591	0.000000	0.000000-M
		+M	0.000000	0.000000	0.000000	0.000000	.001174	0.000000	0.000000	0.000000	+M
1.7648E+02	.0000	-M	0.000000	0.000000	-.034447	0.000000	0.000000	0.000000	0.000000	0.000000	.998813-M
		+M	0.000000	0.000000	-.034447	0.000000	0.000000	0.000000	0.000000	0.000000	+M
1.4889E+02	.1248	-M	0.000000	0.000000	0.000000	-.045406	0.000000	0.000000	0.000000	0.000000	0.000000-M
		+M	0.000000	-.042678	0.000000	0.000000	0.000000	0.000000	0.000000	.998057	+M
1.4889E+02	-.1248	-M	0.000000	-.042678	0.000000	0.000000	0.000000	0.000000	0.000000	.998057	0.000000-M
		+M	0.000000	0.000000	0.000000	-.045406	0.000000	0.000000	0.000000	0.000000	+M
7.3285E+01	.2448	-M	0.000000	0.000000	0.000000	0.000000	-.086497	0.000000	0.000000	0.000000	0.000000-M
		+M	.023604	0.000000	0.000000	0.000000	0.000000	0.000000	.995972	0.000000	+M
7.3285E+01	-.2448	-M	.023604	0.000000	0.000000	0.000000	0.000000	0.000000	.995972	0.000000	0.000000-M
		+M	0.000000	0.000000	0.000000	0.000000	-.086497	0.000000	0.000000	0.000000	+M
-1.2840E+01	0.0000	-M	0.000000	0.000000	0.000000	0.000000	0.000000	.707107	0.000000	0.000000	0.000000-M
		+M	0.000000	0.000000	0.000000	0.000000	0.000000	-.707107	0.000000	0.000000	+M
-2.6519E+01	-.8736	-M	0.000000	.999087	0.000000	0.000000	0.000000	0.000000	0.000000	.042536	0.000000-M
		+M	0.000000	0.000000	0.000000	-.004078	0.000000	0.000000	0.000000	0.000000	+M
-2.6519E+01	.8736	-M	0.000000	0.000000	0.000000	-.004078	0.000000	0.000000	0.000000	0.000000	0.000000-M
		+M	0.000000	.999087	0.000000	0.000000	0.000000	0.000000	0.000000	.042536	+M
-4.9800E+01	0.0000	-M	0.000000	0.000000	0.000000	0.000000	0.000000	.707107	0.000000	0.000000	0.000000-M
		+M	0.000000	0.000000	0.000000	0.000000	0.000000	.707107	0.000000	0.000000	+M
-1.3113E+02	-.4944	-M	0.000000	0.000000	0.000000	0.000000	.996251	0.000000	0.000000	0.000000	0.000000-M
		+M	.000871	0.000000	0.000000	0.000000	0.000000	0.000000	.086501	0.000000	+M
-1.3113E+02	.4944	-M	.000871	0.000000	0.000000	0.000000	0.000000	0.000000	.086501	0.000000	0.000000-M
		+M	0.000000	0.000000	0.000000	0.000000	.996251	0.000000	0.000000	0.000000	+M
-1.5984E+02	-.0000	-M	0.000000	0.000000	-.707107	0.000000	0.000000	0.000000	0.000000	0.000000	-.000000-M
		+M	0.000000	0.000000	.707107	0.000000	0.000000	0.000000	0.000000	0.000000	+M
-1.6064E+02	.0000	-M	0.000000	0.000000	.706267	0.000000	0.000000	0.000000	0.000000	0.000000	.048715-M
		+M	0.000000	0.000000	.706267	0.000000	0.000000	0.000000	0.000000	0.000000	+M
-1.8561E+02	.6234	-M	0.000000	.002139	0.000000	0.000000	0.000000	0.000000	0.000000	.045539	0.000000-M
		+M	0.000000	0.000000	0.000000	.998960	0.000000	0.000000	0.000000	0.000000	+M
-1.8561E+02	-.6234	-M	0.000000	0.000000	0.000000	.998960	0.000000	0.000000	0.000000	0.000000	0.000000-M
		+M	0.000000	.002139	0.000000	0.000000	0.000000	0.000000	0.000000	.045539	+M
ENERGY	HPAL		M=8	M=7	M=6	M=5	M=4	M=3	M=2	M=1	M=0

Y=-.80 X=-.60 OAS=3.1359E+02*W246 B4F4/B2F2= 2.40E+00 B4F4/B6F6=-1.50E+00 B2F2/B6F6=-6.25E-01 B4F4= 4.80E-01

ENERGY	HPAL		M=8	M=7	M=6	M=5	M=4	M=3	M=2	M=1	M=0
1.5866E+02	-.0000	-M	0.000000	0.000000	-.087742	0.000000	0.000000	0.000000	0.000000	0.000000	.992271-M
		+M	0.000000	0.000000	-.087742	0.000000	0.000000	0.000000	0.000000	0.000000	+M
1.4999E+02	.9954	-M	0.000000	0.000000	0.000000	0.000000	.009904	0.000000	0.000000	0.000000	0.000000-M
		+M	.996999	0.000000	0.000000	0.000000	0.000000	0.000000	-.076776	0.000000	+M
1.4999E+02	-.9954	-M	.996999	0.000000	0.000000	0.000000	0.000000	0.000000	-.076776	0.000000	0.000000-M
		+M	0.000000	0.000000	0.000000	0.000000	.009904	0.000000	0.000000	0.000000	+M
1.3194E+02	.1287	-M	0.000000	0.000000	0.000000	-.105639	0.000000	0.000000	0.000000	0.000000	0.000000-M
		+M	0.000000	-.126731	0.000000	0.000000	0.000000	0.000000	0.000000	.986296	+M
1.3194E+02	-.1287	-M	0.000000	-.126731	0.000000	0.000000	0.000000	0.000000	0.000000	.986296	0.000000-M
		+M	0.000000	0.000000	0.000000	-.105639	0.000000	0.000000	0.000000	0.000000	+M
5.8741E+01	.2275	-M	0.000000	0.000000	0.000000	0.000000	-.189587	0.000000	0.000000	0.000000	0.000000-M
		+M	.077259	0.000000	0.000000	0.000000	0.000000	0.000000	.978820	0.000000	+M
5.8741E+01	-.2275	-M	.077259	0.000000	0.000000	0.000000	0.000000	0.000000	.978820	0.000000	0.000000-M
		+M	0.000000	0.000000	0.000000	0.000000	-.189587	0.000000	0.000000	0.000000	+M
1.3513E+01	-.8625	-M	0.000000	.991892	0.000000	0.000000	0.000000	0.000000	0.000000	.124988	0.000000-M
		+M	0.000000	0.000000	0.000000	-.022986	0.000000	0.000000	0.000000	0.000000	+M
1.3513E+01	.8625	-M	0.000000	0.000000	0.000000	-.022986	0.000000	0.000000	0.000000	0.000000	0.000000-M
		+M	0.000000	.991892	0.000000	0.000000	0.000000	0.000000	0.000000	.124988	+M
-5.8800E+00	0.0000	-M	0.000000	0.000000	0.000000	0.000000	0.000000	.707107	0.000000	0.000000	0.000000-M
		+M	0.000000	0.000000	0.000000	0.000000	0.000000	-.707107	0.000000	0.000000	+M
-7.9800E+01	0.0000	-M	0.000000	0.000000	0.000000	0.000000	0.000000	.707107	0.000000	0.000000	0.000000-M
		+M	0.000000	0.000000	0.000000	0.000000	0.000000	.707107	0.000000	0.000000	+M
-1.0368E+02	.0000	-M	0.000000	0.000000	-.707107	0.000000	0.000000	0.000000	0.000000	0.000000	.000000-M
		+M	0.000000	0.000000	.707107	0.000000	0.000000	0.000000	0.000000	0.000000	+M
-1.0778E+02	-.0000	-M	0.000000	0.000000	.701642	0.000000	0.000000	0.000000	0.000000	0.000000	.124086-M
		+M	0.000000	0.000000	.701642	0.000000	0.000000	0.000000	0.000000	0.000000	+M
-1.3001E+02	-.4730	-M	0.000000	0.000000	0.000000	0.000000	.981814	0.000000	0.000000	0.000000	0.000000-M
		+M	.004862	0.000000	0.000000	0.000000	0.000000	0.000000	.189783	0.000000	+M
-1.3001E+02	.4730	-M	.004862	0.000000	0.000000	0.000000	0.000000	0.000000	.189783	0.000000	0.000000-M
		+M	0.000000	0.000000	0.000000	0.000000	.981814	0.000000	0.000000	0.000000	+M
-1.5493E+02	.6162	-M	0.000000	.009468	0.000000	0.000000	0.000000	0.000000	0.000000	.107696	0.000000-M
		+M	0.000000	0.000000	0.000000	.994139	0.000000	0.000000	0.000000	0.000000	+M
-1.5493E+02	-.6162	-M	0.000000	0.000000	0.000000	.994139	0.000000	0.000000	0.000000	0.000000	0.000000-M
		+M	0.000000	.009468	0.000000	0.000000	0.000000	0.000000	0.000000	.107696	+M
ENERGY	HPAL		M=8	M=7	M=6	M=5	M=4	M=3	M=2	M=1	M=0

Y=-.80 X=-.40 OAS=2.7957E+02*W246 B4F4/B2F2= 1.60E+00 B4F4/B6F6=-6.67E-01 B2F2/B6F6=-4.17E-01 B4F4= 3.20E-01

ENERGY	HPAL		M=8	M=7	M=6	M=5	M=4	M=3	M=2	M=1	M=0
1.4596E+02	.0000	-M	0.000000	0.000000	-.174300	0.000000	0.000000	0.000000	0.000000	0.000000	.969143-M
		+M	0.000000	0.000000	-.174300	0.000000	0.000000	0.000000	0.000000	0.000000	+M
1.2203E+02	.1791	-M	0.000000	0.000000	0.000000	-.176331	0.000000	0.000000	0.000000	0.000000	0.000000-M
		+M	0.000000	-.321346	0.000000	0.000000	0.000000	0.000000	0.000000	.930400	+M
1.2203E+02	-.1791	-M	0.000000	-.321346	0.000000	0.000000	0.000000	0.000000	0.000000	.930400	0.000000-M
		+M	0.000000	0.000000	0.000000	-.176331	0.000000	0.000000	0.000000	0.000000	+M
7.8297E+01	.9078	-M	0.000000	0.000000	0.000000	0.000000	.088705	0.000000	0.000000	0.000000	0.000000-M
		+M	.940715	0.000000	0.000000	0.000000	0.000000	0.000000	-.327394	0.000000	+M
7.8297E+01	-.9078	-M	.940715	0.000000	0.000000	0.000000	0.000000	0.000000	-.327394	0.000000	0.000000-M
		+M	0.000000	0.000000	0.000000	0.000000	.088705	0.000000	0.000000	0.000000	+M
4.9622E+01	.7918	-M	0.000000	0.000000	0.000000	-.084343	0.000000	0.000000	0.000000	0.000000	0.000000-M
		+M	0.000000	.946667	0.000000	0.000000	0.000000	0.000000	0.000000	.310980	+M
4.9622E+01	-.7918	-M	0.000000	.946667	0.000000	0.000000	0.000000	0.000000	0.000000	.310980	0.000000-M
		+M	0.000000	0.000000	0.000000	-.084343	0.000000	0.000000	0.000000	0.000000	+M
4.6197E+01	-.2729	-M	.338838	0.000000	0.000000	0.000000	0.000000	0.000000	.894959	0.000000	0.000000-M
		+M	0.000000	0.000000	0.000000	0.000000	-.290236	0.000000	0.000000	0.000000	+M
4.6197E+01	.2729	-M	0.000000	0.000000	0.000000	0.000000	-.290236	0.000000	0.000000	0.000000	0.000000-M
		+M	.338838	0.000000	0.000000	0.000000	0.000000	0.000000	.894959	0.000000	+M
1.0800E+00	0.0000	-M	0.000000	0.000000	0.000000	0.000000	0.000000	.707107	0.000000	0.000000	0.000000-M
		+M	0.000000	0.000000	0.000000	0.000000	0.000000	-.707107	0.000000	0.000000	+M
-4.7520E+01	-.0000	-M	0.000000	0.000000	-.707107	0.000000	0.000000	0.000000	0.000000	0.000000	-.000000-M
		+M	0.000000	0.000000	.707107	0.000000	0.000000	0.000000	0.000000	0.000000	+M
-6.0636E+01	.0000	-M	0.000000	0.000000	.685288	0.000000	0.000000	0.000000	0.000000	0.000000	.246498-M
		+M	0.000000	0.000000	.685288	0.000000	0.000000	0.000000	0.000000	0.000000	+M
-1.0980E+02	0.0000	-M	0.000000	0.000000	0.000000	0.000000	0.000000	.707107	0.000000	0.000000	0.000000-M
		+M	0.000000	0.000000	0.000000	0.000000	0.000000	.707107	0.000000	0.000000	+M
-1.2737E+02	.5959	-M	0.000000	.023638	0.000000	0.000000	0.000000	0.000000	0.000000	.194030	0.000000-M
		+M	0.000000	0.000000	0.000000	.980711	0.000000	0.000000	0.000000	0.000000	+M
-1.2737E+02	-.5959	-M	0.000000	0.000000	0.000000	.980711	0.000000	0.000000	0.000000	0.000000	0.000000-M
		+M	0.000000	.023638	0.000000	0.000000	0.000000	0.000000	0.000000	.194030	+M
-1.3361E+02	-.4307	-M	0.000000	0.000000	0.000000	0.000000	.952835	0.000000	0.000000	0.000000	0.000000-M
		+M	.015634	0.000000	0.000000	0.000000	0.000000	0.000000	.303086	0.000000	+M
-1.3361E+02	.4307	-M	.015634	0.000000	0.000000	0.000000	0.000000	0.000000	.303086	0.000000	0.000000-M
		+M	0.000000	0.000000	0.000000	0.000000	.952835	0.000000	0.000000	0.000000	+M
ENERGY	HPAL		M=8	M=7	M=6	M=5	M=4	M=3	M=2	M=1	M=0

Y=-.80 X=-.20 OAS=2.8659E+02*W246 B4F4/B2F2= 6.00E-01 B4F4/B6F6=-2.50E-01 B2F2/B6F6=-3.12E-01 B4F4= 1.60E-01

ENERGY	HPAL		M=8	M=7	M=6	M=5	M=4	M=3	M=2	M=1	M=0
1.4410E+02	.0000	-M	0.000000	0.000000	-.308245	0.000000	0.000000	0.000000	0.000000	0.000000	.899983-M
		+M	0.000000	0.000000	-.308245	0.000000	0.000000	0.000000	0.000000	0.000000	+M
1.3012E+02	.4364	-M	0.000000	0.000000	0.000000	-.200485	0.000000	0.000000	0.000000	0.000000	0.000000-M
		+M	0.000000	-.674804	0.000000	0.000000	0.000000	0.000000	0.000000	.710243	+M
1.3012E+02	-.4364	-M	0.000000	.674804	0.000000	0.000000	0.000000	0.000000	0.000000	-.710243	0.000000-M
		+M	0.000000	0.000000	0.000000	.200485	0.000000	0.000000	0.000000	0.000000	+M
7.2932E+01	.4877	-M	0.000000	0.000000	0.000000	-.242702	0.000000	0.000000	0.000000	0.000000	0.000000-M
		+M	0.000000	.736574	0.000000	0.000000	0.000000	0.000000	0.000000	.631312	+M
7.2932E+01	-.4877	-M	0.000000	.736574	0.000000	0.000000	0.000000	0.000000	0.000000	.631312	0.000000-M
		+M	0.000000	0.000000	0.000000	-.242702	0.000000	0.000000	0.000000	0.000000	+M
4.9062E+01	.1848	-M	0.000000	0.000000	0.000000	0.000000	-.390752	0.000000	0.000000	0.000000	0.000000-M
		+M	-.256332	0.000000	0.000000	0.000000	0.000000	0.000000	.884085	0.000000	+M
4.9062E+01	-.1848	-M	-.256332	0.000000	0.000000	0.000000	0.000000	0.000000	.884085	0.000000	0.000000-M
		+M	0.000000	0.000000	0.000000	0.000000	-.390752	0.000000	0.000000	0.000000	+M
8.6400E+00	-.0000	-M	0.000000	0.000000	-.707107	0.000000	0.000000	0.000000	0.000000	0.000000	.000000-M
		+M	0.000000	0.000000	.707107	0.000000	0.000000	0.000000	0.000000	0.000000	+M
8.0400E+00	0.0000	-M	0.000000	0.000000	0.000000	0.000000	0.000000	.707107	0.000000	0.000000	0.000000-M
		+M	0.000000	0.000000	0.000000	0.000000	0.000000	-.707107	0.000000	0.000000	+M
-3.5344E+00	-.9341	-M	.965689	0.000000	0.000000	0.000000	0.000000	0.000000	.216797	0.000000	0.000000-M
		+M	0.000000	0.000000	0.000000	0.000000	-.142982	0.000000	0.000000	0.000000	+M
-3.5344E+00	.9341	-M	0.000000	0.000000	0.000000	0.000000	-.142982	0.000000	0.000000	0.000000	0.000000-M
		+M	.965689	0.000000	0.000000	0.000000	0.000000	0.000000	.216797	0.000000	+M
-2.3141E+01	-.0000	-M	0.000000	0.000000	.636384	0.000000	0.000000	0.000000	0.000000	0.000000	.435925-M
		+M	0.000000	0.000000	.636384	0.000000	0.000000	0.000000	0.000000	0.000000	+M
-1.0501E+02	.5491	-M	0.000000	.045809	0.000000	0.000000	0.000000	0.000000	0.000000	.311448	0.000000-M
		+M	0.000000	0.000000	0.000000	.949158	0.000000	0.000000	0.000000	0.000000	+M
-1.0501E+02	-.5491	-M	0.000000	0.000000	0.000000	.949158	0.000000	0.000000	0.000000	0.000000	0.000000-M
		+M	0.000000	.045809	0.000000	0.000000	0.000000	0.000000	0.000000	.311448	+M
-1.3980E+02	0.0000	-M	0.000000	0.000000	0.000000	0.000000	0.000000	.707107	0.000000	0.000000	0.000000-M
		+M	0.000000	0.000000	0.000000	0.000000	0.000000	.707107	0.000000	0.000000	+M
-1.4249E+02	-.3688	-M	0.000000	0.000000	0.000000	0.000000	.909323	0.000000	0.000000	0.000000	0.000000-M
		+M	.041694	0.000000	0.000000	0.000000	0.000000	0.000000	.413996	0.000000	+M
-1.4249E+02	.3688	-M	.041694	0.000000	0.000000	0.000000	0.000000	0.000000	.413996	0.000000	0.000000-M
		+M	0.000000	0.000000	0.000000	0.000000	.909323	0.000000	0.000000	0.000000	+M
ENERGY	HPAL		M=8	M=7	M=6	M=5	M=4	M=3	M=2	M=1	M=0

Y=-.80 X=0.00 OAS=3.3087E+02*W246 B4F4/B2F2= 0. B4F4/B6F6= 0. B2F2/B6F6=-2.50E-01 B4F4= 0.

ENERGY	HPAL		M=8	M=7	M=6	M=5	M=4	M=3	M=2	M=1	M=0
1.6107E+02	.0000	-M	0.000000	0.000000	-.458558	0.000000	0.000000	0.000000	0.000000	0.000000	.761216-M
		+M	0.000000	0.000000	-.458558	0.000000	0.000000	0.000000	0.000000	0.000000	+M
1.6030E+02	.6700	-M	0.000000	0.000000	0.000000	.166983	0.000000	0.000000	0.000000	0.000000	0.000000-M
		+M	0.000000	.868653	0.000000	0.000000	0.000000	0.000000	0.000000	-.466431	+M
1.6030E+02	-.6700	-M	0.000000	.868653	0.000000	0.000000	0.000000	0.000000	0.000000	-.466431	0.000000-M
		+M	0.000000	0.000000	0.000000	.166983	0.000000	0.000000	0.000000	0.000000	+M
8.2238E+01	-.1665	-M	0.000000	.489898	0.000000	0.000000	0.000000	0.000000	0.000000	.758522	0.000000-M
		+M	0.000000	0.000000	0.000000	-.429703	0.000000	0.000000	0.000000	0.000000	+M
8.2238E+01	.1665	-M	0.000000	0.000000	0.000000	-.429703	0.000000	0.000000	0.000000	0.000000	0.000000-M
		+M	0.000000	.489898	0.000000	0.000000	0.000000	0.000000	0.000000	.758522	+M
6.4800E+01	-.0000	-M	0.000000	0.000000	-.707107	0.000000	0.000000	0.000000	0.000000	0.000000	-.000000-M
		+M	0.000000	0.000000	.707107	0.000000	0.000000	0.000000	0.000000	0.000000	+M
4.8047E+01	.0761	-M	0.000000	0.000000	0.000000	0.000000	-.497436	0.000000	0.000000	0.000000	0.000000-M
		+M	-.124668	0.000000	0.000000	0.000000	0.000000	0.000000	.858496	0.000000	+M
4.8047E+01	-.0761	-M	-.124668	0.000000	0.000000	0.000000	0.000000	0.000000	.858496	0.000000	0.000000-M
		+M	0.000000	0.000000	0.000000	0.000000	-.497436	0.000000	0.000000	0.000000	+M
1.5000E+01	0.0000	-M	0.000000	0.000000	0.000000	0.000000	0.000000	.707107	0.000000	0.000000	0.000000-M
		+M	0.000000	0.000000	0.000000	0.000000	0.000000	-.707107	0.000000	0.000000	+M
-5.0720E+00	.0000	-M	0.000000	0.000000	.538261	0.000000	0.000000	0.000000	0.000000	0.000000	.648499-M
		+M	0.000000	0.000000	.538261	0.000000	0.000000	0.000000	0.000000	0.000000	+M
-7.6118E+01	-.9598	-M	.985854	0.000000	0.000000	0.000000	0.000000	0.000000	.050572	0.000000	0.000000-M
		+M	0.000000	0.000000	0.000000	0.000000	-.159796	0.000000	0.000000	0.000000	+M
-7.6118E+01	.9598	-M	0.000000	0.000000	0.000000	0.000000	-.159796	0.000000	0.000000	0.000000	0.000000-M
		+M	.985854	0.000000	0.000000	0.000000	0.000000	0.000000	.050572	0.000000	+M
-9.0739E+01	-.4615	-M	0.000000	0.000000	0.000000	.887396	0.000000	0.000000	0.000000	0.000000	0.000000-M
		+M	0.000000	.073767	0.000000	0.000000	0.000000	0.000000	0.000000	.455068	+M
-9.0739E+01	.4615	-M	0.000000	.073767	0.000000	0.000000	0.000000	0.000000	0.000000	.455068	0.000000-M
		+M	0.000000	0.000000	0.000000	.887396	0.000000	0.000000	0.000000	0.000000	+M
-1.5673E+02	.2859	-M	.112028	0.000000	0.000000	0.000000	0.000000	0.000000	.510321	0.000000	0.000000-M
		+M	0.000000	0.000000	0.000000	0.000000	.852656	0.000000	0.000000	0.000000	+M
-1.5673E+02	-.2859	-M	0.000000	0.000000	0.000000	0.000000	.852656	0.000000	0.000000	0.000000	0.000000-M
		+M	.112028	0.000000	0.000000	0.000000	0.000000	0.000000	.510321	0.000000	+M
-1.6980E+02	0.0000	-M	0.000000	0.000000	0.000000	0.000000	0.000000	.707107	0.000000	0.000000	0.000000-M
		+M	0.000000	0.000000	0.000000	0.000000	0.000000	.707107	0.000000	0.000000	+M
ENERGY	HPAL		M=8	M=7	M=6	M=5	M=4	M=3	M=2	M=1	M=0

Y=-.80 X= .20 OAS=2.7027E+02*W246 B4F4/B2F2=-8.00E-01 B4F4/B6F6= 2.50E-01 B2F2/B6F6=-3.12E-01 B4F4=-1.60E-01

ENERGY	HPAL		M=8	M=7	M=6	M=5	M=4	M=3	M=2	M=1	M=0
1.3719E+02	.7810	-M	0.000000	0.000000	0.000000	.134471	0.000000	0.000000	0.000000	0.000000	0.000000-M
		+M	0.000000	.944846	0.000000	0.000000	0.000000	0.000000	0.000000	-.298635	+M
1.3719E+02	-.7810	-M	0.000000	.944846	0.000000	0.000000	0.000000	0.000000	0.000000	-.298635	0.000000-M
		+M	0.000000	0.000000	0.000000	.134471	0.000000	0.000000	0.000000	0.000000	+M
1.3691E+02	.0000	-M	0.000000	0.000000	.600025	0.000000	0.000000	0.000000	0.000000	0.000000	-.529094-M
		+M	0.000000	0.000000	.600025	0.000000	0.000000	0.000000	0.000000	0.000000	+M
9.6000E+01	-.0000	-M	0.000000	0.000000	-.707107	0.000000	0.000000	0.000000	0.000000	0.000000	.000000-M
		+M	0.000000	0.000000	.707107	0.000000	0.000000	0.000000	0.000000	0.000000	+M
6.2547E+01	.1460	-M	0.000000	.307916	0.000000	0.000000	0.000000	0.000000	0.000000	.670083	0.000000-M
		+M	0.000000	0.000000	0.000000	-.675409	0.000000	0.000000	0.000000	0.000000	+M
6.2547E+01	-.1460	-M	0.000000	0.000000	0.000000	-.675409	0.000000	0.000000	0.000000	0.000000	0.000000-M
		+M	0.000000	.307916	0.000000	0.000000	0.000000	0.000000	0.000000	.670083	+M
3.2671E+01	-.0320	-M	0.000000	0.000000	0.000000	0.000000	-.617745	0.000000	0.000000	0.000000	0.000000-M
		+M	-.074985	0.000000	0.000000	0.000000	0.000000	0.000000	.782795	0.000000	+M
3.2671E+01	.0320	-M	-.074985	0.000000	0.000000	0.000000	0.000000	0.000000	.782795	0.000000	0.000000-M
		+M	0.000000	0.000000	0.000000	0.000000	-.617745	0.000000	0.000000	0.000000	+M
1.4760E+01	0.0000	-M	0.000000	0.000000	0.000000	0.000000	0.000000	.707107	0.000000	0.000000	0.000000-M
		+M	0.000000	0.000000	0.000000	0.000000	0.000000	-.707107	0.000000	0.000000	+M
-9.2310E+00	.0000	-M	0.000000	0.000000	.374126	0.000000	0.000000	0.000000	0.000000	0.000000	.848564-M
		+M	0.000000	0.000000	.374126	0.000000	0.000000	0.000000	0.000000	0.000000	+M
-5.4653E+01	-.2600	-M	0.000000	0.000000	0.000000	.725080	0.000000	0.000000	0.000000	0.000000	0.000000-M
		+M	0.000000	.111594	0.000000	0.000000	0.000000	0.000000	0.000000	.679563	+M
-5.4653E+01	.2600	-M	0.000000	.111594	0.000000	0.000000	0.000000	0.000000	0.000000	.679563	0.000000-M
		+M	0.000000	0.000000	0.000000	.725080	0.000000	0.000000	0.000000	0.000000	+M
-1.0599E+02	.2057	-M	0.000000	0.000000	0.000000	0.000000	.652133	0.000000	0.000000	0.000000	0.000000-M
		+M	-.605126	0.000000	0.000000	0.000000	0.000000	0.000000	.456667	0.000000	+M
-1.0599E+02	-.2057	-M	-.605126	0.000000	0.000000	0.000000	0.000000	0.000000	.456667	0.000000	0.000000-M
		+M	0.000000	0.000000	0.000000	0.000000	.652133	0.000000	0.000000	0.000000	+M
-1.2444E+02	.5763	-M	0.000000	0.000000	0.000000	0.000000	.439447	0.000000	0.000000	0.000000	0.000000-M
		+M	.792591	0.000000	0.000000	0.000000	0.000000	0.000000	.422714	0.000000	+M
-1.2444E+02	-.5763	-M	.792591	0.000000	0.000000	0.000000	0.000000	0.000000	.422714	0.000000	0.000000-M
		+M	0.000000	0.000000	0.000000	0.000000	.439447	0.000000	0.000000	0.000000	+M
-1.3308E+02	0.0000	-M	0.000000	0.000000	0.000000	0.000000	0.000000	.707107	0.000000	0.000000	0.000000-M
		+M	0.000000	0.000000	0.000000	0.000000	0.000000	.707107	0.000000	0.000000	+M
ENERGY	HPAL		M=8	M=7	M=6	M=5	M=4	M=3	M=2	M=1	M=0

Y=-.80 X= .40 OAS=3.0108E+02*W246 B4F4/B2F2=-1.60E+00 B4F4/B6F6= 6.67E-01 B2F2/B6F6=-4.17E-01 B4F4=-3.20E-01

ENERGY	HPAL		M=8	M=7	M=6	M=5	M=4	M=3	M=2	M=1	M=0	
1.4150E+02	.0000	-M	0.000000	0.000000	.679017	0.000000	0.000000	0.000000	0.000000	0.000000	-.279054	-M
		+M	0.000000	0.000000	.679017	0.000000	0.000000	0.000000	0.000000	0.000000		+M
1.2720E+02	-.0000	-M	0.000000	0.000000	-.707107	0.000000	0.000000	0.000000	0.000000	0.000000	.000000	-M
		+M	0.000000	0.000000	.707107	0.000000	0.000000	0.000000	0.000000	0.000000		+M
1.1945E+02	.8239	-M	0.000000	0.000000	0.000000	.131916	0.000000	0.000000	0.000000	0.000000	0.000000	-M
		+M	0.000000	.974283	0.000000	0.000000	0.000000	0.000000	0.000000	-.182679		+M
1.1945E+02	-.8239	-M	0.000000	.974283	0.000000	0.000000	0.000000	0.000000	0.000000	-.182679	0.000000	-M
		+M	0.000000	0.000000	0.000000	.131916	0.000000	0.000000	0.000000	0.000000		+M
7.3776E+01	.4787	-M	0.000000	-.189903	0.000000	0.000000	0.000000	0.000000	0.000000	-.350568	0.000000	-M
		+M	0.000000	0.000000	0.000000	.917082	0.000000	0.000000	0.000000	0.000000		+M
7.3776E+01	-.4787	-M	0.000000	0.000000	0.000000	.917082	0.000000	0.000000	0.000000	0.000000	0.000000	-M
		+M	0.000000	-.189903	0.000000	0.000000	0.000000	0.000000	0.000000	-.350568		+M
2.9544E+01	.2358	-M	.033803	0.000000	0.000000	0.000000	0.000000	0.000000	-.591620	0.000000	0.000000	-M
		+M	0.000000	0.000000	0.000000	0.000000	.805508	0.000000	0.000000	0.000000		+M
2.9544E+01	-.2358	-M	0.000000	0.000000	0.000000	0.000000	.805508	0.000000	0.000000	0.000000	0.000000	-M
		+M	.033803	0.000000	0.000000	0.000000	0.000000	0.000000	-.591620	0.000000		+M
1.4520E+01	0.0000	-M	0.000000	0.000000	0.000000	0.000000	0.000000	.707107	0.000000	0.000000	0.000000	-M
		+M	0.000000	0.000000	0.000000	0.000000	0.000000	-.707107	0.000000	0.000000		+M
-4.2140E+01	.0000	-M	0.000000	0.000000	.197321	0.000000	0.000000	0.000000	0.000000	0.000000	.960275	-M
		+M	0.000000	0.000000	.197321	0.000000	0.000000	0.000000	0.000000	0.000000		+M
-5.4867E+01	-.0299	-M	0.000000	.121286	0.000000	0.000000	0.000000	0.000000	0.000000	.918548	0.000000	-M
		+M	0.000000	0.000000	0.000000	.376244	0.000000	0.000000	0.000000	0.000000		+M
-5.4867E+01	.0299	-M	0.000000	0.000000	0.000000	.376244	0.000000	0.000000	0.000000	0.000000	0.000000	-M
		+M	0.000000	.121286	0.000000	0.000000	0.000000	0.000000	0.000000	.918548		+M
-8.0687E+01	.0031	-M	-.110383	0.000000	0.000000	0.000000	0.000000	0.000000	.798828	0.000000	0.000000	-M
		+M	0.000000	0.000000	0.000000	0.000000	.591346	0.000000	0.000000	0.000000		+M
-8.0687E+01	-.0031	-M	0.000000	0.000000	0.000000	0.000000	.591346	0.000000	0.000000	0.000000	0.000000	-M
		+M	-.110383	0.000000	0.000000	0.000000	0.000000	0.000000	.798828	0.000000		+M
-9.6360E+01	0.0000	-M	0.000000	0.000000	0.000000	0.000000	0.000000	.707107	0.000000	0.000000	0.000000	-M
		+M	0.000000	0.000000	0.000000	0.000000	0.000000	.707107	0.000000	0.000000		+M
-1.5958E+02	-.9889	-M	.993314	0.000000	0.000000	0.000000	0.000000	0.000000	.108903	0.000000	0.000000	-M
		+M	0.000000	0.000000	0.000000	0.000000	.038302	0.000000	0.000000	0.000000		+M
-1.5958E+02	.9889	-M	0.000000	0.000000	0.000000	0.000000	.038302	0.000000	0.000000	0.000000	0.000000	-M
		+M	.993314	0.000000	0.000000	0.000000	0.000000	0.000000	.108903	0.000000		+M
ENERGY	HPAL		M=8	M=7	M=6	M=5	M=4	M=3	M=2	M=1	M=0	

Y=-.80 X= .60 OAS=3.6309E+02*W246 B4F4/B2F2=-2.40E+00 B4F4/B6F6= 1.50E+00 B2F2/B6F6=-6.25E-01 B4F4=-4.80E-01

ENERGY	HPAL		M=8	M=7	M=6	M=5	M=4	M=3	M=2	M=1	M=0	
1.6270E+02	.0000	-M	0.000000	0.000000	.701099	0.000000	0.000000	0.000000	0.000000	0.000000	-.130078	-M
		+M	0.000000	0.000000	.701099	0.000000	0.000000	0.000000	0.000000	0.000000		+M
1.5840E+02	-.0000	-M	0.000000	0.000000	-.707107	0.000000	0.000000	0.000000	0.000000	0.000000	.000000	-M
		+M	0.000000	0.000000	.707107	0.000000	0.000000	0.000000	0.000000	0.000000		+M
1.1558E+02	.5501	-M	0.000000	.191200	0.000000	0.000000	0.000000	0.000000	0.000000	-.163636	0.000000	-M
		+M	0.000000	0.000000	0.000000	.967815	0.000000	0.000000	0.000000	0.000000		+M
1.1558E+02	-.5501	-M	0.000000	0.000000	0.000000	.967815	0.000000	0.000000	0.000000	0.000000	0.000000	-M
		+M	0.000000	.191200	0.000000	0.000000	0.000000	0.000000	0.000000	-.163636		+M
1.0347E+02	-.8127	-M	0.000000	.978479	0.000000	0.000000	0.000000	0.000000	0.000000	-.046174	0.000000	-M
		+M	0.000000	0.000000	0.000000	-.201114	0.000000	0.000000	0.000000	0.000000		+M
1.0347E+02	.8127	-M	0.000000	0.000000	0.000000	-.201114	0.000000	0.000000	0.000000	0.000000	0.000000	-M
		+M	0.000000	.978479	0.000000	0.000000	0.000000	0.000000	0.000000	-.046174		+M
4.9284E+01	.4310	-M	.008701	0.000000	0.000000	0.000000	0.000000	0.000000	-.302983	0.000000	0.000000	-M
		+M	0.000000	0.000000	0.000000	0.000000	.952956	0.000000	0.000000	0.000000		+M
4.9284E+01	-.4310	-M	0.000000	0.000000	0.000000	0.000000	.952956	0.000000	0.000000	0.000000	0.000000	-M
		+M	.008701	0.000000	0.000000	0.000000	0.000000	0.000000	-.302983	0.000000		+M
1.4280E+01	0.0000	-M	0.000000	0.000000	0.000000	0.000000	0.000000	.707107	0.000000	0.000000	0.000000	-M
		+M	0.000000	0.000000	0.000000	0.000000	0.000000	-.707107	0.000000	0.000000		+M
-5.9640E+01	0.0000	-M	0.000000	0.000000	0.000000	0.000000	0.000000	.707107	0.000000	0.000000	0.000000	-M
		+M	0.000000	0.000000	0.000000	0.000000	0.000000	.707107	0.000000	0.000000		+M
-7.2580E+01	-.1833	-M	-.053459	0.000000	0.000000	0.000000	0.000000	0.000000	.951489	0.000000	0.000000	-M
		+M	0.000000	0.000000	0.000000	0.000000	.303004	0.000000	0.000000	0.000000		+M
-7.2580E+01	.1833	-M	0.000000	0.000000	0.000000	0.000000	.303004	0.000000	0.000000	0.000000	0.000000	-M
		+M	-.053459	0.000000	0.000000	0.000000	0.000000	0.000000	.951489	0.000000		+M
-8.7402E+01	-.1124	-M	0.000000	.077597	0.000000	0.000000	0.000000	0.000000	0.000000	.985440	0.000000	-M
		+M	0.000000	0.000000	0.000000	.151286	0.000000	0.000000	0.000000	0.000000		+M
-8.7402E+01	.1124	-M	0.000000	0.000000	0.000000	.151286	0.000000	0.000000	0.000000	0.000000	0.000000	-M
		+M	0.000000	.077597	0.000000	0.000000	0.000000	0.000000	0.000000	.985440		+M
-9.1664E+01	-.0000	-M	0.000000	0.000000	.091979	0.000000	0.000000	0.000000	0.000000	0.000000	.991504	-M
		+M	0.000000	0.000000	.091979	0.000000	0.000000	0.000000	0.000000	0.000000		+M
-2.0038E+02	-.9978	-M	.998532	0.000000	0.000000	0.000000	0.000000	0.000000	.053580	0.000000	0.000000	-M
		+M	0.000000	0.000000	0.000000	0.000000	.007918	0.000000	0.000000	0.000000		+M
-2.0038E+02	.9978	-M	0.000000	0.000000	0.000000	0.000000	.007918	0.000000	0.000000	0.000000	0.000000	-M
		+M	.998532	0.000000	0.000000	0.000000	0.000000	0.000000	.053580	0.000000		+M
ENERGY	HPAL		M=8	M=7	M=6	M=5	M=4	M=3	M=2	M=1	M=0	

Y=-.80 X= .80 OAS=4.3208E+02*W246 B4F4/B2F2=-3.20E+00 B4F4/B6F6= 4.00E+00 B2F2/B6F6=-1.25E+00 B4F4=-6.40E-01

ENERGY	HPAL		M=8	M=7	M=6	M=5	M=4	M=3	M=2	M=1	M=0	
1.9040E+02	-.0000	-M	0.000000	0.000000	.706272	0.000000	0.000000	0.000000	0.000000	0.000000	-.048577	-M
		+M	0.000000	0.000000	.706272	0.000000	0.000000	0.000000	0.000000	0.000000		+M
1.8960E+02	.0000	-M	0.000000	0.000000	-.707107	0.000000	0.000000	0.000000	0.000000	0.000000	-.000000	-M
		+M	0.000000	0.000000	.707107	0.000000	0.000000	0.000000	0.000000	0.000000		+M
1.6530E+02	-.6230	-M	0.000000	0.000000	0.000000	.998659	0.000000	0.000000	0.000000	0.000000	0.000000	-M
		+M	0.000000	.005140	0.000000	0.000000	0.000000	0.000000	0.000000	-.051509		+M
1.6530E+02	.6230	-M	0.000000	.005140	0.000000	0.000000	0.000000	0.000000	0.000000	-.051509	0.000000	-M
		+M	0.000000	0.000000	0.000000	.998659	0.000000	0.000000	0.000000	0.000000		+M
9.0531E+01	.8741	-M	0.000000	0.000000	0.000000	-.006871	0.000000	0.000000	0.000000	0.000000	0.000000	-M
		+M	0.000000	.999416	0.000000	0.000000	0.000000	0.000000	0.000000	-.033475		+M
9.0531E+01	-.8741	-M	0.000000	.999416	0.000000	0.000000	0.000000	0.000000	0.000000	-.033475	0.000000	-M
		+M	0.000000	0.000000	0.000000	-.006871	0.000000	0.000000	0.000000	0.000000		+M
8.7291E+01	.4918	-M	.001137	0.000000	0.000000	0.000000	0.000000	0.000000	-.104474	0.000000	0.000000	-M
		+M	0.000000	0.000000	0.000000	0.000000	.994527	0.000000	0.000000	0.000000		+M
8.7291E+01	-.4918	-M	0.000000	0.000000	0.000000	0.000000	.994527	0.000000	0.000000	0.000000	0.000000	-M
		+M	.001137	0.000000	0.000000	0.000000	0.000000	0.000000	-.104474	0.000000		+M
1.4040E+01	0.0000	-M	0.000000	0.000000	0.000000	0.000000	0.000000	.707107	0.000000	0.000000	0.000000	-M
		+M	0.000000	0.000000	0.000000	0.000000	0.000000	-.707107	0.000000	0.000000		+M
-2.2920E+01	0.0000	-M	0.000000	0.000000	0.000000	0.000000	0.000000	.707107	0.000000	0.000000	0.000000	-M
		+M	0.000000	0.000000	0.000000	0.000000	0.000000	.707107	0.000000	0.000000		+M
-8.2251E+01	-.2422	-M	-.022335	0.000000	0.000000	0.000000	0.000000	0.000000	.994277	0.000000	0.000000	-M
		+M	0.000000	0.000000	0.000000	0.000000	.104473	0.000000	0.000000	0.000000		+M
-8.2251E+01	.2422	-M	0.000000	0.000000	0.000000	0.000000	.104473	0.000000	0.000000	0.000000	0.000000	-M
		+M	-.022335	0.000000	0.000000	0.000000	0.000000	0.000000	.994277	0.000000		+M
-1.3091E+02	-.1239	-M	0.000000	.033784	0.000000	0.000000	0.000000	0.000000	0.000000	.998111	0.000000	-M
		+M	0.000000	0.000000	0.000000	.051307	0.000000	0.000000	0.000000	0.000000		+M
-1.3091E+02	.1239	-M	0.000000	0.000000	0.000000	.051307	0.000000	0.000000	0.000000	0.000000	0.000000	-M
		+M	0.000000	.033784	0.000000	0.000000	0.000000	0.000000	0.000000	.998111		+M
-1.4768E+02	.0000	-M	0.000000	0.000000	.034349	0.000000	0.000000	0.000000	0.000000	0.000000	.998819	-M
		+M	0.000000	0.000000	.034349	0.000000	0.000000	0.000000	0.000000	0.000000		+M
-2.4168E+02	-.9996	-M	.999750	0.000000	0.000000	0.000000	0.000000	0.000000	.022331	0.000000	0.000000	-M
		+M	0.000000	0.000000	0.000000	0.000000	.001203	0.000000	0.000000	0.000000		+M
-2.4168E+02	.9996	-M	0.000000	0.000000	0.000000	0.000000	.001203	0.000000	0.000000	0.000000	0.000000	-M
		+M	.999750	0.000000	0.000000	0.000000	0.000000	0.000000	.022331	0.000000		+M
ENERGY	HPAL		M=8	M=7	M=6	M=5	M=4	M=3	M=2	M=1	M=0	

```
Y=-.60  X=-.80    OAS=3.1727E+02*W246    B4F4/B2F2= 1.20E+00  B4F4/B6F6=-4.00E+00  B2F2/B6F6=-3.33E+00   B4F4= 4.80E-01
   ENERGY      HPAL        M=8       M=7       M=6       M=5       M=4       M=3       M=2       M=1       M=0
 1.7830E+02    .9997   -M 0.000000  0.000000  0.000000  0.000000   .000992  0.000000  0.000000  0.000000  0.000000-M
                       +M  .999782  0.000000  0.000000  0.000000  0.000000  0.000000  -.020841  0.000000        +M
 1.7830E+02   -.9997   -M  .999782  0.000000  0.000000  0.000000  0.000000  0.000000  -.020841  0.000000  0.000000-M
                       +M 0.000000  0.000000  0.000000  0.000000   .000992  0.000000  0.000000  0.000000        +M
 1.2638E+02    .0000   -M 0.000000  0.000000  -.035715  0.000000  0.000000  0.000000  0.000000  0.000000   .998724-M
                       +M 0.000000  0.000000  -.035715  0.000000  0.000000  0.000000  0.000000  0.000000        +M
 1.0596E+02   -.1250   -M 0.000000  -.046944  0.000000  0.000000  0.000000  0.000000  0.000000   .997814  0.000000-M
                       +M 0.000000  0.000000  0.000000  -.046503  0.000000  0.000000  0.000000  0.000000        +M
 1.0596E+02    .1250   -M 0.000000  0.000000  0.000000  -.046503  0.000000  0.000000  0.000000  0.000000  0.000000-M
                       +M 0.000000  -.046944  0.000000  0.000000  0.000000  0.000000  0.000000   .997814        +M
 4.9994E+01    .2445   -M 0.000000  0.000000  0.000000  0.000000  -.088211  0.000000  0.000000  0.000000  0.000000-M
                       +M  .020847  0.000000  0.000000  0.000000  0.000000  0.000000   .995884  0.000000        +M
 4.9994E+01   -.2445   -M  .020847  0.000000  0.000000  0.000000  0.000000  0.000000   .995884  0.000000  0.000000-M
                       +M 0.000000  0.000000  0.000000  0.000000  -.088211  0.000000  0.000000  0.000000        +M
-1.3380E+01   0.0000   -M 0.000000  0.000000  0.000000  0.000000  0.000000   .707107  0.000000  0.000000  0.000000-M
                       +M 0.000000  0.000000  0.000000  0.000000  0.000000  -.707107  0.000000  0.000000        +M
-1.3663E+01   -.8733   -M 0.000000   .998895  0.000000  0.000000  0.000000  0.000000  0.000000   .046796  0.000000-M
                       +M 0.000000  0.000000  0.000000  -.004272  0.000000  0.000000  0.000000  0.000000        +M
-1.3663E+01    .8733   -M 0.000000  0.000000  0.000000  -.004272  0.000000  0.000000  0.000000  0.000000  0.000000-M
                       +M 0.000000   .998895  0.000000  0.000000  0.000000  0.000000  0.000000   .046796        +M
-4.1100E+01   0.0000   -M 0.000000  0.000000  0.000000  0.000000  0.000000   .707107  0.000000  0.000000  0.000000-M
                       +M 0.000000  0.000000  0.000000  0.000000  0.000000   .707107  0.000000  0.000000        +M
-1.0037E+02   -.4942   -M 0.000000  0.000000  0.000000  0.000000   .996101  0.000000  0.000000  0.000000  0.000000-M
                       +M  .000850  0.000000  0.000000  0.000000  0.000000  0.000000   .088213  0.000000        +M
-1.0037E+02    .4942   -M  .000850  0.000000  0.000000  0.000000  0.000000  0.000000   .088213  0.000000  0.000000-M
                       +M 0.000000  0.000000  0.000000  0.000000   .996101  0.000000  0.000000  0.000000        +M
-1.1688E+02    .0000   -M 0.000000  0.000000  -.707107  0.000000  0.000000  0.000000  0.000000  0.000000   .000000-M
                       +M 0.000000  0.000000   .707107  0.000000  0.000000  0.000000  0.000000  0.000000        +M
-1.1750E+02   -.0000   -M 0.000000  0.000000   .706204  0.000000  0.000000  0.000000  0.000000  0.000000   .050508-M
                       +M 0.000000  0.000000   .706204  0.000000  0.000000  0.000000  0.000000  0.000000        +M
-1.3897E+02   -.6234   -M 0.000000  0.000000  0.000000   .998909  0.000000  0.000000  0.000000  0.000000  0.000000-M
                       +M 0.000000   .002086  0.000000  0.000000  0.000000  0.000000  0.000000   .046652        +M
-1.3897E+02    .6234   -M 0.000000   .002086  0.000000  0.000000  0.000000  0.000000  0.000000   .046652  0.000000-M
                       +M 0.000000  0.000000  0.000000   .998909  0.000000  0.000000  0.000000  0.000000        +M
   ENERGY      HPAL        M=8       M=7       M=6       M=5       M=4       M=3       M=2       M=1       M=0

Y=-.60  X=-.60    OAS=2.3844E+02*W246    B4F4/B2F2= 9.00E-01  B4F4/B6F6=-1.50E+00  B2F2/B6F6=-1.67E+00   B4F4= 3.60E-01
   ENERGY      HPAL        M=8       M=7       M=6       M=5       M=4       M=3       M=2       M=1       M=0
 1.2242E+02    .9969   -M 0.000000  0.000000  0.000000  0.000000   .007699  0.000000  0.000000  0.000000  0.000000-M
                       +M  .997974  0.000000  0.000000  0.000000  0.000000  0.000000  -.063160  0.000000        +M
 1.2242E+02   -.9969   -M  .997974  0.000000  0.000000  0.000000  0.000000  0.000000  -.063160  0.000000  0.000000-M
                       +M 0.000000  0.000000  0.000000  0.000000   .007699  0.000000  0.000000  0.000000        +M
 1.1314E+02   -.0000   -M 0.000000  0.000000  -.091809  0.000000  0.000000  0.000000  0.000000  0.000000   .991535-M
                       +M 0.000000  0.000000  -.091809  0.000000  0.000000  0.000000  0.000000  0.000000        +M
 9.3493E+01   -.1322   -M 0.000000  -.145901  0.000000  0.000000  0.000000  0.000000  0.000000   .983361  0.000000-M
                       +M 0.000000  0.000000  0.000000  -.108230  0.000000  0.000000  0.000000  0.000000        +M
 9.3493E+01    .1322   -M 0.000000  0.000000  0.000000  -.108230  0.000000  0.000000  0.000000  0.000000  0.000000-M
                       +M 0.000000  -.145901  0.000000  0.000000  0.000000  0.000000  0.000000   .983361        +M
 3.9239E+01    .2249   -M 0.000000  0.000000  0.000000  0.000000  -.193622  0.000000  0.000000  0.000000  0.000000-M
                       +M  .063454  0.000000  0.000000  0.000000  0.000000  0.000000   .979022  0.000000        +M
 3.9239E+01   -.2249   -M  .063454  0.000000  0.000000  0.000000  0.000000  0.000000   .979022  0.000000  0.000000-M
                       +M 0.000000  0.000000  0.000000  0.000000  -.193622  0.000000  0.000000  0.000000        +M
 1.6165E+01    .8585   -M 0.000000  0.000000  0.000000  -.025300  0.000000  0.000000  0.000000  0.000000  0.000000-M
                       +M 0.000000   .989256  0.000000  0.000000  0.000000  0.000000  0.000000   .143991        +M
 1.6165E+01   -.8585   -M 0.000000   .989256  0.000000  0.000000  0.000000  0.000000  0.000000   .143991  0.000000-M
                       +M 0.000000  0.000000  0.000000  -.025300  0.000000  0.000000  0.000000  0.000000        +M
-8.1600E+00   0.0000   -M 0.000000  0.000000  0.000000  0.000000  0.000000   .707107  0.000000  0.000000  0.000000-M
                       +M 0.000000  0.000000  0.000000  0.000000  0.000000  -.707107  0.000000  0.000000        +M
-6.3600E+01   0.0000   -M 0.000000  0.000000  0.000000  0.000000  0.000000   .707107  0.000000  0.000000  0.000000-M
                       +M 0.000000  0.000000  0.000000  0.000000  0.000000   .707107  0.000000  0.000000        +M
-7.4760E+01   -.0000   -M 0.000000  0.000000  -.707107  0.000000  0.000000  0.000000  0.000000  0.000000  -.000000-M
                       +M 0.000000  0.000000   .707107  0.000000  0.000000  0.000000  0.000000  0.000000        +M
-7.7982E+01    .0000   -M 0.000000  0.000000   .701121  0.000000  0.000000  0.000000  0.000000  0.000000   .129837-M
                       +M 0.000000  0.000000   .701121  0.000000  0.000000  0.000000  0.000000  0.000000        +M
-9.9619E+01   -.4718   -M 0.000000  0.000000  0.000000  0.000000   .981046  0.000000  0.000000  0.000000  0.000000-M
                       +M  .004692  0.000000  0.000000  0.000000  0.000000  0.000000   .193719  0.000000        +M
-9.9619E+01    .4718   -M  .004692  0.000000  0.000000  0.000000  0.000000  0.000000   .193719  0.000000  0.000000-M
                       +M 0.000000  0.000000  0.000000  0.000000   .981046  0.000000  0.000000  0.000000        +M
-1.1602E+02    .6157   -M 0.000000   .009295  0.000000  0.000000  0.000000  0.000000  0.000000   .110759  0.000000-M
                       +M 0.000000  0.000000  0.000000   .993804  0.000000  0.000000  0.000000  0.000000        +M
-1.1602E+02   -.6157   -M 0.000000  0.000000  0.000000   .993804  0.000000  0.000000  0.000000  0.000000  0.000000-M
                       +M 0.000000   .009295  0.000000  0.000000  0.000000  0.000000  0.000000   .110759        +M
   ENERGY      HPAL        M=8       M=7       M=6       M=5       M=4       M=3       M=2       M=1       M=0

Y=-.60  X=-.40    OAS=2.0653E+02*W246    B4F4/B2F2= 6.00E-01  B4F4/B6F6=-6.67E-01  B2F2/B6F6=-1.11E+00   B4F4= 2.40E-01
   ENERGY      HPAL        M=8       M=7       M=6       M=5       M=4       M=3       M=2       M=1       M=0
 1.0405E+02    .0000   -M 0.000000  0.000000  -.184329  0.000000  0.000000  0.000000  0.000000  0.000000   .965425-M
                       +M 0.000000  0.000000  -.184329  0.000000  0.000000  0.000000  0.000000  0.000000        +M
 8.7487E+01    .2189   -M 0.000000  0.000000  0.000000  -.175100  0.000000  0.000000  0.000000  0.000000  0.000000-M
                       +M 0.000000  -.394779  0.000000  0.000000  0.000000  0.000000  0.000000   .901936        +M
 8.7487E+01   -.2189   -M 0.000000  -.394779  0.000000  0.000000  0.000000  0.000000  0.000000   .901936  0.000000-M
                       +M 0.000000  0.000000  0.000000  -.175100  0.000000  0.000000  0.000000  0.000000        +M
 6.7636E+01    .9634   -M 0.000000  0.000000  0.000000  0.000000   .052470  0.000000  0.000000  0.000000  0.000000-M
                       +M  .976699  0.000000  0.000000  0.000000  0.000000  0.000000  -.208100  0.000000        +M
 6.7636E+01   -.9634   -M  .976699  0.000000  0.000000  0.000000  0.000000  0.000000  -.208100  0.000000  0.000000-M
                       +M 0.000000  0.000000  0.000000  0.000000   .052470  0.000000  0.000000  0.000000        +M
 4.1983E+01    .7502   -M 0.000000  0.000000  0.000000  -.100169  0.000000  0.000000  0.000000  0.000000  0.000000-M
                       +M 0.000000   .918479  0.000000  0.000000  0.000000  0.000000  0.000000   .382573        +M
 4.1983E+01   -.7502   -M 0.000000   .918479  0.000000  0.000000  0.000000  0.000000  0.000000   .382573  0.000000-M
                       +M 0.000000  0.000000  0.000000  -.100169  0.000000  0.000000  0.000000  0.000000        +M
 3.1013E+01    .2146   -M 0.000000  0.000000  0.000000  0.000000  -.305019  0.000000  0.000000  0.000000  0.000000-M
                       +M  .214103  0.000000  0.000000  0.000000  0.000000  0.000000   .927967  0.000000        +M
 3.1013E+01   -.2146   -M  .214103  0.000000  0.000000  0.000000  0.000000  0.000000   .927967  0.000000  0.000000-M
                       +M 0.000000  0.000000  0.000000  0.000000  -.305019  0.000000  0.000000  0.000000        +M
-2.9400E+00   0.0000   -M 0.000000  0.000000  0.000000  0.000000  0.000000   .707107  0.000000  0.000000  0.000000-M
                       +M 0.000000  0.000000  0.000000  0.000000  0.000000  -.707107  0.000000  0.000000        +M
-3.2640E+01   -.0000   -M 0.000000  0.000000  -.707107  0.000000  0.000000  0.000000  0.000000  0.000000  -.000000-M
                       +M 0.000000  0.000000   .707107  0.000000  0.000000  0.000000  0.000000  0.000000        +M
-4.2606E+01    .0000   -M 0.000000  0.000000   .682659  0.000000  0.000000  0.000000  0.000000  0.000000   .260681-M
                       +M 0.000000  0.000000   .682659  0.000000  0.000000  0.000000  0.000000  0.000000        +M
-8.6100E+01   0.0000   -M 0.000000  0.000000  0.000000  0.000000  0.000000   .707107  0.000000  0.000000  0.000000-M
                       +M 0.000000  0.000000  0.000000  0.000000  0.000000   .707107  0.000000  0.000000        +M
-9.5510E+01   -.5941   -M 0.000000  0.000000  0.000000   .979442  0.000000  0.000000  0.000000  0.000000  0.000000-M
                       +M 0.000000   .023358  0.000000  0.000000  0.000000  0.000000  0.000000   .200371        +M
-9.5510E+01    .5941   -M 0.000000   .023358  0.000000  0.000000  0.000000  0.000000  0.000000   .200371  0.000000-M
                       +M 0.000000  0.000000  0.000000   .979442  0.000000  0.000000  0.000000  0.000000        +M
-1.0249E+02    .4280   -M  .014784  0.000000  0.000000  0.000000  0.000000  0.000000   .309146  0.000000  0.000000-M
                       +M 0.000000  0.000000  0.000000  0.000000   .950900  0.000000  0.000000  0.000000        +M
-1.0249E+02   -.4280   -M 0.000000  0.000000  0.000000  0.000000   .950900  0.000000  0.000000  0.000000  0.000000-M
                       +M  .014784  0.000000  0.000000  0.000000  0.000000  0.000000   .309146  0.000000        +M
   ENERGY      HPAL        M=8       M=7       M=6       M=5       M=4       M=3       M=2       M=1       M=0
```

Y=-.60 X=-.20 OAS=2.1326E+02*W246 B4F4/B2F2= 3.00E-01 B4F4/B6F6=-2.50E-01 B2F2/B6F6=-8.33E-01 B4F4= 1.20E-01

ENERGY	HPAL		M=8	M=7	M=6	M=5	M=4	M=3	M=2	M=1	M=0
1.0389E+02	-.0000	-M	0.000000	0.000000	-.326836	0.000000	0.000000	0.000000	0.000000	0.000000	.886768-M
		+M	0.000000	0.000000	-.326836	0.000000	0.000000	0.000000	0.000000	0.000000	+M
9.8314E+01	.5430	-M	0.000000	0.000000	0.000000	.173988	0.000000	0.000000	0.000000	0.000000	0.000000-M
		+M	0.000000	.766535	0.000000	0.000000	0.000000	0.000000	0.000000	-.618184	+M
9.8314E+01	-.5430	-M	0.000000	.766535	0.000000	0.000000	0.000000	0.000000	0.000000	-.618184	0.000000-M
		+M	0.000000	0.000000	0.000000	.173988	0.000000	0.000000	0.000000	0.000000	+M
5.5066E+01	.3759	-M	0.000000	0.000000	0.000000	-.275308	0.000000	0.000000	0.000000	0.000000	0.000000-M
		+M	0.000000	.640590	0.000000	0.000000	0.000000	0.000000	0.000000	.716833	+M
5.5066E+01	-.3759	-M	0.000000	.640590	0.000000	0.000000	0.000000	0.000000	0.000000	.716833	0.000000-M
		+M	0.000000	0.000000	0.000000	-.275308	0.000000	0.000000	0.000000	0.000000	+M
3.3712E+01	.2511	-M	0.000000	0.000000	0.000000	0.000000	-.376972	0.000000	0.000000	0.000000	0.000000-M
		+M	-.378922	0.000000	0.000000	0.000000	0.000000	0.000000	.845169	0.000000	+M
3.3712E+01	-.2511	-M	.378922	0.000000	0.000000	0.000000	0.000000	0.000000	-.845169	0.000000	0.000000-M
		+M	0.000000	0.000000	0.000000	0.000000	.376972	0.000000	0.000000	0.000000	+M
9.4800E+00	-.0000	-M	0.000000	0.000000	-.707107	0.000000	0.000000	0.000000	0.000000	0.000000	-.000000-M
		+M	0.000000	0.000000	.707107	0.000000	0.000000	0.000000	0.000000	0.000000	+M
5.9385E+00	.8637	-M	0.000000	0.000000	0.000000	0.000000	-.191692	0.000000	0.000000	0.000000	0.000000-M
		+M	.924649	0.000000	0.000000	0.000000	0.000000	0.000000	.329056	0.000000	+M
5.9385E+00	-.8637	-M	.924649	0.000000	0.000000	0.000000	0.000000	0.000000	.329056	0.000000	0.000000-M
		+M	0.000000	0.000000	0.000000	0.000000	-.191692	0.000000	0.000000	0.000000	+M
2.2800E+00	0.0000	-M	0.000000	0.000000	0.000000	0.000000	0.000000	.707107	0.000000	0.000000	0.000000-M
		+M	0.000000	0.000000	0.000000	0.000000	0.000000	-.707107	0.000000	0.000000	+M
-1.6170E+01	.0000	-M	0.000000	0.000000	.627039	0.000000	0.000000	0.000000	0.000000	0.000000	.462216-M
		+M	0.000000	0.000000	.627039	0.000000	0.000000	0.000000	0.000000	0.000000	+M
-7.9100E+01	.5439	-M	0.000000	.045471	0.000000	0.000000	0.000000	0.000000	0.000000	.322489	0.000000-M
		+M	0.000000	0.000000	0.000000	.945481	0.000000	0.000000	0.000000	0.000000	+M
-7.9100E+01	-.5439	-M	0.000000	0.000000	0.000000	.945481	0.000000	0.000000	0.000000	0.000000	0.000000-M
		+M	0.000000	.045471	0.000000	0.000000	0.000000	0.000000	0.000000	.322489	+M
-1.0860E+02	0.0000	-M	0.000000	0.000000	0.000000	0.000000	0.000000	.707107	0.000000	0.000000	0.000000-M
		+M	0.000000	0.000000	0.000000	0.000000	0.000000	.707107	0.000000	0.000000	+M
-1.0937E+02	.3648	-M	.037968	0.000000	0.000000	0.000000	0.000000	0.000000	.421203	0.000000	0.000000-M
		+M	0.000000	0.000000	0.000000	0.000000	.906171	0.000000	0.000000	0.000000	+M
-1.0937E+02	-.3648	-M	0.000000	0.000000	0.000000	0.000000	.906171	0.000000	0.000000	0.000000	0.000000-M
		+M	.037968	0.000000	0.000000	0.000000	0.000000	0.000000	.421203	0.000000	+M
ENERGY	HPAL		M=8	M=7	M=6	M=5	M=4	M=3	M=2	M=1	M=0

Y=-.60 X=0.00 OAS=2.5515E+02*W246 B4F4/B2F2= 0. B4F4/B6F6= 0. B2F2/B6F6=-6.67E-01 B4F4= 0.

ENERGY	HPAL		M=8	M=7	M=6	M=5	M=4	M=3	M=2	M=1	M=0
1.2405E+02	.7206	-M	0.000000	0.000000	0.000000	.142373	0.000000	0.000000	0.000000	0.000000	0.000000-M
		+M	0.000000	.902434	0.000000	0.000000	0.000000	0.000000	0.000000	-.406624	+M
1.2405E+02	-.7206	-M	0.000000	.902434	0.000000	0.000000	0.000000	0.000000	0.000000	-.406624	0.000000-M
		+M	0.000000	0.000000	0.000000	.142373	0.000000	0.000000	0.000000	0.000000	+M
1.1875E+02	-.0000	-M	0.000000	0.000000	-.477637	0.000000	0.000000	0.000000	0.000000	0.000000	.737377-M
		+M	0.000000	0.000000	-.477637	0.000000	0.000000	0.000000	0.000000	0.000000	+M
5.9606E+01	-.1055	-M	0.000000	.424568	0.000000	0.000000	0.000000	0.000000	0.000000	.783229	0.000000-M
		+M	0.000000	0.000000	0.000000	-.454195	0.000000	0.000000	0.000000	0.000000	+M
5.9606E+01	.1055	-M	0.000000	0.000000	0.000000	-.454195	0.000000	0.000000	0.000000	0.000000	0.000000-M
		+M	0.000000	.424568	0.000000	0.000000	0.000000	0.000000	0.000000	.783229	+M
5.1600E+01	.0000	-M	0.000000	0.000000	-.707107	0.000000	0.000000	0.000000	0.000000	0.000000	.000000-M
		+M	0.000000	0.000000	.707107	0.000000	0.000000	0.000000	0.000000	0.000000	+M
3.2058E+01	.0766	-M	0.000000	0.000000	0.000000	0.000000	-.502498	0.000000	0.000000	0.000000	0.000000-M
		+M	-.145800	0.000000	0.000000	0.000000	0.000000	0.000000	.852196	0.000000	+M
3.2058E+01	-.0766	-M	-.145800	0.000000	0.000000	0.000000	0.000000	0.000000	.852196	0.000000	0.000000-M
		+M	0.000000	0.000000	0.000000	0.000000	-.502498	0.000000	0.000000	0.000000	+M
7.5000E+00	0.0000	-M	0.000000	0.000000	0.000000	0.000000	0.000000	.707107	0.000000	0.000000	0.000000-M
		+M	0.000000	0.000000	0.000000	0.000000	0.000000	-.707107	0.000000	0.000000	+M
-4.7490E+00	-.0000	-M	0.000000	0.000000	.521405	0.000000	0.000000	0.000000	0.000000	0.000000	.675481-M
		+M	0.000000	0.000000	.521405	0.000000	0.000000	0.000000	0.000000	0.000000	+M
-4.7445E+01	-.9592	-M	.984827	0.000000	0.000000	0.000000	0.000000	0.000000	.076699	0.000000	0.000000-M
		+M	0.000000	0.000000	0.000000	0.000000	-.155673	0.000000	0.000000	0.000000	+M
-4.7445E+01	.9592	-M	0.000000	0.000000	0.000000	0.000000	-.155673	0.000000	0.000000	0.000000	0.000000-M
		+M	.984827	0.000000	0.000000	0.000000	0.000000	0.000000	.076699	0.000000	+M
-6.9056E+01	.4511	-M	0.000000	.073176	0.000000	0.000000	0.000000	0.000000	0.000000	.470328	0.000000-M
		+M	0.000000	0.000000	0.000000	.879452	0.000000	0.000000	0.000000	0.000000	+M
-6.9056E+01	-.4511	-M	0.000000	0.000000	0.000000	.879452	0.000000	0.000000	0.000000	0.000000	0.000000-M
		+M	0.000000	.073176	0.000000	0.000000	0.000000	0.000000	0.000000	.470328	+M
-1.2021E+02	.2858	-M	.094123	0.000000	0.000000	0.000000	0.000000	0.000000	.517571	0.000000	0.000000-M
		+M	0.000000	0.000000	0.000000	0.000000	.850448	0.000000	0.000000	0.000000	+M
-1.2021E+02	-.2858	-M	0.000000	0.000000	0.000000	0.000000	.850448	0.000000	0.000000	0.000000	0.000000-M
		+M	.094123	0.000000	0.000000	0.000000	0.000000	0.000000	.517571	0.000000	+M
-1.3110E+02	0.0000	-M	0.000000	0.000000	0.000000	0.000000	0.000000	.707107	0.000000	0.000000	0.000000-M
		+M	0.000000	0.000000	0.000000	0.000000	0.000000	.707107	0.000000	0.000000	+M
ENERGY	HPAL		M=8	M=7	M=6	M=5	M=4	M=3	M=2	M=1	M=0

Y=-.60 X= .20 OAS=2.1168E+02*W246 B4F4/B2F2=-3.00E-01 B4F4/B6F6= 2.50E-01 B2F2/B6F6=-8.33E-01 B4F4=-1.20E-01

ENERGY	HPAL		M=8	M=7	M=6	M=5	M=4	M=3	M=2	M=1	M=0
1.0812E+02	.8058	-M	0.000000	0.000000	0.000000	.111403	0.000000	0.000000	0.000000	0.000000	0.000000-M
		+M	0.000000	.959275	0.000000	0.000000	0.000000	0.000000	0.000000	-.259579	+M
1.0812E+02	-.8058	-M	0.000000	.959275	0.000000	0.000000	0.000000	0.000000	0.000000	-.259579	0.000000-M
		+M	0.000000	0.000000	0.000000	.111403	0.000000	0.000000	0.000000	0.000000	+M
1.0331E+02	.0000	-M	0.000000	0.000000	.612933	0.000000	0.000000	0.000000	0.000000	0.000000	-.498625-M
		+M	0.000000	0.000000	.612933	0.000000	0.000000	0.000000	0.000000	0.000000	+M
7.5000E+01	-.0000	-M	0.000000	0.000000	-.707107	0.000000	0.000000	0.000000	0.000000	0.000000	.000000-M
		+M	0.000000	0.000000	.707107	0.000000	0.000000	0.000000	0.000000	0.000000	+M
4.4974E+01	-.1935	-M	0.000000	0.000000	0.000000	-.701950	0.000000	0.000000	0.000000	0.000000	0.000000-M
		+M	0.000000	.260855	0.000000	0.000000	0.000000	0.000000	0.000000	.662737	+M
4.4974E+01	.1935	-M	0.000000	-.260855	0.000000	0.000000	0.000000	0.000000	0.000000	-.662737	0.000000-M
		+M	0.000000	0.000000	0.000000	.701950	0.000000	0.000000	0.000000	0.000000	+M
2.0761E+01	.0397	-M	-.084522	0.000000	0.000000	0.000000	0.000000	0.000000	.774197	0.000000	0.000000-M
		+M	0.000000	0.000000	0.000000	0.000000	-.627275	0.000000	0.000000	0.000000	+M
2.0761E+01	-.0397	-M	0.000000	0.000000	0.000000	0.000000	-.627275	0.000000	0.000000	0.000000	0.000000-M
		+M	-.084522	0.000000	0.000000	0.000000	0.000000	0.000000	.774197	0.000000	+M
7.3200E+00	0.0000	-M	0.000000	0.000000	0.000000	0.000000	0.000000	.707107	0.000000	0.000000	0.000000-M
		+M	0.000000	0.000000	0.000000	0.000000	0.000000	-.707107	0.000000	0.000000	+M
-1.0547E+01	0.0000	-M	0.000000	0.000000	.352581	0.000000	0.000000	0.000000	0.000000	0.000000	.866818-M
		+M	0.000000	0.000000	.352581	0.000000	0.000000	0.000000	0.000000	0.000000	+M
-4.3532E+01	.2373	-M	0.000000	.108381	0.000000	0.000000	0.000000	0.000000	0.000000	.702423	0.000000-M
		+M	0.000000	0.000000	0.000000	.703460	0.000000	0.000000	0.000000	0.000000	+M
-4.3532E+01	-.2373	-M	0.000000	0.000000	0.000000	.703460	0.000000	0.000000	0.000000	0.000000	0.000000-M
		+M	0.000000	.108381	0.000000	0.000000	0.000000	0.000000	0.000000	.702423	+M
-7.4768E+01	.7114	-M	0.000000	0.000000	0.000000	0.000000	-.406552	0.000000	0.000000	0.000000	0.000000-M
		+M	.883431	0.000000	0.000000	0.000000	0.000000	0.000000	-.232951	0.000000	+M
-7.4768E+01	-.7114	-M	.883431	0.000000	0.000000	0.000000	0.000000	0.000000	-.232951	0.000000	0.000000-M
		+M	0.000000	0.000000	0.000000	0.000000	-.406552	0.000000	0.000000	0.000000	+M
-9.1313E+01	.0784	-M	0.000000	0.000000	0.000000	0.000000	.664260	0.000000	0.000000	0.000000	0.000000-M
		+M	.460876	0.000000	0.000000	0.000000	0.000000	0.000000	.588517	0.000000	+M
-9.1313E+01	-.0784	-M	.460876	0.000000	0.000000	0.000000	0.000000	0.000000	.588517	0.000000	0.000000-M
		+M	0.000000	0.000000	0.000000	0.000000	.664260	0.000000	0.000000	0.000000	+M
-1.0356E+02	0.0000	-M	0.000000	0.000000	0.000000	0.000000	0.000000	.707107	0.000000	0.000000	0.000000-M
		+M	0.000000	0.000000	0.000000	0.000000	0.000000	.707107	0.000000	0.000000	+M
ENERGY	HPAL		M=8	M=7	M=6	M=5	M=4	M=3	M=2	M=1	M=0

Y=-.60 X= .40 OAS=2.1840E+02*W246 B4F4/B2F2=-6.00E-01 B4F4/B6F6= 6.67E-01 B2F2/B6F6=-1.11E+00 B4F4=-2.40E-01

ENERGY	HPAL	M=8	M=7	M=6	M=5	M=4	M=3	M=2	M=1	M=0
1.0846E+02	.0000	-M 0.000000	0.000000	.682197	0.000000	0.000000	0.000000	0.000000	0.000000	-.263087-M
		+M 0.000000	0.000000	.682197	0.000000	0.000000	0.000000	0.000000	0.000000	+M
9.8400E+01	-.0000	-M 0.000000	0.000000	-.707107	0.000000	0.000000	0.000000	0.000000	0.000000	.000000-M
		+M 0.000000	0.000000	.707107	0.000000	0.000000	0.000000	0.000000	0.000000	+M
9.5408E+01	.8402	-M 0.000000	0.000000	0.000000	.102745	0.000000	0.000000	0.000000	0.000000	0.000000-M
		+M 0.000000	.981913	0.000000	0.000000	0.000000	0.000000	0.000000	-.159031	+M
9.5408E+01	-.8402	-M 0.000000	.981913	0.000000	0.000000	0.000000	0.000000	0.000000	-.159031	0.000000-M
		+M 0.000000	0.000000	0.000000	.102745	0.000000	0.000000	0.000000	0.000000	+M
5.5040E+01	.5036	-M 0.000000	-.152177	0.000000	0.000000	0.000000	0.000000	0.000000	-.340023	0.000000-M
		+M 0.000000	0.000000	0.000000	.928023	0.000000	0.000000	0.000000	0.000000	+M
5.5040E+01	-.5036	-M 0.000000	0.000000	0.000000	.928023	0.000000	0.000000	0.000000	0.000000	0.000000-M
		+M 0.000000	-.152177	0.000000	0.000000	0.000000	0.000000	0.000000	-.340023	+M
1.9147E+01	.2475	-M .036382	0.000000	0.000000	0.000000	0.000000	0.000000	-.577976	0.000000	0.000000-M
		+M 0.000000	0.000000	0.000000	0.000000	.815243	0.000000	0.000000	0.000000	+M
1.9147E+01	-.2475	-M 0.000000	0.000000	0.000000	0.000000	.815243	0.000000	0.000000	0.000000	0.000000-M
		+M .036382	0.000000	0.000000	0.000000	0.000000	0.000000	-.577976	0.000000	+M
7.1400E+00	0.0000	-M 0.000000	0.000000	0.000000	0.000000	0.000000	.707107	0.000000	0.000000	0.000000-M
		+M 0.000000	0.000000	0.000000	0.000000	0.000000	-.707107	0.000000	0.000000	+M
-3.6944E+01	-.0000	-M 0.000000	0.000000	.186030	0.000000	0.000000	0.000000	0.000000	0.000000	.964772-M
		+M 0.000000	0.000000	.186030	0.000000	0.000000	0.000000	0.000000	0.000000	+M
-4.5928E+01	-.0384	-M 0.000000	.112649	0.000000	0.000000	0.000000	0.000000	0.000000	.926873	0.000000-M
		+M 0.000000	0.000000	0.000000	.358074	0.000000	0.000000	0.000000	0.000000	+M
-4.5928E+01	.0384	-M 0.000000	0.000000	0.000000	.358074	0.000000	0.000000	0.000000	0.000000	0.000000-M
		+M 0.000000	.112649	0.000000	0.000000	0.000000	0.000000	0.000000	.926873	+M
-6.4251E+01	-.0166	-M -.145360	0.000000	0.000000	0.000000	0.000000	0.000000	.804046	0.000000	0.000000-M
		+M 0.000000	0.000000	0.000000	0.000000	.576525	0.000000	0.000000	0.000000	+M
-6.4251E+01	.0166	-M 0.000000	0.000000	0.000000	0.000000	.576525	0.000000	0.000000	0.000000	0.000000-M
		+M -.145360	0.000000	0.000000	0.000000	0.000000	0.000000	.804046	0.000000	+M
-7.6020E+01	0.0000	-M 0.000000	0.000000	0.000000	0.000000	0.000000	.707107	0.000000	0.000000	0.000000-M
		+M 0.000000	0.000000	0.000000	0.000000	0.000000	.707107	0.000000	0.000000	+M
-1.0994E+02	-.9809	-M .988710	0.000000	0.000000	0.000000	0.000000	0.000000	.139478	0.000000	0.000000-M
		+M 0.000000	0.000000	0.000000	0.000000	.054762	0.000000	0.000000	0.000000	+M
-1.0994E+02	.9809	-M 0.000000	0.000000	0.000000	0.000000	.054762	0.000000	0.000000	0.000000	0.000000-M
		+M .988710	0.000000	0.000000	0.000000	0.000000	0.000000	.139478	0.000000	+M
ENERGY	HPAL	M=8	M=7	M=6	M=5	M=4	M=3	M=2	M=1	M=0

Y=-.60 X= .60 OAS=2.6522E+02*W246 B4F4/B2F2=-9.00E-01 B4F4/B6F6= 1.50E+00 B2F2/B6F6=-1.67E+00 B4F4=-3.60E-01

ENERGY	HPAL	M=8	M=7	M=6	M=5	M=4	M=3	M=2	M=1	M=0
1.2488E+02	-.0000	-M 0.000000	0.000000	.701622	0.000000	0.000000	0.000000	0.000000	0.000000	-.124307-M
		+M 0.000000	0.000000	.701622	0.000000	0.000000	0.000000	0.000000	0.000000	+M
1.2180E+02	.0000	-M 0.000000	0.000000	-.707107	0.000000	0.000000	0.000000	0.000000	0.000000	-.000000-M
		+M 0.000000	0.000000	.707107	0.000000	0.000000	0.000000	0.000000	0.000000	+M
8.7311E+01	.2919	-M 0.000000	.457048	0.000000	0.000000	0.000000	0.000000	0.000000	-.162441	0.000000-M
		+M 0.000000	0.000000	0.000000	.874483	0.000000	0.000000	0.000000	0.000000	+M
8.7311E+01	-.2919	-M 0.000000	0.000000	0.000000	.874483	0.000000	0.000000	0.000000	0.000000	0.000000-M
		+M 0.000000	.457048	0.000000	0.000000	0.000000	0.000000	0.000000	-.162441	+M
8.3271E+01	-.5539	-M 0.000000	.886542	0.000000	0.000000	0.000000	0.000000	0.000000	.003877	0.000000-M
		+M 0.000000	0.000000	0.000000	-.462631	0.000000	0.000000	0.000000	0.000000	+M
8.3271E+01	.5539	-M 0.000000	0.000000	0.000000	-.462631	0.000000	0.000000	0.000000	0.000000	0.000000-M
		+M 0.000000	.886542	0.000000	0.000000	0.000000	0.000000	0.000000	.003877	+M
3.4697E+01	.4350	-M .009043	0.000000	0.000000	0.000000	0.000000	0.000000	-.294217	0.000000	0.000000-M
		+M 0.000000	0.000000	0.000000	0.000000	.955696	0.000000	0.000000	0.000000	+M
3.4697E+01	-.4350	-M 0.000000	0.000000	0.000000	0.000000	.955696	0.000000	0.000000	0.000000	0.000000-M
		+M .009043	0.000000	0.000000	0.000000	0.000000	0.000000	-.294217	0.000000	+M
6.9600E+00	0.0000	-M 0.000000	0.000000	0.000000	0.000000	0.000000	.707107	0.000000	0.000000	0.000000-M
		+M 0.000000	0.000000	0.000000	0.000000	0.000000	-.707107	0.000000	0.000000	+M
-4.8480E+01	0.0000	-M 0.000000	0.000000	0.000000	0.000000	0.000000	.707107	0.000000	0.000000	0.000000-M
		+M 0.000000	0.000000	0.000000	0.000000	0.000000	.707107	0.000000	0.000000	+M
-5.9117E+01	-.1881	-M -.063306	0.000000	0.000000	0.000000	0.000000	0.000000	.953649	0.000000	0.000000-M
		+M 0.000000	0.000000	0.000000	0.000000	.294186	0.000000	0.000000	0.000000	+M
-5.9117E+01	.1881	-M 0.000000	0.000000	0.000000	0.000000	.294186	0.000000	0.000000	0.000000	0.000000-M
		+M -.063306	0.000000	0.000000	0.000000	0.000000	0.000000	.953649	0.000000	+M
-7.1102E+01	.1129	-M 0.000000	0.000000	0.000000	.145782	0.000000	0.000000	0.000000	0.000000	0.000000-M
		+M 0.000000	.071760	0.000000	0.000000	0.000000	0.000000	0.000000	.986711	+M
-7.1102E+01	-.1129	-M 0.000000	.071760	0.000000	0.000000	0.000000	0.000000	0.000000	.986711	0.000000-M
		+M 0.000000	0.000000	0.000000	.145782	0.000000	0.000000	0.000000	0.000000	+M
-7.4602E+01	-.0000	-M 0.000000	0.000000	.087898	0.000000	0.000000	0.000000	0.000000	0.000000	.992244-M
		+M 0.000000	0.000000	.087898	0.000000	0.000000	0.000000	0.000000	0.000000	+M
-1.4034E+02	-.9969	-M .997953	0.000000	0.000000	0.000000	0.000000	0.000000	.063161	0.000000	0.000000-M
		+M 0.000000	0.000000	0.000000	0.000000	.010002	0.000000	0.000000	0.000000	+M
-1.4034E+02	.9969	-M 0.000000	0.000000	0.000000	0.000000	.010002	0.000000	0.000000	0.000000	0.000000-M
		+M .997953	0.000000	0.000000	0.000000	0.000000	0.000000	.063161	0.000000	+M
ENERGY	HPAL	M=8	M=7	M=6	M=5	M=4	M=3	M=2	M=1	M=0

Y=-.60 X= .80 OAS=3.1705E+02*W246 B4F4/B2F2=-1.20E+00 B4F4/B6F6= 4.00E+00 B2F2/B6F6=-3.33E+00 B4F4=-4.80E-01

ENERGY	HPAL	M=8	M=7	M=6	M=5	M=4	M=3	M=2	M=1	M=0
1.4578E+02	-.0000	-M 0.000000	0.000000	.706328	0.000000	0.000000	0.000000	0.000000	0.000000	-.046915-M
		+M 0.000000	0.000000	.706328	0.000000	0.000000	0.000000	0.000000	0.000000	+M
1.4520E+02	.0000	-M 0.000000	0.000000	-.707107	0.000000	0.000000	0.000000	0.000000	0.000000	-.000000-M
		+M 0.000000	0.000000	.707107	0.000000	0.000000	0.000000	0.000000	0.000000	+M
1.2421E+02	.6231	-M 0.000000	.005606	0.000000	0.000000	0.000000	0.000000	0.000000	-.050168	0.000000-M
		+M 0.000000	0.000000	0.000000	.998725	0.000000	0.000000	0.000000	0.000000	+M
1.2421E+02	-.6231	-M 0.000000	0.000000	0.000000	.998725	0.000000	0.000000	0.000000	0.000000	0.000000-M
		+M 0.000000	.005606	0.000000	0.000000	0.000000	0.000000	0.000000	-.050168	+M
7.4135E+01	.8742	-M 0.000000	0.000000	0.000000	-.007177	0.000000	0.000000	0.000000	0.000000	0.000000-M
		+M 0.000000	.999488	0.000000	0.000000	0.000000	0.000000	0.000000	-.031195	+M
7.4135E+01	-.8742	-M 0.000000	.999488	0.000000	0.000000	0.000000	0.000000	0.000000	-.031195	0.000000-M
		+M 0.000000	0.000000	0.000000	-.007177	0.000000	0.000000	0.000000	0.000000	+M
6.3436E+01	.4922	-M .001168	0.000000	0.000000	0.000000	0.000000	0.000000	-.102092	0.000000	0.000000-M
		+M 0.000000	0.000000	0.000000	0.000000	.994774	0.000000	0.000000	0.000000	+M
6.3436E+01	-.4922	-M 0.000000	0.000000	0.000000	0.000000	.994774	0.000000	0.000000	0.000000	0.000000-M
		+M .001168	0.000000	0.000000	0.000000	0.000000	0.000000	-.102092	0.000000	+M
6.7800E+00	0.0000	-M 0.000000	0.000000	0.000000	0.000000	0.000000	.707107	0.000000	0.000000	0.000000-M
		+M 0.000000	0.000000	0.000000	0.000000	0.000000	-.707107	0.000000	0.000000	+M
-2.0940E+01	0.0000	-M 0.000000	0.000000	0.000000	0.000000	0.000000	.707107	0.000000	0.000000	0.000000-M
		+M 0.000000	0.000000	0.000000	0.000000	0.000000	.707107	0.000000	0.000000	+M
-6.6648E+01	-.2427	-M -.025535	0.000000	0.000000	0.000000	0.000000	0.000000	.994448	0.000000	0.000000-M
		+M 0.000000	0.000000	0.000000	0.000000	.102088	0.000000	0.000000	0.000000	+M
-6.6648E+01	.2427	-M 0.000000	0.000000	0.000000	0.000000	.102088	0.000000	0.000000	0.000000	0.000000-M
		+M -.025535	0.000000	0.000000	0.000000	0.000000	0.000000	.994448	0.000000	+M
-1.0391E+02	-.1239	-M 0.000000	.031515	0.000000	0.000000	0.000000	0.000000	0.000000	.998253	0.000000-M
		+M 0.000000	0.000000	0.000000	.049968	0.000000	0.000000	0.000000	0.000000	+M
-1.0391E+02	.1239	-M 0.000000	0.000000	0.000000	.049968	0.000000	0.000000	0.000000	0.000000	0.000000-M
		+M 0.000000	.031515	0.000000	0.000000	0.000000	0.000000	0.000000	.998253	+M
-1.1674E+02	.0000	-M 0.000000	0.000000	.033174	0.000000	0.000000	0.000000	0.000000	0.000000	.998899-M
		+M 0.000000	0.000000	.033174	0.000000	0.000000	0.000000	0.000000	0.000000	+M
-1.7127E+02	-.9995	-M .999673	0.000000	0.000000	0.000000	0.000000	0.000000	.025521	0.000000	0.000000-M
		+M 0.000000	0.000000	0.000000	0.000000	.001445	0.000000	0.000000	0.000000	+M
-1.7127E+02	.9995	-M 0.000000	0.000000	0.000000	0.000000	.001445	0.000000	0.000000	0.000000	0.000000-M
		+M .999673	0.000000	0.000000	0.000000	0.000000	0.000000	.025521	0.000000	+M
ENERGY	HPAL	M=8	M=7	M=6	M=5	M=4	M=3	M=2	M=1	M=0

```
Y=-.40  X=-.80   OAS=2.2452E+02*W246   B4F4/B2F2= 5.33E-01  B4F4/B6F6=-4.00E+00  B2F2/B6F6=-7.50E+00   B4F4= 3.20E-01
  ENERGY       HPAL          M=8        M=7        M=6        M=5        M=4        M=3        M=2        M=1        M=0
 1.3219E+02    .9998    -M 0.000000  0.000000   0.000000   0.000000    .000741   0.000000   0.000000   0.000000   0.000000-M
                        +M  .999857  0.000000   0.000000   0.000000   0.000000   0.000000   -.016900   0.000000          +M
 1.3219E+02   -.9998    -M  .999857  0.000000   0.000000   0.000000   0.000000   0.000000   -.016900   0.000000   0.000000-M
                        +M 0.000000  0.000000   0.000000   0.000000    .000741   0.000000   0.000000   0.000000          +M
 7.6268E+01   0.0000    -M 0.000000  0.000000   -.038552   0.000000   0.000000   0.000000   0.000000   0.000000    .998513-M
                        +M 0.000000  0.000000   -.038552   0.000000   0.000000   0.000000   0.000000   0.000000          +M
 6.3032E+01   -.1258    -M 0.000000  -.058659   0.000000   0.000000   0.000000   0.000000   0.000000    .997082   0.000000-M
                        +M 0.000000  0.000000   0.000000   -.048848   0.000000   0.000000   0.000000   0.000000          +M
 6.3032E+01    .1258    -M 0.000000  0.000000   0.000000   -.048848   0.000000   0.000000   0.000000   0.000000   0.000000-M
                        +M 0.000000  -.058659   0.000000   0.000000   0.000000   0.000000   0.000000    .997082          +M
 2.6702E+01    .2439    -M 0.000000  0.000000   0.000000   0.000000   -.091846   0.000000   0.000000   0.000000   0.000000-M
                        +M  .016897  0.000000   0.000000   0.000000   0.000000   0.000000    .995630   0.000000          +M
 2.6702E+01   -.2439    -M  .016897  0.000000   0.000000   0.000000   0.000000   0.000000    .995630   0.000000   0.000000-M
                        +M 0.000000  0.000000   0.000000   0.000000   -.091846   0.000000   0.000000   0.000000          +M
-6.1934E-01    .8724    -M 0.000000  0.000000   0.000000   -.004873   0.000000   0.000000   0.000000   0.000000   0.000000-M
                        +M 0.000000   .998276   0.000000   0.000000   0.000000   0.000000   0.000000    .058491          +M
-6.1934E-01   -.8724    -M 0.000000   .998276   0.000000   0.000000   0.000000   0.000000   0.000000    .058491   0.000000-M
                        +M 0.000000  0.000000   0.000000   -.004873   0.000000   0.000000   0.000000   0.000000          +M
-1.3920E+01   0.0000    -M 0.000000  0.000000   0.000000   0.000000   0.000000    .707107   0.000000   0.000000   0.000000-M
                        +M 0.000000  0.000000   0.000000   0.000000   0.000000   -.707107   0.000000   0.000000          +M
-3.2400E+01   0.0000    -M 0.000000  0.000000   0.000000   0.000000   0.000000    .707107   0.000000   0.000000   0.000000-M
                        +M 0.000000  0.000000   0.000000   0.000000   0.000000    .707107   0.000000   0.000000          +M
-6.9613E+01    .4937    -M  .000815  0.000000   0.000000   0.000000   0.000000   0.000000    .091845   0.000000   0.000000-M
                        +M 0.000000  0.000000   0.000000   0.000000    .995773   0.000000   0.000000   0.000000          +M
-6.9613E+01   -.4937    -M 0.000000  0.000000   0.000000   0.000000    .995773   0.000000   0.000000   0.000000   0.000000-M
                        +M  .000815  0.000000   0.000000   0.000000   0.000000   0.000000    .091845   0.000000          +M
-7.3920E+01    .0000    -M 0.000000  0.000000   -.707107   0.000000   0.000000   0.000000   0.000000   0.000000    .000000-M
                        +M 0.000000  0.000000    .707107   0.000000   0.000000   0.000000   0.000000   0.000000          +M
-7.4368E+01   -.0000    -M 0.000000  0.000000    .706055   0.000000   0.000000   0.000000   0.000000   0.000000    .054521-M
                        +M 0.000000  0.000000    .706055   0.000000   0.000000   0.000000   0.000000   0.000000          +M
-9.2333E+01    .6232    -M 0.000000   .002002   0.000000   0.000000   0.000000   0.000000   0.000000    .049050   0.000000-M
                        +M 0.000000  0.000000   0.000000    .998794   0.000000   0.000000   0.000000   0.000000          +M
-9.2333E+01   -.6232    -M 0.000000  0.000000   0.000000    .998794   0.000000   0.000000   0.000000   0.000000   0.000000-M
                        +M 0.000000   .002002   0.000000   0.000000   0.000000   0.000000   0.000000    .049050          +M
  ENERGY       HPAL          M=8        M=7        M=6        M=5        M=4        M=3        M=2        M=1        M=0

Y=-.40  X=-.60   OAS=1.7200E+02*W246   B4F4/B2F2= 4.00E-01  B4F4/B6F6=-1.50E+00  B2F2/B6F6=-3.75E+00   B4F4= 2.40E-01
  ENERGY       HPAL          M=8        M=7        M=6        M=5        M=4        M=3        M=2        M=1        M=0
 9.4887E+01    .9983    -M 0.000000  0.000000   0.000000   0.000000    .005117   0.000000   0.000000   0.000000   0.000000-M
                        +M  .998900  0.000000   0.000000   0.000000   0.000000   0.000000   -.046610   0.000000          +M
 9.4887E+01   -.9983    -M  .998900  0.000000   0.000000   0.000000   0.000000   0.000000   -.046610   0.000000   0.000000-M
                        +M 0.000000  0.000000   0.000000   0.000000    .005117   0.000000   0.000000   0.000000          +M
 6.7651E+01   -.0000    -M 0.000000  0.000000   -.101150   0.000000   0.000000   0.000000   0.000000   0.000000    .989716-M
                        +M 0.000000  0.000000   -.101150   0.000000   0.000000   0.000000   0.000000   0.000000          +M
 5.5246E+01    .1477    -M 0.000000  0.000000   0.000000   -.113018   0.000000   0.000000   0.000000   0.000000   0.000000-M
                        +M 0.000000  -.207583   0.000000   0.000000   0.000000   0.000000   0.000000    .971667          +M
 5.5246E+01   -.1477    -M 0.000000  -.207583   0.000000   0.000000   0.000000   0.000000   0.000000    .971667   0.000000-M
                        +M 0.000000  0.000000   0.000000   -.113018   0.000000   0.000000   0.000000   0.000000          +M
 1.9708E+01   -.2210    -M  .046682  0.000000   0.000000   0.000000   0.000000   0.000000    .978261   0.000000   0.000000-M
                        +M 0.000000  0.000000   0.000000   0.000000   -.202054   0.000000   0.000000   0.000000          +M
 1.9708E+01    .2210    -M 0.000000  0.000000   0.000000   0.000000   -.202054   0.000000   0.000000   0.000000   0.000000-M
                        +M  .046682  0.000000   0.000000   0.000000   0.000000   0.000000    .978261   0.000000          +M
 1.8630E+01    .8418    -M 0.000000  0.000000   0.000000   -.033156   0.000000   0.000000   0.000000   0.000000   0.000000-M
                        +M 0.000000   .978176   0.000000   0.000000   0.000000   0.000000   0.000000    .205117          +M
 1.8630E+01   -.8418    -M 0.000000   .978176   0.000000   0.000000   0.000000   0.000000   0.000000    .205117   0.000000-M
                        +M 0.000000  0.000000   0.000000   -.033156   0.000000   0.000000   0.000000   0.000000          +M
-1.0440E+01   0.0000    -M 0.000000  0.000000   0.000000   0.000000   0.000000    .707107   0.000000   0.000000   0.000000-M
                        +M 0.000000  0.000000   0.000000   0.000000   0.000000   -.707107   0.000000   0.000000          +M
-4.5840E+01   -.0000    -M 0.000000  0.000000   -.707107   0.000000   0.000000   0.000000   0.000000   0.000000   -.000000-M
                        +M 0.000000  0.000000    .707107   0.000000   0.000000   0.000000   0.000000   0.000000          +M
-4.7400E+01   0.0000    -M 0.000000  0.000000   0.000000   0.000000   0.000000    .707107   0.000000   0.000000   0.000000-M
                        +M 0.000000  0.000000   0.000000   0.000000   0.000000    .707107   0.000000   0.000000          +M
-4.8211E+01    .0000    -M 0.000000  0.000000    .699835   0.000000   0.000000   0.000000   0.000000   0.000000    .143048-M
                        +M 0.000000  0.000000    .699835   0.000000   0.000000   0.000000   0.000000   0.000000          +M
-6.9236E+01   -.4693    -M 0.000000  0.000000   0.000000   0.000000    .979361   0.000000   0.000000   0.000000   0.000000-M
                        +M  .004412  0.000000   0.000000   0.000000   0.000000   0.000000    .202071   0.000000          +M
-6.9236E+01    .4693    -M  .004412  0.000000   0.000000   0.000000   0.000000   0.000000    .202071   0.000000   0.000000-M
                        +M 0.000000  0.000000   0.000000   0.000000    .979361   0.000000   0.000000   0.000000          +M
-7.7116E+01    .6145    -M 0.000000   .009035   0.000000   0.000000   0.000000   0.000000   0.000000    .117434   0.000000-M
                        +M 0.000000  0.000000   0.000000    .993040   0.000000   0.000000   0.000000   0.000000          +M
-7.7116E+01   -.6145    -M 0.000000  0.000000   0.000000    .993040   0.000000   0.000000   0.000000   0.000000   0.000000-M
                        +M 0.000000   .009035   0.000000   0.000000   0.000000   0.000000   0.000000    .117434          +M
  ENERGY       HPAL          M=8        M=7        M=6        M=5        M=4        M=3        M=2        M=1        M=0

Y=-.40  X=-.40   OAS=1.3367E+02*W246   B4F4/B2F2= 2.67E-01  B4F4/B6F6=-6.67E-01  B2F2/B6F6=-2.50E+00   B4F4= 1.60E-01
  ENERGY       HPAL          M=8        M=7        M=6        M=5        M=4        M=3        M=2        M=1        M=0
 6.2283E+01   0.0000    -M 0.000000  0.000000   -.207767   0.000000   0.000000   0.000000   0.000000   0.000000    .955859-M
                        +M 0.000000  0.000000   -.207767   0.000000   0.000000   0.000000   0.000000   0.000000          +M
 5.7913E+01    .9887    -M 0.000000  0.000000   0.000000   0.000000    .025710   0.000000   0.000000   0.000000   0.000000-M
                        +M  .992795  0.000000   0.000000   0.000000   0.000000   0.000000   -.117035   0.000000          +M
 5.7913E+01   -.9887    -M  .992795  0.000000   0.000000   0.000000   0.000000   0.000000   -.117035   0.000000   0.000000-M
                        +M 0.000000  0.000000   0.000000   0.000000    .025710   0.000000   0.000000   0.000000          +M
 5.4937E+01   -.3954    -M 0.000000  -.619941   0.000000   0.000000   0.000000   0.000000   0.000000    .769338   0.000000-M
                        +M 0.000000  0.000000   0.000000   -.154249   0.000000   0.000000   0.000000   0.000000          +M
 5.4937E+01    .3954    -M 0.000000  0.000000   0.000000   -.154249   0.000000   0.000000   0.000000   0.000000   0.000000-M
                        +M 0.000000  -.619941   0.000000   0.000000   0.000000   0.000000   0.000000    .769338          +M
 3.2382E+01    .5694    -M 0.000000  0.000000   0.000000   -.150549   0.000000   0.000000   0.000000   0.000000   0.000000-M
                        +M 0.000000   .784312   0.000000   0.000000   0.000000   0.000000   0.000000    .601822          +M
 3.2382E+01   -.5694    -M 0.000000   .784312   0.000000   0.000000   0.000000   0.000000   0.000000    .601822   0.000000-M
                        +M 0.000000  0.000000   0.000000   -.150549   0.000000   0.000000   0.000000   0.000000          +M
 1.4915E+01    .1833    -M 0.000000  0.000000   0.000000   0.000000   -.321170   0.000000   0.000000   0.000000   0.000000-M
                        +M  .119070  0.000000   0.000000   0.000000   0.000000   0.000000    .939506   0.000000          +M
 1.4915E+01   -.1833    -M  .119070  0.000000   0.000000   0.000000   0.000000   0.000000    .939506   0.000000   0.000000-M
                        +M 0.000000  0.000000   0.000000   0.000000   -.321170   0.000000   0.000000   0.000000          +M
-6.9600E+00   0.0000    -M 0.000000  0.000000   0.000000   0.000000   0.000000    .707107   0.000000   0.000000   0.000000-M
                        +M 0.000000  0.000000   0.000000   0.000000   0.000000   -.707107   0.000000   0.000000          +M
-1.7760E+01   -.0000    -M 0.000000  0.000000   -.707107   0.000000   0.000000   0.000000   0.000000   0.000000   -.000000-M
                        +M 0.000000  0.000000    .707107   0.000000   0.000000   0.000000   0.000000   0.000000          +M
-2.5323E+01    .0000    -M 0.000000  0.000000    .675894   0.000000   0.000000   0.000000   0.000000   0.000000    .293826-M
                        +M 0.000000  0.000000    .675894   0.000000   0.000000   0.000000   0.000000   0.000000          +M
-6.2400E+01   0.0000    -M 0.000000  0.000000   0.000000   0.000000   0.000000    .707107   0.000000   0.000000   0.000000-M
                        +M 0.000000  0.000000   0.000000   0.000000   0.000000    .707107   0.000000   0.000000          +M
-6.3679E+01    .5898    -M 0.000000   .022992   0.000000   0.000000   0.000000   0.000000   0.000000    .214311   0.000000-M
                        +M 0.000000  0.000000   0.000000    .976495   0.000000   0.000000   0.000000   0.000000          +M
-6.3679E+01   -.5898    -M 0.000000  0.000000   0.000000    .976495   0.000000   0.000000   0.000000   0.000000   0.000000-M
                        +M 0.000000   .022992   0.000000   0.000000   0.000000   0.000000   0.000000    .214311          +M
-7.1388E+01   -.4220    -M 0.000000  0.000000   0.000000   0.000000    .946672   0.000000   0.000000   0.000000   0.000000-M
                        +M  .013433  0.000000   0.000000   0.000000   0.000000   0.000000    .321917   0.000000          +M
-7.1388E+01    .4220    -M  .013433  0.000000   0.000000   0.000000   0.000000   0.000000    .321917   0.000000   0.000000-M
                        +M 0.000000  0.000000   0.000000   0.000000    .946672   0.000000   0.000000   0.000000          +M
  ENERGY       HPAL          M=8        M=7        M=6        M=5        M=4        M=3        M=2        M=1        M=0
```

Y=-.40 X=-.20 OAS=1.4663E+02*W246 B4F4/B2F2= 1.33E-01 B4F4/B6F6=-2.50E-01 B2F2/B6F6=-1.87E+00 B4F4= 8.00E-02

ENERGY	HPAL		M=8	M=7	M=6	M=5	M=4	M=3	M=2	M=1	M=0
6.9230E+01	.7072	-M	0.000000	0.000000	0.000000	.120443	0.000000	0.000000	0.000000	0.000000	0.000000-M
		+M	0.000000	.889272	0.000000	0.000000	0.000000	0.000000	0.000000	-.441236	+M
6.9230E+01	-.7072	-M	0.000000	.889272	0.000000	0.000000	0.000000	0.000000	0.000000	-.441236	0.000000-M
		+M	0.000000	0.000000	0.000000	.120443	0.000000	0.000000	0.000000	0.000000	+M
6.4153E+01	-.0000	-M	0.000000	0.000000	-.367975	0.000000	0.000000	0.000000	0.000000	0.000000	.853925-M
		+M	0.000000	0.000000	-.367975	0.000000	0.000000	0.000000	0.000000	0.000000	+M
3.4568E+01	.1996	-M	0.000000	0.000000	0.000000	-.328173	0.000000	0.000000	0.000000	0.000000	0.000000-M
		+M	0.000000	.455150	0.000000	0.000000	0.000000	0.000000	0.000000	.827732	+M
3.4568E+01	-.1996	-M	0.000000	.455150	0.000000	0.000000	0.000000	0.000000	0.000000	.827732	0.000000-M
		+M	0.000000	0.000000	0.000000	-.328173	0.000000	0.000000	0.000000	0.000000	+M
2.4031E+01	.7513	-M	0.000000	0.000000	0.000000	0.000000	.209162	0.000000	0.000000	0.000000	0.000000-M
		+M	.843926	0.000000	0.000000	0.000000	0.000000	0.000000	-.494005	0.000000	+M
2.4031E+01	-.7513	-M	.843926	0.000000	0.000000	0.000000	0.000000	0.000000	-.494005	0.000000	0.000000-M
		+M	0.000000	0.000000	0.000000	0.000000	.209162	0.000000	0.000000	0.000000	+M
1.0320E+01	.0000	-M	0.000000	0.000000	-.707107	0.000000	0.000000	0.000000	0.000000	0.000000	.000000-M
		+M	0.000000	0.000000	.707107	0.000000	0.000000	0.000000	0.000000	0.000000	+M
9.7849E+00	.3543	-M	0.000000	0.000000	0.000000	0.000000	-.384201	0.000000	0.000000	0.000000	0.000000-M
		+M	.535476	0.000000	0.000000	0.000000	0.000000	0.000000	.752101	0.000000	+M
9.7849E+00	-.3543	-M	.535476	0.000000	0.000000	0.000000	0.000000	0.000000	.752101	0.000000	0.000000-M
		+M	0.000000	0.000000	0.000000	0.000000	-.384201	0.000000	0.000000	0.000000	+M
-3.4800E+00	0.0000	-M	0.000000	0.000000	0.000000	0.000000	0.000000	.707107	0.000000	0.000000	0.000000-M
		+M	0.000000	0.000000	0.000000	0.000000	0.000000	-.707107	0.000000	0.000000	+M
-9.6729E+00	.0000	-M	0.000000	0.000000	.603816	0.000000	0.000000	0.000000	0.000000	0.000000	.520396-M
		+M	0.000000	0.000000	.603816	0.000000	0.000000	0.000000	0.000000	0.000000	+M
-5.3278E+01	.5318	-M	0.000000	.045107	0.000000	0.000000	0.000000	0.000000	0.000000	.346655	0.000000-M
		+M	0.000000	0.000000	0.000000	.936908	0.000000	0.000000	0.000000	0.000000	+M
-5.3278E+01	-.5318	-M	0.000000	0.000000	0.000000	.936908	0.000000	0.000000	0.000000	0.000000	0.000000-M
		+M	0.000000	.045107	0.000000	0.000000	0.000000	0.000000	0.000000	.346655	+M
-7.6295E+01	-.3557	-M	0.000000	0.000000	0.000000	0.000000	.899245	0.000000	0.000000	0.000000	0.000000-M
		+M	.032486	0.000000	0.000000	0.000000	0.000000	0.000000	.436238	0.000000	+M
-7.6295E+01	.3557	-M	.032486	0.000000	0.000000	0.000000	0.000000	0.000000	.436238	0.000000	0.000000-M
		+M	0.000000	0.000000	0.000000	0.000000	.899245	0.000000	0.000000	0.000000	+M
-7.7400E+01	0.0000	-M	0.000000	0.000000	0.000000	0.000000	0.000000	.707107	0.000000	0.000000	0.000000-M
		+M	0.000000	0.000000	0.000000	0.000000	0.000000	.707107	0.000000	0.000000	+M
ENERGY	HPAL		M=8	M=7	M=6	M=5	M=4	M=3	M=2	M=1	M=0

Y=-.40 X=0.00 OAS=1.8125E+02*W246 B4F4/B2F2= 0. B4F4/B6F6= 0. B2F2/B6F6=-1.50E+00 B4F4= 0.

ENERGY	HPAL		M=8	M=7	M=6	M=5	M=4	M=3	M=2	M=1	M=0
8.8852E+01	.7841	-M	0.000000	0.000000	0.000000	.104762	0.000000	0.000000	0.000000	0.000000	0.000000-M
		+M	0.000000	.943272	0.000000	0.000000	0.000000	0.000000	0.000000	-.315080	+M
8.8852E+01	-.7841	-M	0.000000	.943272	0.000000	0.000000	0.000000	0.000000	0.000000	-.315080	0.000000-M
		+M	0.000000	0.000000	0.000000	.104762	0.000000	0.000000	0.000000	0.000000	+M
7.7078E+01	.0000	-M	0.000000	0.000000	.514399	0.000000	0.000000	0.000000	0.000000	0.000000	-.686140-M
		+M	0.000000	0.000000	.514399	0.000000	0.000000	0.000000	0.000000	0.000000	+M
3.8400E+01	-.0000	-M	0.000000	0.000000	-.707107	0.000000	0.000000	0.000000	0.000000	0.000000	.000000-M
		+M	0.000000	0.000000	.707107	0.000000	0.000000	0.000000	0.000000	0.000000	+M
3.6102E+01	-.0185	-M	0.000000	.324056	0.000000	0.000000	0.000000	0.000000	0.000000	.805007	0.000000-M
		+M	0.000000	0.000000	0.000000	-.496942	0.000000	0.000000	0.000000	0.000000	+M
3.6102E+01	.0185	-M	0.000000	0.000000	0.000000	-.496942	0.000000	0.000000	0.000000	0.000000	0.000000-M
		+M	0.000000	.324056	0.000000	0.000000	0.000000	0.000000	0.000000	.805007	+M
1.6370E+01	.0932	-M	0.000000	0.000000	0.000000	0.000000	-.507253	0.000000	0.000000	0.000000	0.000000-M
		+M	-.219536	0.000000	0.000000	0.000000	0.000000	0.000000	.833366	0.000000	+M
1.6370E+01	-.0932	-M	-.219536	0.000000	0.000000	0.000000	0.000000	0.000000	.833366	0.000000	0.000000-M
		+M	0.000000	0.000000	0.000000	0.000000	-.507253	0.000000	0.000000	0.000000	+M
0.	0.0000	-M	0.000000	0.000000	0.000000	0.000000	0.000000	.707107	0.000000	0.000000	0.000000-M
		+M	0.000000	0.000000	0.000000	0.000000	0.000000	-.707107	0.000000	0.000000	+M
-5.0785E+00	.0000	-M	0.000000	0.000000	.485174	0.000000	0.000000	0.000000	0.000000	0.000000	.727470-M
		+M	0.000000	0.000000	.485174	0.000000	0.000000	0.000000	0.000000	0.000000	+M
-1.8967E+01	-.9365	-M	.972947	0.000000	0.000000	0.000000	0.000000	0.000000	.148668	0.000000	0.000000-M
		+M	0.000000	0.000000	0.000000	0.000000	-.176839	0.000000	0.000000	0.000000	+M
-1.8967E+01	.9365	-M	0.000000	0.000000	0.000000	0.000000	-.176839	0.000000	0.000000	0.000000	0.000000-M
		+M	.972947	0.000000	0.000000	0.000000	0.000000	0.000000	.148668	0.000000	+M
-4.7555E+01	-.4276	-M	0.000000	0.000000	0.000000	.861444	0.000000	0.000000	0.000000	0.000000	0.000000-M
		+M	0.000000	.072291	0.000000	0.000000	0.000000	0.000000	0.000000	.502681	+M
-4.7555E+01	.4276	-M	0.000000	.072291	0.000000	0.000000	0.000000	0.000000	0.000000	.502681	0.000000-M
		+M	0.000000	0.000000	0.000000	.861444	0.000000	0.000000	0.000000	0.000000	+M
-8.3802E+01	.2797	-M	.071960	0.000000	0.000000	0.000000	0.000000	0.000000	.532353	0.000000	0.000000-M
		+M	0.000000	0.000000	0.000000	0.000000	.843459	0.000000	0.000000	0.000000	+M
-8.3802E+01	-.2797	-M	0.000000	0.000000	0.000000	0.000000	.843459	0.000000	0.000000	0.000000	0.000000-M
		+M	.071960	0.000000	0.000000	0.000000	0.000000	0.000000	.532353	0.000000	+M
-9.2400E+01	0.0000	-M	0.000000	0.000000	0.000000	0.000000	0.000000	.707107	0.000000	0.000000	0.000000-M
		+M	0.000000	0.000000	0.000000	0.000000	0.000000	.707107	0.000000	0.000000	+M
ENERGY	HPAL		M=8	M=7	M=6	M=5	M=4	M=3	M=2	M=1	M=0

Y=-.40 X= .20 OAS=1.5353E+02*W246 B4F4/B2F2=-1.33E-01 B4F4/B6F6= 2.50E-01 B2F2/B6F6=-1.87E+00 B4F4=-8.00E-02

ENERGY	HPAL		M=8	M=7	M=6	M=5	M=4	M=3	M=2	M=1	M=0
7.9492E+01	.8343	-M	0.000000	0.000000	0.000000	.079609	0.000000	0.000000	0.000000	0.000000	0.000000-M
		+M	0.000000	.975718	0.000000	0.000000	0.000000	0.000000	0.000000	-.204052	+M
7.9492E+01	-.8343	-M	0.000000	.975718	0.000000	0.000000	0.000000	0.000000	0.000000	-.204052	0.000000-M
		+M	0.000000	0.000000	0.000000	.079609	0.000000	0.000000	0.000000	0.000000	+M
7.0215E+01	.0000	-M	0.000000	0.000000	.633904	0.000000	0.000000	0.000000	0.000000	0.000000	-.443093-M
		+M	0.000000	0.000000	.633904	0.000000	0.000000	0.000000	0.000000	0.000000	+M
5.4000E+01	-.0000	-M	0.000000	0.000000	-.707107	0.000000	0.000000	0.000000	0.000000	0.000000	-.000000-M
		+M	0.000000	0.000000	.707107	0.000000	0.000000	0.000000	0.000000	0.000000	+M
2.7349E+01	.2674	-M	0.000000	-.193652	0.000000	0.000000	0.000000	0.000000	0.000000	-.633836	0.000000-M
		+M	0.000000	0.000000	0.000000	.748833	0.000000	0.000000	0.000000	0.000000	+M
2.7349E+01	-.2674	-M	0.000000	0.000000	0.000000	.748833	0.000000	0.000000	0.000000	0.000000	0.000000-M
		+M	0.000000	-.193652	0.000000	0.000000	0.000000	0.000000	0.000000	-.633836	+M
8.9862E+00	-.0520	-M	0.000000	0.000000	0.000000	0.000000	-.644789	0.000000	0.000000	0.000000	0.000000-M
		+M	-.114182	0.000000	0.000000	0.000000	0.000000	0.000000	.755784	0.000000	+M
8.9862E+00	.0520	-M	-.114182	0.000000	0.000000	0.000000	0.000000	0.000000	.755784	0.000000	0.000000-M
		+M	0.000000	0.000000	0.000000	0.000000	-.644789	0.000000	0.000000	0.000000	+M
-1.2000E-01	0.0000	-M	0.000000	0.000000	0.000000	0.000000	0.000000	.707107	0.000000	0.000000	0.000000-M
		+M	0.000000	0.000000	0.000000	0.000000	0.000000	-.707107	0.000000	0.000000	+M
-1.2375E+01	.0000	-M	0.000000	0.000000	.313314	0.000000	0.000000	0.000000	0.000000	0.000000	.896475-M
		+M	0.000000	0.000000	.313314	0.000000	0.000000	0.000000	0.000000	0.000000	+M
-3.2800E+01	-.1918	-M	0.000000	0.000000	0.000000	.657960	0.000000	0.000000	0.000000	0.000000	0.000000-M
		+M	0.000000	.102342	0.000000	0.000000	0.000000	0.000000	0.000000	.746066	+M
-3.2800E+01	.1918	-M	0.000000	.102342	0.000000	0.000000	0.000000	0.000000	0.000000	.746066	0.000000-M
		+M	0.000000	0.000000	0.000000	.657960	0.000000	0.000000	0.000000	0.000000	+M
-3.8140E+01	-.9301	-M	-.976115	0.000000	0.000000	0.000000	0.000000	0.000000	.035403	0.000000	0.000000-M
		+M	0.000000	0.000000	0.000000	0.000000	.214352	0.000000	0.000000	0.000000	+M
-3.8140E+01	.9301	-M	0.000000	0.000000	0.000000	0.000000	-.214352	0.000000	0.000000	0.000000	0.000000-M
		+M	.976115	0.000000	0.000000	0.000000	0.000000	0.000000	-.035403	0.000000	+M
-6.3726E+01	.1281	-M	.184831	0.000000	0.000000	0.000000	0.000000	0.000000	.653864	0.000000	0.000000-M
		+M	0.000000	0.000000	0.000000	0.000000	.733689	0.000000	0.000000	0.000000	+M
-6.3726E+01	-.1281	-M	0.000000	0.000000	0.000000	0.000000	.733689	0.000000	0.000000	0.000000	0.000000-M
		+M	.184831	0.000000	0.000000	0.000000	0.000000	0.000000	.653864	0.000000	+M
-7.4040E+01	0.0000	-M	0.000000	0.000000	0.000000	0.000000	0.000000	.707107	0.000000	0.000000	0.000000-M
		+M	0.000000	0.000000	0.000000	0.000000	0.000000	.707107	0.000000	0.000000	+M
ENERGY	HPAL		M=8	M=7	M=6	M=5	M=4	M=3	M=2	M=1	M=0

```
Y=-.40  X= .40   OAS=1.3655E+02*W246   B4F4/B2F2=-2.67E-01  B4F4/B6F6= 6.67E-01  B2F2/B6F6=-2.50E+00   B4F4=-1.60E-01
   ENERGY      HPAL         M=8      M=7      M=6      M=5      M=4      M=3      M=2      M=1      M=0
 7.5565E+01    .0000  -M 0.000000 0.000000  .687198 0.000000 0.000000 0.000000 0.000000 0.000000 -.235622-M
                      +M 0.000000 0.000000  .687198 0.000000 0.000000 0.000000 0.000000 0.000000         +M
 7.1560E+01    .8561  -M 0.000000 0.000000 0.000000  .067417 0.000000 0.000000 0.000000 0.000000 0.000000-M
                      +M 0.000000  .989621 0.000000 0.000000 0.000000 0.000000 0.000000 -.126904         +M
 7.1560E+01   -.8561  -M 0.000000  .989621 0.000000 0.000000 0.000000 0.000000 0.000000 -.126904 0.000000-M
                      +M 0.000000 0.000000 0.000000  .067417 0.000000 0.000000 0.000000 0.000000         +M
 6.9600E+01   -.0000  -M 0.000000 0.000000 -.707107 0.000000 0.000000 0.000000 0.000000 0.000000  .000000-M
                      +M 0.000000 0.000000  .707107 0.000000 0.000000 0.000000 0.000000 0.000000         +M
 3.6295E+01    .5340  -M 0.000000 -.104687 0.000000 0.000000 0.000000 0.000000 0.000000 -.315302 0.000000-M
                      +M 0.000000 0.000000 0.000000  .943200 0.000000 0.000000 0.000000 0.000000         +M
 3.6295E+01   -.5340  -M 0.000000 0.000000 0.000000  .943200 0.000000 0.000000 0.000000 0.000000 0.000000-M
                      +M 0.000000 -.104687 0.000000 0.000000 0.000000 0.000000 0.000000 -.315302         +M
 8.8480E+00    .2692  -M  .043499 0.000000 0.000000 0.000000 0.000000 0.000000 -.551290 0.000000 0.000000-M
                      +M 0.000000 0.000000 0.000000 0.000000  .833179 0.000000 0.000000 0.000000         +M
 8.8480E+00   -.2692  -M 0.000000 0.000000 0.000000 0.000000  .833179 0.000000 0.000000 0.000000 0.000000-M
                      +M  .043499 0.000000 0.000000 0.000000 0.000000 0.000000 -.551290 0.000000         +M
-2.4000E-01   0.0000  -M 0.000000 0.000000 0.000000 0.000000 0.000000  .707107 0.000000 0.000000 0.000000-M
                      +M 0.000000 0.000000 0.000000 0.000000 0.000000 -.707107 0.000000 0.000000         +M
-3.1885E+01   -.0000  -M 0.000000 0.000000  .166610 0.000000 0.000000 0.000000 0.000000 0.000000  .971845-M
                      +M 0.000000 0.000000  .166610 0.000000 0.000000 0.000000 0.000000 0.000000         +M
-3.7175E+01    .0529  -M 0.000000 0.000000 0.000000  .325315 0.000000 0.000000 0.000000 0.000000 0.000000-M
                      +M 0.000000  .098439 0.000000 0.000000 0.000000 0.000000 0.000000  .940468         +M
-3.7175E+01   -.0529  -M 0.000000  .098439 0.000000 0.000000 0.000000 0.000000 0.000000  .940468 0.000000-M
                      +M 0.000000 0.000000 0.000000  .325315 0.000000 0.000000 0.000000 0.000000         +M
-4.7220E+01   -.1295  -M -.346795 0.000000 0.000000 0.000000 0.000000 0.000000  .773779 0.000000 0.000000-M
                      +M 0.000000 0.000000 0.000000 0.000000  .530093 0.000000 0.000000 0.000000         +M
-4.7220E+01    .1295  -M 0.000000 0.000000 0.000000 0.000000  .530093 0.000000 0.000000 0.000000 0.000000-M
                      +M -.346795 0.000000 0.000000 0.000000 0.000000 0.000000  .773779 0.000000         +M
-5.5680E+01   0.0000  -M 0.000000 0.000000 0.000000 0.000000 0.000000  .707107 0.000000 0.000000 0.000000-M
                      +M 0.000000 0.000000 0.000000 0.000000 0.000000  .707107 0.000000 0.000000         +M
-6.0988E+01   -.8898  -M  .936932 0.000000 0.000000 0.000000 0.000000 0.000000  .312001 0.000000 0.000000-M
                      +M 0.000000 0.000000 0.000000 0.000000  .157526 0.000000 0.000000 0.000000         +M
-6.0988E+01    .8898  -M 0.000000 0.000000 0.000000 0.000000  .157526 0.000000 0.000000 0.000000 0.000000-M
                      +M  .936932 0.000000 0.000000 0.000000 0.000000 0.000000  .312001 0.000000         +M
   ENERGY      HPAL         M=8      M=7      M=6      M=5      M=4      M=3      M=2      M=1      M=0

Y=-.40  X= .60   OAS=1.6744E+02*W246   B4F4/B2F2=-4.00E-01  B4F4/B6F6= 1.50E+00  B2F2/B6F6=-3.75E+00   B4F4=-2.40E-01
   ENERGY      HPAL         M=8      M=7      M=6      M=5      M=4      M=3      M=2      M=1      M=0
 6.7085E+01   -.0000  -M 0.000000 0.000000  .702485 0.000000 0.000000 0.000000 0.000000 0.000000 -.114147-M
                      +M 0.000000 0.000000  .702485 0.000000 0.000000 0.000000 0.000000 0.000000         +M
 8.5200E+01    .0000  -M 0.000000 0.000000 -.707107 0.000000 0.000000 0.000000 0.000000 0.000000 -.000000-M
                      +M 0.000000 0.000000  .707107 0.000000 0.000000 0.000000 0.000000 0.000000         +M
 6.4499E+01    .8401  -M 0.000000 0.000000 0.000000  .141176 0.000000 0.000000 0.000000 0.000000 0.000000-M
                      +M 0.000000  .986615 0.000000 0.000000 0.000000 0.000000 0.000000 -.081614         +M
 6.4499E+01   -.8401  -M 0.000000  .986615 0.000000 0.000000 0.000000 0.000000 0.000000 -.081614 0.000000-M
                      +M 0.000000 0.000000 0.000000  .141176 0.000000 0.000000 0.000000 0.000000         +M
 5.7659E+01   -.5792  -M 0.000000 0.000000 0.000000  .980624 0.000000 0.000000 0.000000 0.000000 0.000000-M
                      +M 0.000000 -.150675 0.000000 0.000000 0.000000 0.000000 0.000000 -.125198         +M
 5.7659E+01    .5792  -M 0.000000 -.150675 0.000000 0.000000 0.000000 0.000000 0.000000 -.125198 0.000000-M
                      +M 0.000000 0.000000 0.000000  .980624 0.000000 0.000000 0.000000 0.000000         +M
 2.0139E+01    .4419  -M  .009936 0.000000 0.000000 0.000000 0.000000 0.000000 -.277953 0.000000 0.000000-M
                      +M 0.000000 0.000000 0.000000 0.000000  .960543 0.000000 0.000000 0.000000         +M
 2.0139E+01   -.4419  -M 0.000000 0.000000 0.000000 0.000000  .960543 0.000000 0.000000 0.000000 0.000000-M
                      +M  .009936 0.000000 0.000000 0.000000 0.000000 0.000000 -.277953 0.000000         +M
-3.6000E-01   0.0000  -M 0.000000 0.000000 0.000000 0.000000 0.000000  .707107 0.000000 0.000000 0.000000-M
                      +M 0.000000 0.000000 0.000000 0.000000 0.000000 -.707107 0.000000 0.000000         +M
-3.7320E+01   0.0000  -M 0.000000 0.000000 0.000000 0.000000 0.000000  .707107 0.000000 0.000000 0.000000-M
                      +M 0.000000 0.000000 0.000000 0.000000 0.000000  .707107 0.000000 0.000000         +M
-4.5625E+01   -.1997  -M -.099502 0.000000 0.000000 0.000000 0.000000 0.000000  .955548 0.000000 0.000000-M
                      +M 0.000000 0.000000 0.000000 0.000000  .277537 0.000000 0.000000 0.000000         +M
-4.5625E+01    .1997  -M 0.000000 0.000000 0.000000 0.000000  .277537 0.000000 0.000000 0.000000 0.000000-M
                      +M -.099502 0.000000 0.000000 0.000000 0.000000 0.000000  .955548 0.000000         +M
-5.4838E+01   -.1141  -M 0.000000  .062357 0.000000 0.000000 0.000000 0.000000 0.000000  .988769 0.000000-M
                      +M 0.000000 0.000000 0.000000  .135820 0.000000 0.000000 0.000000 0.000000         +M
-5.4838E+01    .1141  -M 0.000000 0.000000 0.000000  .135820 0.000000 0.000000 0.000000 0.000000 0.000000-M
                      +M 0.000000  .062357 0.000000 0.000000 0.000000 0.000000 0.000000  .988769         +M
-5.7565E+01   -.0000  -M 0.000000 0.000000  .080714 0.000000 0.000000 0.000000 0.000000 0.000000  .993464-M
                      +M 0.000000 0.000000  .080714 0.000000 0.000000 0.000000 0.000000 0.000000         +M
-8.0354E+01   -.9923  -M  .994988 0.000000 0.000000 0.000000 0.000000 0.000000  .098333 0.000000 0.000000-M
                      +M 0.000000 0.000000 0.000000 0.000000  .018163 0.000000 0.000000 0.000000         +M
-8.0354E+01    .9923  -M 0.000000 0.000000 0.000000 0.000000  .018163 0.000000 0.000000 0.000000 0.000000-M
                      +M  .994988 0.000000 0.000000 0.000000 0.000000 0.000000  .098333 0.000000         +M
   ENERGY      HPAL         M=8      M=7      M=6      M=5      M=4      M=3      M=2      M=1      M=0

Y=-.40  X= .80   OAS=2.0202E+02*W246   B4F4/B2F2=-5.33E-01  B4F4/B6F6= 4.00E+00  B2F2/B6F6=-7.50E+00   B4F4=-3.20E-01
   ENERGY      HPAL         M=8      M=7      M=6      M=5      M=4      M=3      M=2      M=1      M=0
 1.0116E+02    .0000  -M 0.000000 0.000000  .706425 0.000000 0.000000 0.000000 0.000000 0.000000 -.043911-M
                      +M 0.000000 0.000000  .706425 0.000000 0.000000 0.000000 0.000000 0.000000         +M
 1.0080E+02   -.0000  -M 0.000000 0.000000 -.707107 0.000000 0.000000 0.000000 0.000000 0.000000  .000000-M
                      +M 0.000000 0.000000  .707107 0.000000 0.000000 0.000000 0.000000 0.000000         +M
 8.3123E+01    .6232  -M 0.000000  .007007 0.000000 0.000000 0.000000 0.000000 0.000000 -.047697 0.000000-M
                      +M 0.000000 0.000000 0.000000  .998837 0.000000 0.000000 0.000000 0.000000         +M
 8.3123E+01   -.6232  -M 0.000000 0.000000 0.000000  .998837 0.000000 0.000000 0.000000 0.000000 0.000000-M
                      +M 0.000000  .007007 0.000000 0.000000 0.000000 0.000000 0.000000 -.047697         +M
 5.7743E+01    .8743  -M 0.000000 0.000000 0.000000 -.008321 0.000000 0.000000 0.000000 0.000000 0.000000-M
                      +M 0.000000  .999589 0.000000 0.000000 0.000000 0.000000 0.000000 -.027416         +M
 5.7743E+01   -.8743  -M 0.000000  .999589 0.000000 0.000000 0.000000 0.000000 0.000000 -.027416 0.000000-M
                      +M 0.000000 0.000000 0.000000 -.008321 0.000000 0.000000 0.000000 0.000000         +M
 3.9584E+01    .4928  -M  .001245 0.000000 0.000000 0.000000 0.000000 0.000000 -.097636 0.000000 0.000000-M
                      +M 0.000000 0.000000 0.000000 0.000000  .995221 0.000000 0.000000 0.000000         +M
 3.9584E+01   -.4928  -M 0.000000 0.000000 0.000000 0.000000  .995221 0.000000 0.000000 0.000000 0.000000-M
                      +M  .001245 0.000000 0.000000 0.000000 0.000000 0.000000 -.097636 0.000000         +M
-4.8000E-01   0.0000  -M 0.000000 0.000000 0.000000 0.000000 0.000000  .707107 0.000000 0.000000 0.000000-M
                      +M 0.000000 0.000000 0.000000 0.000000 0.000000 -.707107 0.000000 0.000000         +M
-1.8960E+01   0.0000  -M 0.000000 0.000000 0.000000 0.000000 0.000000  .707107 0.000000 0.000000 0.000000-M
                      +M 0.000000 0.000000 0.000000 0.000000 0.000000  .707107 0.000000 0.000000         +M
-5.1040E+01    .2438  -M 0.000000 0.000000 0.000000 0.000000  .097618 0.000000 0.000000 0.000000 0.000000-M
                      +M -.035773 0.000000 0.000000 0.000000 0.000000 0.000000  .994581 0.000000         +M
-5.1040E+01   -.2438  -M -.035773 0.000000 0.000000 0.000000 0.000000 0.000000  .994581 0.000000 0.000000-M
                      +M 0.000000 0.000000 0.000000 0.000000  .097618 0.000000 0.000000 0.000000         +M
-7.6905E+01   -.1239  -M 0.000000  .027781 0.000000 0.000000 0.000000 0.000000 0.000000  .998486 0.000000-M
                      +M 0.000000 0.000000 0.000000  .047485 0.000000 0.000000 0.000000 0.000000         +M
-7.6905E+01    .1239  -M 0.000000 0.000000 0.000000  .047485 0.000000 0.000000 0.000000 0.000000 0.000000-M
                      +M 0.000000  .027781 0.000000 0.000000 0.000000 0.000000 0.000000  .998486         +M
-8.5800E+01    .0000  -M 0.000000 0.000000  .031050 0.000000 0.000000 0.000000 0.000000 0.000000  .999035-M
                      +M 0.000000 0.000000  .031050 0.000000 0.000000 0.000000 0.000000 0.000000         +M
-1.0086E+02   -.9990  -M  .999359 0.000000 0.000000 0.000000 0.000000 0.000000  .035724 0.000000 0.000000-M
                      +M 0.000000 0.000000 0.000000 0.000000  .002255 0.000000 0.000000 0.000000         +M
-1.0086E+02    .9990  -M 0.000000 0.000000 0.000000 0.000000  .002255 0.000000 0.000000 0.000000 0.000000-M
                      +M  .999359 0.000000 0.000000 0.000000 0.000000 0.000000  .035724 0.000000         +M
   ENERGY      HPAL         M=8      M=7      M=6      M=5      M=4      M=3      M=2      M=1      M=0
```

Y=-.20 X=-.80 OAS=1.3179E+02*W246 B4F4/B2F2= 2.00E-01 B4F4/B6F6=-4.00E+00 B2F2/B6F6=-2.00E+01 B4F4= 1.60E-01

ENERGY	HPAL		M=8	M=7	M=6	M=5	M=4	M=3	M=2	M=1	M=0
8.6090E+01	.9999	-M	0.000000	0.000000	0.000000	0.000000	.000382	0.000000	0.000000	0.000000	0.000000-M
		+M	.999942	0.000000	0.000000	0.000000	0.000000	0.000000	-.010784	0.000000	+M
8.6090E+01	-.9999	-M	.999942	0.000000	0.000000	0.000000	0.000000	0.000000	-.010784	0.000000	0.000000-M
		+M	0.000000	0.000000	0.000000	0.000000	.000382	0.000000	0.000000	0.000000	+M
2.6214E+01	-.0000	-M	0.000000	0.000000	-.050587	0.000000	0.000000	0.000000	0.000000	0.000000	.997438-M
		+M	0.000000	0.000000	-.050587	0.000000	0.000000	0.000000	0.000000	0.000000	+M
2.0368E+01	.1599	-M	0.000000	0.000000	0.000000	-.056112	0.000000	0.000000	0.000000	0.000000	0.000000-M
		+M	0.000000	-.223020	0.000000	0.000000	0.000000	0.000000	0.000000	.973197	+M
2.0368E+01	-.1599	-M	0.000000	-.223020	0.000000	0.000000	0.000000	0.000000	0.000000	.973197	0.000000-M
		+M	0.000000	0.000000	0.000000	-.056112	0.000000	0.000000	0.000000	0.000000	+M
1.1773E+01	-.8375	-M	0.000000	.974812	0.000000	0.000000	0.000000	0.000000	0.000000	.222540	0.000000-M
		+M	0.000000	0.000000	0.000000	-.014757	0.000000	0.000000	0.000000	0.000000	+M
1.1773E+01	.8375	-M	0.000000	0.000000	0.000000	-.014757	0.000000	0.000000	0.000000	0.000000	0.000000-M
		+M	0.000000	.974812	0.000000	0.000000	0.000000	0.000000	0.000000	.222540	+M
3.4144E+00	-.2419	-M	.010765	0.000000	0.000000	0.000000	0.000000	0.000000	.994439	0.000000	0.000000-M
		+M	0.000000	0.000000	0.000000	0.000000	-.104761	0.000000	0.000000	0.000000	+M
3.4144E+00	.2419	-M	0.000000	0.000000	0.000000	0.000000	-.104761	0.000000	0.000000	0.000000	0.000000-M
		+M	.010765	0.000000	0.000000	0.000000	0.000000	0.000000	.994439	0.000000	+M
-1.4460E+01	0.0000	-M	0.000000	0.000000	0.000000	0.000000	0.000000	.707107	0.000000	0.000000	0.000000-M
		+M	0.000000	0.000000	0.000000	0.000000	0.000000	-.707107	0.000000	0.000000	+M
-2.3700E+01	0.0000	-M	0.000000	0.000000	0.000000	0.000000	0.000000	.707107	0.000000	0.000000	0.000000-M
		+M	0.000000	0.000000	0.000000	0.000000	0.000000	.707107	0.000000	0.000000	+M
-3.0960E+01	.0000	-M	0.000000	0.000000	-.707107	0.000000	0.000000	0.000000	0.000000	0.000000	.000000-M
		+M	0.000000	0.000000	.707107	0.000000	0.000000	0.000000	0.000000	0.000000	+M
-3.1254E+01	-.0000	-M	0.000000	0.000000	.705295	0.000000	0.000000	0.000000	0.000000	0.000000	.071541-M
		+M	0.000000	0.000000	.705295	0.000000	0.000000	0.000000	0.000000	0.000000	+M
-3.8864E+01	-.4918	-M	0.000000	0.000000	0.000000	0.000000	.994497	0.000000	0.000000	0.000000	0.000000-M
		+M	.000750	0.000000	0.000000	0.000000	0.000000	0.000000	.104759	0.000000	+M
-3.8864E+01	.4918	-M	.000750	0.000000	0.000000	0.000000	0.000000	0.000000	.104759	0.000000	0.000000-M
		+M	0.000000	0.000000	0.000000	0.000000	.994497	0.000000	0.000000	0.000000	+M
-4.5701E+01	-.6225	-M	0.000000	0.000000	0.000000	.998315	0.000000	0.000000	0.000000	0.000000	0.000000-M
		+M	0.000000	.001875	0.000000	0.000000	0.000000	0.000000	0.000000	.057990	+M
-4.5701E+01	.6225	-M	0.000000	.001875	0.000000	0.000000	0.000000	0.000000	0.000000	.057990	0.000000-M
		+M	0.000000	0.000000	0.000000	.998315	0.000000	0.000000	0.000000	0.000000	+M
ENERGY	HPAL		M=8	M=7	M=6	M=5	M=4	M=3	M=2	M=1	M=0

Y=-.20 X=-.60 OAS=1.0631E+02*W246 B4F4/B2F2= 1.50E-01 B4F4/B6F6=-1.50E+00 B2F2/B6F6=-1.00E+01 B4F4= 1.20E-01

ENERGY	HPAL		M=8	M=7	M=6	M=5	M=4	M=3	M=2	M=1	M=0
6.7407E+01	-.9995	-M	.999657	0.000000	0.000000	0.000000	0.000000	0.000000	-.026103	0.000000	0.000000-M
		+M	0.000000	0.000000	0.000000	0.000000	.002207	0.000000	0.000000	0.000000	+M
6.7407E+01	.9995	-M	0.000000	0.000000	0.000000	0.000000	.002207	0.000000	0.000000	0.000000	0.000000-M
		+M	.999657	0.000000	0.000000	0.000000	0.000000	0.000000	-.026103	0.000000	+M
2.4191E+01	.7571	-M	0.000000	0.000000	0.000000	.048343	0.000000	0.000000	0.000000	0.000000	0.000000-M
		+M	0.000000	.919294	0.000000	0.000000	0.000000	0.000000	0.000000	-.390591	+M
2.4191E+01	-.7571	-M	0.000000	.919294	0.000000	0.000000	0.000000	0.000000	0.000000	-.390591	0.000000-M
		+M	0.000000	0.000000	0.000000	.048343	0.000000	0.000000	0.000000	0.000000	+M
2.2353E+01	.0000	-M	0.000000	0.000000	-.144552	0.000000	0.000000	0.000000	0.000000	0.000000	.978882-M
		+M	0.000000	0.000000	-.144552	0.000000	0.000000	0.000000	0.000000	0.000000	+M
1.3948E+01	-.2274	-M	0.000000	.393473	0.000000	0.000000	0.000000	0.000000	0.000000	.909355	0.000000-M
		+M	0.000000	0.000000	0.000000	-.135104	0.000000	0.000000	0.000000	0.000000	+M
1.3948E+01	.2274	-M	0.000000	0.000000	0.000000	-.135104	0.000000	0.000000	0.000000	0.000000	0.000000-M
		+M	0.000000	.393473	0.000000	0.000000	0.000000	0.000000	0.000000	.909355	+M
1.7166E-01	-.2102	-M	.025904	0.000000	0.000000	0.000000	0.000000	0.000000	.972437	0.000000	0.000000-M
		+M	0.000000	0.000000	0.000000	0.000000	-.231722	0.000000	0.000000	0.000000	+M
1.7166E-01	.2102	-M	0.000000	0.000000	0.000000	0.000000	-.231722	0.000000	0.000000	0.000000	0.000000-M
		+M	.025904	0.000000	0.000000	0.000000	0.000000	0.000000	.972437	0.000000	+M
-1.2720E+01	0.0000	-M	0.000000	0.000000	0.000000	0.000000	0.000000	.707107	0.000000	0.000000	0.000000-M
		+M	0.000000	0.000000	0.000000	0.000000	0.000000	-.707107	0.000000	0.000000	+M
-1.6920E+01	-.0000	-M	0.000000	0.000000	-.707107	0.000000	0.000000	0.000000	0.000000	0.000000	-.000000-M
		+M	0.000000	0.000000	.707107	0.000000	0.000000	0.000000	0.000000	0.000000	+M
-1.8633E+01	.0000	-M	0.000000	0.000000	.692174	0.000000	0.000000	0.000000	0.000000	0.000000	.204428-M
		+M	0.000000	0.000000	.692174	0.000000	0.000000	0.000000	0.000000	0.000000	+M
-3.1200E+01	0.0000	-M	0.000000	0.000000	0.000000	0.000000	0.000000	.707107	0.000000	0.000000	0.000000-M
		+M	0.000000	0.000000	0.000000	0.000000	0.000000	.707107	0.000000	0.000000	+M
-3.8259E+01	-.6095	-M	0.000000	0.000000	0.000000	.989651	0.000000	0.000000	0.000000	0.000000	0.000000-M
		+M	0.000000	.008810	0.000000	0.000000	0.000000	0.000000	0.000000	.143222	+M
-3.8259E+01	.6095	-M	0.000000	.008810	0.000000	0.000000	0.000000	0.000000	0.000000	.143222	0.000000-M
		+M	0.000000	0.000000	0.000000	.989651	0.000000	0.000000	0.000000	0.000000	+M
-3.8898E+01	-.4597	-M	0.000000	0.000000	0.000000	0.000000	.972779	0.000000	0.000000	0.000000	0.000000-M
		+M	.003903	0.000000	0.000000	0.000000	0.000000	0.000000	.231700	0.000000	+M
-3.8898E+01	.4597	-M	.003903	0.000000	0.000000	0.000000	0.000000	0.000000	.231700	0.000000	0.000000-M
		+M	0.000000	0.000000	0.000000	0.000000	.972779	0.000000	0.000000	0.000000	+M
ENERGY	HPAL		M=8	M=7	M=6	M=5	M=4	M=3	M=2	M=1	M=0

Y=-.20 X=-.40 OAS=8.9173E+01*W246 B4F4/B2F2= 1.00E-01 B4F4/B6F6=-6.67E-01 B2F2/B6F6=-6.67E+00 B4F4= 8.00E-02

ENERGY	HPAL		M=8	M=7	M=6	M=5	M=4	M=3	M=2	M=1	M=0
4.8775E+01	.9980	-M	0.000000	0.000000	0.000000	0.000000	.007899	0.000000	0.000000	0.000000	0.000000-M
		+M	.998708	0.000000	0.000000	0.000000	0.000000	0.000000	-.050195	0.000000	+M
4.8775E+01	-.9980	-M	.998708	0.000000	0.000000	0.000000	0.000000	0.000000	-.050195	0.000000	0.000000-M
		+M	0.000000	0.000000	0.000000	0.000000	.007899	0.000000	0.000000	0.000000	+M
3.4362E+01	.8307	-M	0.000000	0.000000	0.000000	.042444	0.000000	0.000000	0.000000	0.000000	0.000000-M
		+M	0.000000	.970943	0.000000	0.000000	0.000000	0.000000	0.000000	-.235518	+M
3.4362E+01	-.8307	-M	0.000000	.970943	0.000000	0.000000	0.000000	0.000000	0.000000	-.235518	0.000000-M
		+M	0.000000	0.000000	0.000000	.042444	0.000000	0.000000	0.000000	0.000000	+M
2.1554E+01	-.0000	-M	0.000000	0.000000	-.317992	0.000000	0.000000	0.000000	0.000000	0.000000	.893175-M
		+M	0.000000	0.000000	-.317992	0.000000	0.000000	0.000000	0.000000	0.000000	+M
1.0985E+01	-.1141	-M	0.000000	.238179	0.000000	0.000000	0.000000	0.000000	0.000000	.933783	0.000000-M
		+M	0.000000	0.000000	0.000000	-.267059	0.000000	0.000000	0.000000	0.000000	+M
1.0985E+01	.1141	-M	0.000000	0.000000	0.000000	-.267059	0.000000	0.000000	0.000000	0.000000	0.000000-M
		+M	0.000000	.238179	0.000000	0.000000	0.000000	0.000000	0.000000	.933783	+M
-1.6570E+00	-.1513	-M	.049599	0.000000	0.000000	0.000000	0.000000	0.000000	.929253	0.000000	0.000000-M
		+M	0.000000	0.000000	0.000000	0.000000	-.366099	0.000000	0.000000	0.000000	+M
-1.6570E+00	.1513	-M	0.000000	0.000000	0.000000	0.000000	-.366099	0.000000	0.000000	0.000000	0.000000-M
		+M	.049599	0.000000	0.000000	0.000000	0.000000	0.000000	.929253	0.000000	+M
-2.8800E+00	-.0000	-M	0.000000	0.000000	-.707107	0.000000	0.000000	0.000000	0.000000	0.000000	-.000000-M
		+M	0.000000	0.000000	.707107	0.000000	0.000000	0.000000	0.000000	0.000000	+M
-9.0742E+00	.0000	-M	0.000000	0.000000	.631570	0.000000	0.000000	0.000000	0.000000	0.000000	.449708-M
		+M	0.000000	0.000000	.631570	0.000000	0.000000	0.000000	0.000000	0.000000	+M
-1.0980E+01	0.0000	-M	0.000000	0.000000	0.000000	0.000000	0.000000	.707107	0.000000	0.000000	0.000000-M
		+M	0.000000	0.000000	0.000000	0.000000	0.000000	-.707107	0.000000	0.000000	+M
-3.2027E+01	-.5698	-M	0.000000	0.000000	0.000000	.962745	0.000000	0.000000	0.000000	0.000000	0.000000-M
		+M	0.000000	.023264	0.000000	0.000000	0.000000	0.000000	0.000000	.269408	+M
-3.2027E+01	.5698	-M	0.000000	.023264	0.000000	0.000000	0.000000	0.000000	0.000000	.269408	0.000000-M
		+M	0.000000	0.000000	0.000000	.962745	0.000000	0.000000	0.000000	0.000000	+M
-3.8700E+01	0.0000	-M	0.000000	0.000000	0.000000	0.000000	0.000000	.707107	0.000000	0.000000	0.000000-M
		+M	0.000000	0.000000	0.000000	0.000000	0.000000	.707107	0.000000	0.000000	+M
-4.0398E+01	-.3993	-M	0.000000	0.000000	0.000000	0.000000	.930542	0.000000	0.000000	0.000000	0.000000-M
		+M	.011036	0.000000	0.000000	0.000000	0.000000	0.000000	.366018	0.000000	+M
-4.0398E+01	.3993	-M	.011036	0.000000	0.000000	0.000000	0.000000	0.000000	.366018	0.000000	0.000000-M
		+M	0.000000	0.000000	0.000000	0.000000	.930542	0.000000	0.000000	0.000000	+M
ENERGY	HPAL		M=8	M=7	M=6	M=5	M=4	M=3	M=2	M=1	M=0

Y=-.20 X=-.20 OAS=9.1095E+01*W246 B4F4/B2F2= 5.00E-02 B4F4/B6F6=-2.50E-01 B2F2/B6F6=-5.00E+00 B4F4= 4.00E-02

ENERGY	HPAL		M=8	M=7	M=6	M=5	M=4	M=3	M=2	M=1	M=0
4.4895E+01	.8432	-M	0.000000	0.000000	0.000000	.045419	0.000000	0.000000	0.000000	0.000000	0.000000-M
		+M	0.000000	.979623	0.000000	0.000000	0.000000	0.000000	0.000000	-.195641	+M
4.4895E+01	-.8432	-M	0.000000	.979623	0.000000	0.000000	0.000000	0.000000	0.000000	-.195641	0.000000-M
		+M	0.000000	0.000000	0.000000	.045419	0.000000	0.000000	0.000000	0.000000	+M
3.0267E+01	.9920	-M	0.000000	0.000000	0.000000	0.000000	.026581	0.000000	0.000000	0.000000	0.000000-M
		+M	.994993	0.000000	0.000000	0.000000	0.000000	0.000000	-.096349	0.000000	+M
3.0267E+01	-.9920	-M	.994993	0.000000	0.000000	0.000000	0.000000	0.000000	-.096349	0.000000	0.000000-M
		+M	0.000000	0.000000	0.000000	0.000000	.026581	0.000000	0.000000	0.000000	+M
2.7032E+01	.0000	-M	0.000000	0.000000	.508159	0.000000	0.000000	0.000000	0.000000	0.000000	-.695377-M
		+M	0.000000	0.000000	.508159	0.000000	0.000000	0.000000	0.000000	0.000000	+M
1.1160E+01	-.0000	-M	0.000000	0.000000	-.707107	0.000000	0.000000	0.000000	0.000000	0.000000	-.000000-M
		+M	0.000000	0.000000	.707107	0.000000	0.000000	0.000000	0.000000	0.000000	+M
9.8140E+00	-.0091	-M	0.000000	.195490	0.000000	0.000000	0.000000	0.000000	0.000000	.876944	0.000000-M
		+M	0.000000	0.000000	0.000000	-.439036	0.000000	0.000000	0.000000	0.000000	+M
9.8140E+00	.0091	-M	0.000000	0.000000	0.000000	-.439036	0.000000	0.000000	0.000000	0.000000	0.000000-M
		+M	0.000000	.195490	0.000000	0.000000	0.000000	0.000000	0.000000	.876944	+M
-2.1031E+00	-.0800	-M	.097091	0.000000	0.000000	0.000000	0.000000	0.000000	.868598	0.000000	0.000000-M
		+M	0.000000	0.000000	0.000000	0.000000	-.485912	0.000000	0.000000	0.000000	+M
-2.1031E+00	.0800	-M	0.000000	0.000000	0.000000	0.000000	-.485912	0.000000	0.000000	0.000000	0.000000-M
		+M	.097091	0.000000	0.000000	0.000000	0.000000	0.000000	.868598	0.000000	+M
-5.7922E+00	.0000	-M	0.000000	0.000000	.491706	0.000000	0.000000	0.000000	0.000000	0.000000	.718645-M
		+M	0.000000	0.000000	.491706	0.000000	0.000000	0.000000	0.000000	0.000000	+M
-9.2400E+00	0.0000	-M	0.000000	0.000000	0.000000	0.000000	0.000000	.707107	0.000000	0.000000	0.000000-M
		+M	0.000000	0.000000	0.000000	0.000000	0.000000	-.707107	0.000000	0.000000	+M
-2.7949E+01	-.4773	-M	0.000000	0.000000	0.000000	.897321	0.000000	0.000000	0.000000	0.000000	0.000000-M
		+M	0.000000	.046063	0.000000	0.000000	0.000000	0.000000	0.000000	.438969	+M
-2.7949E+01	.4773	-M	0.000000	.046063	0.000000	0.000000	0.000000	0.000000	0.000000	.438969	0.000000-M
		+M	0.000000	0.000000	0.000000	.897321	0.000000	0.000000	0.000000	0.000000	+M
-4.3403E+01	-.3220	-M	0.000000	0.000000	0.000000	0.000000	.873604	0.000000	0.000000	0.000000	0.000000-M
		+M	.023728	0.000000	0.000000	0.000000	0.000000	0.000000	.486059	0.000000	+M
-4.3403E+01	.3220	-M	.023728	0.000000	0.000000	0.000000	0.000000	0.000000	.486059	0.000000	0.000000-M
		+M	0.000000	0.000000	0.000000	0.000000	.873604	0.000000	0.000000	0.000000	+M
-4.6200E+01	0.0000	-M	0.000000	0.000000	0.000000	0.000000	0.000000	.707107	0.000000	0.000000	0.000000-M
		+M	0.000000	0.000000	0.000000	0.000000	0.000000	.707107	0.000000	0.000000	+M
ENERGY	HPAL		M=8	M=7	M=6	M=5	M=4	M=3	M=2	M=1	M=0

Y=-.20 X=0.00 OAS=1.0920E+02*W246 B4F4/B2F2= 0. B4F4/B6F6= 0. B2F2/B6F6=-4.00E+00 B4F4= 0.

ENERGY	HPAL		M=8	M=7	M=6	M=5	M=4	M=3	M=2	M=1	M=0
5.5498E+01	.8474	-M	0.000000	0.000000	0.000000	.050003	0.000000	0.000000	0.000000	0.000000	0.000000-M
		+M	0.000000	.982701	0.000000	0.000000	0.000000	0.000000	0.000000	-.178321	+M
5.5498E+01	-.8474	-M	0.000000	.982701	0.000000	0.000000	0.000000	0.000000	0.000000	-.178321	0.000000-M
		+M	0.000000	0.000000	0.000000	.050003	0.000000	0.000000	0.000000	0.000000	+M
3.7901E+01	.0000	-M	0.000000	0.000000	.601123	0.000000	0.000000	0.000000	0.000000	0.000000	-.526596-M
		+M	0.000000	0.000000	.601123	0.000000	0.000000	0.000000	0.000000	0.000000	+M
2.5200E+01	-.0000	-M	0.000000	0.000000	-.707107	0.000000	0.000000	0.000000	0.000000	0.000000	.000000-M
		+M	0.000000	0.000000	.707107	0.000000	0.000000	0.000000	0.000000	0.000000	+M
1.2326E+01	.9343	-M	0.000000	0.000000	0.000000	0.000000	.120368	0.000000	0.000000	0.000000	0.000000-M
		+M	.962720	0.000000	0.000000	0.000000	0.000000	0.000000	-.242244	0.000000	+M
1.2326E+01	-.9343	-M	.962720	0.000000	0.000000	0.000000	0.000000	0.000000	-.242244	0.000000	0.000000-M
		+M	0.000000	0.000000	0.000000	0.000000	.120368	0.000000	0.000000	0.000000	+M
1.1639E+01	.1356	-M	0.000000	.171088	0.000000	0.000000	0.000000	0.000000	0.000000	.770742	0.000000-M
		+M	0.000000	0.000000	0.000000	-.613747	0.000000	0.000000	0.000000	0.000000	+M
1.1639E+01	-.1356	-M	0.000000	0.000000	0.000000	-.613747	0.000000	0.000000	0.000000	0.000000	0.000000-M
		+M	0.000000	.171088	0.000000	0.000000	0.000000	0.000000	0.000000	.770742	+M
-1.8467E+00	-.0611	-M	.266894	0.000000	0.000000	0.000000	0.000000	0.000000	.778228	0.000000	0.000000-M
		+M	0.000000	0.000000	0.000000	0.000000	-.568444	0.000000	0.000000	0.000000	+M
-1.8467E+00	.0611	-M	0.000000	0.000000	0.000000	0.000000	-.568444	0.000000	0.000000	0.000000	0.000000-M
		+M	.266894	0.000000	0.000000	0.000000	0.000000	0.000000	.778228	0.000000	+M
-7.5000E+00	0.0000	-M	0.000000	0.000000	0.000000	0.000000	0.000000	.707107	0.000000	0.000000	0.000000-M
		+M	0.000000	0.000000	0.000000	0.000000	0.000000	-.707107	0.000000	0.000000	+M
-7.9011E+00	-.0000	-M	0.000000	0.000000	.372359	0.000000	0.000000	0.000000	0.000000	0.000000	.850116-M
		+M	0.000000	0.000000	.372359	0.000000	0.000000	0.000000	0.000000	0.000000	+M
-2.6937E+01	-.3368	-M	0.000000	0.000000	0.000000	.787917	0.000000	0.000000	0.000000	0.000000	0.000000-M
		+M	0.000000	.070905	0.000000	0.000000	0.000000	0.000000	0.000000	.611685	+M
-2.6937E+01	.3368	-M	0.000000	.070905	0.000000	0.000000	0.000000	0.000000	0.000000	.611685	0.000000-M
		+M	0.000000	0.000000	0.000000	.787917	0.000000	0.000000	0.000000	0.000000	+M
-4.7679E+01	.2453	-M	.044028	0.000000	0.000000	0.000000	0.000000	0.000000	.579378	0.000000	0.000000-M
		+M	0.000000	0.000000	0.000000	0.000000	.813869	0.000000	0.000000	0.000000	+M
-4.7679E+01	-.2453	-M	0.000000	0.000000	0.000000	0.000000	.813869	0.000000	0.000000	0.000000	0.000000-M
		+M	.044028	0.000000	0.000000	0.000000	0.000000	0.000000	.579378	0.000000	+M
-5.3700E+01	0.0000	-M	0.000000	0.000000	0.000000	0.000000	0.000000	.707107	0.000000	0.000000	0.000000-M
		+M	0.000000	0.000000	0.000000	0.000000	0.000000	.707107	0.000000	0.000000	+M
ENERGY	HPAL		M=8	M=7	M=6	M=5	M=4	M=3	M=2	M=1	M=0

Y=-.20 X= .20 OAS=9.6127E+01*W246 B4F4/B2F2=-5.00E-02 B4F4/B6F6= 2.50E-01 B2F2/B6F6=-5.00E+00 B4F4=-4.00E-02

ENERGY	HPAL		M=8	M=7	M=6	M=5	M=4	M=3	M=2	M=1	M=0
5.1607E+01	.8616	-M	0.000000	0.000000	0.000000	.037065	0.000000	0.000000	0.000000	0.000000	0.000000-M
		+M	0.000000	.991748	0.000000	0.000000	0.000000	0.000000	0.000000	-.122731	+M
5.1607E+01	-.8616	-M	0.000000	.991748	0.000000	0.000000	0.000000	0.000000	0.000000	-.122731	0.000000-M
		+M	0.000000	0.000000	0.000000	.037065	0.000000	0.000000	0.000000	0.000000	+M
3.8535E+01	.0000	-M	0.000000	0.000000	.669996	0.000000	0.000000	0.000000	0.000000	0.000000	-.319706-M
		+M	0.000000	0.000000	.669996	0.000000	0.000000	0.000000	0.000000	0.000000	+M
3.3000E+01	-.0000	-M	0.000000	0.000000	-.707107	0.000000	0.000000	0.000000	0.000000	0.000000	.000000-M
		+M	0.000000	0.000000	.707107	0.000000	0.000000	0.000000	0.000000	0.000000	+M
1.0434E+01	.4146	-M	0.000000	-.095289	0.000000	0.000000	0.000000	0.000000	0.000000	-.512222	0.000000-M
		+M	0.000000	0.000000	0.000000	.853551	0.000000	0.000000	0.000000	0.000000	+M
1.0434E+01	-.4146	-M	0.000000	0.000000	0.000000	.853551	0.000000	0.000000	0.000000	0.000000	0.000000-M
		+M	0.000000	-.095289	0.000000	0.000000	0.000000	0.000000	0.000000	-.512222	+M
2.1707E+00	.8005	-M	0.000000	0.000000	0.000000	0.000000	.271565	0.000000	0.000000	0.000000	0.000000-M
		+M	.898777	0.000000	0.000000	0.000000	0.000000	0.000000	-.344168	0.000000	+M
2.1707E+00	-.8005	-M	.898777	0.000000	0.000000	0.000000	0.000000	0.000000	-.344168	0.000000	0.000000-M
		+M	0.000000	0.000000	0.000000	0.000000	.271565	0.000000	0.000000	0.000000	+M
-4.1963E+00	-.0564	-M	-.433479	0.000000	0.000000	0.000000	0.000000	0.000000	-.605057	0.000000	0.000000-M
		+M	0.000000	0.000000	0.000000	0.000000	.667834	0.000000	0.000000	0.000000	+M
-4.1963E+00	.0564	-M	0.000000	0.000000	0.000000	0.000000	.667834	0.000000	0.000000	0.000000	0.000000-M
		+M	-.433479	0.000000	0.000000	0.000000	0.000000	0.000000	-.605057	0.000000	+M
-7.5600E+00	0.0000	-M	0.000000	0.000000	0.000000	0.000000	0.000000	.707107	0.000000	0.000000	0.000000-M
		+M	0.000000	0.000000	0.000000	0.000000	0.000000	-.707107	0.000000	0.000000	+M
-1.5615E+01	.0000	-M	0.000000	0.000000	.226066	0.000000	0.000000	0.000000	0.000000	0.000000	.947517-M
		+M	0.000000	0.000000	.226066	0.000000	0.000000	0.000000	0.000000	0.000000	+M
-2.3520E+01	.0720	-M	0.000000	.085772	0.000000	0.000000	0.000000	0.000000	0.000000	.850039	0.000000-M
		+M	0.000000	0.000000	0.000000	.519690	0.000000	0.000000	0.000000	0.000000	+M
-2.3520E+01	-.0720	-M	0.000000	0.000000	0.000000	.519690	0.000000	0.000000	0.000000	0.000000	0.000000-M
		+M	0.000000	.085772	0.000000	0.000000	0.000000	0.000000	0.000000	.850039	+M
-3.8414E+01	-.1070	-M	0.000000	0.000000	0.000000	0.000000	.693001	0.000000	0.000000	0.000000	0.000000-M
		+M	.065535	0.000000	0.000000	0.000000	0.000000	0.000000	.717952	0.000000	+M
-3.8414E+01	.1070	-M	.065535	0.000000	0.000000	0.000000	0.000000	0.000000	.717952	0.000000	0.000000-M
		+M	0.000000	0.000000	0.000000	0.000000	.693001	0.000000	0.000000	0.000000	+M
-4.4520E+01	0.0000	-M	0.000000	0.000000	0.000000	0.000000	0.000000	.707107	0.000000	0.000000	0.000000-M
		+M	0.000000	0.000000	0.000000	0.000000	0.000000	.707107	0.000000	0.000000	+M
ENERGY	HPAL		M=8	M=7	M=6	M=5	M=4	M=3	M=2	M=1	M=0

Y=-.20 X= .40 OAS=8.3353E+01*W246 B4F4/B2F2=-1.00E-01 B4F4/B6F6= 6.67E-01 B2F2/B6F6=-6.67E+00 B4F4=-8.00E-02

ENERGY	HPAL		M=8	M=7	M=6	M=5	M=4	M=3	M=2	M=1	M=0
4.8013E+01	-.8690	-M	0.000000	.996388	0.000000	0.000000	0.000000	0.000000	0.000000	-.080317	0.000000-M
		+M	0.000000	0.000000	0.000000	.027577	0.000000	0.000000	0.000000	0.000000	+M
4.8013E+01	.8690	-M	0.000000	0.000000	0.000000	.027577	0.000000	0.000000	0.000000	0.000000	0.000000-M
		+M	0.000000	.996388	0.000000	0.000000	0.000000	0.000000	0.000000	-.080317	+M
4.3026E+01	-.0000	-M	0.000000	0.000000	.695806	0.000000	0.000000	0.000000	0.000000	0.000000	-.178064-M
		+M	0.000000	0.000000	.695806	0.000000	0.000000	0.000000	0.000000	0.000000	+M
4.0800E+01	.0000	-M	0.000000	0.000000	-.707107	0.000000	0.000000	0.000000	0.000000	0.000000	.000000-M
		+M	0.000000	0.000000	.707107	0.000000	0.000000	0.000000	0.000000	0.000000	+M
1.7763E+01	.5755	-M	0.000000	-.046794	0.000000	0.000000	0.000000	0.000000	0.000000	-.248309	0.000000-M
		+M	0.000000	0.000000	0.000000	.967550	0.000000	0.000000	0.000000	0.000000	+M
1.7763E+01	-.5755	-M	0.000000	0.000000	0.000000	.967550	0.000000	0.000000	0.000000	0.000000	0.000000-M
		+M	0.000000	-.046794	0.000000	0.000000	0.000000	0.000000	0.000000	-.248309	+M
-1.0148E+00	.2948	-M	.149583	0.000000	0.000000	0.000000	0.000000	0.000000	-.478354	0.000000	0.000000-M
		+M	0.000000	0.000000	0.000000	0.000000	.865334	0.000000	0.000000	0.000000	+M
-1.0148E+00	-.2948	-M	0.000000	0.000000	0.000000	0.000000	.865334	0.000000	0.000000	0.000000	0.000000-M
		+M	.149583	0.000000	0.000000	0.000000	0.000000	0.000000	-.478354	0.000000	+M
-7.6200E+00	0.0000	-M	0.000000	0.000000	0.000000	0.000000	0.000000	.707107	0.000000	0.000000	0.000000-M
		+M	0.000000	0.000000	0.000000	0.000000	0.000000	-.707107	0.000000	0.000000	+M
-9.5543E+00	-.9516	-M	.983653	0.000000	0.000000	0.000000	0.000000	0.000000	-.016746	0.000000	0.000000-M
		+M	0.000000	0.000000	0.000000	0.000000	-.179293	0.000000	0.000000	0.000000	+M
-9.5543E+00	.9516	-M	0.000000	0.000000	0.000000	0.000000	-.179293	0.000000	0.000000	0.000000	0.000000-M
		+M	.983653	0.000000	0.000000	0.000000	0.000000	0.000000	-.016746	0.000000	+M
-2.7186E+01	-.0000	-M	0.000000	0.000000	.125910	0.000000	0.000000	0.000000	0.000000	0.000000	.984019-M
		+M	0.000000	0.000000	.125910	0.000000	0.000000	0.000000	0.000000	0.000000	+M
-2.8936E+01	-.0815	-M	0.000000	.070863	0.000000	0.000000	0.000000	0.000000	0.000000	.965345	0.000000-M
		+M	0.000000	0.000000	0.000000	.251171	0.000000	0.000000	0.000000	0.000000	+M
-2.8936E+01	.0815	-M	0.000000	0.000000	0.000000	.251171	0.000000	0.000000	0.000000	0.000000	0.000000-M
		+M	0.000000	.070863	0.000000	0.000000	0.000000	0.000000	0.000000	.965345	+M
-3.3111E+01	-.0933	-M	.100256	0.000000	0.000000	0.000000	0.000000	0.000000	.878008	0.000000	0.000000-M
		+M	0.000000	0.000000	0.000000	0.000000	.468029	0.000000	0.000000	0.000000	+M
-3.3111E+01	.0933	-M	0.000000	0.000000	0.000000	0.000000	.468029	0.000000	0.000000	0.000000	0.000000-M
		+M	.100256	0.000000	0.000000	0.000000	0.000000	0.000000	.878008	0.000000	+M
-3.5340E+01	0.0000	-M	0.000000	0.000000	0.000000	0.000000	0.000000	.707107	0.000000	0.000000	0.000000-M
		+M	0.000000	0.000000	0.000000	0.000000	0.000000	.707107	0.000000	0.000000	+M
ENERGY	HPAL		M=8	M=7	M=6	M=5	M=4	M=3	M=2	M=1	M=0

Y=-.20 X= .60 OAS=8.9948E+01*W246 B4F4/B2F2=-1.50E-01 B4F4/B6F6= 1.50E+00 B2F2/B6F6=-1.00E+01 B4F4=-1.20E-01

ENERGY	HPAL		M=8	M=7	M=6	M=5	M=4	M=3	M=2	M=1	M=0
4.9354E+01	.0000	-M	0.000000	0.000000	.704136	0.000000	0.000000	0.000000	0.000000	0.000000	-.091567-M
		+M	0.000000	0.000000	.704136	0.000000	0.000000	0.000000	0.000000	0.000000	+M
4.8600E+01	-.0000	-M	0.000000	0.000000	-.707107	0.000000	0.000000	0.000000	0.000000	0.000000	.000000-M
		+M	0.000000	0.000000	.707107	0.000000	0.000000	0.000000	0.000000	0.000000	+M
4.4618E+01	-.8726	-M	0.000000	.998625	0.000000	0.000000	0.000000	0.000000	0.000000	-.047495	0.000000-M
		+M	0.000000	0.000000	0.000000	.022182	0.000000	0.000000	0.000000	0.000000	+M
4.4618E+01	.8726	-M	0.000000	0.000000	0.000000	.022182	0.000000	0.000000	0.000000	0.000000	0.000000-M
		+M	0.000000	.998625	0.000000	0.000000	0.000000	0.000000	0.000000	-.047495	+M
2.9211E+01	.6146	-M	0.000000	-.027360	0.000000	0.000000	0.000000	0.000000	0.000000	-.111305	0.000000-M
		+M	0.000000	0.000000	0.000000	.993410	0.000000	0.000000	0.000000	0.000000	+M
2.9211E+01	-.6146	-M	0.000000	0.000000	0.000000	.993410	0.000000	0.000000	0.000000	0.000000	0.000000-M
		+M	0.000000	-.027360	0.000000	0.000000	0.000000	0.000000	0.000000	-.111305	+M
5.6765E+00	.4572	-M	.016570	0.000000	0.000000	0.000000	0.000000	0.000000	-.237726	0.000000	0.000000-M
		+M	0.000000	0.000000	0.000000	0.000000	.971191	0.000000	0.000000	0.000000	+M
5.6765E+00	-.4572	-M	0.000000	0.000000	0.000000	0.000000	.971191	0.000000	0.000000	0.000000	0.000000-M
		+M	.016570	0.000000	0.000000	0.000000	0.000000	0.000000	-.237726	0.000000	+M
-7.6800E+00	0.0000	-M	0.000000	0.000000	0.000000	0.000000	0.000000	.707107	0.000000	0.000000	0.000000-M
		+M	0.000000	0.000000	0.000000	0.000000	0.000000	-.707107	0.000000	0.000000	+M
-1.9771E+01	-.9845	-M	.990790	0.000000	0.000000	0.000000	0.000000	0.000000	-.126651	0.000000	0.000000-M
		+M	0.000000	0.000000	0.000000	0.000000	-.047906	0.000000	0.000000	0.000000	+M
-1.9771E+01	.9845	-M	0.000000	0.000000	0.000000	0.000000	-.047906	0.000000	0.000000	0.000000	0.000000-M
		+M	.990790	0.000000	0.000000	0.000000	0.000000	0.000000	-.126651	0.000000	+M
-2.6160E+01	0.0000	-M	0.000000	0.000000	0.000000	0.000000	0.000000	.707107	0.000000	0.000000	0.000000-M
		+M	0.000000	0.000000	0.000000	0.000000	0.000000	.707107	0.000000	0.000000	+M
-3.2825E+01	.2227	-M	0.000000	0.000000	0.000000	0.000000	.233438	0.000000	0.000000	0.000000	0.000000-M
		+M	.134391	0.000000	0.000000	0.000000	0.000000	0.000000	.963040	0.000000	+M
-3.2825E+01	-.2227	-M	.134391	0.000000	0.000000	0.000000	0.000000	0.000000	.963040	0.000000	0.000000-M
		+M	0.000000	0.000000	0.000000	0.000000	.233438	0.000000	0.000000	0.000000	+M
-3.8669E+01	.1170	-M	0.000000	0.000000	0.000000	.112451	0.000000	0.000000	0.000000	0.000000	0.000000-M
		+M	0.000000	.044713	0.000000	0.000000	0.000000	0.000000	0.000000	.992651	+M
-3.8669E+01	-.1170	-M	0.000000	.044713	0.000000	0.000000	0.000000	0.000000	0.000000	.992651	0.000000-M
		+M	0.000000	0.000000	0.000000	.112451	0.000000	0.000000	0.000000	0.000000	+M
-4.0594E+01	-.0000	-M	0.000000	0.000000	.064747	0.000000	0.000000	0.000000	0.000000	0.000000	.995799-M
		+M	0.000000	0.000000	.064747	0.000000	0.000000	0.000000	0.000000	0.000000	+M
ENERGY	HPAL		M=8	M=7	M=6	M=5	M=4	M=3	M=2	M=1	M=0

Y=-.20 X= .80 OAS=1.1142E+02*W246 B4F4/B2F2=-2.00E-01 B4F4/B6F6= 4.00E+00 B2F2/B6F6=-2.00E+01 B4F4=-1.60E-01

ENERGY	HPAL		M=8	M=7	M=6	M=5	M=4	M=3	M=2	M=1	M=0
5.6551E+01	-.0000	-M	0.000000	0.000000	.706627	0.000000	0.000000	0.000000	0.000000	0.000000	-.036829-M
		+M	0.000000	0.000000	.706627	0.000000	0.000000	0.000000	0.000000	0.000000	+M
5.6400E+01	.0000	-M	0.000000	0.000000	-.707107	0.000000	0.000000	0.000000	0.000000	0.000000	-.000000-M
		+M	0.000000	0.000000	.707107	0.000000	0.000000	0.000000	0.000000	0.000000	+M
4.2046E+01	.6048	-M	0.000000	.111808	0.000000	0.000000	0.000000	0.000000	0.000000	-.043355	0.000000-M
		+M	0.000000	0.000000	0.000000	.992784	0.000000	0.000000	0.000000	0.000000	+M
4.2046E+01	-.6048	-M	0.000000	0.000000	0.000000	.992784	0.000000	0.000000	0.000000	0.000000	0.000000-M
		+M	0.000000	.111808	0.000000	0.000000	0.000000	0.000000	0.000000	-.043355	+M
4.1350E+01	-.8558	-M	0.000000	.993518	0.000000	0.000000	0.000000	0.000000	0.000000	-.015727	0.000000-M
		+M	0.000000	0.000000	0.000000	-.112578	0.000000	0.000000	0.000000	0.000000	+M
4.1350E+01	.8558	-M	0.000000	0.000000	0.000000	-.112578	0.000000	0.000000	0.000000	0.000000	0.000000-M
		+M	0.000000	.993518	0.000000	0.000000	0.000000	0.000000	0.000000	-.015727	+M
1.5742E+01	.4944	-M	.001674	0.000000	0.000000	0.000000	0.000000	0.000000	-.086314	0.000000	0.000000-M
		+M	0.000000	0.000000	0.000000	0.000000	.996267	0.000000	0.000000	0.000000	+M
1.5742E+01	-.4944	-M	0.000000	0.000000	0.000000	0.000000	.996267	0.000000	0.000000	0.000000	0.000000-M
		+M	.001674	0.000000	0.000000	0.000000	0.000000	0.000000	-.086314	0.000000	+M
-7.7400E+00	0.0000	-M	0.000000	0.000000	0.000000	0.000000	0.000000	.707107	0.000000	0.000000	0.000000-M
		+M	0.000000	0.000000	0.000000	0.000000	0.000000	-.707107	0.000000	0.000000	+M
-1.6980E+01	0.0000	-M	0.000000	0.000000	0.000000	0.000000	0.000000	.707107	0.000000	0.000000	0.000000-M
		+M	0.000000	0.000000	0.000000	0.000000	0.000000	.707107	0.000000	0.000000	+M
-3.0249E+01	-.9788	-M	.985899	0.000000	0.000000	0.000000	0.000000	0.000000	-.166569	0.000000	0.000000-M
		+M	0.000000	0.000000	0.000000	0.000000	-.016088	0.000000	0.000000	0.000000	+M
-3.0249E+01	.9788	-M	0.000000	0.000000	0.000000	0.000000	-.016088	0.000000	0.000000	0.000000	0.000000-M
		+M	.985899	0.000000	0.000000	0.000000	0.000000	0.000000	-.166569	0.000000	+M
-3.5653E+01	-.2656	-M	.167336	0.000000	0.000000	0.000000	0.000000	0.000000	.982245	0.000000	0.000000-M
		+M	0.000000	0.000000	0.000000	0.000000	.084818	0.000000	0.000000	0.000000	+M
-3.5653E+01	.2656	-M	0.000000	0.000000	0.000000	0.000000	.084818	0.000000	0.000000	0.000000	0.000000-M
		+M	.167336	0.000000	0.000000	0.000000	0.000000	0.000000	.982245	0.000000	+M
-4.9915E+01	.1240	-M	0.000000	0.000000	0.000000	.041316	0.000000	0.000000	0.000000	0.000000	0.000000-M
		+M	0.000000	.020494	0.000000	0.000000	0.000000	0.000000	0.000000	.998936	+M
-4.9915E+01	-.1240	-M	0.000000	.020494	0.000000	0.000000	0.000000	0.000000	0.000000	.998936	0.000000-M
		+M	0.000000	0.000000	0.000000	.041316	0.000000	0.000000	0.000000	0.000000	+M
-5.4871E+01	.0000	-M	0.000000	0.000000	.026042	0.000000	0.000000	0.000000	0.000000	0.000000	.999322-M
		+M	0.000000	0.000000	.026042	0.000000	0.000000	0.000000	0.000000	0.000000	+M
ENERGY	HPAL		M=8	M=7	M=6	M=5	M=4	M=3	M=2	M=1	M=0

Y= .20 X=-.80 OAS=1.1477E+02*W246 B4F4/B2F2=-2.00E-01 B4F4/B6F6=-4.00E+00 B2F2/B6F6= 2.00E+01 B4F4=-1.60E-01

ENERGY	HPAL		M=8	M=7	M=6	M=5	M=4	M=3	M=2	M=1	M=0
5.0307E+01	-.0000	-M	0.000000	0.000000	.706655	0.000000	0.000000	0.000000	0.000000	0.000000	.035753-M
		+M	0.000000	0.000000	.706655	0.000000	0.000000	0.000000	0.000000	0.000000	+M
5.0160E+01	.0000	-M	0.000000	0.000000	-.707107	0.000000	0.000000	0.000000	0.000000	0.000000	.000000-M
		+M	0.000000	0.000000	.707107	0.000000	0.000000	0.000000	0.000000	0.000000	+M
4.7219E+01	.6240	-M	0.000000	.003531	0.000000	0.000000	0.000000	0.000000	0.000000	.036637	0.000000-M
		+M	0.000000	0.000000	0.000000	.999322	0.000000	0.000000	0.000000	0.000000	+M
4.7219E+01	-.6240	-M	0.000000	0.000000	0.000000	.999322	0.000000	0.000000	0.000000	0.000000	0.000000-M
		+M	0.000000	.003531	0.000000	0.000000	0.000000	0.000000	0.000000	.036637	+M
2.7841E+01	.8746	-M	0.000000	0.000000	0.000000	-.004339	0.000000	0.000000	0.000000	0.000000	0.000000-M
		+M	0.000000	.999749	0.000000	0.000000	0.000000	0.000000	0.000000	.021987	+M
2.7841E+01	-.8746	-M	0.000000	.999749	0.000000	0.000000	0.000000	0.000000	0.000000	.021987	0.000000-M
		+M	0.000000	0.000000	0.000000	-.004339	0.000000	0.000000	0.000000	0.000000	+M
2.5919E+01	-.4961	-M	0.000000	0.000000	0.000000	0.000000	.997390	0.000000	0.000000	0.000000	0.000000-M
		+M	.001346	0.000000	0.000000	0.000000	0.000000	0.000000	.072188	0.000000	+M
2.5919E+01	.4961	-M	.001346	0.000000	0.000000	0.000000	0.000000	0.000000	.072188	0.000000	0.000000-M
		+M	0.000000	0.000000	0.000000	0.000000	.997390	0.000000	0.000000	0.000000	+M
-3.0000E-01	0.0000	-M	0.000000	0.000000	0.000000	0.000000	0.000000	.707107	0.000000	0.000000	0.000000-M
		+M	0.000000	0.000000	0.000000	0.000000	0.000000	.707107	0.000000	0.000000	+M
-9.5400E+00	0.0000	-M	0.000000	0.000000	0.000000	0.000000	0.000000	-.707107	0.000000	0.000000	0.000000-M
		+M	0.000000	0.000000	0.000000	0.000000	0.000000	.707107	0.000000	0.000000	+M
-2.2020E+01	-.9966	-M	.997741	0.000000	0.000000	0.000000	0.000000	0.000000	.066893	0.000000	0.000000-M
		+M	0.000000	0.000000	0.000000	0.000000	-.006188	0.000000	0.000000	0.000000	+M
-2.2020E+01	.9966	-M	0.000000	0.000000	0.000000	0.000000	-.006188	0.000000	0.000000	0.000000	0.000000-M
		+M	.997741	0.000000	0.000000	0.000000	0.000000	0.000000	.066893	0.000000	+M
-3.5339E+01	-.2495	-M	-.067165	0.000000	0.000000	0.000000	0.000000	0.000000	.995145	0.000000	0.000000-M
		+M	0.000000	0.000000	0.000000	0.000000	-.071935	0.000000	0.000000	0.000000	+M
-3.5339E+01	.2495	-M	0.000000	0.000000	0.000000	0.000000	-.071935	0.000000	0.000000	0.000000	0.000000-M
		+M	-.067165	0.000000	0.000000	0.000000	0.000000	0.000000	.995145	0.000000	+M
-5.6700E+01	.1244	-M	0.000000	0.000000	0.000000	-.036550	0.000000	0.000000	0.000000	0.000000	0.000000-M
		+M	0.000000	-.022131	0.000000	0.000000	0.000000	0.000000	0.000000	.999087	+M
-5.6700E+01	-.1244	-M	0.000000	-.022131	0.000000	0.000000	0.000000	0.000000	0.000000	.999087	0.000000-M
		+M	0.000000	0.000000	0.000000	-.036550	0.000000	0.000000	0.000000	0.000000	+M
-6.4467E+01	0.0000	-M	0.000000	0.000000	-.025281	0.000000	0.000000	0.000000	0.000000	0.000000	.999361-M
		+M	0.000000	0.000000	-.025281	0.000000	0.000000	0.000000	0.000000	0.000000	+M
ENERGY	HPAL		M=8	M=7	M=6	M=5	M=4	M=3	M=2	M=1	M=0

Y= .20 X=-.60 OAS=9.9142E+01*W246 B4F4/B2F2=-1.50E-01 B4F4/B6F6=-1.50E+00 B2F2/B6F6= 1.00E+01 B4F4=-1.20E-01

ENERGY	HPAL		M=8	M=7	M=6	M=5	M=4	M=3	M=2	M=1	M=0
3.9401E+01	.6194	-M	0.000000	.014303	0.000000	0.000000	0.000000	0.000000	0.000000	.084059	0.000000-M
		+M	0.000000	0.000000	0.000000	.996358	0.000000	0.000000	0.000000	0.000000	+M
3.9401E+01	-.6194	-M	0.000000	0.000000	0.000000	.996358	0.000000	0.000000	0.000000	0.000000	0.000000-M
		+M	0.000000	.014303	0.000000	0.000000	0.000000	0.000000	0.000000	.084059	+M
3.6822E+01	-.0000	-M	0.000000	0.000000	.704533	0.000000	0.000000	0.000000	0.000000	0.000000	.085246-M
		+M	0.000000	0.000000	.704533	0.000000	0.000000	0.000000	0.000000	0.000000	+M
3.6120E+01	.0000	-M	0.000000	0.000000	-.707107	0.000000	0.000000	0.000000	0.000000	0.000000	.000000-M
		+M	0.000000	0.000000	.707107	0.000000	0.000000	0.000000	0.000000	0.000000	+M
2.5401E+01	.4814	-M	.009771	0.000000	0.000000	0.000000	0.000000	0.000000	.157018	0.000000	0.000000-M
		+M	0.000000	0.000000	0.000000	0.000000	.987547	0.000000	0.000000	0.000000	+M
2.5401E+01	-.4814	-M	0.000000	0.000000	0.000000	0.000000	.987547	0.000000	0.000000	0.000000	0.000000-M
		+M	.009771	0.000000	0.000000	0.000000	0.000000	0.000000	.157018	0.000000	+M
1.7596E+01	.8724	-M	0.000000	0.000000	0.000000	-.018740	0.000000	0.000000	0.000000	0.000000	0.000000-M
		+M	0.000000	.998459	0.000000	0.000000	0.000000	0.000000	0.000000	.052232	+M
1.7596E+01	-.8724	-M	0.000000	.998459	0.000000	0.000000	0.000000	0.000000	0.000000	.052232	0.000000-M
		+M	0.000000	0.000000	0.000000	-.018740	0.000000	0.000000	0.000000	0.000000	+M
7.2000E+00	0.0000	-M	0.000000	0.000000	0.000000	0.000000	0.000000	.707107	0.000000	0.000000	0.000000-M
		+M	0.000000	0.000000	0.000000	0.000000	0.000000	.707107	0.000000	0.000000	+M
-3.2522E+00	-.9967	-M	.998003	0.000000	0.000000	0.000000	0.000000	0.000000	.060108	0.000000	0.000000-M
		+M	0.000000	0.000000	0.000000	0.000000	-.019432	0.000000	0.000000	0.000000	+M
-3.2522E+00	.9967	-M	0.000000	0.000000	0.000000	0.000000	-.019432	0.000000	0.000000	0.000000	0.000000-M
		+M	.998003	0.000000	0.000000	0.000000	0.000000	0.000000	.060108	0.000000	+M
-1.1280E+01	0.0000	-M	0.000000	0.000000	0.000000	0.000000	0.000000	-.707107	0.000000	0.000000	0.000000-M
		+M	0.000000	0.000000	0.000000	0.000000	0.000000	.707107	0.000000	0.000000	+M
-3.1629E+01	-.2346	-M	-.062411	0.000000	0.000000	0.000000	0.000000	0.000000	.985765	0.000000	0.000000-M
		+M	0.000000	0.000000	0.000000	0.000000	-.156117	0.000000	0.000000	0.000000	+M
-3.1629E+01	.2346	-M	0.000000	0.000000	0.000000	0.000000	-.156117	0.000000	0.000000	0.000000	0.000000-M
		+M	-.062411	0.000000	0.000000	0.000000	0.000000	0.000000	.985765	0.000000	+M
-5.2076E+01	-.1220	-M	0.000000	-.053617	0.000000	0.000000	0.000000	0.000000	0.000000	.995091	0.000000-M
		+M	0.000000	0.000000	0.000000	-.083182	0.000000	0.000000	0.000000	0.000000	+M
-5.2076E+01	.1220	-M	0.000000	0.000000	0.000000	-.083182	0.000000	0.000000	0.000000	0.000000	0.000000-M
		+M	0.000000	-.053617	0.000000	0.000000	0.000000	0.000000	0.000000	.995091	+M
-5.9742E+01	.0000	-M	0.000000	0.000000	-.060278	0.000000	0.000000	0.000000	0.000000	0.000000	.996360-M
		+M	0.000000	0.000000	-.060278	0.000000	0.000000	0.000000	0.000000	0.000000	+M
ENERGY	HPAL		M=8	M=7	M=6	M=5	M=4	M=3	M=2	M=1	M=0

Y= .20 X=-.40 OAS=8.7834E+01*W246 B4F4/B2F2=-1.00E-01 B4F4/B6F6=-6.67E-01 B2F2/B6F6= 6.67E+00 B4F4=-8.00E-02

ENERGY	HPAL		M=8	M=7	M=6	M=5	M=4	M=3	M=2	M=1	M=0
3.2128E+01	.6073	-M	0.000000	.032751	0.000000	0.000000	0.000000	0.000000	0.000000	.146560	0.000000-M
		+M	0.000000	0.000000	0.000000	.988660	0.000000	0.000000	0.000000	0.000000	+M
3.2128E+01	-.6073	-M	0.000000	0.000000	0.000000	.988660	0.000000	0.000000	0.000000	0.000000	0.000000-M
		+M	0.000000	.032751	0.000000	0.000000	0.000000	0.000000	0.000000	.146560	+M
2.5861E+01	.4458	-M	.064638	0.000000	0.000000	0.000000	0.000000	0.000000	.252840	0.000000	0.000000-M
		+M	0.000000	0.000000	0.000000	0.000000	.965346	0.000000	0.000000	0.000000	+M
2.5861E+01	-.4458	-M	0.000000	0.000000	0.000000	0.000000	.965346	0.000000	0.000000	0.000000	0.000000-M
		+M	.064638	0.000000	0.000000	0.000000	0.000000	0.000000	.252840	0.000000	+M
2.4026E+01	-.0000	-M	0.000000	0.000000	.698425	0.000000	0.000000	0.000000	0.000000	0.000000	.156217-M
		+M	0.000000	0.000000	.698425	0.000000	0.000000	0.000000	0.000000	0.000000	+M
2.2080E+01	.0000	-M	0.000000	0.000000	-.707107	0.000000	0.000000	0.000000	0.000000	0.000000	-.000000-M
		+M	0.000000	0.000000	.707107	0.000000	0.000000	0.000000	0.000000	0.000000	+M
1.5469E+01	-.9898	-M	.996181	0.000000	0.000000	0.000000	0.000000	0.000000	.040554	0.000000	0.000000-M
		+M	0.000000	0.000000	0.000000	0.000000	-.077325	0.000000	0.000000	0.000000	+M
1.5469E+01	.9898	-M	0.000000	0.000000	0.000000	0.000000	-.077325	0.000000	0.000000	0.000000	0.000000-M
		+M	.996181	0.000000	0.000000	0.000000	0.000000	0.000000	.040554	0.000000	+M
1.4700E+01	0.0000	-M	0.000000	0.000000	0.000000	0.000000	0.000000	.707107	0.000000	0.000000	0.000000-M
		+M	0.000000	0.000000	0.000000	0.000000	0.000000	.707107	0.000000	0.000000	+M
7.5330E+00	.8650	-M	0.000000	0.000000	0.000000	-.046935	0.000000	0.000000	0.000000	0.000000	0.000000-M
		+M	0.000000	.994428	0.000000	0.000000	0.000000	0.000000	0.000000	.094394	+M
7.5330E+00	-.8650	-M	0.000000	.994428	0.000000	0.000000	0.000000	0.000000	0.000000	.094394	0.000000-M
		+M	0.000000	0.000000	0.000000	-.046935	0.000000	0.000000	0.000000	0.000000	+M
-1.3020E+01	0.0000	-M	0.000000	0.000000	0.000000	0.000000	0.000000	-.707107	0.000000	0.000000	0.000000-M
		+M	0.000000	0.000000	0.000000	0.000000	0.000000	.707107	0.000000	0.000000	+M
-2.8851E+01	.2060	-M	0.000000	0.000000	0.000000	0.000000	-.249254	0.000000	0.000000	0.000000	0.000000-M
		+M	-.058699	0.000000	0.000000	0.000000	0.000000	0.000000	.966658	0.000000	+M
-2.8851E+01	-.2060	-M	-.058699	0.000000	0.000000	0.000000	0.000000	0.000000	.966658	0.000000	0.000000-M
		+M	0.000000	0.000000	0.000000	0.000000	-.249254	0.000000	0.000000	0.000000	+M
-4.8181E+01	-.1173	-M	0.000000	-.100202	0.000000	0.000000	0.000000	0.000000	0.000000	.984688	0.000000-M
		+M	0.000000	0.000000	0.000000	-.142652	0.000000	0.000000	0.000000	0.000000	+M
-4.8181E+01	.1173	-M	0.000000	0.000000	0.000000	-.142652	0.000000	0.000000	0.000000	0.000000	0.000000-M
		+M	0.000000	-.100202	0.000000	0.000000	0.000000	0.000000	0.000000	.984688	+M
-5.5706E+01	-.0000	-M	0.000000	0.000000	-.110462	0.000000	0.000000	0.000000	0.000000	0.000000	.987723-M
		+M	0.000000	0.000000	-.110462	0.000000	0.000000	0.000000	0.000000	0.000000	+M
ENERGY	HPAL		M=8	M=7	M=6	M=5	M=4	M=3	M=2	M=1	M=0

Y= .20 X=-.20 OAS=8.7373E+01*W246 B4F4/B2F2=-5.00E-02 B4F4/B6F6=-2.50E-01 B2F2/B6F6= 5.00E+00 B4F4=-4.00E-02

ENERGY	HPAL		M=8	M=7	M=6	M=5	M=4	M=3	M=2	M=1	M=0
3.4478E+01	.9613	-M	0.000000	0.000000	0.000000	0.000000	.140828	0.000000	0.000000	0.000000	0.000000-M
		+M	.983980	0.000000	0.000000	0.000000	0.000000	0.000000	.109319	0.000000	+M
3.4478E+01	-.9613	-M	.983980	0.000000	0.000000	0.000000	0.000000	0.000000	.109319	0.000000	0.000000-M
		+M	0.000000	0.000000	0.000000	0.000000	.140828	0.000000	0.000000	0.000000	+M
2.7089E+01	-.3745	-M	0.000000	0.000000	0.000000	0.000000	.928148	0.000000	0.000000	0.000000	0.000000-M
		+M	-.169644	0.000000	0.000000	0.000000	0.000000	0.000000	.331304	0.000000	+M
2.7089E+01	.3745	-M	-.169644	0.000000	0.000000	0.000000	0.000000	0.000000	.331304	0.000000	0.000000-M
		+M	0.000000	0.000000	0.000000	0.000000	.928148	0.000000	0.000000	0.000000	+M
2.5698E+01	.5805	-M	0.000000	.058908	0.000000	0.000000	0.000000	0.000000	0.000000	.228947	0.000000-M
		+M	0.000000	0.000000	0.000000	.971655	0.000000	0.000000	0.000000	0.000000	+M
2.5698E+01	-.5805	-M	0.000000	0.000000	0.000000	.971655	0.000000	0.000000	0.000000	0.000000	0.000000-M
		+M	0.000000	.058908	0.000000	0.000000	0.000000	0.000000	0.000000	.228947	+M
2.2200E+01	0.0000	-M	0.000000	0.000000	0.000000	0.000000	0.000000	.707107	0.000000	0.000000	0.000000-M
		+M	0.000000	0.000000	0.000000	0.000000	0.000000	.707107	0.000000	0.000000	+M
1.2456E+01	.0000	-M	0.000000	0.000000	.682800	0.000000	0.000000	0.000000	0.000000	0.000000	.259938-M
		+M	0.000000	0.000000	.682800	0.000000	0.000000	0.000000	0.000000	0.000000	+M
8.0400E+00	-.0000	-M	0.000000	0.000000	-.707107	0.000000	0.000000	0.000000	0.000000	0.000000	-.000000-M
		+M	0.000000	0.000000	.707107	0.000000	0.000000	0.000000	0.000000	0.000000	+M
-2.2302E+00	.8435	-M	0.000000	0.000000	0.000000	-.095829	0.000000	0.000000	0.000000	0.000000	0.000000-M
		+M	0.000000	.983467	0.000000	0.000000	0.000000	0.000000	0.000000	.153654	+M
-2.2302E+00	-.8435	-M	0.000000	.983467	0.000000	0.000000	0.000000	0.000000	0.000000	.153654	0.000000-M
		+M	0.000000	0.000000	0.000000	-.095829	0.000000	0.000000	0.000000	0.000000	+M
-1.4760E+01	0.0000	-M	0.000000	0.000000	0.000000	0.000000	0.000000	-.707107	0.000000	0.000000	0.000000-M
		+M	0.000000	0.000000	0.000000	0.000000	0.000000	.707107	0.000000	0.000000	+M
-2.7127E+01	-.1632	-M	-.054807	0.000000	0.000000	0.000000	0.000000	0.000000	.937170	0.000000	0.000000-M
		+M	0.000000	0.000000	0.000000	0.000000	-.344542	0.000000	0.000000	0.000000	+M
-2.7127E+01	.1632	-M	0.000000	0.000000	0.000000	0.000000	-.344542	0.000000	0.000000	0.000000	0.000000-M
		+M	-.054807	0.000000	0.000000	0.000000	0.000000	0.000000	.937170	0.000000	+M
-4.5428E+01	.1120	-M	0.000000	0.000000	0.000000	-.216110	0.000000	0.000000	0.000000	0.000000	0.000000-M
		+M	0.000000	-.171238	0.000000	0.000000	0.000000	0.000000	0.000000	.961236	+M
-4.5428E+01	-.1120	-M	0.000000	-.171238	0.000000	0.000000	0.000000	0.000000	0.000000	.961236	0.000000-M
		+M	0.000000	0.000000	0.000000	-.216110	0.000000	0.000000	0.000000	0.000000	+M
-5.2896E+01	.0000	-M	0.000000	0.000000	-.183804	0.000000	0.000000	0.000000	0.000000	0.000000	.965625-M
		+M	0.000000	0.000000	-.183804	0.000000	0.000000	0.000000	0.000000	0.000000	+M
ENERGY	HPAL		M=8	M=7	M=6	M=5	M=4	M=3	M=2	M=1	M=0

Y= .20 X=0.00 OAS=1.0545E+02*W246 B4F4/B2F2= 0. B4F4/B6F6= 0. B2F2/B6F6= 4.00E+00 B4F4= 0.

ENERGY	HPAL		M=8	M=7	M=6	M=5	M=4	M=3	M=2	M=1	M=0
5.3167E+01	-.9908	-M	.995245	0.000000	0.000000	0.000000	0.000000	0.000000	.081729	0.000000	0.000000-M
		+M	0.000000	0.000000	0.000000	0.000000	.052994	0.000000	0.000000	0.000000	+M
5.3167E+01	.9908	-M	0.000000	0.000000	0.000000	0.000000	.052994	0.000000	0.000000	0.000000	0.000000-M
		+M	.995245	0.000000	0.000000	0.000000	0.000000	0.000000	.081729	0.000000	+M
2.9706E+01	.3516	-M	-.083126	0.000000	0.000000	0.000000	0.000000	0.000000	.429046	0.000000	0.000000-M
		+M	0.000000	0.000000	0.000000	0.000000	.899449	0.000000	0.000000	0.000000	+M
2.9706E+01	-.3516	-M	0.000000	0.000000	0.000000	0.000000	.899449	0.000000	0.000000	0.000000	0.000000-M
		+M	-.083126	0.000000	0.000000	0.000000	0.000000	0.000000	.429046	0.000000	+M
2.9700E+01	0.0000	-M	0.000000	0.000000	0.000000	0.000000	0.000000	.707107	0.000000	0.000000	0.000000-M
		+M	0.000000	0.000000	0.000000	0.000000	0.000000	.707107	0.000000	0.000000	+M
2.0548E+01	.5292	-M	0.000000	.090888	0.000000	0.000000	0.000000	0.000000	0.000000	.333576	0.000000-M
		+M	0.000000	0.000000	0.000000	.938332	0.000000	0.000000	0.000000	0.000000	+M
2.0548E+01	-.5292	-M	0.000000	0.000000	0.000000	.938332	0.000000	0.000000	0.000000	0.000000	0.000000-M
		+M	0.000000	.090888	0.000000	0.000000	0.000000	0.000000	0.000000	.333576	+M
3.0836E+00	.0000	-M	0.000000	0.000000	.646506	0.000000	0.000000	0.000000	0.000000	0.000000	.405044-M
		+M	0.000000	0.000000	.646506	0.000000	0.000000	0.000000	0.000000	0.000000	+M
-6.0000E+00	-.0000	-M	0.000000	0.000000	-.707107	0.000000	0.000000	0.000000	0.000000	0.000000	.000000-M
		+M	0.000000	0.000000	.707107	0.000000	0.000000	0.000000	0.000000	0.000000	+M
-1.1503E+01	-.7873	-M	0.000000	.956026	0.000000	0.000000	0.000000	0.000000	0.000000	.234602	0.000000-M
		+M	0.000000	0.000000	0.000000	-.176003	0.000000	0.000000	0.000000	0.000000	+M
-1.1503E+01	.7873	-M	0.000000	0.000000	0.000000	-.176003	0.000000	0.000000	0.000000	0.000000	0.000000-M
		+M	0.000000	.956026	0.000000	0.000000	0.000000	0.000000	0.000000	.234602	+M
-1.6500E+01	0.0000	-M	0.000000	0.000000	0.000000	0.000000	0.000000	-.707107	0.000000	0.000000	0.000000-M
		+M	0.000000	0.000000	0.000000	0.000000	0.000000	.707107	0.000000	0.000000	+M
-2.6474E+01	-.1108	-M	-.050774	0.000000	0.000000	0.000000	0.000000	0.000000	.899578	0.000000	0.000000-M
		+M	0.000000	0.000000	0.000000	0.000000	-.433800	0.000000	0.000000	0.000000	+M
-2.6474E+01	.1108	-M	0.000000	0.000000	0.000000	0.000000	-.433800	0.000000	0.000000	0.000000	0.000000-M
		+M	-.050774	0.000000	0.000000	0.000000	0.000000	0.000000	.899578	0.000000	+M
-4.4445E+01	.1169	-M	0.000000	0.000000	0.000000	-.297585	0.000000	0.000000	0.000000	0.000000	0.000000-M
		+M	0.000000	-.278845	0.000000	0.000000	0.000000	0.000000	0.000000	.913066	+M
-4.4445E+01	-.1169	-M	0.000000	-.278845	0.000000	0.000000	0.000000	0.000000	0.000000	.913066	0.000000-M
		+M	0.000000	0.000000	0.000000	-.297585	0.000000	0.000000	0.000000	0.000000	+M
-5.2284E+01	.0000	-M	0.000000	0.000000	-.286409	0.000000	0.000000	0.000000	0.000000	0.000000	.914297-M
		+M	0.000000	0.000000	-.286409	0.000000	0.000000	0.000000	0.000000	0.000000	+M
ENERGY	HPAL		M=8	M=7	M=6	M=5	M=4	M=3	M=2	M=1	M=0

Y= .20 X= .20 OAS=1.0238E+02*W246 B4F4/B2F2= 5.00E-02 B4F4/B6F6= 2.50E-01 B2F2/B6F6= 5.00E+00 B4F4= 4.00E-02

ENERGY	HPAL		M=8	M=7	M=6	M=5	M=4	M=3	M=2	M=1	M=0
6.3386E+01	-.9976	-M	.998515	0.000000	0.000000	0.000000	0.000000	0.000000	.051977	0.000000	0.000000-M
		+M	0.000000	0.000000	0.000000	0.000000	.016346	0.000000	0.000000	0.000000	+M
6.3386E+01	.9976	-M	0.000000	0.000000	0.000000	0.000000	.016346	0.000000	0.000000	0.000000	0.000000-M
		+M	.998515	0.000000	0.000000	0.000000	0.000000	0.000000	.051977	0.000000	+M
2.0520E+01	0.0000	-M	0.000000	0.000000	0.000000	0.000000	0.000000	.707107	0.000000	0.000000	0.000000-M
		+M	0.000000	0.000000	0.000000	0.000000	0.000000	.707107	0.000000	0.000000	+M
1.7993E+01	.2976	-M	-.040878	0.000000	0.000000	0.000000	0.000000	0.000000	.516258	0.000000	0.000000-M
		+M	0.000000	0.000000	0.000000	0.000000	.855457	0.000000	0.000000	0.000000	+M
1.7993E+01	-.2976	-M	0.000000	0.000000	0.000000	0.000000	.855457	0.000000	0.000000	0.000000	0.000000-M
		+M	-.040878	0.000000	0.000000	0.000000	0.000000	0.000000	.516258	0.000000	+M
8.3191E+00	.4169	-M	0.000000	.181312	0.000000	0.000000	0.000000	0.000000	0.000000	.460095	0.000000-M
		+M	0.000000	0.000000	0.000000	.869159	0.000000	0.000000	0.000000	0.000000	+M
8.3191E+00	-.4169	-M	0.000000	0.000000	0.000000	.869159	0.000000	0.000000	0.000000	0.000000	0.000000-M
		+M	0.000000	.181312	0.000000	0.000000	0.000000	0.000000	0.000000	.460095	+M
-3.1219E+00	.0000	-M	0.000000	0.000000	.592605	0.000000	0.000000	0.000000	0.000000	0.000000	.545562-M
		+M	0.000000	0.000000	.592605	0.000000	0.000000	0.000000	0.000000	0.000000	+M
-9.2194E+00	-.7214	-M	0.000000	.938513	0.000000	0.000000	0.000000	0.000000	0.000000	.183087	0.000000-M
		+M	0.000000	0.000000	0.000000	-.292698	0.000000	0.000000	0.000000	0.000000	+M
-9.2194E+00	.7214	-M	0.000000	0.000000	0.000000	-.292698	0.000000	0.000000	0.000000	0.000000	0.000000-M
		+M	0.000000	.938513	0.000000	0.000000	0.000000	0.000000	0.000000	.183087	+M
-1.3800E+01	-.0000	-M	0.000000	0.000000	-.707107	0.000000	0.000000	0.000000	0.000000	0.000000	.000000-M
		+M	0.000000	0.000000	.707107	0.000000	0.000000	0.000000	0.000000	0.000000	+M
-1.6440E+01	0.0000	-M	0.000000	0.000000	0.000000	0.000000	0.000000	-.707107	0.000000	0.000000	0.000000-M
		+M	0.000000	0.000000	0.000000	0.000000	0.000000	.707107	0.000000	0.000000	+M
-2.1740E+01	-.0500	-M	-.036025	0.000000	0.000000	0.000000	0.000000	0.000000	.854854	0.000000	0.000000-M
		+M	0.000000	0.000000	0.000000	0.000000	-.517616	0.000000	0.000000	0.000000	+M
-2.1740E+01	.0500	-M	0.000000	0.000000	0.000000	0.000000	-.517616	0.000000	0.000000	0.000000	0.000000-M
		+M	-.036025	0.000000	0.000000	0.000000	0.000000	0.000000	.854854	0.000000	+M
-3.2820E+01	.0706	-M	0.000000	0.000000	0.000000	-.398610	0.000000	0.000000	0.000000	0.000000	0.000000-M
		+M	0.000000	-.293801	0.000000	0.000000	0.000000	0.000000	0.000000	.868787	+M
-3.2820E+01	-.0706	-M	0.000000	-.293801	0.000000	0.000000	0.000000	0.000000	0.000000	.868787	0.000000-M
		+M	0.000000	0.000000	0.000000	-.398610	0.000000	0.000000	0.000000	0.000000	+M
-3.8998E+01	.0000	-M	0.000000	0.000000	-.385770	0.000000	0.000000	0.000000	0.000000	0.000000	.838071-M
		+M	0.000000	0.000000	-.385770	0.000000	0.000000	0.000000	0.000000	0.000000	+M
ENERGY	HPAL		M=8	M=7	M=6	M=5	M=4	M=3	M=2	M=1	M=0

Y= .20 X= .40 OAS=1.0417E+02*W246 B4F4/B2F2= 1.00E-01 B4F4/B6F6= 6.67E-01 B2F2/B6F6= 6.67E+00 B4F4= 8.00E-02

ENERGY	HPAL		M=8	M=7	M=6	M=5	M=4	M=3	M=2	M=1	M=0
7.3693E+01	-.9991	-M	.999387	0.000000	0.000000	0.000000	0.000000	0.000000	.034530	0.000000	0.000000-M
		+M	0.000000	0.000000	0.000000	0.000000	.005837	0.000000	0.000000	0.000000	+M
7.3693E+01	.9991	-M	0.000000	0.000000	0.000000	0.000000	.005837	0.000000	0.000000	0.000000	0.000000-M
		+M	.999387	0.000000	0.000000	0.000000	0.000000	0.000000	.034530	0.000000	+M
1.1340E+01	0.0000	-M	0.000000	0.000000	0.000000	0.000000	0.000000	.707107	0.000000	0.000000	0.000000-M
		+M	0.000000	0.000000	0.000000	0.000000	0.000000	.707107	0.000000	0.000000	+M
7.8332E+00	.1506	-M	-.027817	0.000000	0.000000	0.000000	0.000000	0.000000	.681440	0.000000	0.000000-M
		+M	0.000000	0.000000	0.000000	0.000000	.731346	0.000000	0.000000	0.000000	+M
7.8332E+00	-.1506	-M	0.000000	0.000000	0.000000	0.000000	.731346	0.000000	0.000000	0.000000	0.000000-M
		+M	-.027817	0.000000	0.000000	0.000000	0.000000	0.000000	.681440	0.000000	+M
1.2708E+00	-.0464	-M	0.000000	.447351	0.000000	0.000000	0.000000	0.000000	0.000000	.703521	0.000000-M
		+M	0.000000	0.000000	0.000000	.552210	0.000000	0.000000	0.000000	0.000000	+M
1.2708E+00	.0464	-M	0.000000	0.000000	0.000000	.552210	0.000000	0.000000	0.000000	0.000000	0.000000-M
		+M	0.000000	.447351	0.000000	0.000000	0.000000	0.000000	0.000000	.703521	+M
-4.5586E+00	.0000	-M	0.000000	0.000000	.413890	0.000000	0.000000	0.000000	0.000000	0.000000	.810796-M
		+M	0.000000	0.000000	.413890	0.000000	0.000000	0.000000	0.000000	0.000000	+M
-8.7532E+00	-.5230	-M	0.000000	-.865127	0.000000	0.000000	0.000000	0.000000	0.000000	.183828	0.000000-M
		+M	0.000000	0.000000	0.000000	.466650	0.000000	0.000000	0.000000	0.000000	+M
-8.7532E+00	.5230	-M	0.000000	0.000000	0.000000	-.466650	0.000000	0.000000	0.000000	0.000000	0.000000-M
		+M	0.000000	.865127	0.000000	0.000000	0.000000	0.000000	0.000000	-.183828	+M
-1.6380E+01	0.0000	-M	0.000000	0.000000	0.000000	0.000000	0.000000	-.707107	0.000000	0.000000	0.000000-M
		+M	0.000000	0.000000	0.000000	0.000000	0.000000	.707107	0.000000	0.000000	+M
-1.8646E+01	.0985	-M	-.021276	0.000000	0.000000	0.000000	0.000000	0.000000	.731059	0.000000	0.000000-M
		+M	0.000000	0.000000	0.000000	0.000000	-.681982	0.000000	0.000000	0.000000	+M
-1.8646E+01	-.0985	-M	0.000000	0.000000	0.000000	0.000000	-.681982	0.000000	0.000000	0.000000	0.000000-M
		+M	-.021276	0.000000	0.000000	0.000000	0.000000	0.000000	.731059	0.000000	+M
-2.1600E+01	-.0000	-M	0.000000	0.000000	-.707107	0.000000	0.000000	0.000000	0.000000	0.000000	.000000-M
		+M	0.000000	0.000000	.707107	0.000000	0.000000	0.000000	0.000000	0.000000	+M
-2.4558E+01	.1944	-M	0.000000	.226786	0.000000	0.000000	0.000000	0.000000	0.000000	-.686488	0.000000-M
		+M	0.000000	0.000000	0.000000	.690870	0.000000	0.000000	0.000000	0.000000	+M
-2.4558E+01	-.1944	-M	0.000000	0.000000	0.000000	.690870	0.000000	0.000000	0.000000	0.000000	0.000000-M
		+M	0.000000	.226786	0.000000	0.000000	0.000000	0.000000	0.000000	-.686488	+M
-3.0481E+01	.0000	-M	0.000000	0.000000	.573320	0.000000	0.000000	0.000000	0.000000	0.000000	-.585329-M
		+M	0.000000	0.000000	.573320	0.000000	0.000000	0.000000	0.000000	0.000000	+M
ENERGY	HPAL		M=8	M=7	M=6	M=5	M=4	M=3	M=2	M=1	M=0

Y= .20 X= .60 OAS=1.1548E+02*W246 B4F4/B2F2= 1.50E-01 B4F4/B6F6= 1.50E+00 B2F2/B6F6= 1.00E+01 B4F4= 1.20E-01

ENERGY	HPAL		M=8	M=7	M=6	M=5	M=4	M=3	M=2	M=1	M=0
8.4038E+01	-.9997	-M	.999776	0.000000	0.000000	0.000000	0.000000	0.000000	.021095	0.000000	0.000000-M
		+M	0.000000	0.000000	0.000000	0.000000	.001852	0.000000	0.000000	0.000000	+M
8.4038E+01	.9997	-M	0.000000	0.000000	0.000000	0.000000	.001852	0.000000	0.000000	0.000000	0.000000-M
		+M	.999776	0.000000	0.000000	0.000000	0.000000	0.000000	.021095	0.000000	+M
4.2998E+00	.1996	-M	0.000000	0.000000	0.000000	.219257	0.000000	0.000000	0.000000	0.000000	0.000000-M
		+M	0.000000	.384152	0.000000	0.000000	0.000000	0.000000	0.000000	.896858	+M
4.2998E+00	-.1996	-M	0.000000	.384152	0.000000	0.000000	0.000000	0.000000	0.000000	.896858	0.000000-M
		+M	0.000000	0.000000	0.000000	.219257	0.000000	0.000000	0.000000	0.000000	+M
3.4855E+00	.0000	-M	0.000000	0.000000	.171111	0.000000	0.000000	0.000000	0.000000	0.000000	.970279-M
		+M	0.000000	0.000000	.171111	0.000000	0.000000	0.000000	0.000000	0.000000	+M
2.5204E+00	.1158	-M	0.000000	0.000000	0.000000	0.000000	.423505	0.000000	0.000000	0.000000	0.000000-M
		+M	-.019894	0.000000	0.000000	0.000000	0.000000	0.000000	.905675	0.000000	+M
2.5204E+00	-.1158	-M	-.019894	0.000000	0.000000	0.000000	0.000000	0.000000	.905675	0.000000	0.000000-M
		+M	0.000000	0.000000	0.000000	0.000000	.423505	0.000000	0.000000	0.000000	+M
2.1600E+00	0.0000	-M	0.000000	0.000000	0.000000	0.000000	0.000000	.707107	0.000000	0.000000	0.000000-M
		+M	0.000000	0.000000	0.000000	0.000000	0.000000	.707107	0.000000	0.000000	+M
-5.9124E+00	-.7501	-M	0.000000	.922447	0.000000	0.000000	0.000000	0.000000	0.000000	-.362807	0.000000-M
		+M	0.000000	0.000000	0.000000	-.132144	0.000000	0.000000	0.000000	0.000000	+M
-5.9124E+00	.7501	-M	0.000000	0.000000	0.000000	-.132144	0.000000	0.000000	0.000000	0.000000	0.000000-M
		+M	0.000000	.922447	0.000000	0.000000	0.000000	0.000000	0.000000	-.362807	+M
-1.6320E+01	0.0000	-M	0.000000	0.000000	0.000000	0.000000	0.000000	-.707107	0.000000	0.000000	0.000000-M
		+M	0.000000	0.000000	0.000000	0.000000	0.000000	.707107	0.000000	0.000000	+M
-2.0438E+01	-.3654	-M	0.000000	0.000000	0.000000	0.000000	.905892	0.000000	0.000000	0.000000	0.000000-M
		+M	.007257	0.000000	0.000000	0.000000	0.000000	0.000000	-.423447	0.000000	+M
-2.0438E+01	.3654	-M	.007257	0.000000	0.000000	0.000000	0.000000	0.000000	-.423447	0.000000	0.000000-M
		+M	0.000000	0.000000	0.000000	0.000000	.905892	0.000000	0.000000	0.000000	+M
-2.8747E+01	.5747	-M	0.000000	.038966	0.000000	0.000000	0.000000	0.000000	0.000000	-.253016	0.000000-M
		+M	0.000000	0.000000	0.000000	.966677	0.000000	0.000000	0.000000	0.000000	+M
-2.8747E+01	-.5747	-M	0.000000	0.000000	0.000000	.966677	0.000000	0.000000	0.000000	0.000000	0.000000-M
		+M	0.000000	.038966	0.000000	0.000000	0.000000	0.000000	0.000000	-.253016	+M
-2.9400E+01	-.0000	-M	0.000000	0.000000	-.707107	0.000000	0.000000	0.000000	0.000000	0.000000	.000000-M
		+M	0.000000	0.000000	.707107	0.000000	0.000000	0.000000	0.000000	0.000000	+M
-3.1445E+01	.0000	-M	0.000000	0.000000	.686091	0.000000	0.000000	0.000000	0.000000	0.000000	-.241988-M
		+M	0.000000	0.000000	.686091	0.000000	0.000000	0.000000	0.000000	0.000000	+M
ENERGY	HPAL		M=8	M=7	M=6	M=5	M=4	M=3	M=2	M=1	M=0

Y= .20 X= .80 OAS=1.3496E+02*W246 B4F4/B2F2= 2.00E-01 B4F4/B6F6= 4.00E+00 B2F2/B6F6= 2.00E+01 B4F4= 1.60E-01

ENERGY	HPAL		M=8	M=7	M=6	M=5	M=4	M=3	M=2	M=1	M=0
9.4409E+01	.9999	-M	0.000000	0.000000	0.000000	0.000000	.000353	0.000000	0.000000	0.000000	0.000000-M
		+M	.999952	0.000000	0.000000	0.000000	0.000000	0.000000	.009819	0.000000	+M
9.4409E+01	-.9999	-M	.999952	0.000000	0.000000	0.000000	0.000000	0.000000	.009819	0.000000	0.000000-M
		+M	0.000000	0.000000	0.000000	0.000000	.000353	0.000000	0.000000	0.000000	+M
1.6632E+01	.0000	-M	0.000000	0.000000	.053710	0.000000	0.000000	0.000000	0.000000	0.000000	.997111-M
		+M	0.000000	0.000000	.053710	0.000000	0.000000	0.000000	0.000000	0.000000	+M
1.3426E+01	-.1332	-M	0.000000	.125615	0.000000	0.000000	0.000000	0.000000	0.000000	.989609	0.000000-M
		+M	0.000000	0.000000	0.000000	.069958	0.000000	0.000000	0.000000	0.000000	+M
1.3426E+01	.1332	-M	0.000000	0.000000	0.000000	.069958	0.000000	0.000000	0.000000	0.000000	0.000000-M
		+M	0.000000	.125615	0.000000	0.000000	0.000000	0.000000	0.000000	.989609	+M
3.7198E+00	.2360	-M	0.000000	0.000000	0.000000	0.000000	.136868	0.000000	0.000000	0.000000	0.000000-M
		+M	-.009775	0.000000	0.000000	0.000000	0.000000	0.000000	.990541	0.000000	+M
3.7198E+00	-.2360	-M	-.009775	0.000000	0.000000	0.000000	0.000000	0.000000	.990541	0.000000	0.000000-M
		+M	0.000000	0.000000	0.000000	0.000000	.136868	0.000000	0.000000	0.000000	+M
-1.5560E+00	-.8630	-M	0.000000	.992073	0.000000	0.000000	0.000000	0.000000	0.000000	-.125061	0.000000-M
		+M	0.000000	0.000000	0.000000	-.012261	0.000000	0.000000	0.000000	0.000000	+M
-1.5560E+00	.8630	-M	0.000000	0.000000	0.000000	-.012261	0.000000	0.000000	0.000000	0.000000	0.000000-M
		+M	0.000000	.992073	0.000000	0.000000	0.000000	0.000000	0.000000	-.125061	+M
-7.0200E+00	0.0000	-M	0.000000	0.000000	0.000000	0.000000	0.000000	.707107	0.000000	0.000000	0.000000-M
		+M	0.000000	0.000000	0.000000	0.000000	0.000000	.707107	0.000000	0.000000	+M
-1.6260E+01	0.0000	-M	0.000000	0.000000	0.000000	0.000000	0.000000	-.707107	0.000000	0.000000	0.000000-M
		+M	0.000000	0.000000	0.000000	0.000000	0.000000	.707107	0.000000	0.000000	+M
-2.8769E+01	-.4859	-M	0.000000	0.000000	0.000000	0.000000	.990589	0.000000	0.000000	0.000000	0.000000-M
		+M	.000994	0.000000	0.000000	0.000000	0.000000	0.000000	-.136865	0.000000	+M
-2.8769E+01	.4859	-M	.000994	0.000000	0.000000	0.000000	0.000000	0.000000	-.136865	0.000000	0.000000-M
		+M	0.000000	0.000000	0.000000	0.000000	.990589	0.000000	0.000000	0.000000	+M
-3.7200E+01	.0000	-M	0.000000	0.000000	-.707107	0.000000	0.000000	0.000000	0.000000	0.000000	-.000000-M
		+M	0.000000	0.000000	.707107	0.000000	0.000000	0.000000	0.000000	0.000000	+M
-3.7512E+01	-.0000	-M	0.000000	0.000000	.705064	0.000000	0.000000	0.000000	0.000000	0.000000	-.075958-M
		+M	0.000000	0.000000	.705064	0.000000	0.000000	0.000000	0.000000	0.000000	+M
-4.0550E+01	-.6212	-M	0.000000	0.000000	0.000000	.997475	0.000000	0.000000	0.000000	0.000000	0.000000-M
		+M	0.000000	.003385	0.000000	0.000000	0.000000	0.000000	0.000000	-.070944	+M
-4.0550E+01	.6212	-M	0.000000	.003385	0.000000	0.000000	0.000000	0.000000	0.000000	-.070944	0.000000-M
		+M	0.000000	0.000000	0.000000	.997475	0.000000	0.000000	0.000000	0.000000	+M
ENERGY	HPAL		M=8	M=7	M=6	M=5	M=4	M=3	M=2	M=1	M=0

Y= .40 X=-.80 OAS=1.9846E+02*W246 B4F4/B2F2=-5.33E-01 B4F4/B6F6=-4.00E+00 B2F2/B6F6= 7.50E+00 B4F4=-3.20E-01

ENERGY	HPAL		M=8	M=7	M=6	M=5	M=4	M=3	M=2	M=1	M=0
9.3475E+01	.6237	-M	0.000000	.002464	0.000000	0.000000	0.000000	0.000000	0.000000	.041392	0.000000-M
		+M	0.000000	0.000000	0.000000	.999140	0.000000	0.000000	0.000000	0.000000	+M
9.3475E+01	-.6237	-M	0.000000	0.000000	0.000000	.999140	0.000000	0.000000	0.000000	0.000000	0.000000-M
		+M	0.000000	.002464	0.000000	0.000000	0.000000	0.000000	0.000000	.041392	+M
8.8668E+01	-.0000	-M	0.000000	0.000000	.706471	0.000000	0.000000	0.000000	0.000000	0.000000	.042390-M
		+M	0.000000	0.000000	.706471	0.000000	0.000000	0.000000	0.000000	0.000000	+M
8.8320E+01	.0000	-M	0.000000	0.000000	-.707107	0.000000	0.000000	0.000000	0.000000	0.000000	.000000-M
		+M	0.000000	0.000000	.707107	0.000000	0.000000	0.000000	0.000000	0.000000	+M
5.9907E+01	.4952	-M	.000994	0.000000	0.000000	0.000000	0.000000	0.000000	.079976	0.000000	0.000000-M
		+M	0.000000	0.000000	0.000000	0.000000	.996796	0.000000	0.000000	0.000000	+M
5.9907E+01	-.4952	-M	0.000000	0.000000	0.000000	0.000000	.996796	0.000000	0.000000	0.000000	0.000000-M
		+M	.000994	0.000000	0.000000	0.000000	0.000000	0.000000	.079976	0.000000	+M
3.0715E+01	.8743	-M	0.000000	0.000000	0.000000	-.003739	0.000000	0.000000	0.000000	0.000000	0.000000-M
		+M	0.000000	.999520	0.000000	0.000000	0.000000	0.000000	0.000000	.030751	+M
3.0715E+01	-.8743	-M	0.000000	.999520	0.000000	0.000000	0.000000	0.000000	0.000000	.030751	0.000000-M
		+M	0.000000	0.000000	0.000000	-.003739	0.000000	0.000000	0.000000	0.000000	+M
1.4400E+01	0.0000	-M	0.000000	0.000000	0.000000	0.000000	0.000000	.707107	0.000000	0.000000	0.000000-M
		+M	0.000000	0.000000	0.000000	0.000000	0.000000	.707107	0.000000	0.000000	+M
-4.0800E+00	0.0000	-M	0.000000	0.000000	0.000000	0.000000	0.000000	-.707107	0.000000	0.000000	0.000000-M
		+M	0.000000	0.000000	0.000000	0.000000	0.000000	.707107	0.000000	0.000000	+M
-5.0532E+01	-.2473	-M	.052977	0.000000	0.000000	0.000000	0.000000	0.000000	.995393	0.000000	0.000000-M
		+M	0.000000	0.000000	0.000000	0.000000	-.079917	0.000000	0.000000	0.000000	+M
-5.0532E+01	.2473	-M	0.000000	0.000000	0.000000	0.000000	-.079917	0.000000	0.000000	0.000000	0.000000-M
		+M	.052977	0.000000	0.000000	0.000000	0.000000	0.000000	.995393	0.000000	+M
-8.4255E+01	-.9979	-M	.998595	0.000000	0.000000	0.000000	0.000000	0.000000	-.052887	0.000000	0.000000-M
		+M	0.000000	0.000000	0.000000	0.000000	.003248	0.000000	0.000000	0.000000	+M
-8.4255E+01	.9979	-M	0.000000	0.000000	0.000000	0.000000	.003248	0.000000	0.000000	0.000000	0.000000-M
		+M	.998595	0.000000	0.000000	0.000000	0.000000	0.000000	-.052887	0.000000	+M
-9.0470E+01	-.1244	-M	0.000000	-.030879	0.000000	0.000000	0.000000	0.000000	0.000000	.998670	0.000000-M
		+M	0.000000	0.000000	0.000000	-.041296	0.000000	0.000000	0.000000	0.000000	+M
-9.0470E+01	.1244	-M	0.000000	0.000000	0.000000	-.041296	0.000000	0.000000	0.000000	0.000000	0.000000-M
		+M	0.000000	-.030879	0.000000	0.000000	0.000000	0.000000	0.000000	.998670	+M
-1.0499E+02	.0000	-M	0.000000	0.000000	-.029974	0.000000	0.000000	0.000000	0.000000	0.000000	.999101-M
		+M	0.000000	0.000000	-.029974	0.000000	0.000000	0.000000	0.000000	0.000000	+M
ENERGY	HPAL		M=8	M=7	M=6	M=5	M=4	M=3	M=2	M=1	M=0

Y= .40 X=-.60 OAS=1.7380E+02*W246 B4F4/B2F2=-4.00E-01 B4F4/B6F6=-1.50E+00 B2F2/B6F6= 3.75E+00 B4F4=-2.40E-01

ENERGY	HPAL		M=8	M=7	M=6	M=5	M=4	M=3	M=2	M=1	M=0
7.7992E+01	-.6178	-M	0.000000	0.000000	0.000000	.995278	0.000000	0.000000	0.000000	0.000000	0.000000-M
		+M	0.000000	.010594	0.000000	0.000000	0.000000	0.000000	0.000000	.096488	+M
7.7992E+01	.6178	-M	0.000000	.010594	0.000000	0.000000	0.000000	0.000000	0.000000	.096488	0.000000-M
		+M	0.000000	0.000000	0.000000	.995278	0.000000	0.000000	0.000000	0.000000	+M
6.1964E+01	-.0000	-M	0.000000	0.000000	.703232	0.000000	0.000000	0.000000	0.000000	0.000000	.104544-M
		+M	0.000000	0.000000	.703232	0.000000	0.000000	0.000000	0.000000	0.000000	+M
6.0240E+01	.0000	-M	0.000000	0.000000	-.707107	0.000000	0.000000	0.000000	0.000000	0.000000	.000000-M
		+M	0.000000	0.000000	.707107	0.000000	0.000000	0.000000	0.000000	0.000000	+M
5.9128E+01	.4770	-M	.005911	0.000000	0.000000	0.000000	0.000000	0.000000	.174800	0.000000	0.000000-M
		+M	0.000000	0.000000	0.000000	0.000000	.984586	0.000000	0.000000	0.000000	+M
5.9128E+01	-.4770	-M	0.000000	0.000000	0.000000	0.000000	.984586	0.000000	0.000000	0.000000	0.000000-M
		+M	.005911	0.000000	0.000000	0.000000	0.000000	0.000000	.174800	0.000000	+M
2.9400E+01	0.0000	-M	0.000000	0.000000	0.000000	0.000000	0.000000	.707107	0.000000	0.000000	0.000000-M
		+M	0.000000	0.000000	0.000000	0.000000	0.000000	.707107	0.000000	0.000000	+M
1.0400E+01	-.8697	-M	0.000000	.996632	0.000000	0.000000	0.000000	0.000000	0.000000	.079918	0.000000-M
		+M	0.000000	0.000000	0.000000	-.018356	0.000000	0.000000	0.000000	0.000000	+M
1.0400E+01	.8697	-M	0.000000	0.000000	0.000000	-.018356	0.000000	0.000000	0.000000	0.000000	0.000000-M
		+M	0.000000	.996632	0.000000	0.000000	0.000000	0.000000	0.000000	.079918	+M
-7.5600E+00	0.0000	-M	0.000000	0.000000	0.000000	0.000000	0.000000	-.707107	0.000000	0.000000	0.000000-M
		+M	0.000000	0.000000	0.000000	0.000000	0.000000	.707107	0.000000	0.000000	+M
-4.1140E+01	-.4466	-M	.533749	0.000000	0.000000	0.000000	0.000000	0.000000	.832066	0.000000	0.000000-M
		+M	0.000000	0.000000	0.000000	0.000000	-.150927	0.000000	0.000000	0.000000	+M
-4.1140E+01	.4466	-M	0.000000	0.000000	0.000000	0.000000	-.150927	0.000000	0.000000	0.000000	0.000000-M
		+M	.533749	0.000000	0.000000	0.000000	0.000000	0.000000	.832066	0.000000	+M
-4.8948E+01	-.7804	-M	.845622	0.000000	0.000000	0.000000	0.000000	0.000000	-.526414	0.000000	0.000000-M
		+M	0.000000	0.000000	0.000000	0.000000	.088381	0.000000	0.000000	0.000000	+M
-4.8948E+01	.7804	-M	0.000000	0.000000	0.000000	0.000000	.088381	0.000000	0.000000	0.000000	0.000000-M
		+M	.845622	0.000000	0.000000	0.000000	0.000000	0.000000	-.526414	0.000000	+M
-8.1553E+01	.1231	-M	0.000000	0.000000	0.000000	-.095316	0.000000	0.000000	0.000000	0.000000	0.000000-M
		+M	0.000000	-.081312	0.000000	0.000000	0.000000	0.000000	0.000000	.992121	+M
-8.1553E+01	-.1231	-M	0.000000	-.081312	0.000000	0.000000	0.000000	0.000000	0.000000	.992121	0.000000-M
		+M	0.000000	0.000000	0.000000	-.095316	0.000000	0.000000	0.000000	0.000000	+M
-9.5804E+01	.0000	-M	0.000000	0.000000	-.073923	0.000000	0.000000	0.000000	0.000000	0.000000	.994520-M
		+M	0.000000	0.000000	-.073923	0.000000	0.000000	0.000000	0.000000	0.000000	+M
ENERGY	HPAL		M=8	M=7	M=6	M=5	M=4	M=3	M=2	M=1	M=0

Y= .40 X=-.40 OAS=1.5237E+02*W246 B4F4/B2F2=-2.67E-01 B4F4/B6F6=-6.67E-01 B2F2/B6F6= 2.50E+00 B4F4=-1.60E-01

ENERGY	HPAL		M=8	M=7	M=6	M=5	M=4	M=3	M=2	M=1	M=0
6.3839E+01	.6020	-M	0.000000	.025686	0.000000	0.000000	0.000000	0.000000	0.000000	.171171	0.000000-M
		+M	0.000000	0.000000	0.000000	.984907	0.000000	0.000000	0.000000	0.000000	+M
6.3839E+01	-.6020	-M	0.000000	0.000000	0.000000	.984907	0.000000	0.000000	0.000000	0.000000	0.000000-M
		+M	0.000000	.025686	0.000000	0.000000	0.000000	0.000000	0.000000	.171171	+M
6.0506E+01	.4406	-M	.021530	0.000000	0.000000	0.000000	0.000000	0.000000	.279834	0.000000	0.000000-M
		+M	0.000000	0.000000	0.000000	0.000000	.959807	0.000000	0.000000	0.000000	+M
6.0506E+01	-.4406	-M	0.000000	0.000000	0.000000	0.000000	.959807	0.000000	0.000000	0.000000	0.000000-M
		+M	.021530	0.000000	0.000000	0.000000	0.000000	0.000000	.279834	0.000000	+M
4.4400E+01	0.0000	-M	0.000000	0.000000	0.000000	0.000000	0.000000	.707107	0.000000	0.000000	0.000000-M
		+M	0.000000	0.000000	0.000000	0.000000	0.000000	.707107	0.000000	0.000000	+M
3.7176E+01	-.0000	-M	0.000000	0.000000	.692856	0.000000	0.000000	0.000000	0.000000	0.000000	.199751-M
		+M	0.000000	0.000000	.692856	0.000000	0.000000	0.000000	0.000000	0.000000	+M
3.2160E+01	.0000	-M	0.000000	0.000000	-.707107	0.000000	0.000000	0.000000	0.000000	0.000000	.000000-M
		+M	0.000000	0.000000	.707107	0.000000	0.000000	0.000000	0.000000	0.000000	+M
-8.4207E+00	.9744	-M	0.000000	0.000000	0.000000	0.000000	-.068065	0.000000	0.000000	0.000000	0.000000-M
		+M	.985145	0.000000	0.000000	0.000000	0.000000	0.000000	.157662	0.000000	+M
-8.4207E+00	-.9744	-M	.985145	0.000000	0.000000	0.000000	0.000000	0.000000	.157662	0.000000	0.000000-M
		+M	0.000000	0.000000	0.000000	0.000000	-.068065	0.000000	0.000000	0.000000	+M
-9.1263E+00	-.8503	-M	0.000000	.984914	0.000000	0.000000	0.000000	0.000000	0.000000	.164323	0.000000-M
		+M	0.000000	0.000000	0.000000	-.054245	0.000000	0.000000	0.000000	0.000000	+M
-9.1263E+00	.8503	-M	0.000000	0.000000	0.000000	-.054245	0.000000	0.000000	0.000000	0.000000	0.000000-M
		+M	0.000000	.984914	0.000000	0.000000	0.000000	0.000000	0.000000	.164323	+M
-1.1040E+01	0.0000	-M	0.000000	0.000000	0.000000	0.000000	0.000000	-.707107	0.000000	0.000000	0.000000-M
		+M	0.000000	0.000000	0.000000	0.000000	0.000000	.707107	0.000000	0.000000	+M
-3.9125E+01	.2162	-M	0.000000	0.000000	0.000000	0.000000	-.272282	0.000000	0.000000	0.000000	0.000000-M
		+M	-.170372	0.000000	0.000000	0.000000	0.000000	0.000000	.947014	0.000000	+M
-3.9125E+01	-.2162	-M	-.170372	0.000000	0.000000	0.000000	0.000000	0.000000	.947014	0.000000	0.000000-M
		+M	0.000000	0.000000	0.000000	0.000000	-.272282	0.000000	0.000000	0.000000	+M
-7.4753E+01	.1267	-M	0.000000	0.000000	0.000000	-.164367	0.000000	0.000000	0.000000	0.000000	0.000000-M
		+M	0.000000	-.171128	0.000000	0.000000	0.000000	0.000000	0.000000	.971442	+M
-7.4753E+01	-.1267	-M	0.000000	-.171128	0.000000	0.000000	0.000000	0.000000	0.000000	.971442	0.000000-M
		+M	0.000000	0.000000	0.000000	-.164367	0.000000	0.000000	0.000000	0.000000	+M
-8.8536E+01	.0000	-M	0.000000	0.000000	-.141245	0.000000	0.000000	0.000000	0.000000	0.000000	.979847-M
		+M	0.000000	0.000000	-.141245	0.000000	0.000000	0.000000	0.000000	0.000000	+M
ENERGY	HPAL		M=8	M=7	M=6	M=5	M=4	M=3	M=2	M=1	M=0

```
Y= .40  X=-.20   OAS=1.4944E+02*W246   B4F4/B2F2=-1.33E-01  B4F4/B6F6=-2.50E-01  B2F2/B6F6= 1.87E+00   B4F4=-8.00E-02
  ENERGY      HPAL         M=8       M=7       M=6       M=5       M=4       M=3       M=2       M=1       M=0
 6.4406E+01   .3794   -M  .076284  0.000000  0.000000  0.000000  0.000000  0.000000   .386230  0.000000  0.000000-M
                      +M 0.000000  0.000000  0.000000  0.000000   .919243  0.000000  0.000000  0.000000          +M
 6.4406E+01  -.3794   -M 0.000000  0.000000  0.000000  0.000000   .919243  0.000000  0.000000  0.000000  0.000000-M
                      +M  .076284  0.000000  0.000000  0.000000  0.000000  0.000000   .386230  0.000000          +M
 5.9400E+01  0.0000   -M 0.000000  0.000000  0.000000  0.000000  0.000000   .707107  0.000000  0.000000  0.000000-M
                      +M 0.000000  0.000000  0.000000  0.000000  0.000000   .707107  0.000000  0.000000          +M
 5.1822E+01  -.5661   -M 0.000000  0.000000  0.000000   .961188  0.000000  0.000000  0.000000  0.000000  0.000000-M
                      +M 0.000000   .048631  0.000000  0.000000  0.000000  0.000000  0.000000   .271574          +M
 5.1822E+01   .5661   -M 0.000000   .048631  0.000000  0.000000  0.000000  0.000000  0.000000   .271574  0.000000-M
                      +M 0.000000  0.000000  0.000000   .961188  0.000000  0.000000  0.000000  0.000000          +M
 2.8631E+01  -.9787   -M  .991813  0.000000  0.000000  0.000000  0.000000  0.000000   .065299  0.000000  0.000000-M
                      +M 0.000000  0.000000  0.000000  0.000000  -.109742  0.000000  0.000000  0.000000          +M
 2.8631E+01   .9787   -M 0.000000  0.000000  0.000000  0.000000  -.109742  0.000000  0.000000  0.000000  0.000000-M
                      +M  .991813  0.000000  0.000000  0.000000  0.000000  0.000000   .065299  0.000000          +M
 1.6157E+01   .0000   -M 0.000000  0.000000   .663572  0.000000  0.000000  0.000000  0.000000  0.000000   .345464-M
                      +M 0.000000  0.000000   .663572  0.000000  0.000000  0.000000  0.000000  0.000000          +M
 4.0800E+00  -.0000   -M 0.000000  0.000000  -.707107  0.000000  0.000000  0.000000  0.000000  0.000000  -.000000-M
                      +M 0.000000  0.000000   .707107  0.000000  0.000000  0.000000  0.000000  0.000000          +M
-1.4520E+01  0.0000   -M 0.000000  0.000000  0.000000  0.000000  0.000000  -.707107  0.000000  0.000000  0.000000-M
                      +M 0.000000  0.000000  0.000000  0.000000  0.000000   .707107  0.000000  0.000000          +M
-2.6789E+01  -.7730   -M 0.000000   .939448  0.000000  0.000000  0.000000  0.000000  0.000000   .314393  0.000000-M
                      +M 0.000000  0.000000  0.000000  -.136359  0.000000  0.000000  0.000000  0.000000          +M
-2.6789E+01   .7730   -M 0.000000  0.000000  0.000000  -.136359  0.000000  0.000000  0.000000  0.000000  0.000000-M
                      +M 0.000000   .939448  0.000000  0.000000  0.000000  0.000000  0.000000   .314393          +M
-3.6158E+01  -.1507   -M -.102412  0.000000  0.000000  0.000000  0.000000  0.000000   .920088  0.000000  0.000000-M
                      +M 0.000000  0.000000  0.000000  0.000000  -.378087  0.000000  0.000000  0.000000          +M
-3.6158E+01   .1507   -M 0.000000  0.000000  0.000000  0.000000  -.378087  0.000000  0.000000  0.000000  0.000000-M
                      +M -.102412  0.000000  0.000000  0.000000  0.000000  0.000000   .920088  0.000000          +M
-7.1953E+01  -.1682   -M 0.000000  -.339222  0.000000  0.000000  0.000000  0.000000  0.000000   .909618  0.000000-M
                      +M 0.000000  0.000000  0.000000  -.239840  0.000000  0.000000  0.000000  0.000000          +M
-7.1953E+01   .1682   -M 0.000000  0.000000  0.000000  -.239840  0.000000  0.000000  0.000000  0.000000  0.000000-M
                      +M 0.000000  -.339222  0.000000  0.000000  0.000000  0.000000  0.000000   .909618          +M
-8.5037E+01   .0000   -M 0.000000  0.000000  -.244280  0.000000  0.000000  0.000000  0.000000  0.000000   .938432-M
                      +M 0.000000  0.000000  -.244280  0.000000  0.000000  0.000000  0.000000  0.000000          +M
  ENERGY      HPAL         M=8       M=7       M=6       M=5       M=4       M=3       M=2       M=1       M=0

Y= .40  X=0.00   OAS=1.6288E+02*W246   B4F4/B2F2= 0.        B4F4/B6F6= 0.         B2F2/B6F6= 1.50E+00   B4F4= 0.
  ENERGY      HPAL         M=8       M=7       M=6       M=5       M=4       M=3       M=2       M=1       M=0
 7.4400E+01  0.0000   -M 0.000000  0.000000  0.000000  0.000000  0.000000   .707107  0.000000  0.000000  0.000000-M
                      +M 0.000000  0.000000  0.000000  0.000000  0.000000   .707107  0.000000  0.000000          +M
 7.2918E+01  -.0696   -M  .533469  0.000000  0.000000  0.000000  0.000000  0.000000   .436230  0.000000  0.000000-M
                      +M 0.000000  0.000000  0.000000  0.000000   .724648  0.000000  0.000000  0.000000          +M
 7.2918E+01   .0696   -M 0.000000  0.000000  0.000000  0.000000   .724648  0.000000  0.000000  0.000000  0.000000-M
                      +M  .533469  0.000000  0.000000  0.000000  0.000000  0.000000   .436230  0.000000          +M
 6.3525E+01  -.5927   -M  .842249  0.000000  0.000000  0.000000  0.000000  0.000000  -.195332  0.000000  0.000000-M
                      +M 0.000000  0.000000  0.000000  0.000000  -.502457  0.000000  0.000000  0.000000          +M
 6.3525E+01   .5927   -M 0.000000  0.000000  0.000000  0.000000  -.502457  0.000000  0.000000  0.000000  0.000000-M
                      +M  .842249  0.000000  0.000000  0.000000  0.000000  0.000000  -.195332  0.000000          +M
 4.3119E+01  -.4973   -M 0.000000  0.000000  0.000000   .914186  0.000000  0.000000  0.000000  0.000000  0.000000-M
                      +M 0.000000   .077781  0.000000  0.000000  0.000000  0.000000  0.000000   .397762          +M
 4.3119E+01   .4973   -M 0.000000   .077781  0.000000  0.000000  0.000000  0.000000  0.000000   .397762  0.000000-M
                      +M 0.000000  0.000000  0.000000   .914186  0.000000  0.000000  0.000000  0.000000          +M
 2.0805E+00  -.0000   -M 0.000000  0.000000   .596662  0.000000  0.000000  0.000000  0.000000  0.000000   .536645-M
                      +M 0.000000  0.000000   .596662  0.000000  0.000000  0.000000  0.000000  0.000000          +M
-1.8000E+01  0.0000   -M 0.000000  0.000000  0.000000  0.000000  0.000000  -.707107  0.000000  0.000000  0.000000-M
                      +M 0.000000  0.000000  0.000000  0.000000  0.000000   .707107  0.000000  0.000000          +M
-2.4000E+01   .0000   -M 0.000000  0.000000  -.707107  0.000000  0.000000  0.000000  0.000000  0.000000   .000000-M
                      +M 0.000000  0.000000   .707107  0.000000  0.000000  0.000000  0.000000  0.000000          +M
-3.5642E+01   .0877   -M 0.000000  0.000000  0.000000  0.000000  -.471618  0.000000  0.000000  0.000000  0.000000-M
                      +M -.077639  0.000000  0.000000  0.000000  0.000000  0.000000   .878379  0.000000          +M
-3.5642E+01  -.0877   -M -.077639  0.000000  0.000000  0.000000  0.000000  0.000000   .878379  0.000000  0.000000-M
                      +M 0.000000  0.000000  0.000000  0.000000  -.471618  0.000000  0.000000  0.000000          +M
-4.0453E+01  -.5419   -M 0.000000   .801636  0.000000  0.000000  0.000000  0.000000  0.000000   .520212  0.000000-M
                      +M 0.000000  0.000000  0.000000  -.294549  0.000000  0.000000  0.000000  0.000000          +M
-4.0453E+01   .5419   -M 0.000000  0.000000  0.000000  -.294549  0.000000  0.000000  0.000000  0.000000  0.000000-M
                      +M 0.000000   .801636  0.000000  0.000000  0.000000  0.000000  0.000000   .520212          +M
-7.6466E+01  -.3304   -M 0.000000   .592731  0.000000  0.000000  0.000000  0.000000  0.000000  -.755755  0.000000-M
                      +M 0.000000  0.000000  0.000000   .278397  0.000000  0.000000  0.000000  0.000000          +M
-7.6466E+01   .3304   -M 0.000000  0.000000  0.000000  -.278397  0.000000  0.000000  0.000000  0.000000  0.000000-M
                      +M 0.000000  -.592731  0.000000  0.000000  0.000000  0.000000  0.000000   .755755          +M
-8.8480E+01  -.0000   -M 0.000000  0.000000  -.379465  0.000000  0.000000  0.000000  0.000000  0.000000   .843808-M
                      +M 0.000000  0.000000  -.379465  0.000000  0.000000  0.000000  0.000000  0.000000          +M
  ENERGY      HPAL         M=8       M=7       M=6       M=5       M=4       M=3       M=2       M=1       M=0

Y= .40  X= .20   OAS=1.5625E+02*W246   B4F4/B2F2= 1.33E-01  B4F4/B6F6= 2.50E-01  B2F2/B6F6= 1.87E+00   B4F4= 8.00E-02
  ENERGY      HPAL         M=8       M=7       M=6       M=5       M=4       M=3       M=2       M=1       M=0
 8.7135E+01   .9875   -M 0.000000  0.000000  0.000000  0.000000   .055812  0.000000  0.000000  0.000000  0.000000-M
                      +M  .993215  0.000000  0.000000  0.000000  0.000000  0.000000   .102020  0.000000          +M
 8.7135E+01  -.9875   -M  .993215  0.000000  0.000000  0.000000  0.000000  0.000000   .102020  0.000000  0.000000-M
                      +M 0.000000  0.000000  0.000000  0.000000   .055812  0.000000  0.000000  0.000000          +M
 5.6040E+01  0.0000   -M 0.000000  0.000000  0.000000  0.000000  0.000000   .707107  0.000000  0.000000  0.000000-M
                      +M 0.000000  0.000000  0.000000  0.000000  0.000000   .707107  0.000000  0.000000          +M
 4.7323E+01   .2406   -M -.104304  0.000000  0.000000  0.000000  0.000000  0.000000   .569331  0.000000  0.000000-M
                      +M 0.000000  0.000000  0.000000  0.000000   .815465  0.000000  0.000000  0.000000          +M
 4.7323E+01  -.2406   -M 0.000000  0.000000  0.000000  0.000000   .815465  0.000000  0.000000  0.000000  0.000000-M
                      +M -.104304  0.000000  0.000000  0.000000  0.000000  0.000000   .569331  0.000000          +M
 2.1572E+01  -.3463   -M 0.000000  0.000000  0.000000   .803033  0.000000  0.000000  0.000000  0.000000  0.000000-M
                      +M 0.000000   .128282  0.000000  0.000000  0.000000  0.000000  0.000000   .581963          +M
 2.1572E+01   .3463   -M 0.000000   .128282  0.000000  0.000000  0.000000  0.000000  0.000000   .581963  0.000000-M
                      +M 0.000000  0.000000  0.000000   .803033  0.000000  0.000000  0.000000  0.000000          +M
-3.1293E+00  -.0000   -M 0.000000  0.000000   .472894  0.000000  0.000000  0.000000  0.000000  0.000000   .743466-M
                      +M 0.000000  0.000000   .472894  0.000000  0.000000  0.000000  0.000000  0.000000          +M
-1.7880E+01  0.0000   -M 0.000000  0.000000  0.000000  0.000000  0.000000  -.707107  0.000000  0.000000  0.000000-M
                      +M 0.000000  0.000000  0.000000  0.000000  0.000000   .707107  0.000000  0.000000          +M
-2.7179E+01  -.0031   -M -.051418  0.000000  0.000000  0.000000  0.000000  0.000000   .815754  0.000000  0.000000-M
                      +M 0.000000  0.000000  0.000000  0.000000  -.576109  0.000000  0.000000  0.000000          +M
-2.7179E+01   .0031   -M 0.000000  0.000000  0.000000  0.000000  -.576109  0.000000  0.000000  0.000000  0.000000-M
                      +M -.051418  0.000000  0.000000  0.000000  0.000000  0.000000   .815754  0.000000          +M
-3.3689E+01  -.2555   -M 0.000000   .657454  0.000000  0.000000  0.000000  0.000000  0.000000   .556320  0.000000-M
                      +M 0.000000  0.000000  0.000000  -.508195  0.000000  0.000000  0.000000  0.000000          +M
-3.3689E+01   .2555   -M 0.000000  0.000000  0.000000  -.508195  0.000000  0.000000  0.000000  0.000000  0.000000-M
                      +M 0.000000   .657454  0.000000  0.000000  0.000000  0.000000  0.000000   .556320          +M
-3.9600E+01  -.0000   -M 0.000000  0.000000  -.707107  0.000000  0.000000  0.000000  0.000000  0.000000   .000000-M
                      +M 0.000000  0.000000   .707107  0.000000  0.000000  0.000000  0.000000  0.000000          +M
-5.8322E+01  -.4658   -M 0.000000   .742494  0.000000  0.000000  0.000000  0.000000  0.000000  -.593150  0.000000-M
                      +M 0.000000  0.000000  0.000000   .311249  0.000000  0.000000  0.000000  0.000000          +M
-5.8322E+01   .4658   -M 0.000000  0.000000  0.000000   .311249  0.000000  0.000000  0.000000  0.000000  0.000000-M
                      +M 0.000000   .742494  0.000000  0.000000  0.000000  0.000000  0.000000  -.593150          +M
-6.9111E+01   .0000   -M 0.000000  0.000000   .525710  0.000000  0.000000  0.000000  0.000000  0.000000  -.668773-M
                      +M 0.000000  0.000000   .525710  0.000000  0.000000  0.000000  0.000000  0.000000          +M
  ENERGY      HPAL         M=8       M=7       M=6       M=5       M=4       M=3       M=2       M=1       M=0
```

Y= .40 X= .40 OAS=1.7246E+02*W246 B4F4/B2F2= 2.67E-01 B4F4/B6F6= 6.67E-01 B2F2/B6F6= 2.50E+00 B4F4= 1.60E-01

ENERGY	HPAL		M=8	M=7	M=6	M=5	M=4	M=3	M=2	M=1	M=0
1.0751E+02	.9972	-M	0.000000	0.000000	0.000000	0.000000	.013961	0.000000	0.000000	0.000000	0.000000-M
		+M	.998260	0.000000	0.000000	0.000000	0.000000	0.000000	.057296	0.000000	+M
1.0751E+02	-.9972	-M	.998260	0.000000	0.000000	0.000000	0.000000	0.000000	.057296	0.000000	0.000000-M
		+M	0.000000	0.000000	0.000000	0.000000	.013961	0.000000	0.000000	0.000000	+M
3.7680E+01	0.0000	-M	0.000000	0.000000	0.000000	0.000000	0.000000	.707107	0.000000	0.000000	0.000000-M
		+M	0.000000	0.000000	0.000000	0.000000	0.000000	.707107	0.000000	0.000000	+M
2.9833E+01	-.0652	-M	0.000000	0.000000	0.000000	0.000000	.650404	0.000000	0.000000	0.000000	0.000000-M
		+M	-.052589	0.000000	0.000000	0.000000	0.000000	0.000000	.757765	0.000000	+M
2.9833E+01	.0652	-M	-.052589	0.000000	0.000000	0.000000	0.000000	0.000000	.757765	0.000000	0.000000-M
		+M	0.000000	0.000000	0.000000	0.000000	.650404	0.000000	0.000000	0.000000	+M
1.5018E+01	-.0138	-M	0.000000	0.000000	0.000000	.464998	0.000000	0.000000	0.000000	0.000000	0.000000-M
		+M	0.000000	.176694	0.000000	0.000000	0.000000	0.000000	0.000000	.867500	+M
1.5018E+01	.0138	-M	0.000000	.176694	0.000000	0.000000	0.000000	0.000000	0.000000	.867500	0.000000-M
		+M	0.000000	0.000000	0.000000	.464998	0.000000	0.000000	0.000000	0.000000	+M
6.8731E+00	-.0000	-M	0.000000	0.000000	.260564	0.000000	0.000000	0.000000	0.000000	0.000000	.929630-M
		+M	0.000000	0.000000	.260564	0.000000	0.000000	0.000000	0.000000	0.000000	+M
-1.7760E+01	0.0000	-M	0.000000	0.000000	0.000000	0.000000	0.000000	-.707107	0.000000	0.000000	0.000000-M
		+M	0.000000	0.000000	0.000000	0.000000	0.000000	.707107	0.000000	0.000000	+M
-2.3581E+01	.1821	-M	.026686	0.000000	0.000000	0.000000	0.000000	0.000000	-.650007	0.000000	0.000000-M
		+M	0.000000	0.000000	0.000000	0.000000	.759460	0.000000	0.000000	0.000000	+M
-2.3581E+01	-.1821	-M	0.000000	0.000000	0.000000	0.000000	.759460	0.000000	0.000000	0.000000	0.000000-M
		+M	.026686	0.000000	0.000000	0.000000	0.000000	0.000000	-.650007	0.000000	+M
-3.5592E+01	-.0386	-M	0.000000	-.639623	0.000000	0.000000	0.000000	0.000000	0.000000	-.257876	0.000000-M
		+M	0.000000	0.000000	0.000000	.724143	0.000000	0.000000	0.000000	0.000000	+M
-3.5592E+01	.0386	-M	0.000000	0.000000	0.000000	-.724143	0.000000	0.000000	0.000000	0.000000	0.000000-M
		+M	0.000000	.639623	0.000000	0.000000	0.000000	0.000000	0.000000	.257876	+M
-4.6506E+01	-.3502	-M	0.000000	.748106	0.000000	0.000000	0.000000	0.000000	0.000000	-.425375	0.000000-M
		+M	0.000000	0.000000	0.000000	.509307	0.000000	0.000000	0.000000	0.000000	+M
-4.6506E+01	.3502	-M	0.000000	0.000000	0.000000	.509307	0.000000	0.000000	0.000000	0.000000	0.000000-M
		+M	0.000000	.748106	0.000000	0.000000	0.000000	0.000000	0.000000	-.425375	+M
-5.5200E+01	-.0000	-M	0.000000	0.000000	-.707107	0.000000	0.000000	0.000000	0.000000	0.000000	.000000-M
		+M	0.000000	0.000000	.707107	0.000000	0.000000	0.000000	0.000000	0.000000	+M
-6.4953E+01	.0000	-M	0.000000	0.000000	.657348	0.000000	0.000000	0.000000	0.000000	0.000000	-.368493-M
		+M	0.000000	0.000000	.657348	0.000000	0.000000	0.000000	0.000000	0.000000	+M
ENERGY	HPAL		M=8	M=7	M=6	M=5	M=4	M=3	M=2	M=1	M=0

Y= .40 X= .60 OAS=2.0160E+02*W246 B4F4/B2F2= 4.00E-01 B4F4/B6F6= 1.50E+00 B2F2/B6F6= 3.75E+00 B4F4= 2.40E-01

ENERGY	HPAL		M=8	M=7	M=6	M=5	M=4	M=3	M=2	M=1	M=0
1.2812E+02	.9992	-M	0.000000	0.000000	0.000000	0.000000	.003779	0.000000	0.000000	0.000000	0.000000-M
		+M	.999456	0.000000	0.000000	0.000000	0.000000	0.000000	.032760	0.000000	+M
1.2812E+02	-.9992	-M	.999456	0.000000	0.000000	0.000000	0.000000	0.000000	.032760	0.000000	0.000000-M
		+M	0.000000	0.000000	0.000000	0.000000	.003779	0.000000	0.000000	0.000000	+M
2.9561E+01	-.0000	-M	0.000000	0.000000	.114058	0.000000	0.000000	0.000000	0.000000	0.000000	.986905-M
		+M	0.000000	0.000000	.114058	0.000000	0.000000	0.000000	0.000000	0.000000	+M
2.8363E+01	.1113	-M	0.000000	0.000000	0.000000	.178979	0.000000	0.000000	0.000000	0.000000	0.000000-M
		+M	0.000000	.117498	0.000000	0.000000	0.000000	0.000000	0.000000	.976811	+M
2.8363E+01	-.1113	-M	0.000000	.117498	0.000000	0.000000	0.000000	0.000000	0.000000	.976811	0.000000-M
		+M	0.000000	0.000000	0.000000	.178979	0.000000	0.000000	0.000000	0.000000	+M
2.3242E+01	.1614	-M	0.000000	0.000000	0.000000	0.000000	.345172	0.000000	0.000000	0.000000	0.000000-M
		+M	-.032050	0.000000	0.000000	0.000000	0.000000	0.000000	.937992	0.000000	+M
2.3242E+01	-.1614	-M	-.032050	0.000000	0.000000	0.000000	0.000000	0.000000	.937992	0.000000	0.000000-M
		+M	0.000000	0.000000	0.000000	0.000000	.345172	0.000000	0.000000	0.000000	+M
1.9320E+01	0.0000	-M	0.000000	0.000000	0.000000	0.000000	0.000000	.707107	0.000000	0.000000	0.000000-M
		+M	0.000000	0.000000	0.000000	0.000000	0.000000	.707107	0.000000	0.000000	+M
-1.7640E+01	0.0000	-M	0.000000	0.000000	0.000000	0.000000	0.000000	-.707107	0.000000	0.000000	0.000000-M
		+M	0.000000	0.000000	0.000000	0.000000	0.000000	.707107	0.000000	0.000000	+M
-3.1119E+01	-.4106	-M	0.000000	0.000000	0.000000	0.000000	.938532	0.000000	0.000000	0.000000	0.000000-M
		+M	.007763	0.000000	0.000000	0.000000	0.000000	0.000000	-.345105	0.000000	+M
-3.1119E+01	.4106	-M	.007763	0.000000	0.000000	0.000000	0.000000	0.000000	-.345105	0.000000	0.000000-M
		+M	0.000000	0.000000	0.000000	0.000000	.938532	0.000000	0.000000	0.000000	+M
-3.4670E+01	-.8572	-M	0.000000	.991293	0.000000	0.000000	0.000000	0.000000	0.000000	-.104580	0.000000-M
		+M	0.000000	0.000000	0.000000	-.080010	0.000000	0.000000	0.000000	0.000000	+M
-3.4670E+01	.8572	-M	0.000000	0.000000	0.000000	-.080010	0.000000	0.000000	0.000000	0.000000	0.000000-M
		+M	0.000000	.991293	0.000000	0.000000	0.000000	0.000000	0.000000	-.104580	+M
-5.7413E+01	-.5935	-M	0.000000	0.000000	0.000000	.980594	0.000000	0.000000	0.000000	0.000000	0.000000-M
		+M	0.000000	.059437	0.000000	0.000000	0.000000	0.000000	0.000000	-.186822	+M
-5.7413E+01	.5935	-M	0.000000	.059437	0.000000	0.000000	0.000000	0.000000	0.000000	-.186822	0.000000-M
		+M	0.000000	0.000000	0.000000	.980594	0.000000	0.000000	0.000000	0.000000	+M
-7.0800E+01	-.0000	-M	0.000000	0.000000	-.707107	0.000000	0.000000	0.000000	0.000000	0.000000	.000000-M
		+M	0.000000	0.000000	.707107	0.000000	0.000000	0.000000	0.000000	0.000000	+M
-7.3481E+01	.0000	-M	0.000000	0.000000	.697847	0.000000	0.000000	0.000000	0.000000	0.000000	-.161303-M
		+M	0.000000	0.000000	.697847	0.000000	0.000000	0.000000	0.000000	0.000000	+M
ENERGY	HPAL		M=8	M=7	M=6	M=5	M=4	M=3	M=2	M=1	M=0

Y= .40 X= .80 OAS=2.3569E+02*W246 B4F4/B2F2= 5.33E-01 B4F4/B6F6= 4.00E+00 B2F2/B6F6= 7.50E+00 B4F4= 3.20E-01

ENERGY	HPAL		M=8	M=7	M=6	M=5	M=4	M=3	M=2	M=1	M=0
1.4883E+02	-.9998	-M	.999893	0.000000	0.000000	0.000000	0.000000	0.000000	.014645	0.000000	0.000000-M
		+M	0.000000	0.000000	0.000000	0.000000	.000654	0.000000	0.000000	0.000000	+M
1.4883E+02	.9998	-M	0.000000	0.000000	0.000000	0.000000	.000654	0.000000	0.000000	0.000000	0.000000-M
		+M	.999893	0.000000	0.000000	0.000000	0.000000	0.000000	.014645	0.000000	+M
5.7109E+01	0.0000	-M	0.000000	0.000000	.040346	0.000000	0.000000	0.000000	0.000000	0.000000	.998371-M
		+M	0.000000	0.000000	.040346	0.000000	0.000000	0.000000	0.000000	0.000000	+M
4.9462E+01	.1243	-M	0.000000	0.000000	0.000000	.057789	0.000000	0.000000	0.000000	0.000000	0.000000-M
		+M	0.000000	.048415	0.000000	0.000000	0.000000	0.000000	0.000000	.997154	+M
4.9462E+01	-.1243	-M	0.000000	.048415	0.000000	0.000000	0.000000	0.000000	0.000000	.997154	0.000000-M
		+M	0.000000	0.000000	0.000000	.057789	0.000000	0.000000	0.000000	0.000000	+M
2.7241E+01	.2401	-M	0.000000	0.000000	0.000000	0.000000	.115798	0.000000	0.000000	0.000000	0.000000-M
		+M	-.014623	0.000000	0.000000	0.000000	0.000000	0.000000	.993165	0.000000	+M
2.7241E+01	-.2401	-M	-.014623	0.000000	0.000000	0.000000	0.000000	0.000000	.993165	0.000000	0.000000-M
		+M	0.000000	0.000000	0.000000	0.000000	.115798	0.000000	0.000000	0.000000	+M
9.6000E-01	0.0000	-M	0.000000	0.000000	0.000000	0.000000	0.000000	.707107	0.000000	0.000000	0.000000-M
		+M	0.000000	0.000000	0.000000	0.000000	0.000000	.707107	0.000000	0.000000	+M
-1.7520E+01	0.0000	-M	0.000000	0.000000	0.000000	0.000000	0.000000	-.707107	0.000000	0.000000	0.000000-M
		+M	0.000000	0.000000	0.000000	0.000000	0.000000	.707107	0.000000	0.000000	+M
-2.7820E+01	-.8732	-M	0.000000	.998819	0.000000	0.000000	0.000000	0.000000	0.000000	-.048102	0.000000-M
		+M	0.000000	0.000000	0.000000	-.006797	0.000000	0.000000	0.000000	0.000000	+M
-2.7820E+01	.8732	-M	0.000000	0.000000	0.000000	-.006797	0.000000	0.000000	0.000000	0.000000	0.000000-M
		+M	0.000000	.998819	0.000000	0.000000	0.000000	0.000000	0.000000	-.048102	+M
-4.9347E+01	-.4899	-M	0.000000	0.000000	0.000000	0.000000	.993273	0.000000	0.000000	0.000000	0.000000-M
		+M	.001046	0.000000	0.000000	0.000000	0.000000	0.000000	-.115795	0.000000	+M
-4.9347E+01	.4899	-M	.001046	0.000000	0.000000	0.000000	0.000000	0.000000	-.115795	0.000000	0.000000-M
		+M	0.000000	0.000000	0.000000	0.000000	.993273	0.000000	0.000000	0.000000	+M
-8.2002E+01	-.6224	-M	0.000000	0.000000	0.000000	.998306	0.000000	0.000000	0.000000	0.000000	0.000000-M
		+M	0.000000	.003998	0.000000	0.000000	0.000000	0.000000	0.000000	-.058050	+M
-8.2002E+01	.6224	-M	0.000000	.003998	0.000000	0.000000	0.000000	0.000000	0.000000	-.058050	0.000000-M
		+M	0.000000	0.000000	0.000000	.998306	0.000000	0.000000	0.000000	0.000000	+M
-8.6400E+01	.0000	-M	0.000000	0.000000	-.707107	0.000000	0.000000	0.000000	0.000000	0.000000	-.000000-M
		+M	0.000000	0.000000	.707107	0.000000	0.000000	0.000000	0.000000	0.000000	+M
-8.6869E+01	-.0000	-M	0.000000	0.000000	.705955	0.000000	0.000000	0.000000	0.000000	0.000000	-.057058-M
		+M	0.000000	0.000000	.705955	0.000000	0.000000	0.000000	0.000000	0.000000	+M
ENERGY	HPAL		M=8	M=7	M=6	M=5	M=4	M=3	M=2	M=1	M=0

Y= .60 X=-.80 OAS=2.8606E+02*W246 B4F4/B2F2=-1.20E+00 B4F4/B6F6=-4.00E+00 B2F2/B6F6= 3.33E+00 B4F4=-4.80E-01

ENERGY	HPAL		M=8	M=7	M=6	M=5	M=4	M=3	M=2	M=1	M=0
1.3973E+02	.6236	-M	0.000000	.002285	0.000000	0.000000	0.000000	0.000000	0.000000	.043267	0.000000-M
		+M	0.000000	0.000000	0.000000	.999061	0.000000	0.000000	0.000000	0.000000	+M
1.3973E+02	-.6236	-M	0.000000	0.000000	0.000000	.999061	0.000000	0.000000	0.000000	0.000000	0.000000-M
		+M	0.000000	.002285	0.000000	0.000000	0.000000	0.000000	0.000000	.043267	+M
1.2704E+02	.0000	-M	0.000000	0.000000	.706385	0.000000	0.000000	0.000000	0.000000	0.000000	.045184-M
		+M	0.000000	0.000000	.706385	0.000000	0.000000	0.000000	0.000000	0.000000	+M
1.2648E+02	-.0000	-M	0.000000	0.000000	-.707107	0.000000	0.000000	0.000000	0.000000	0.000000	-.000000-M
		+M	0.000000	0.000000	.707107	0.000000	0.000000	0.000000	0.000000	0.000000	+M
9.3900E+01	.4948	-M	.000927	0.000000	0.000000	0.000000	0.000000	0.000000	.082957	0.000000	0.000000-M
		+M	0.000000	0.000000	0.000000	0.000000	.996553	0.000000	0.000000	0.000000	+M
9.3900E+01	-.4948	-M	0.000000	0.000000	0.000000	0.000000	.996553	0.000000	0.000000	0.000000	0.000000-M
		+M	.000927	0.000000	0.000000	0.000000	0.000000	0.000000	.082957	0.000000	+M
3.3599E+01	.8740	-M	0.000000	0.000000	0.000000	-.003820	0.000000	0.000000	0.000000	0.000000	0.000000-M
		+M	0.000000	.999365	0.000000	0.000000	0.000000	0.000000	0.000000	.035430	+M
3.3599E+01	-.8740	-M	0.000000	.999365	0.000000	0.000000	0.000000	0.000000	0.000000	.035430	0.000000-M
		+M	0.000000	0.000000	0.000000	-.003820	0.000000	0.000000	0.000000	0.000000	+M
2.9100E+01	0.0000	-M	0.000000	0.000000	0.000000	0.000000	0.000000	.707107	0.000000	0.000000	0.000000-M
		+M	0.000000	0.000000	0.000000	0.000000	0.000000	.707107	0.000000	0.000000	+M
1.3800E+00	0.0000	-M	0.000000	0.000000	0.000000	0.000000	0.000000	-.707107	0.000000	0.000000	0.000000-M
		+M	0.000000	0.000000	0.000000	0.000000	0.000000	.707107	0.000000	0.000000	+M
-6.5891E+01	.2457	-M	0.000000	0.000000	0.000000	0.000000	-.082942	0.000000	0.000000	0.000000	0.000000-M
		+M	.033278	0.000000	0.000000	0.000000	0.000000	0.000000	.995999	0.000000	+M
-6.5891E+01	-.2457	-M	.033278	0.000000	0.000000	0.000000	0.000000	0.000000	.995999	0.000000	0.000000-M
		+M	0.000000	0.000000	0.000000	0.000000	-.082942	0.000000	0.000000	0.000000	+M
-1.2425E+02	-.1246	-M	0.000000	-.035562	0.000000	0.000000	0.000000	0.000000	0.000000	.998435	0.000000-M
		+M	0.000000	0.000000	0.000000	-.043159	0.000000	0.000000	0.000000	0.000000	+M
-1.2425E+02	.1246	-M	0.000000	0.000000	0.000000	-.043159	0.000000	0.000000	0.000000	0.000000	0.000000-M
		+M	0.000000	-.035562	0.000000	0.000000	0.000000	0.000000	0.000000	.998435	+M
-1.4552E+02	.0000	-M	0.000000	0.000000	-.031950	0.000000	0.000000	0.000000	0.000000	0.000000	.998979-M
		+M	0.000000	0.000000	-.031950	0.000000	0.000000	0.000000	0.000000	0.000000	+M
-1.4633E+02	.9992	-M	0.000000	0.000000	0.000000	0.000000	.001837	0.000000	0.000000	0.000000	0.000000-M
		+M	.999446	0.000000	0.000000	0.000000	0.000000	0.000000	-.033240	0.000000	+M
-1.4633E+02	-.9992	-M	.999446	0.000000	0.000000	0.000000	0.000000	0.000000	-.033240	0.000000	0.000000-M
		+M	0.000000	0.000000	0.000000	0.000000	.001837	0.000000	0.000000	0.000000	+M
ENERGY	HPAL		M=8	M=7	M=6	M=5	M=4	M=3	M=2	M=1	M=0

Y= .60 X=-.60 OAS=2.4852E+02*W246 B4F4/B2F2=-9.00E-01 B4F4/B6F6=-1.50E+00 B2F2/B6F6= 1.67E+00 B4F4=-3.60E-01

ENERGY	HPAL		M=8	M=7	M=6	M=5	M=4	M=3	M=2	M=1	M=0
1.1660E+02	.6171	-M	0.000000	.009966	0.000000	0.000000	0.000000	0.000000	0.000000	.101517	0.000000-M
		+M	0.000000	0.000000	0.000000	.994784	0.000000	0.000000	0.000000	0.000000	+M
1.1660E+02	-.6171	-M	0.000000	0.000000	0.000000	.994784	0.000000	0.000000	0.000000	0.000000	0.000000-M
		+M	0.000000	.009966	0.000000	0.000000	0.000000	0.000000	0.000000	.101517	+M
9.2882E+01	.4752	-M	.005330	0.000000	0.000000	0.000000	0.000000	0.000000	.181641	0.000000	0.000000-M
		+M	0.000000	0.000000	0.000000	0.000000	.983351	0.000000	0.000000	0.000000	+M
9.2882E+01	-.4752	-M	0.000000	0.000000	0.000000	0.000000	.983351	0.000000	0.000000	0.000000	0.000000-M
		+M	.005330	0.000000	0.000000	0.000000	0.000000	0.000000	.181641	0.000000	+M
8.7159E+01	-.0000	-M	0.000000	0.000000	.702575	0.000000	0.000000	0.000000	0.000000	0.000000	.113036-M
		+M	0.000000	0.000000	.702575	0.000000	0.000000	0.000000	0.000000	0.000000	+M
8.4360E+01	.0000	-M	0.000000	0.000000	-.707107	0.000000	0.000000	0.000000	0.000000	0.000000	.000000-M
		+M	0.000000	0.000000	.707107	0.000000	0.000000	0.000000	0.000000	0.000000	+M
5.1600E+01	0.0000	-M	0.000000	0.000000	0.000000	0.000000	0.000000	.707107	0.000000	0.000000	0.000000-M
		+M	0.000000	0.000000	0.000000	0.000000	0.000000	.707107	0.000000	0.000000	+M
3.2902E+00	.8674	-M	0.000000	0.000000	0.000000	-.019827	0.000000	0.000000	0.000000	0.000000	0.000000-M
		+M	0.000000	.995125	0.000000	0.000000	0.000000	0.000000	0.000000	.096605	+M
3.2902E+00	-.8674	-M	0.000000	.995125	0.000000	0.000000	0.000000	0.000000	0.000000	.096605	0.000000-M
		+M	0.000000	0.000000	0.000000	-.019827	0.000000	0.000000	0.000000	0.000000	+M
-3.8400E+00	0.0000	-M	0.000000	0.000000	0.000000	0.000000	0.000000	-.707107	0.000000	0.000000	0.000000-M
		+M	0.000000	0.000000	0.000000	0.000000	0.000000	.707107	0.000000	0.000000	+M
-5.4454E+01	-.2417	-M	.146584	0.000000	0.000000	0.000000	0.000000	0.000000	.972600	0.000000	0.000000-M
		+M	0.000000	0.000000	0.000000	0.000000	-.180449	0.000000	0.000000	0.000000	+M
-5.4454E+01	.2417	-M	0.000000	0.000000	0.000000	0.000000	-.180449	0.000000	0.000000	0.000000	0.000000-M
		+M	.146584	0.000000	0.000000	0.000000	0.000000	0.000000	.972600	0.000000	+M
-9.0868E+01	-.9835	-M	.989184	0.000000	0.000000	0.000000	0.000000	0.000000	-.145105	0.000000	0.000000-M
		+M	0.000000	0.000000	0.000000	0.000000	.021441	0.000000	0.000000	0.000000	+M
-9.0868E+01	.9835	-M	0.000000	0.000000	0.000000	0.000000	.021441	0.000000	0.000000	0.000000	0.000000-M
		+M	.989184	0.000000	0.000000	0.000000	0.000000	0.000000	-.145105	0.000000	+M
-1.1113E+02	-.1247	-M	0.000000	-.098114	0.000000	0.000000	0.000000	0.000000	0.000000	.990132	0.000000-M
		+M	0.000000	0.000000	0.000000	-.100060	0.000000	0.000000	0.000000	0.000000	+M
-1.1113E+02	.1247	-M	0.000000	0.000000	0.000000	-.100060	0.000000	0.000000	0.000000	0.000000	0.000000-M
		+M	0.000000	-.098114	0.000000	0.000000	0.000000	0.000000	0.000000	.990132	+M
-1.3192E+02	0.0000	-M	0.000000	0.000000	-.079928	0.000000	0.000000	0.000000	0.000000	0.000000	.993591-M
		+M	0.000000	0.000000	-.079928	0.000000	0.000000	0.000000	0.000000	0.000000	+M
ENERGY	HPAL		M=8	M=7	M=6	M=5	M=4	M=3	M=2	M=1	M=0

Y= .60 X=-.40 OAS=2.1722E+02*W246 B4F4/B2F2=-6.00E-01 B4F4/B6F6=-6.67E-01 B2F2/B6F6= 1.11E+00 B4F4=-2.40E-01

ENERGY	HPAL		M=8	M=7	M=6	M=5	M=4	M=3	M=2	M=1	M=0
9.5623E+01	-.5994	-M	0.000000	0.000000	0.000000	.983111	0.000000	0.000000	0.000000	0.000000	0.000000-M
		+M	0.000000	.024515	0.000000	0.000000	0.000000	0.000000	0.000000	.181360	+M
9.5623E+01	.5994	-M	0.000000	.024515	0.000000	0.000000	0.000000	0.000000	0.000000	.181360	0.000000-M
		+M	0.000000	0.000000	0.000000	.983111	0.000000	0.000000	0.000000	0.000000	+M
9.5237E+01	-.4362	-M	0.000000	0.000000	0.000000	0.000000	.956711	0.000000	0.000000	0.000000	0.000000-M
		+M	.018113	0.000000	0.000000	0.000000	0.000000	0.000000	.290476	0.000000	+M
9.5237E+01	.4362	-M	.018113	0.000000	0.000000	0.000000	0.000000	0.000000	.290476	0.000000	0.000000-M
		+M	0.000000	0.000000	0.000000	0.000000	.956711	0.000000	0.000000	0.000000	+M
7.4100E+01	0.0000	-M	0.000000	0.000000	0.000000	0.000000	0.000000	.707107	0.000000	0.000000	0.000000-M
		+M	0.000000	0.000000	0.000000	0.000000	0.000000	.707107	0.000000	0.000000	+M
5.0554E+01	-.0000	-M	0.000000	0.000000	.689820	0.000000	0.000000	0.000000	0.000000	0.000000	.219766-M
		+M	0.000000	0.000000	.689820	0.000000	0.000000	0.000000	0.000000	0.000000	+M
4.2240E+01	.0000	-M	0.000000	0.000000	-.707107	0.000000	0.000000	0.000000	0.000000	0.000000	.000000-M
		+M	0.000000	0.000000	.707107	0.000000	0.000000	0.000000	0.000000	0.000000	+M
-9.0600E+00	0.0000	-M	0.000000	0.000000	0.000000	0.000000	0.000000	-.707107	0.000000	0.000000	0.000000-M
		+M	0.000000	0.000000	0.000000	0.000000	0.000000	.707107	0.000000	0.000000	+M
-2.5303E+01	-.8346	-M	0.000000	.974768	0.000000	0.000000	0.000000	0.000000	0.000000	.213916	0.000000-M
		+M	0.000000	0.000000	0.000000	-.063769	0.000000	0.000000	0.000000	0.000000	+M
-2.5303E+01	.8346	-M	0.000000	0.000000	0.000000	-.063769	0.000000	0.000000	0.000000	0.000000	0.000000-M
		+M	0.000000	.974768	0.000000	0.000000	0.000000	0.000000	0.000000	.213916	+M
-3.0529E+01	-.8622	-M	.913337	0.000000	0.000000	0.000000	0.000000	0.000000	.384511	0.000000	0.000000-M
		+M	0.000000	0.000000	0.000000	0.000000	-.134037	0.000000	0.000000	0.000000	+M
-3.0529E+01	.8622	-M	0.000000	0.000000	0.000000	0.000000	-.134037	0.000000	0.000000	0.000000	0.000000-M
		+M	.913337	0.000000	0.000000	0.000000	0.000000	0.000000	.384511	0.000000	+M
-5.1268E+01	-.3241	-M	-.406801	0.000000	0.000000	0.000000	0.000000	0.000000	.876228	0.000000	0.000000-M
		+M	0.000000	0.000000	0.000000	0.000000	-.258337	0.000000	0.000000	0.000000	+M
-5.1268E+01	.3241	-M	0.000000	0.000000	0.000000	0.000000	-.258337	0.000000	0.000000	0.000000	0.000000-M
		+M	-.406801	0.000000	0.000000	0.000000	0.000000	0.000000	.876228	0.000000	+M
-1.0188E+02	.1398	-M	0.000000	0.000000	0.000000	-.171540	0.000000	0.000000	0.000000	0.000000	0.000000-M
		+M	0.000000	-.221868	0.000000	0.000000	0.000000	0.000000	0.000000	.959869	+M
-1.0188E+02	-.1398	-M	0.000000	-.221868	0.000000	0.000000	0.000000	0.000000	0.000000	.959869	0.000000-M
		+M	0.000000	0.000000	0.000000	-.171540	0.000000	0.000000	0.000000	0.000000	+M
-1.2159E+02	.0000	-M	0.000000	0.000000	-.155398	0.000000	0.000000	0.000000	0.000000	0.000000	.975553-M
		+M	0.000000	0.000000	-.155398	0.000000	0.000000	0.000000	0.000000	0.000000	+M
ENERGY	HPAL		M=8	M=7	M=6	M=5	M=4	M=3	M=2	M=1	M=0

Y= .60 X=-.20 OAS=2.1939E+02*W246 B4F4/B2F2=-3.00E-01 B4F4/B6F6=-2.50E-01 B2F2/B6F6= 8.33E-01 B4F4=-1.20E-01

ENERGY	HPAL		M=8	M=7	M=6	M=5	M=4	M=3	M=2	M=1	M=0	
1.0144E+02	-.3763	-M	0.000000	0.000000	0.000000	0.000000	.915393	0.000000	0.000000	0.000000	0.000000	-M
		+M	.054090	0.000000	0.000000	0.000000	0.000000	0.000000	.398911	0.000000		+M
1.0144E+02	.3763	-M	.054090	0.000000	0.000000	0.000000	0.000000	0.000000	.398911	0.000000	0.000000	-M
		+M	0.000000	0.000000	0.000000	0.000000	.915393	0.000000	0.000000	0.000000		+M
9.6600E+01	0.0000	-M	0.000000	0.000000	0.000000	0.000000	0.000000	.707107	0.000000	0.000000	0.000000	-M
		+M	0.000000	0.000000	0.000000	0.000000	0.000000	.707107	0.000000	0.000000		+M
7.8148E+01	-.5589	-M	0.000000	0.000000	0.000000	.956072	0.000000	0.000000	0.000000	0.000000	0.000000	-M
		+M	0.000000	.046976	0.000000	0.000000	0.000000	0.000000	0.000000	.289344		+M
7.8148E+01	.5589	-M	0.000000	.046976	0.000000	0.000000	0.000000	0.000000	0.000000	.289344	0.000000	-M
		+M	0.000000	0.000000	0.000000	.956072	0.000000	0.000000	0.000000	0.000000		+M
2.3446E+01	.9737	-M	0.000000	0.000000	0.000000	0.000000	-.106745	0.000000	0.000000	0.000000	0.000000	-M
		+M	.988074	0.000000	0.000000	0.000000	0.000000	0.000000	.110974	0.000000		+M
2.3446E+01	-.9737	-M	.988074	0.000000	0.000000	0.000000	0.000000	0.000000	.110974	0.000000	0.000000	-M
		+M	0.000000	0.000000	0.000000	0.000000	-.106745	0.000000	0.000000	0.000000		+M
2.0630E+01	.0000	-M	0.000000	0.000000	.652686	0.000000	0.000000	0.000000	0.000000	0.000000	.384709	-M
		+M	0.000000	0.000000	.652686	0.000000	0.000000	0.000000	0.000000	0.000000		+M
1.2000E-01	-.0000	-M	0.000000	0.000000	-.707107	0.000000	0.000000	0.000000	0.000000	0.000000	-.000000	-M
		+M	0.000000	0.000000	.707107	0.000000	0.000000	0.000000	0.000000	0.000000		+M
-1.4280E+01	0.0000	-M	0.000000	0.000000	0.000000	0.000000	0.000000	-.707107	0.000000	0.000000	0.000000	-M
		+M	0.000000	0.000000	0.000000	0.000000	0.000000	.707107	0.000000	0.000000		+M
-4.5561E+01	-.1526	-M	-.144167	0.000000	0.000000	0.000000	0.000000	0.000000	.910250	0.000000	0.000000	-M
		+M	0.000000	0.000000	0.000000	0.000000	-.388151	0.000000	0.000000	0.000000		+M
-4.5561E+01	.1526	-M	0.000000	0.000000	0.000000	0.000000	-.388151	0.000000	0.000000	0.000000	0.000000	-M
		+M	-.144167	0.000000	0.000000	0.000000	0.000000	0.000000	.910250	0.000000		+M
-4.9059E+01	-.6852	-M	0.000000	.881908	0.000000	0.000000	0.000000	0.000000	0.000000	.437449	0.000000	-M
		+M	0.000000	0.000000	0.000000	-.175720	0.000000	0.000000	0.000000	0.000000		+M
-4.9059E+01	.6852	-M	0.000000	0.000000	0.000000	-.175720	0.000000	0.000000	0.000000	0.000000	0.000000	-M
		+M	0.000000	.881908	0.000000	0.000000	0.000000	0.000000	0.000000	.437449		+M
-1.0097E+02	.2487	-M	0.000000	0.000000	0.000000	-.234625	0.000000	0.000000	0.000000	0.000000	0.000000	-M
		+M	0.000000	-.469076	0.000000	0.000000	0.000000	0.000000	0.000000	.851422		+M
-1.0097E+02	-.2487	-M	0.000000	-.469076	0.000000	0.000000	0.000000	0.000000	0.000000	.851422	0.000000	-M
		+M	0.000000	0.000000	0.000000	-.234625	0.000000	0.000000	0.000000	0.000000		+M
-1.1795E+02	-.0000	-M	0.000000	0.000000	-.272031	0.000000	0.000000	0.000000	0.000000	0.000000	.923038	-M
		+M	0.000000	0.000000	-.272031	0.000000	0.000000	0.000000	0.000000	0.000000		+M
ENERGY	HPAL		M=8	M=7	M=6	M=5	M=4	M=3	M=2	M=1	M=0	

Y= .60 X=0.00 OAS=2.4552E+02*W246 B4F4/B2F2= 0. B4F4/B6F6= 0. B2F2/B6F6= 6.67E-01 B4F4= 0.

ENERGY	HPAL		M=8	M=7	M=6	M=5	M=4	M=3	M=2	M=1	M=0	
1.1910E+02	0.0000	-M	0.000000	0.000000	0.000000	0.000000	0.000000	.707107	0.000000	0.000000	0.000000	-M
		+M	0.000000	0.000000	0.000000	0.000000	0.000000	.707107	0.000000	0.000000		+M
1.1206E+02	-.2595	-M	0.000000	0.000000	0.000000	0.000000	.847358	0.000000	0.000000	0.000000	0.000000	-M
		+M	.196686	0.000000	0.000000	0.000000	0.000000	0.000000	.493253	0.000000		+M
1.1206E+02	.2595	-M	.196686	0.000000	0.000000	0.000000	0.000000	0.000000	.493253	0.000000	0.000000	-M
		+M	0.000000	0.000000	0.000000	0.000000	.847358	0.000000	0.000000	0.000000		+M
7.8192E+01	-.9287	-M	.975895	0.000000	0.000000	0.000000	0.000000	0.000000	-.015130	0.000000	0.000000	-M
		+M	0.000000	0.000000	0.000000	0.000000	-.217715	0.000000	0.000000	0.000000		+M
7.8192E+01	.9287	-M	0.000000	0.000000	0.000000	0.000000	-.217715	0.000000	0.000000	0.000000	0.000000	-M
		+M	.975895	0.000000	0.000000	0.000000	0.000000	0.000000	-.015130	0.000000		+M
6.6131E+01	.4818	-M	0.000000	.075514	0.000000	0.000000	0.000000	0.000000	0.000000	.423693	0.000000	-M
		+M	0.000000	0.000000	0.000000	.902653	0.000000	0.000000	0.000000	0.000000		+M
6.6131E+01	-.4818	-M	0.000000	0.000000	0.000000	.902653	0.000000	0.000000	0.000000	0.000000	0.000000	-M
		+M	0.000000	.075514	0.000000	0.000000	0.000000	0.000000	0.000000	.423693		+M
2.8206E+00	0.0000	-M	0.000000	0.000000	.571490	0.000000	0.000000	0.000000	0.000000	0.000000	.588896	-M
		+M	0.000000	0.000000	.571490	0.000000	0.000000	0.000000	0.000000	0.000000		+M
-1.9500E+01	0.0000	-M	0.000000	0.000000	0.000000	0.000000	0.000000	-.707107	0.000000	0.000000	0.000000	-M
		+M	0.000000	0.000000	0.000000	0.000000	0.000000	.707107	0.000000	0.000000		+M
-4.2000E+01	0.0000	-M	0.000000	0.000000	-.707107	0.000000	0.000000	0.000000	0.000000	0.000000	.000000	-M
		+M	0.000000	0.000000	.707107	0.000000	0.000000	0.000000	0.000000	0.000000		+M
-4.5055E+01	-.0808	-M	-.094568	0.000000	0.000000	0.000000	0.000000	0.000000	.869755	0.000000	0.000000	-M
		+M	0.000000	0.000000	0.000000	0.000000	-.484339	0.000000	0.000000	0.000000		+M
-4.5055E+01	.0808	-M	0.000000	0.000000	0.000000	0.000000	-.484339	0.000000	0.000000	0.000000	0.000000	-M
		+M	-.094568	0.000000	0.000000	0.000000	0.000000	0.000000	.869755	0.000000		+M
-6.3485E+01	-.3532	-M	0.000000	.660610	0.000000	0.000000	0.000000	0.000000	0.000000	.656820	0.000000	-M
		+M	0.000000	0.000000	0.000000	-.363568	0.000000	0.000000	0.000000	0.000000		+M
-6.3485E+01	.3532	-M	0.000000	0.000000	0.000000	-.363568	0.000000	0.000000	0.000000	0.000000	0.000000	-M
		+M	0.000000	.660610	0.000000	0.000000	0.000000	0.000000	0.000000	.656820		+M
-1.1485E+02	-.5036	-M	0.000000	.746922	0.000000	0.000000	0.000000	0.000000	0.000000	-.623756	0.000000	-M
		+M	0.000000	0.000000	0.000000	.230297	0.000000	0.000000	0.000000	0.000000		+M
-1.1485E+02	.5036	-M	0.000000	0.000000	0.000000	.230297	0.000000	0.000000	0.000000	0.000000	0.000000	-M
		+M	0.000000	.746922	0.000000	0.000000	0.000000	0.000000	0.000000	-.623756		+M
-1.2642E+02	0.0000	-M	0.000000	0.000000	-.416412	0.000000	0.000000	0.000000	0.000000	0.000000	.808209	-M
		+M	0.000000	0.000000	-.416412	0.000000	0.000000	0.000000	0.000000	0.000000		+M
ENERGY	HPAL		M=8	M=7	M=6	M=5	M=4	M=3	M=2	M=1	M=0	

Y= .60 X= .20 OAS=2.1375E+02*W246 B4F4/B2F2= 3.00E-01 B4F4/B6F6= 2.50E-01 B2F2/B6F6= 8.33E-01 B4F4= 1.20E-01

ENERGY	HPAL		M=8	M=7	M=6	M=5	M=4	M=3	M=2	M=1	M=0	
1.1135E+02	.9610	-M	0.000000	0.000000	0.000000	0.000000	.115139	0.000000	0.000000	0.000000	0.000000	-M
		+M	.980446	0.000000	0.000000	0.000000	0.000000	0.000000	.159592	0.000000		+M
1.1135E+02	-.9610	-M	.980446	0.000000	0.000000	0.000000	0.000000	0.000000	.159592	0.000000	0.000000	-M
		+M	0.000000	0.000000	0.000000	0.000000	.115139	0.000000	0.000000	0.000000		+M
9.1560E+01	0.0000	-M	0.000000	0.000000	0.000000	0.000000	0.000000	.707107	0.000000	0.000000	0.000000	-M
		+M	0.000000	0.000000	0.000000	0.000000	0.000000	.707107	0.000000	0.000000		+M
7.6494E+01	.1973	-M	-.187316	0.000000	0.000000	0.000000	0.000000	0.000000	.577474	0.000000	0.000000	-M
		+M	0.000000	0.000000	0.000000	0.000000	.794630	0.000000	0.000000	0.000000		+M
7.6494E+01	-.1973	-M	0.000000	0.000000	0.000000	0.000000	.794630	0.000000	0.000000	0.000000	0.000000	-M
		+M	-.187316	0.000000	0.000000	0.000000	0.000000	0.000000	.577474	0.000000		+M
3.6068E+01	-.3076	-M	0.000000	0.000000	0.000000	.768813	0.000000	0.000000	0.000000	0.000000	0.000000	-M
		+M	0.000000	.119471	0.000000	0.000000	0.000000	0.000000	0.000000	.628214		+M
3.6068E+01	.3076	-M	0.000000	.119471	0.000000	0.000000	0.000000	0.000000	0.000000	.628214	0.000000	-M
		+M	0.000000	0.000000	0.000000	.768813	0.000000	0.000000	0.000000	0.000000		+M
4.3311E-02	.0000	-M	0.000000	0.000000	.424968	0.000000	0.000000	0.000000	0.000000	0.000000	.799252	-M
		+M	0.000000	0.000000	.424968	0.000000	0.000000	0.000000	0.000000	0.000000		+M
-1.9320E+01	0.0000	-M	0.000000	0.000000	0.000000	0.000000	0.000000	-.707107	0.000000	0.000000	0.000000	-M
		+M	0.000000	0.000000	0.000000	0.000000	0.000000	.707107	0.000000	0.000000		+M
-3.2922E+01	.0138	-M	-.060326	0.000000	0.000000	0.000000	0.000000	0.000000	.800659	0.000000	0.000000	-M
		+M	0.000000	0.000000	0.000000	0.000000	-.596076	0.000000	0.000000	0.000000		+M
-3.2922E+01	-.0138	-M	0.000000	0.000000	0.000000	0.000000	-.596076	0.000000	0.000000	0.000000	0.000000	-M
		+M	-.060326	0.000000	0.000000	0.000000	0.000000	0.000000	.800659	0.000000		+M
-5.0302E+01	-.0110	-M	0.000000	.460644	0.000000	0.000000	0.000000	0.000000	0.000000	.650895	0.000000	-M
		+M	0.000000	0.000000	0.000000	-.603443	0.000000	0.000000	0.000000	0.000000		+M
-5.0302E+01	.0110	-M	0.000000	0.000000	0.000000	-.603443	0.000000	0.000000	0.000000	0.000000	0.000000	-M
		+M	0.000000	.460644	0.000000	0.000000	0.000000	0.000000	0.000000	.650895		+M
-6.5400E+01	-.0000	-M	0.000000	0.000000	-.707107	0.000000	0.000000	0.000000	0.000000	0.000000	-.000000	-M
		+M	0.000000	0.000000	.707107	0.000000	0.000000	0.000000	0.000000	0.000000		+M
-9.2926E+01	-.6716	-M	0.000000	.879508	0.000000	0.000000	0.000000	0.000000	0.000000	-.426243	0.000000	-M
		+M	0.000000	0.000000	0.000000	.211620	0.000000	0.000000	0.000000	0.000000		+M
-9.2926E+01	.6716	-M	0.000000	0.000000	0.000000	.211620	0.000000	0.000000	0.000000	0.000000	0.000000	-M
		+M	0.000000	.879508	0.000000	0.000000	0.000000	0.000000	0.000000	-.426243		+M
-1.0240E+02	.0000	-M	0.000000	0.000000	.565157	0.000000	0.000000	0.000000	0.000000	0.000000	-.600996	-M
		+M	0.000000	0.000000	.565157	0.000000	0.000000	0.000000	0.000000	0.000000		+M
ENERGY	HPAL		M=8	M=7	M=6	M=5	M=4	M=3	M=2	M=1	M=0	

```
Y= .60  X= .40    OAS=2.4270E+02*W246   B4F4/B2F2= 6.00E-01  B4F4/B6F6= 6.67E-01  B2F2/B6F6= 1.11E+00   B4F4= 2.40E-01
  ENERGY        HPAL         M=8        M=7        M=6        M=5        M=4        M=3        M=2        M=1        M=0
 1.4140E+02     .9952    -M 0.000000   0.000000   0.000000   0.000000    .021076   0.000000   0.000000   0.000000   0.000000-M
                         +M  .997050   0.000000   0.000000   0.000000   0.000000   0.000000    .073809   0.000000              +M
 1.4140E+02    -.9952    -M  .997050   0.000000   0.000000   0.000000   0.000000   0.000000    .073809   0.000000   0.000000-M
                         +M 0.000000   0.000000   0.000000   0.000000    .021076   0.000000   0.000000   0.000000              +M
 6.4020E+01    0.0000    -M 0.000000   0.000000   0.000000   0.000000   0.000000    .707107   0.000000   0.000000   0.000000-M
                         +M 0.000000   0.000000   0.000000   0.000000   0.000000    .707107   0.000000   0.000000              +M
 5.2206E+01    -.0368    -M 0.000000   0.000000   0.000000   0.000000    .622435   0.000000   0.000000   0.000000   0.000000-M
                         +M -.070859   0.000000   0.000000   0.000000   0.000000   0.000000    .779457   0.000000              +M
 5.2206E+01     .0368    -M -.070859   0.000000   0.000000   0.000000   0.000000   0.000000    .779457   0.000000   0.000000-M
                         +M 0.000000   0.000000   0.000000   0.000000    .622435   0.000000   0.000000   0.000000              +M
 3.1058E+01    -.0083    -M 0.000000    .145430   0.000000   0.000000   0.000000   0.000000   0.000000    .895573   0.000000-M
                         +M 0.000000   0.000000   0.000000    .420475   0.000000   0.000000   0.000000   0.000000              +M
 3.1058E+01     .0083    -M 0.000000   0.000000   0.000000    .420475   0.000000   0.000000   0.000000   0.000000   0.000000-M
                         +M 0.000000    .145430   0.000000   0.000000   0.000000   0.000000   0.000000    .895573              +M
 2.0179E+01    -.0000    -M 0.000000   0.000000    .226818   0.000000   0.000000   0.000000   0.000000   0.000000    .947157-M
                         +M 0.000000   0.000000    .226818   0.000000   0.000000   0.000000   0.000000   0.000000              +M
-1.9140E+01    0.0000    -M 0.000000   0.000000   0.000000   0.000000   0.000000   -.707107   0.000000   0.000000   0.000000-M
                         +M 0.000000   0.000000   0.000000   0.000000   0.000000    .707107   0.000000   0.000000              +M
-2.8962E+01    -.2084    -M 0.000000   0.000000   0.000000   0.000000    .782388   0.000000   0.000000   0.000000   0.000000-M
                         +M  .029513   0.000000   0.000000   0.000000   0.000000   0.000000   -.622092   0.000000              +M
-2.8962E+01     .2084    -M  .029513   0.000000   0.000000   0.000000   0.000000   0.000000   -.622092   0.000000   0.000000-M
                         +M 0.000000   0.000000   0.000000   0.000000    .782388   0.000000   0.000000   0.000000              +M
-5.5553E+01    -.3522    -M 0.000000   0.000000   0.000000    .869579   0.000000   0.000000   0.000000   0.000000   0.000000-M
                         +M 0.000000   -.346267   0.000000   0.000000   0.000000   0.000000   0.000000   -.352040              +M
-5.5553E+01     .3522    -M 0.000000   -.346267   0.000000   0.000000   0.000000   0.000000   0.000000   -.352040   0.000000-M
                         +M 0.000000   0.000000   0.000000    .869579   0.000000   0.000000   0.000000   0.000000              +M
-7.7625E+01    -.7189    -M 0.000000    .926795   0.000000   0.000000   0.000000   0.000000   0.000000   -.272060   0.000000-M
                         +M 0.000000   0.000000   0.000000    .258910   0.000000   0.000000   0.000000   0.000000              +M
-7.7625E+01     .7189    -M 0.000000   0.000000   0.000000    .258910   0.000000   0.000000   0.000000   0.000000   0.000000-M
                         +M 0.000000    .926795   0.000000   0.000000   0.000000   0.000000   0.000000   -.272060              +M
-8.8800E+01    -.0000    -M 0.000000   0.000000   -.707107   0.000000   0.000000   0.000000   0.000000   0.000000    .000000-M
                         +M 0.000000   0.000000    .707107   0.000000   0.000000   0.000000   0.000000   0.000000              +M
-1.0130E+02     .0000    -M 0.000000   0.000000    .669741   0.000000   0.000000   0.000000   0.000000   0.000000   -.320769-M
                         +M 0.000000   0.000000    .669741   0.000000   0.000000   0.000000   0.000000   0.000000              +M
  ENERGY        HPAL         M=8        M=7        M=6        M=5        M=4        M=3        M=2        M=1        M=0

Y= .60  X= .60    OAS=2.8802E+02*W246   B4F4/B2F2= 9.00E-01  B4F4/B6F6= 1.50E+00  B2F2/B6F6= 1.67E+00   B4F4= 3.60E-01
  ENERGY        HPAL         M=8        M=7        M=6        M=5        M=4        M=3        M=2        M=1        M=0
 1.7222E+02     .9987    -M 0.000000   0.000000   0.000000   0.000000    .005177   0.000000   0.000000   0.000000   0.000000-M
                         +M  .999179   0.000000   0.000000   0.000000   0.000000   0.000000    .040189   0.000000              +M
 1.7222E+02    -.9987    -M  .999179   0.000000   0.000000   0.000000   0.000000   0.000000    .040189   0.000000   0.000000-M
                         +M 0.000000   0.000000   0.000000   0.000000    .005177   0.000000   0.000000   0.000000              +M
 5.5921E+01    -.0000    -M 0.000000   0.000000    .102396   0.000000   0.000000   0.000000   0.000000   0.000000    .989459-M
                         +M 0.000000   0.000000    .102396   0.000000   0.000000   0.000000   0.000000   0.000000              +M
 5.3424E+01    -.1113    -M 0.000000    .094447   0.000000   0.000000   0.000000   0.000000   0.000000    .981793   0.000000-M
                         +M 0.000000   0.000000   0.000000    .164807   0.000000   0.000000   0.000000   0.000000              +M
 5.3424E+01     .1113    -M 0.000000   0.000000   0.000000    .164807   0.000000   0.000000   0.000000   0.000000   0.000000-M
                         +M 0.000000    .094447   0.000000   0.000000   0.000000   0.000000   0.000000    .981793              +M
 4.4153E+01     .1725    -M 0.000000   0.000000   0.000000   0.000000    .323972   0.000000   0.000000   0.000000   0.000000-M
                         +M -.039697   0.000000   0.000000   0.000000   0.000000   0.000000    .945233   0.000000              +M
 4.4153E+01    -.1725    -M -.039697   0.000000   0.000000   0.000000   0.000000   0.000000    .945233   0.000000   0.000000-M
                         +M 0.000000   0.000000   0.000000   0.000000    .323972   0.000000   0.000000   0.000000              +M
 3.6480E+01    0.0000    -M 0.000000   0.000000   0.000000   0.000000   0.000000    .707107   0.000000   0.000000   0.000000-M
                         +M 0.000000   0.000000   0.000000   0.000000   0.000000    .707107   0.000000   0.000000              +M
-1.8960E+01    0.0000    -M 0.000000   0.000000   0.000000   0.000000   0.000000   -.707107   0.000000   0.000000   0.000000-M
                         +M 0.000000   0.000000   0.000000   0.000000   0.000000    .707107   0.000000   0.000000              +M
-4.2009E+01    -.4212    -M 0.000000   0.000000   0.000000   0.000000    .946053   0.000000   0.000000   0.000000   0.000000-M
                         +M  .008127   0.000000   0.000000   0.000000   0.000000   0.000000   -.323911   0.000000              +M
-4.2009E+01     .4212    -M  .008127   0.000000   0.000000   0.000000   0.000000   0.000000   -.323911   0.000000   0.000000-M
                         +M 0.000000   0.000000   0.000000   0.000000    .946053   0.000000   0.000000   0.000000              +M
-6.4212E+01     .8556    -M 0.000000   0.000000   0.000000   -.099291   0.000000   0.000000   0.000000   0.000000   0.000000-M
                         +M 0.000000    .991937   0.000000   0.000000   0.000000   0.000000   0.000000   -.078755              +M
-6.4212E+01    -.8556    -M 0.000000    .991937   0.000000   0.000000   0.000000   0.000000   0.000000   -.078755   0.000000-M
                         +M 0.000000   0.000000   0.000000   -.099291   0.000000   0.000000   0.000000   0.000000              +M
-8.6293E+01    -.5919    -M 0.000000   0.000000   0.000000    .981315   0.000000   0.000000   0.000000   0.000000   0.000000-M
                         +M 0.000000    .084504   0.000000   0.000000   0.000000   0.000000   0.000000   -.172856              +M
-8.6293E+01     .5919    -M 0.000000    .084504   0.000000   0.000000   0.000000   0.000000   0.000000   -.172856   0.000000-M
                         +M 0.000000   0.000000   0.000000    .981315   0.000000   0.000000   0.000000   0.000000              +M
-1.1220E+02    -.0000    -M 0.000000   0.000000   -.707107   0.000000   0.000000   0.000000   0.000000   0.000000    .000000-M
                         +M 0.000000   0.000000    .707107   0.000000   0.000000   0.000000   0.000000   0.000000              +M
-1.1580E+02     .0000    -M 0.000000   0.000000    .699653   0.000000   0.000000   0.000000   0.000000   0.000000   -.144810-M
                         +M 0.000000   0.000000    .699653   0.000000   0.000000   0.000000   0.000000   0.000000              +M
  ENERGY        HPAL         M=8        M=7        M=6        M=5        M=4        M=3        M=2        M=1        M=0

Y= .60  X= .80    OAS=3.3950E+02*W246   B4F4/B2F2= 1.20E+00  B4F4/B6F6= 4.00E+00  B2F2/B6F6= 3.33E+00   B4F4= 4.80E-01
  ENERGY        HPAL         M=8        M=7        M=6        M=5        M=4        M=3        M=2        M=1        M=0
 2.0325E+02     .9998    -M 0.000000   0.000000   0.000000   0.000000    .000852   0.000000   0.000000   0.000000   0.000000-M
                         +M  .999846   0.000000   0.000000   0.000000   0.000000   0.000000    .017516   0.000000              +M
 2.0325E+02    -.9998    -M  .999846   0.000000   0.000000   0.000000   0.000000   0.000000    .017516   0.000000   0.000000-M
                         +M 0.000000   0.000000   0.000000   0.000000    .000852   0.000000   0.000000   0.000000              +M
 9.7609E+01    -.0000    -M 0.000000   0.000000    .037250   0.000000   0.000000   0.000000   0.000000   0.000000    .998611-M
                         +M 0.000000   0.000000    .037250   0.000000   0.000000   0.000000   0.000000   0.000000              +M
 8.5608E+01    -.1240    -M 0.000000    .040141   0.000000   0.000000   0.000000   0.000000   0.000000    .997706   0.000000-M
                         +M 0.000000   0.000000   0.000000    .054516   0.000000   0.000000   0.000000   0.000000              +M
 8.5608E+01     .1240    -M 0.000000   0.000000   0.000000    .054516   0.000000   0.000000   0.000000   0.000000   0.000000-M
                         +M 0.000000    .040141   0.000000   0.000000   0.000000   0.000000   0.000000    .997706              +M
 5.0777E+01     .2411    -M 0.000000   0.000000   0.000000   0.000000    .110120   0.000000   0.000000   0.000000   0.000000-M
                         +M -.017503   0.000000   0.000000   0.000000   0.000000   0.000000    .993764   0.000000              +M
 5.0777E+01    -.2411    -M -.017503   0.000000   0.000000   0.000000   0.000000   0.000000    .993764   0.000000   0.000000-M
                         +M 0.000000   0.000000   0.000000   0.000000    .110120   0.000000   0.000000   0.000000              +M
 8.9400E+00    0.0000    -M 0.000000   0.000000   0.000000   0.000000   0.000000    .707107   0.000000   0.000000   0.000000-M
                         +M 0.000000   0.000000   0.000000   0.000000   0.000000    .707107   0.000000   0.000000              +M
-1.8780E+01    0.0000    -M 0.000000   0.000000   0.000000   0.000000   0.000000   -.707107   0.000000   0.000000   0.000000-M
                         +M 0.000000   0.000000   0.000000   0.000000   0.000000    .707107   0.000000   0.000000              +M
-5.4184E+01    -.8737    -M 0.000000    .999184   0.000000   0.000000   0.000000   0.000000   0.000000   -.039839   0.000000-M
                         +M 0.000000   0.000000   0.000000   -.006609   0.000000   0.000000   0.000000   0.000000              +M
-5.4184E+01     .8737    -M 0.000000   0.000000   0.000000   -.006609   0.000000   0.000000   0.000000   0.000000   0.000000-M
                         +M 0.000000    .999184   0.000000   0.000000   0.000000   0.000000   0.000000   -.039839              +M
-6.9944E+01    -.4909    -M 0.000000   0.000000   0.000000   0.000000    .993918   0.000000   0.000000   0.000000   0.000000-M
                         +M  .001082   0.000000   0.000000   0.000000   0.000000   0.000000   -.110118   0.000000              +M
-6.9944E+01     .4909    -M  .001082   0.000000   0.000000   0.000000   0.000000   0.000000   -.110118   0.000000   0.000000-M
                         +M 0.000000   0.000000   0.000000   0.000000    .993918   0.000000   0.000000   0.000000              +M
-1.2346E+02    -.6227    -M 0.000000   0.000000   0.000000    .998491   0.000000   0.000000   0.000000   0.000000   0.000000-M
                         +M 0.000000    .004422   0.000000   0.000000   0.000000   0.000000   0.000000   -.054737              +M
-1.2346E+02     .6227    -M 0.000000    .004422   0.000000   0.000000   0.000000   0.000000   0.000000   -.054737   0.000000-M
                         +M 0.000000   0.000000   0.000000    .998491   0.000000   0.000000   0.000000   0.000000              +M
-1.3560E+02     .0000    -M 0.000000   0.000000   -.707107   0.000000   0.000000   0.000000   0.000000   0.000000   -.000000-M
                         +M 0.000000   0.000000    .707107   0.000000   0.000000   0.000000   0.000000   0.000000              +M
-1.3625E+02    -.0000    -M 0.000000   0.000000    .706125   0.000000   0.000000   0.000000   0.000000   0.000000   -.052680-M
                         +M 0.000000   0.000000    .706125   0.000000   0.000000   0.000000   0.000000   0.000000              +M
  ENERGY        HPAL         M=8        M=7        M=6        M=5        M=4        M=3        M=2        M=1        M=0
```

```
Y= .80  X=-.80    OAS=3.9441E+02*W246    B4F4/B2F2=-3.20E+00   B4F4/B6F6=-4.00E+00   B2F2/B6F6= 1.25E+00    B4F4=-6.40E-01
  ENERGY        HPAL        M=8        M=7        M=6        M=5        M=4        M=3        M=2        M=1        M=0
 1.8599E+02    .6235    -M 0.000000    .002213   0.000000   0.000000   0.000000   0.000000   0.000000    .044271   0.000000-M
                        +M 0.000000   0.000000   0.000000    .999017   0.000000   0.000000   0.000000   0.000000           +M
 1.8599E+02   -.6235    -M 0.000000   0.000000   0.000000    .999017   0.000000   0.000000   0.000000   0.000000   0.000000-M
                        +M 0.000000    .002213   0.000000   0.000000   0.000000   0.000000   0.000000    .044271           +M
 1.6541E+02    .0000    -M 0.000000   0.000000    .706335   0.000000   0.000000   0.000000   0.000000   0.000000    .046724-M
                        +M 0.000000   0.000000    .706335   0.000000   0.000000   0.000000   0.000000   0.000000           +M
 1.6464E+02   -.0000    -M 0.000000   0.000000   -.707107   0.000000   0.000000   0.000000   0.000000   0.000000   -.000000-M
                        +M 0.000000   0.000000    .707107   0.000000   0.000000   0.000000   0.000000   0.000000           +M
 1.2789E+02    .4946    -M  .000900   0.000000   0.000000   0.000000   0.000000   0.000000    .084532   0.000000   0.000000-M
                        +M 0.000000   0.000000   0.000000   0.000000    .996420   0.000000   0.000000   0.000000           +M
 1.2789E+02   -.4946    -M 0.000000   0.000000   0.000000   0.000000    .996420   0.000000   0.000000   0.000000   0.000000-M
                        +M  .000900   0.000000   0.000000   0.000000   0.000000   0.000000    .084532   0.000000           +M
 4.3800E+01   0.0000    -M 0.000000   0.000000   0.000000   0.000000   0.000000    .707107   0.000000   0.000000   0.000000-M
                        +M 0.000000   0.000000   0.000000   0.000000   0.000000    .707107   0.000000   0.000000           +M
 3.6487E+01    .8739    -M 0.000000   0.000000   0.000000   -.003912   0.000000   0.000000   0.000000   0.000000   0.000000-M
                        +M 0.000000    .999257   0.000000   0.000000   0.000000   0.000000   0.000000    .038344           +M
 3.6487E+01   -.8739    -M 0.000000    .999257   0.000000   0.000000   0.000000   0.000000   0.000000    .038344   0.000000-M
                        +M 0.000000   0.000000   0.000000   -.003912   0.000000   0.000000   0.000000   0.000000           +M
 6.8400E+00   0.0000    -M 0.000000   0.000000   0.000000   0.000000   0.000000   -.707107   0.000000   0.000000   0.000000-M
                        +M 0.000000   0.000000   0.000000   0.000000   0.000000    .707107   0.000000   0.000000           +M
-8.1234E+01   -.2452    -M  .028054   0.000000   0.000000   0.000000   0.000000   0.000000    .996026   0.000000   0.000000-M
                        +M 0.000000   0.000000   0.000000   0.000000   -.084523   0.000000   0.000000   0.000000           +M
-8.1234E+01    .2452    -M 0.000000   0.000000   0.000000   0.000000   -.084523   0.000000   0.000000   0.000000   0.000000-M
                        +M  .028054   0.000000   0.000000   0.000000   0.000000   0.000000    .996026   0.000000           +M
-1.5804E+02    .1246    -M 0.000000   0.000000   0.000000   -.044153   0.000000   0.000000   0.000000   0.000000   0.000000-M
                        +M 0.000000   -.038479   0.000000   0.000000   0.000000   0.000000   0.000000    .998283           +M
-1.5804E+02   -.1246    -M 0.000000   -.038479   0.000000   0.000000   0.000000   0.000000   0.000000    .998283   0.000000-M
                        +M 0.000000   0.000000   0.000000   -.044153   0.000000   0.000000   0.000000   0.000000           +M
-1.8605E+02   0.0000    -M 0.000000   0.000000   -.033039   0.000000   0.000000   0.000000   0.000000   0.000000    .998908-M
                        +M 0.000000   0.000000   -.033039   0.000000   0.000000   0.000000   0.000000   0.000000           +M
-2.0842E+02   -.9994    -M  .999606   0.000000   0.000000   0.000000   0.000000   0.000000   -.028030   0.000000   0.000000-M
                        +M 0.000000   0.000000   0.000000   0.000000    .001475   0.000000   0.000000   0.000000           +M
-2.0842E+02    .9994    -M 0.000000   0.000000   0.000000   0.000000    .001475   0.000000   0.000000   0.000000   0.000000-M
                        +M  .999606   0.000000   0.000000   0.000000   0.000000   0.000000   -.028030   0.000000           +M
   ENERGY       HPAL        M=8        M=7        M=6        M=5        M=4        M=3        M=2        M=1        M=0

Y= .80  X=-.60    OAS=3.2328E+02*W246    B4F4/B2F2=-2.40E+00   B4F4/B6F6=-1.50E+00   B2F2/B6F6= 6.25E-01    B4F4=-4.80E-01
  ENERGY        HPAL        M=8        M=7        M=6        M=5        M=4        M=3        M=2        M=1        M=0
 1.5522E+02   -.6167    -M 0.000000   0.000000   0.000000    .994505   0.000000   0.000000   0.000000   0.000000   0.000000-M
                        +M 0.000000    .009717   0.000000   0.000000   0.000000   0.000000   0.000000    .104236           +M
 1.5522E+02    .6167    -M 0.000000    .009717   0.000000   0.000000   0.000000   0.000000   0.000000    .104236   0.000000-M
                        +M 0.000000   0.000000   0.000000    .994505   0.000000   0.000000   0.000000   0.000000           +M
 1.2664E+02    .4742    -M  .005099   0.000000   0.000000   0.000000   0.000000   0.000000    .185257   0.000000   0.000000-M
                        +M 0.000000   0.000000   0.000000   0.000000    .982677   0.000000   0.000000   0.000000           +M
 1.2664E+02   -.4742    -M 0.000000   0.000000   0.000000   0.000000    .982677   0.000000   0.000000   0.000000   0.000000-M
                        +M  .005099   0.000000   0.000000   0.000000   0.000000   0.000000    .185257   0.000000           +M
 1.1237E+02   -.0000    -M 0.000000   0.000000    .702183   0.000000   0.000000   0.000000   0.000000   0.000000    .117810-M
                        +M 0.000000   0.000000    .702183   0.000000   0.000000   0.000000   0.000000   0.000000           +M
 1.0848E+02    .0000    -M 0.000000   0.000000   -.707107   0.000000   0.000000   0.000000   0.000000   0.000000    .000000-M
                        +M 0.000000   0.000000    .707107   0.000000   0.000000   0.000000   0.000000   0.000000           +M
 7.3800E+01   0.0000    -M 0.000000   0.000000   0.000000   0.000000   0.000000    .707107   0.000000   0.000000   0.000000-M
                        +M 0.000000   0.000000   0.000000   0.000000   0.000000    .707107   0.000000   0.000000           +M
-1.2000E-01   0.0000    -M 0.000000   0.000000   0.000000   0.000000   0.000000   -.707107   0.000000   0.000000   0.000000-M
                        +M 0.000000   0.000000   0.000000   0.000000   0.000000    .707107   0.000000   0.000000           +M
-3.7764E+00   -.8656    -M 0.000000    .993954   0.000000   0.000000   0.000000   0.000000   0.000000    .107771   0.000000-M
                        +M 0.000000   0.000000   0.000000   -.021008   0.000000   0.000000   0.000000   0.000000           +M
-3.7764E+00    .8656    -M 0.000000   0.000000   0.000000   -.021008   0.000000   0.000000   0.000000   0.000000   0.000000-M
                        +M 0.000000    .993954   0.000000   0.000000   0.000000   0.000000   0.000000    .107771           +M
-6.6379E+01   -.2326    -M  .104322   0.000000   0.000000   0.000000   0.000000   0.000000    .977229   0.000000   0.000000-M
                        +M 0.000000   0.000000   0.000000   0.000000   -.184772   0.000000   0.000000   0.000000           +M
-6.6379E+01    .2326    -M 0.000000   0.000000   0.000000   0.000000   -.184772   0.000000   0.000000   0.000000   0.000000-M
                        +M  .104322   0.000000   0.000000   0.000000   0.000000   0.000000    .977229   0.000000           +M
-1.3418E+02   -.9917    -M  .994530   0.000000   0.000000   0.000000   0.000000   0.000000   -.103457   0.000000   0.000000-M
                        +M 0.000000   0.000000   0.000000   0.000000    .014343   0.000000   0.000000   0.000000           +M
-1.3418E+02    .9917    -M 0.000000   0.000000   0.000000   0.000000    .014343   0.000000   0.000000   0.000000   0.000000-M
                        +M  .994530   0.000000   0.000000   0.000000   0.000000   0.000000   -.103457   0.000000           +M
-1.4077E+02   -.1261    -M 0.000000   -.109368   0.000000   0.000000   0.000000   0.000000   0.000000    .988696   0.000000-M
                        +M 0.000000   0.000000   0.000000   -.102558   0.000000   0.000000   0.000000   0.000000           +M
-1.4077E+02    .1261    -M 0.000000   0.000000   0.000000   -.102558   0.000000   0.000000   0.000000   0.000000   0.000000-M
                        +M 0.000000   -.109368   0.000000   0.000000   0.000000   0.000000   0.000000    .988696           +M
-1.6805E+02    .0000    -M 0.000000   0.000000   -.083304   0.000000   0.000000   0.000000   0.000000   0.000000    .993036-M
                        +M 0.000000   0.000000   -.083304   0.000000   0.000000   0.000000   0.000000   0.000000           +M
   ENERGY       HPAL        M=8        M=7        M=6        M=5        M=4        M=3        M=2        M=1        M=0

Y= .80  X=-.40    OAS=2.8472E+02*W246    B4F4/B2F2=-1.60E+00   B4F4/B6F6=-6.67E-01   B2F2/B6F6= 4.17E-01    B4F4=-3.20E-01
  ENERGY        HPAL        M=8        M=7        M=6        M=5        M=4        M=3        M=2        M=1        M=0
 1.2999E+02    .4338    -M  .016863   0.000000   0.000000   0.000000   0.000000   0.000000    .296087   0.000000   0.000000-M
                        +M 0.000000   0.000000   0.000000   0.000000    .955012   0.000000   0.000000   0.000000           +M
 1.2999E+02   -.4338    -M 0.000000   0.000000   0.000000   0.000000    .955012   0.000000   0.000000   0.000000   0.000000-M
                        +M  .016863   0.000000   0.000000   0.000000   0.000000   0.000000    .296087   0.000000           +M
 1.2743E+02   -.5979    -M 0.000000   0.000000   0.000000    .982081   0.000000   0.000000   0.000000   0.000000   0.000000-M
                        +M 0.000000    .024069   0.000000   0.000000   0.000000   0.000000   0.000000    .186914           +M
 1.2743E+02    .5979    -M 0.000000    .024069   0.000000   0.000000   0.000000   0.000000   0.000000    .186914   0.000000-M
                        +M 0.000000   0.000000   0.000000    .982081   0.000000   0.000000   0.000000   0.000000           +M
 1.0380E+02   0.0000    -M 0.000000   0.000000   0.000000   0.000000   0.000000    .707107   0.000000   0.000000   0.000000-M
                        +M 0.000000   0.000000   0.000000   0.000000   0.000000    .707107   0.000000   0.000000           +M
 6.4016E+01   -.0000    -M 0.000000   0.000000    .687944   0.000000   0.000000   0.000000   0.000000   0.000000    .231225-M
                        +M 0.000000   0.000000    .687944   0.000000   0.000000   0.000000   0.000000   0.000000           +M
 5.2320E+01    .0000    -M 0.000000   0.000000   -.707107   0.000000   0.000000   0.000000   0.000000   0.000000    .000000-M
                        +M 0.000000   0.000000    .707107   0.000000   0.000000   0.000000   0.000000   0.000000           +M
-7.0800E+00   0.0000    -M 0.000000   0.000000   0.000000   0.000000   0.000000   -.707107   0.000000   0.000000   0.000000-M
                        +M 0.000000   0.000000   0.000000   0.000000   0.000000    .707107   0.000000   0.000000           +M
-4.1177E+01   -.8204    -M 0.000000    .965537   0.000000   0.000000   0.000000   0.000000   0.000000    .250310   0.000000-M
                        +M 0.000000   0.000000   0.000000   -.071303   0.000000   0.000000   0.000000   0.000000           +M
-4.1177E+01    .8204    -M 0.000000   0.000000   0.000000   -.071303   0.000000   0.000000   0.000000   0.000000   0.000000-M
                        +M 0.000000    .965537   0.000000   0.000000   0.000000   0.000000   0.000000    .250310           +M
-4.7760E+01   -.5653    -M  .686782   0.000000   0.000000   0.000000   0.000000   0.000000    .690744   0.000000   0.000000-M
                        +M 0.000000   0.000000   0.000000   0.000000   -.226282   0.000000   0.000000   0.000000           +M
-4.7760E+01    .5653    -M 0.000000   0.000000   0.000000   0.000000   -.226282   0.000000   0.000000   0.000000   0.000000-M
                        +M  .686782   0.000000   0.000000   0.000000   0.000000   0.000000    .690744   0.000000           +M
-6.8309E+01   -.6185    -M  .726668   0.000000   0.000000   0.000000   0.000000   0.000000   -.659701   0.000000   0.000000-M
                        +M 0.000000   0.000000   0.000000   0.000000    .191699   0.000000   0.000000   0.000000           +M
-6.8309E+01    .6185    -M 0.000000   0.000000   0.000000   0.000000    .191699   0.000000   0.000000   0.000000   0.000000-M
                        +M  .726668   0.000000   0.000000   0.000000   0.000000   0.000000   -.659701   0.000000           +M
-1.2933E+02   -.1525    -M 0.000000   -.259152   0.000000   0.000000   0.000000   0.000000   0.000000    .949952   0.000000-M
                        +M 0.000000   0.000000   0.000000   -.174448   0.000000   0.000000   0.000000   0.000000           +M
-1.2933E+02    .1525    -M 0.000000   0.000000   0.000000   -.174448   0.000000   0.000000   0.000000   0.000000   0.000000-M
                        +M 0.000000   -.259152   0.000000   0.000000   0.000000   0.000000   0.000000    .949952           +M
-1.5474E+02    .0000    -M 0.000000   0.000000   -.163501   0.000000   0.000000   0.000000   0.000000   0.000000    .972900-M
                        +M 0.000000   0.000000   -.163501   0.000000   0.000000   0.000000   0.000000   0.000000           +M
   ENERGY       HPAL        M=8        M=7        M=6        M=5        M=4        M=3        M=2        M=1        M=0
```

Y= .80 X=-.20 OAS=2.8967E+02*W246 B4F4/B2F2=-8.00E-01 B4F4/B6F6=-2.50E-01 B2F2/B6F6= 3.12E-01 B4F4=-1.60E-01

ENERGY	HPAL		M=8	M=7	M=6	M=5	M=4	M=3	M=2	M=1	M=0
1.3852E+02	.3732	-M	.047527	0.000000	0.000000	0.000000	0.000000	0.000000	.405632	0.000000	0.000000-M
		+M	0.000000	0.000000	0.000000	0.000000	.912800	0.000000	0.000000	0.000000	+M
1.3852E+02	-.3732	-M	0.000000	0.000000	0.000000	0.000000	.912800	0.000000	0.000000	0.000000	0.000000-M
		+M	.047527	0.000000	0.000000	0.000000	0.000000	0.000000	.405632	0.000000	+M
1.3380E+02	0.0000	-M	0.000000	0.000000	0.000000	0.000000	0.000000	.707107	0.000000	0.000000	0.000000-M
		+M	0.000000	0.000000	0.000000	0.000000	0.000000	.707107	0.000000	0.000000	+M
1.0454E+02	-.5547	-M	0.000000	0.000000	0.000000	.953114	0.000000	0.000000	0.000000	0.000000	0.000000-M
		+M	0.000000	.046369	0.000000	0.000000	0.000000	0.000000	0.000000	.299036	+M
1.0454E+02	.5547	-M	0.000000	.046369	0.000000	0.000000	0.000000	0.000000	0.000000	.299036	0.000000-M
		+M	0.000000	0.000000	0.000000	.953114	0.000000	0.000000	0.000000	0.000000	+M
2.5386E+01	.0000	-M	0.000000	0.000000	.645928	0.000000	0.000000	0.000000	0.000000	0.000000	.406885-M
		+M	0.000000	0.000000	.645928	0.000000	0.000000	0.000000	0.000000	0.000000	+M
1.8451E+01	.9637	-M	0.000000	0.000000	0.000000	0.000000	-.116139	0.000000	0.000000	0.000000	0.000000-M
		+M	.982408	0.000000	0.000000	0.000000	0.000000	0.000000	.146243	0.000000	+M
1.8451E+01	-.9637	-M	.982408	0.000000	0.000000	0.000000	0.000000	0.000000	.146243	0.000000	0.000000-M
		+M	0.000000	0.000000	0.000000	0.000000	-.116139	0.000000	0.000000	0.000000	+M
-3.8400E+00	-.0000	-M	0.000000	0.000000	-.707107	0.000000	0.000000	0.000000	0.000000	0.000000	.000000-M
		+M	0.000000	0.000000	.707107	0.000000	0.000000	0.000000	0.000000	0.000000	+M
-1.4040E+01	0.0000	-M	0.000000	0.000000	0.000000	0.000000	0.000000	-.707107	0.000000	0.000000	0.000000-M
		+M	0.000000	0.000000	0.000000	0.000000	0.000000	.707107	0.000000	0.000000	+M
-5.5211E+01	-.1595	-M	-.180600	0.000000	0.000000	0.000000	0.000000	0.000000	.902262	0.000000	0.000000-M
		+M	0.000000	0.000000	0.000000	0.000000	-.391545	0.000000	0.000000	0.000000	+M
-5.5211E+01	.1595	-M	0.000000	0.000000	0.000000	0.000000	-.391545	0.000000	0.000000	0.000000	0.000000-M
		+M	-.180600	0.000000	0.000000	0.000000	0.000000	0.000000	.902262	0.000000	+M
-6.9794E+01	-.6095	-M	0.000000	.829116	0.000000	0.000000	0.000000	0.000000	0.000000	.520651	0.000000-M
		+M	0.000000	0.000000	0.000000	-.203689	0.000000	0.000000	0.000000	0.000000	+M
-6.9794E+01	.6095	-M	0.000000	0.000000	0.000000	-.203689	0.000000	0.000000	0.000000	0.000000	0.000000-M
		+M	0.000000	.829116	0.000000	0.000000	0.000000	0.000000	0.000000	.520651	+M
-1.3159E+02	-.3202	-M	0.000000	-.557151	0.000000	0.000000	0.000000	0.000000	0.000000	.799687	0.000000-M
		+M	0.000000	0.000000	0.000000	-.223794	0.000000	0.000000	0.000000	0.000000	+M
-1.3159E+02	.3202	-M	0.000000	0.000000	0.000000	-.223794	0.000000	0.000000	0.000000	0.000000	0.000000-M
		+M	0.000000	-.557151	0.000000	0.000000	0.000000	0.000000	0.000000	.799687	+M
-1.5115E+02	.0000	-M	0.000000	0.000000	-.287711	0.000000	0.000000	0.000000	0.000000	0.000000	.913480-M
		+M	0.000000	0.000000	-.287711	0.000000	0.000000	0.000000	0.000000	0.000000	+M
ENERGY	HPAL		M=8	M=7	M=6	M=5	M=4	M=3	M=2	M=1	M=0

Y= .80 X=0.00 OAS=3.2872E+02*W246 B4F4/B2F2= 0. B4F4/B6F6= 0. B2F2/B6F6= 2.50E-01 B4F4= 0.

ENERGY	HPAL		M=8	M=7	M=6	M=5	M=4	M=3	M=2	M=1	M=0
1.6380E+02	0.0000	-M	0.000000	0.000000	0.000000	0.000000	0.000000	.707107	0.000000	0.000000	0.000000-M
		+M	0.000000	0.000000	0.000000	0.000000	0.000000	.707107	0.000000	0.000000	+M
1.5261E+02	.2793	-M	.146151	0.000000	0.000000	0.000000	0.000000	0.000000	.501506	0.000000	0.000000-M
		+M	0.000000	0.000000	0.000000	0.000000	.852720	0.000000	0.000000	0.000000	+M
1.5261E+02	-.2793	-M	0.000000	0.000000	0.000000	0.000000	.852720	0.000000	0.000000	0.000000	0.000000-M
		+M	.146151	0.000000	0.000000	0.000000	0.000000	0.000000	.501506	0.000000	+M
9.1544E+01	.9513	-M	0.000000	0.000000	0.000000	0.000000	-.179676	0.000000	0.000000	0.000000	0.000000-M
		+M	.983545	0.000000	0.000000	0.000000	0.000000	0.000000	.018878	0.000000	+M
9.1544E+01	-.9513	-M	.983545	0.000000	0.000000	0.000000	0.000000	0.000000	.018878	0.000000	0.000000-M
		+M	0.000000	0.000000	0.000000	0.000000	-.179676	0.000000	0.000000	0.000000	+M
8.9280E+01	.4730	-M	0.000000	.074635	0.000000	0.000000	0.000000	0.000000	0.000000	.437580	0.000000-M
		+M	0.000000	0.000000	0.000000	.896077	0.000000	0.000000	0.000000	0.000000	+M
8.9280E+01	-.4730	-M	0.000000	0.000000	0.000000	.896077	0.000000	0.000000	0.000000	0.000000	0.000000-M
		+M	0.000000	.074635	0.000000	0.000000	0.000000	0.000000	0.000000	.437580	+M
4.1156E+00	-.0000	-M	0.000000	0.000000	.557085	0.000000	0.000000	0.000000	0.000000	0.000000	.615883-M
		+M	0.000000	0.000000	.557085	0.000000	0.000000	0.000000	0.000000	0.000000	+M
-2.1000E+01	0.0000	-M	0.000000	0.000000	0.000000	0.000000	0.000000	-.707107	0.000000	0.000000	0.000000-M
		+M	0.000000	0.000000	0.000000	0.000000	0.000000	.707107	0.000000	0.000000	+M
-5.4557E+01	-.0780	-M	-.106207	0.000000	0.000000	0.000000	0.000000	0.000000	.864948	0.000000	0.000000-M
		+M	0.000000	0.000000	0.000000	0.000000	-.490494	0.000000	0.000000	0.000000	+M
-5.4557E+01	.0780	-M	0.000000	0.000000	0.000000	0.000000	-.490494	0.000000	0.000000	0.000000	0.000000-M
		+M	-.106207	0.000000	0.000000	0.000000	0.000000	0.000000	.864948	0.000000	+M
-6.0000E+01	0.0000	-M	0.000000	0.000000	-.707107	0.000000	0.000000	0.000000	0.000000	0.000000	.000000-M
		+M	0.000000	0.000000	.707107	0.000000	0.000000	0.000000	0.000000	0.000000	+M
-8.4304E+01	-.2602	-M	0.000000	.580436	0.000000	0.000000	0.000000	0.000000	0.000000	.711617	0.000000-M
		+M	0.000000	0.000000	0.000000	-.395848	0.000000	0.000000	0.000000	0.000000	+M
-8.4304E+01	.2602	-M	0.000000	0.000000	0.000000	-.395848	0.000000	0.000000	0.000000	0.000000	0.000000-M
		+M	0.000000	.580436	0.000000	0.000000	0.000000	0.000000	0.000000	.711617	+M
-1.5558E+02	-.5879	-M	0.000000	.810878	0.000000	0.000000	0.000000	0.000000	0.000000	-.549660	0.000000-M
		+M	0.000000	0.000000	0.000000	.200876	0.000000	0.000000	0.000000	0.000000	+M
-1.5558E+02	.5879	-M	0.000000	0.000000	0.000000	.200876	0.000000	0.000000	0.000000	0.000000	0.000000-M
		+M	0.000000	.810878	0.000000	0.000000	0.000000	0.000000	0.000000	-.549660	+M
-1.6492E+02	.0000	-M	0.000000	0.000000	-.435495	0.000000	0.000000	0.000000	0.000000	0.000000	.787837-M
		+M	0.000000	0.000000	-.435495	0.000000	0.000000	0.000000	0.000000	0.000000	+M
ENERGY	HPAL		M=8	M=7	M=6	M=5	M=4	M=3	M=2	M=1	M=0

Y= .80 X= .20 OAS=2.7266E+02*W246 B4F4/B2F2= 8.00E-01 B4F4/B6F6= 2.50E-01 B2F2/B6F6= 3.12E-01 B4F4= 1.60E-01

ENERGY	HPAL		M=8	M=7	M=6	M=5	M=4	M=3	M=2	M=1	M=0
1.3622E+02	.9044	-M	0.000000	0.000000	0.000000	0.000000	.194174	0.000000	0.000000	0.000000	0.000000-M
		+M	.954087	0.000000	0.000000	0.000000	0.000000	0.000000	.228068	0.000000	+M
1.3622E+02	-.9044	-M	.954087	0.000000	0.000000	0.000000	0.000000	0.000000	.228068	0.000000	0.000000-M
		+M	0.000000	0.000000	0.000000	0.000000	.194174	0.000000	0.000000	0.000000	+M
1.2708E+02	0.0000	-M	0.000000	0.000000	0.000000	0.000000	0.000000	.707107	0.000000	0.000000	0.000000-M
		+M	0.000000	0.000000	0.000000	0.000000	0.000000	.707107	0.000000	0.000000	+M
1.0509E+02	-.1323	-M	0.000000	0.000000	0.000000	0.000000	.771401	0.000000	0.000000	0.000000	0.000000-M
		+M	-.292133	0.000000	0.000000	0.000000	0.000000	0.000000	.565331	0.000000	+M
1.0509E+02	.1323	-M	-.292133	0.000000	0.000000	0.000000	0.000000	0.000000	.565331	0.000000	0.000000-M
		+M	0.000000	0.000000	0.000000	0.000000	.771401	0.000000	0.000000	0.000000	+M
5.0923E+01	-.2864	-M	0.000000	0.000000	0.000000	.749658	0.000000	0.000000	0.000000	0.000000	0.000000-M
		+M	0.000000	.115719	0.000000	0.000000	0.000000	0.000000	0.000000	.651630	+M
5.0923E+01	.2864	-M	0.000000	.115719	0.000000	0.000000	0.000000	0.000000	0.000000	.651630	0.000000-M
		+M	0.000000	0.000000	0.000000	.749658	0.000000	0.000000	0.000000	0.000000	+M
3.9604E+00	-.0000	-M	0.000000	0.000000	.401387	0.000000	0.000000	0.000000	0.000000	0.000000	.823272-M
		+M	0.000000	0.000000	.401387	0.000000	0.000000	0.000000	0.000000	0.000000	+M
-2.0760E+01	0.0000	-M	0.000000	0.000000	0.000000	0.000000	0.000000	-.707107	0.000000	0.000000	0.000000-M
		+M	0.000000	0.000000	0.000000	0.000000	0.000000	.707107	0.000000	0.000000	+M
-3.8755E+01	-.0221	-M	0.000000	0.000000	0.000000	0.000000	-.606001	0.000000	0.000000	0.000000	0.000000-M
		+M	-.066160	0.000000	0.000000	0.000000	0.000000	0.000000	.792708	0.000000	+M
-3.8755E+01	.0221	-M	-.066160	0.000000	0.000000	0.000000	0.000000	0.000000	.792708	0.000000	0.000000-M
		+M	0.000000	0.000000	0.000000	0.000000	-.606001	0.000000	0.000000	0.000000	+M
-6.5325E+01	-.0723	-M	0.000000	0.000000	0.000000	-.639243	0.000000	0.000000	0.000000	0.000000	0.000000-M
		+M	0.000000	.381610	0.000000	0.000000	0.000000	0.000000	0.000000	.667639	+M
-6.5325E+01	.0723	-M	0.000000	.381610	0.000000	0.000000	0.000000	0.000000	0.000000	.667639	0.000000-M
		+M	0.000000	0.000000	0.000000	-.639243	0.000000	0.000000	0.000000	0.000000	+M
-9.1200E+01	-.0000	-M	0.000000	0.000000	-.707107	0.000000	0.000000	0.000000	0.000000	0.000000	.000000-M
		+M	0.000000	0.000000	.707107	0.000000	0.000000	0.000000	0.000000	0.000000	+M
-1.2948E+02	-.7337	-M	0.000000	.917051	0.000000	0.000000	0.000000	0.000000	0.000000	-.360049	0.000000-M
		+M	0.000000	0.000000	0.000000	.171410	0.000000	0.000000	0.000000	0.000000	+M
-1.2948E+02	.7337	-M	0.000000	0.000000	0.000000	.171410	0.000000	0.000000	0.000000	0.000000	0.000000-M
		+M	0.000000	.917051	0.000000	0.000000	0.000000	0.000000	0.000000	-.360049	+M
-1.3644E+02	.0000	-M	0.000000	0.000000	.582141	0.000000	0.000000	0.000000	0.000000	0.000000	-.567647-M
		+M	0.000000	0.000000	.582141	0.000000	0.000000	0.000000	0.000000	0.000000	+M
ENERGY	HPAL		M=8	M=7	M=6	M=5	M=4	M=3	M=2	M=1	M=0

```
Y= .80  X= .40   OAS=3.1324E+02*W246   B4F4/B2F2= 1.60E+00  B4F4/B6F6= 6.67E-01  B2F2/B6F6= 4.17E-01   B4F4= 3.20E-01
  ENERGY       HPAL        M=8       M=7       M=6       M=5       M=4       M=3       M=2       M=1        M=0
 1.7533E+02   -.9933   -M  .995895  0.000000  0.000000  0.000000  0.000000  0.000000   .086397  0.000000  0.000000-M
                       +M 0.000000  0.000000  0.000000  0.000000   .026983  0.000000  0.000000  0.000000         +M
 1.7533E+02    .9933   -M 0.000000  0.000000  0.000000  0.000000   .026983  0.000000  0.000000  0.000000  0.000000-M
                       +M  .995895  0.000000  0.000000  0.000000  0.000000  0.000000   .086397  0.000000         +M
 9.0360E+01   0.0000   -M 0.000000  0.000000  0.000000  0.000000  0.000000   .707107  0.000000  0.000000  0.000000-M
                       +M 0.000000  0.000000  0.000000  0.000000  0.000000   .707107  0.000000  0.000000         +M
 7.4640E+01   -.0222   -M 0.000000  0.000000  0.000000  0.000000   .608388  0.000000  0.000000  0.000000  0.000000-M
                       +M -.084939  0.000000  0.000000  0.000000  0.000000  0.000000   .789081  0.000000         +M
 7.4640E+01    .0222   -M -.084939  0.000000  0.000000  0.000000  0.000000  0.000000   .789081  0.000000  0.000000-M
                       +M 0.000000  0.000000  0.000000  0.000000   .608388  0.000000  0.000000  0.000000         +M
 4.7422E+01   -.0186   -M 0.000000   .133432  0.000000  0.000000  0.000000  0.000000  0.000000   .906899  0.000000-M
                       +M 0.000000  0.000000  0.000000   .399661  0.000000  0.000000  0.000000  0.000000         +M
 4.7422E+01    .0186   -M 0.000000  0.000000  0.000000   .399661  0.000000  0.000000  0.000000  0.000000  0.000000-M
                       +M 0.000000   .133432  0.000000  0.000000  0.000000  0.000000  0.000000   .906899         +M
 3.3748E+01    .0000   -M 0.000000  0.000000   .212539  0.000000  0.000000  0.000000  0.000000  0.000000   .953758-M
                       +M 0.000000  0.000000   .212539  0.000000  0.000000  0.000000  0.000000  0.000000         +M
-2.0520E+01   0.0000   -M 0.000000  0.000000  0.000000  0.000000  0.000000  -.707107  0.000000  0.000000  0.000000-M
                       +M 0.000000  0.000000  0.000000  0.000000  0.000000   .707107  0.000000  0.000000         +M
-3.4451E+01   -.2211   -M 0.000000  0.000000  0.000000  0.000000   .793181  0.000000  0.000000  0.000000  0.000000-M
                       +M  .031271  0.000000  0.000000  0.000000  0.000000  0.000000  -.608183  0.000000         +M
-3.4451E+01    .2211   -M  .031271  0.000000  0.000000  0.000000  0.000000  0.000000  -.608183  0.000000  0.000000-M
                       +M 0.000000  0.000000  0.000000  0.000000   .793181  0.000000  0.000000  0.000000         +M
-7.4111E+01   -.4293   -M 0.000000  0.000000  0.000000   .897537  0.000000  0.000000  0.000000  0.000000  0.000000-M
                       +M 0.000000  -.258007  0.000000  0.000000  0.000000  0.000000  0.000000  -.357575         +M
-7.4111E+01    .4293   -M 0.000000  -.258007  0.000000  0.000000  0.000000  0.000000  0.000000  -.357575  0.000000-M
                       +M 0.000000  0.000000  0.000000   .897537  0.000000  0.000000  0.000000  0.000000         +M
-1.1047E+02   -.7857   -M 0.000000   .956885  0.000000  0.000000  0.000000  0.000000  0.000000  -.222875  0.000000-M
                       +M 0.000000  0.000000  0.000000   .186274  0.000000  0.000000  0.000000  0.000000         +M
-1.1047E+02    .7857   -M 0.000000  0.000000  0.000000   .186274  0.000000  0.000000  0.000000  0.000000  0.000000-M
                       +M 0.000000   .956885  0.000000  0.000000  0.000000  0.000000  0.000000  -.222875         +M
-1.2240E+02   -.0000   -M 0.000000  0.000000  -.707107  0.000000  0.000000  0.000000  0.000000  0.000000   .000000-M
                       +M 0.000000  0.000000   .707107  0.000000  0.000000  0.000000  0.000000  0.000000         +M
-1.3791E+02    .0000   -M 0.000000  0.000000   .674409  0.000000  0.000000  0.000000  0.000000  0.000000  -.300575-M
                       +M 0.000000  0.000000   .674409  0.000000  0.000000  0.000000  0.000000  0.000000         +M
  ENERGY       HPAL        M=8       M=7       M=6       M=5       M=4       M=3       M=2       M=1        M=0
```

```
Y= .80  X= .60   OAS=3.7449E+02*W246   B4F4/B2F2= 2.40E+00  B4F4/B6F6= 1.50E+00  B2F2/B6F6= 6.25E-01   B4F4= 4.80E-01
  ENERGY       HPAL        M=8       M=7       M=6       M=5       M=4       M=3       M=2       M=1        M=0
 2.1632E+02    .9984   -M 0.000000  0.000000  0.000000  0.000000   .006202  0.000000  0.000000  0.000000  0.000000-M
                       +M  .998953  0.000000  0.000000  0.000000  0.000000  0.000000   .045336  0.000000         +M
 2.1632E+02   -.9984   -M  .998953  0.000000  0.000000  0.000000  0.000000  0.000000   .045336  0.000000  0.000000-M
                       +M 0.000000  0.000000  0.000000  0.000000   .006202  0.000000  0.000000  0.000000         +M
 8.2322E+01   -.0000   -M 0.000000  0.000000   .097391  0.000000  0.000000  0.000000  0.000000  0.000000   .990470-M
                       +M 0.000000  0.000000   .097391  0.000000  0.000000  0.000000  0.000000  0.000000         +M
 7.8560E+01   -.1117   -M 0.000000   .085973  0.000000  0.000000  0.000000  0.000000  0.000000   .983624  0.000000-M
                       +M 0.000000  0.000000  0.000000   .158407  0.000000  0.000000  0.000000  0.000000         +M
 7.8560E+01    .1117   -M 0.000000  0.000000  0.000000   .158407  0.000000  0.000000  0.000000  0.000000  0.000000-M
                       +M 0.000000   .085973  0.000000  0.000000  0.000000  0.000000  0.000000   .983624         +M
 6.5095E+01    .1775   -M 0.000000  0.000000  0.000000  0.000000   .314156  0.000000  0.000000  0.000000  0.000000-M
                       +M -.044988  0.000000  0.000000  0.000000  0.000000  0.000000   .948305  0.000000         +M
 6.5095E+01   -.1775   -M -.044988  0.000000  0.000000  0.000000  0.000000  0.000000   .948305  0.000000  0.000000-M
                       +M 0.000000  0.000000  0.000000  0.000000   .314156  0.000000  0.000000  0.000000         +M
 5.3640E+01   0.0000   -M 0.000000  0.000000  0.000000  0.000000  0.000000   .707107  0.000000  0.000000  0.000000-M
                       +M 0.000000  0.000000  0.000000  0.000000  0.000000   .707107  0.000000  0.000000         +M
-2.0280E+01   0.0000   -M 0.000000  0.000000  0.000000  0.000000  0.000000  -.707107  0.000000  0.000000  0.000000-M
                       +M 0.000000  0.000000  0.000000  0.000000  0.000000   .707107  0.000000  0.000000         +M
-5.2940E+01   -.4259   -M 0.000000  0.000000  0.000000  0.000000   .949351  0.000000  0.000000  0.000000  0.000000-M
                       +M  .008361  0.000000  0.000000  0.000000  0.000000  0.000000  -.314106  0.000000         +M
-5.2940E+01    .4259   -M  .008361  0.000000  0.000000  0.000000  0.000000  0.000000  -.314106  0.000000  0.000000-M
                       +M 0.000000  0.000000  0.000000  0.000000   .949351  0.000000  0.000000  0.000000         +M
-9.3766E+01    .8485   -M 0.000000  0.000000  0.000000  -.124179  0.000000  0.000000  0.000000  0.000000  0.000000-M
                       +M 0.000000   .990027  0.000000  0.000000  0.000000  0.000000  0.000000  -.066534         +M
-9.3766E+01   -.8485   -M 0.000000   .990027  0.000000  0.000000  0.000000  0.000000  0.000000  -.066534  0.000000-M
                       +M 0.000000  0.000000  0.000000  -.124179  0.000000  0.000000  0.000000  0.000000         +M
-1.1523E+02   -.5853   -M 0.000000  0.000000  0.000000   .979534  0.000000  0.000000  0.000000  0.000000  0.000000-M
                       +M 0.000000   .111606  0.000000  0.000000  0.000000  0.000000  0.000000  -.167503         +M
-1.1523E+02    .5853   -M 0.000000   .111606  0.000000  0.000000  0.000000  0.000000  0.000000  -.167503  0.000000-M
                       +M 0.000000  0.000000  0.000000   .979534  0.000000  0.000000  0.000000  0.000000         +M
-1.5360E+02    .0000   -M 0.000000  0.000000  -.707107  0.000000  0.000000  0.000000  0.000000  0.000000   .000000-M
                       +M 0.000000  0.000000   .707107  0.000000  0.000000  0.000000  0.000000  0.000000         +M
-1.5816E+02   -.0000   -M 0.000000  0.000000   .700368  0.000000  0.000000  0.000000  0.000000  0.000000  -.137732-M
                       +M 0.000000  0.000000   .700368  0.000000  0.000000  0.000000  0.000000  0.000000         +M
  ENERGY       HPAL        M=8       M=7       M=6       M=5       M=4       M=3       M=2       M=1        M=0
```

```
Y= .80  X= .80   OAS=4.4330E+02*W246   B4F4/B2F2= 3.20E+00  B4F4/B6F6= 4.00E+00  B2F2/B6F6= 1.25E+00   B4F4= 6.40E-01
  ENERGY       HPAL        M=8       M=7       M=6       M=5       M=4       M=3       M=2       M=1        M=0
 2.5767E+02    .9997   -M 0.000000  0.000000  0.000000  0.000000   .000988  0.000000  0.000000  0.000000  0.000000-M
                       +M  .999811  0.000000  0.000000  0.000000  0.000000  0.000000   .019419  0.000000         +M
 2.5767E+02   -.9997   -M  .999811  0.000000  0.000000  0.000000  0.000000  0.000000   .019419  0.000000  0.000000-M
                       +M 0.000000  0.000000  0.000000  0.000000   .000988  0.000000  0.000000  0.000000         +M
 1.3811E+02   0.0000   -M 0.000000  0.000000   .035873  0.000000  0.000000  0.000000  0.000000  0.000000   .998712-M
                       +M 0.000000  0.000000   .035873  0.000000  0.000000  0.000000  0.000000  0.000000         +M
 1.2176E+02   -.1239   -M 0.000000   .036979  0.000000  0.000000  0.000000  0.000000  0.000000   .997909  0.000000-M
                       +M 0.000000  0.000000  0.000000   .053010  0.000000  0.000000  0.000000  0.000000         +M
 1.2176E+02    .1239   -M 0.000000  0.000000  0.000000   .053010  0.000000  0.000000  0.000000  0.000000  0.000000-M
                       +M 0.000000   .036979  0.000000  0.000000  0.000000  0.000000  0.000000   .997909         +M
 7.4315E+01    .2416   -M 0.000000  0.000000  0.000000  0.000000   .107481  0.000000  0.000000  0.000000  0.000000-M
                       +M -.019413  0.000000  0.000000  0.000000  0.000000  0.000000   .994018  0.000000         +M
 7.4315E+01   -.2416   -M -.019413  0.000000  0.000000  0.000000  0.000000  0.000000   .994018  0.000000  0.000000-M
                       +M 0.000000  0.000000  0.000000  0.000000   .107481  0.000000  0.000000  0.000000         +M
 1.6920E+01   0.0000   -M 0.000000  0.000000  0.000000  0.000000  0.000000   .707107  0.000000  0.000000  0.000000-M
                       +M 0.000000  0.000000  0.000000  0.000000  0.000000   .707107  0.000000  0.000000         +M
-2.0040E+01   0.0000   -M 0.000000  0.000000  0.000000  0.000000  0.000000  -.707107  0.000000  0.000000  0.000000-M
                       +M 0.000000  0.000000  0.000000  0.000000  0.000000   .707107  0.000000  0.000000         +M
-8.0555E+01   -.8739   -M 0.000000   .999305  0.000000  0.000000  0.000000  0.000000  0.000000  -.036677  0.000000-M
                       +M 0.000000  0.000000  0.000000  -.006665  0.000000  0.000000  0.000000  0.000000         +M
-8.0555E+01    .8739   -M 0.000000  0.000000  0.000000  -.006665  0.000000  0.000000  0.000000  0.000000  0.000000-M
                       +M 0.000000   .999305  0.000000  0.000000  0.000000  0.000000  0.000000  -.036677         +M
-9.0545E+01   -.4913   -M 0.000000  0.000000  0.000000  0.000000   .994207  0.000000  0.000000  0.000000  0.000000-M
                       +M  .001105  0.000000  0.000000  0.000000  0.000000  0.000000  -.107480  0.000000         +M
-9.0545E+01    .4913   -M  .001105  0.000000  0.000000  0.000000  0.000000  0.000000  -.107480  0.000000  0.000000-M
                       +M 0.000000  0.000000  0.000000  0.000000   .994207  0.000000  0.000000  0.000000         +M
-1.6493E+02   -.6228   -M 0.000000  0.000000  0.000000   .998572  0.000000  0.000000  0.000000  0.000000  0.000000-M
                       +M 0.000000   .004707  0.000000  0.000000  0.000000  0.000000  0.000000  -.053220         +M
-1.6493E+02    .6228   -M 0.000000   .004707  0.000000  0.000000  0.000000  0.000000  0.000000  -.053220  0.000000-M
                       +M 0.000000  0.000000  0.000000   .998572  0.000000  0.000000  0.000000  0.000000         +M
-1.8480E+02   -.0000   -M 0.000000  0.000000  -.707107  0.000000  0.000000  0.000000  0.000000  0.000000   .000000-M
                       +M 0.000000  0.000000   .707107  0.000000  0.000000  0.000000  0.000000  0.000000         +M
-1.8563E+02    .0000   -M 0.000000  0.000000   .706196  0.000000  0.000000  0.000000  0.000000  0.000000  -.050732-M
                       +M 0.000000  0.000000   .706196  0.000000  0.000000  0.000000  0.000000  0.000000         +M
  ENERGY       HPAL        M=8       M=7       M=6       M=5       M=4       M=3       M=2       M=1        M=0
```

RESULTS FOR $J = 15/2$

Y=-.80 X=-.80 OAS=3.1025E+02*W246 B4F4/B2F2= 3.20E+00 B4F4/B6F6=-4.00E+00 B2F2/B6F6=-1.25E+00

ENERGY	HEPAL	HEPER	HE		M=15/2	M=13/2	M=11/2	M=9/2	M=7/2	M=5/2	M=3/2	M=1/2	
1.7141E+02	.9992	0.0000	.9992	-M	0.000000	0.000000	0.000000	.001153	0.000000	0.000000	0.000000	0.000000	-M
				+M	.999486	0.000000	0.000000	0.000000	0.000000	0.000000	-.032050	0.000000	+M
1.7141E+02	-.9992	0.0000	.9992	-M	.999486	0.000000	0.000000	0.000000	0.000000	0.000000	-.032050	0.000000	-M
				+M	0.000000	0.000000	0.000000	.001153	0.000000	0.000000	0.000000	0.000000	+M
1.2926E+02	.0668	.5329	.5371	-M	0.000000	0.000000	-.033342	0.000000	0.000000	0.000000	0.000000	0.000000	-M
				+M	0.000000	-.035197	0.000000	0.000000	0.000000	0.000000	0.000000	.998824	+M
1.2926E+02	-.0668	.5329	.5371	-M	0.000000	-.035197	0.000000	0.000000	0.000000	0.000000	0.000000	.998824	-M
				+M	0.000000	0.000000	-.033342	0.000000	0.000000	0.000000	0.000000	0.000000	+M
8.3230E+01	.1988	0.0000	.1988	-M	0.000000	0.000000	0.000000	-.050219	0.000000	0.000000	0.000000	0.000000	-M
				+M	.032067	0.000000	0.000000	0.000000	0.000000	0.000000	.998223	0.000000	+M
8.3230E+01	-.1988	0.0000	.1988	-M	.032067	0.000000	0.000000	0.000000	0.000000	0.000000	.998223	0.000000	-M
				+M	0.000000	0.000000	0.000000	-.050219	0.000000	0.000000	0.000000	0.000000	+M
6.3110E+00	.3171	-.1393	.3464	-M	0.000000	0.000000	0.000000	0.000000	-.142290	0.000000	0.000000	0.000000	-M
				+M	0.000000	0.000000	0.000000	0.000000	0.000000	.989825	0.000000	0.000000	+M
6.3110E+00	-.3171	-.1393	.3464	-M	0.000000	0.000000	0.000000	0.000000	0.000000	.989825	0.000000	0.000000	-M
				+M	0.000000	0.000000	0.000000	0.000000	-.142290	0.000000	0.000000	0.000000	+M
-3.5524E+01	-.8657	-.0016	.8657	-M	0.000000	.999378	0.000000	0.000000	0.000000	0.000000	0.000000	.035112	-M
				+M	0.000000	0.000000	-.003145	0.000000	0.000000	0.000000	0.000000	0.000000	+M
-3.5524E+01	.8657	-.0016	.8657	-M	0.000000	0.000000	-.003145	0.000000	0.000000	0.000000	0.000000	0.000000	-M
				+M	0.000000	.999378	0.000000	0.000000	0.000000	0.000000	0.000000	.035112	+M
-8.2151E+01	-.4505	.1393	.4715	-M	0.000000	0.000000	0.000000	0.000000	.989825	0.000000	0.000000	0.000000	-M
				+M	0.000000	0.000000	0.000000	0.000000	0.000000	.142290	0.000000	0.000000	+M
-8.2151E+01	.4505	.1393	.4715	-M	0.000000	0.000000	0.000000	0.000000	0.000000	.142290	0.000000	0.000000	-M
				+M	0.000000	0.000000	0.000000	0.000000	.989825	0.000000	0.000000	0.000000	+M
-1.3369E+02	-.7324	.0020	.7324	-M	0.000000	0.000000	.999439	0.000000	0.000000	0.000000	0.000000	0.000000	-M
				+M	0.000000	.001971	0.000000	0.000000	0.000000	0.000000	0.000000	.033432	+M
-1.3369E+02	.7324	.0020	.7324	-M	0.000000	.001971	0.000000	0.000000	0.000000	0.000000	0.000000	.033432	-M
				+M	0.000000	0.000000	.999439	0.000000	0.000000	0.000000	0.000000	0.000000	+M
-1.3884E+02	.5980	0.0000	.5980	-M	.000458	0.000000	0.000000	0.000000	0.000000	0.000000	.050230	0.000000	-M
				+M	0.000000	0.000000	0.000000	.998738	0.000000	0.000000	0.000000	0.000000	+M
-1.3884E+02	-.5980	0.0000	.5980	-M	0.000000	0.000000	0.000000	.998738	0.000000	0.000000	0.000000	0.000000	-M
				+M	.000458	0.000000	0.000000	0.000000	0.000000	0.000000	.050230	0.000000	+M
ENERGY	HEPAL	HEPER	HE		M=15/2	M=13/2	M=11/2	M=9/2	M=7/2	M=5/2	M=3/2	M=1/2	

Y=-.80 X=-.60 OAS=2.3614E+02*W246 B4F4/B2F2= 2.40E+00 B4F4/B6F6=-1.50E+00 B2F2/B6F6=-6.25E-01

ENERGY	HEPAL	HEPER	HE		M=15/2	M=13/2	M=11/2	M=9/2	M=7/2	M=5/2	M=3/2	M=1/2	
1.1789E+02	.9895	0.0000	.9895	-M	0.000000	0.000000	0.000000	.010815	0.000000	0.000000	0.000000	0.000000	-M
				+M	.993504	0.000000	0.000000	0.000000	0.000000	0.000000	-.113280	0.000000	+M
1.1789E+02	-.9895	0.0000	.9895	-M	.993504	0.000000	0.000000	0.000000	0.000000	0.000000	-.113280	0.000000	-M
				+M	0.000000	0.000000	0.000000	.010815	0.000000	0.000000	0.000000	0.000000	+M
1.1319E+02	-.0689	.5302	.5346	-M	0.000000	-.099794	0.000000	0.000000	0.000000	0.000000	0.000000	.991387	-M
				+M	0.000000	0.000000	-.084809	0.000000	0.000000	0.000000	0.000000	0.000000	+M
1.1319E+02	.0689	.5302	.5346	-M	0.000000	0.000000	-.084809	0.000000	0.000000	0.000000	0.000000	0.000000	-M
				+M	0.000000	-.099794	0.000000	0.000000	0.000000	0.000000	0.000000	.991387	+M
6.8171E+01	-.1989	0.0000	.1989	-M	.113758	0.000000	0.000000	0.000000	0.000000	0.000000	.986278	0.000000	-M
				+M	0.000000	0.000000	0.000000	-.119642	0.000000	0.000000	0.000000	0.000000	+M
6.8171E+01	.1989	0.0000	.1989	-M	0.000000	0.000000	0.000000	-.119642	0.000000	0.000000	0.000000	0.000000	-M
				+M	.113758	0.000000	0.000000	0.000000	0.000000	0.000000	.986278	0.000000	+M
1.4441E+00	.2619	-.2820	.3849	-M	0.000000	0.000000	0.000000	0.000000	-.298872	0.000000	0.000000	0.000000	-M
				+M	0.000000	0.000000	0.000000	0.000000	0.000000	.954293	0.000000	0.000000	+M
1.4441E+00	-.2619	-.2820	.3849	-M	0.000000	0.000000	0.000000	0.000000	0.000000	.954293	0.000000	0.000000	-M
				+M	0.000000	0.000000	0.000000	0.000000	-.298872	0.000000	0.000000	0.000000	+M
-3.1881E+00	.8583	-.0085	.8583	-M	0.000000	0.000000	-.019490	0.000000	0.000000	0.000000	0.000000	0.000000	-M
				+M	0.000000	.994948	0.000000	0.000000	0.000000	0.000000	0.000000	.098485	+M
-3.1881E+00	-.8583	-.0085	.8583	-M	0.000000	.994948	0.000000	0.000000	0.000000	0.000000	0.000000	.098485	-M
				+M	0.000000	0.000000	-.019490	0.000000	0.000000	0.000000	0.000000	0.000000	+M
-8.5924E+01	-.3952	.2820	.4855	-M	0.000000	0.000000	0.000000	0.000000	.954293	0.000000	0.000000	0.000000	-M
				+M	0.000000	0.000000	0.000000	0.000000	0.000000	.298872	0.000000	0.000000	+M
-8.5924E+01	.3952	.2820	.4855	-M	0.000000	0.000000	0.000000	0.000000	0.000000	.298872	0.000000	0.000000	-M
				+M	0.000000	0.000000	0.000000	0.000000	.954293	0.000000	0.000000	0.000000	+M
-9.3320E+01	-.7272	.0117	.7273	-M	0.000000	0.000000	.996207	0.000000	0.000000	0.000000	0.000000	0.000000	-M
				+M	0.000000	.010969	0.000000	0.000000	0.000000	0.000000	0.000000	.086326	+M
-9.3320E+01	.7272	.0117	.7273	-M	0.000000	.010969	0.000000	0.000000	0.000000	0.000000	0.000000	.086326	-M
				+M	0.000000	0.000000	.996207	0.000000	0.000000	0.000000	0.000000	0.000000	+M
-1.1826E+02	-.5884	0.0000	.5884	-M	0.000000	0.000000	0.000000	.992758	0.000000	0.000000	0.000000	0.000000	-M
				+M	.002886	0.000000	0.000000	0.000000	0.000000	0.000000	.120095	0.000000	+M
-1.1826E+02	.5884	0.0000	.5884	-M	.002886	0.000000	0.000000	0.000000	0.000000	0.000000	.120095	0.000000	-M
				+M	0.000000	0.000000	0.000000	.992758	0.000000	0.000000	0.000000	0.000000	+M
ENERGY	HEPAL	HEPER	HE		M=15/2	M=13/2	M=11/2	M=9/2	M=7/2	M=5/2	M=3/2	M=1/2	

Y=-.80 X=-.40 OAS=2.0150E+02*W246 B4F4/B2F2= 1.60E+00 B4F4/B6F6=-6.67E-01 B2F2/B6F6=-4.17E-01

ENERGY	HEPAL	HEPER	HE		M=15/2	M=13/2	M=11/2	M=9/2	M=7/2	M=5/2	M=3/2	M=1/2	
1.0118E+02	-.0899	.5163	.5241	-M	0.000000	-.238047	0.000000	0.000000	0.000000	0.000000	0.000000	.956915	-M
				+M	0.000000	0.000000	-.166275	0.000000	0.000000	0.000000	0.000000	0.000000	+M
1.0118E+02	.0899	.5163	.5241	-M	0.000000	0.000000	-.166275	0.000000	0.000000	0.000000	0.000000	0.000000	-M
				+M	0.000000	-.238047	0.000000	0.000000	0.000000	0.000000	0.000000	.956915	+M
6.8914E+01	.7303	0.0000	.7303	-M	0.000000	0.000000	0.000000	.115192	0.000000	0.000000	0.000000	0.000000	-M
				+M	.822254	0.000000	0.000000	0.000000	0.000000	0.000000	-.557342	0.000000	+M
6.8914E+01	-.7303	0.0000	.7303	-M	.822254	0.000000	0.000000	0.000000	0.000000	0.000000	-.557342	0.000000	-M
				+M	0.000000	0.000000	0.000000	.115192	0.000000	0.000000	0.000000	0.000000	+M
5.1209E+01	-.4314	0.0000	.4314	-M	.569008	0.000000	0.000000	0.000000	0.000000	0.000000	.801041	0.000000	-M
				+M	0.000000	0.000000	0.000000	-.185910	0.000000	0.000000	0.000000	0.000000	+M
5.1209E+01	.4314	0.0000	.4314	-M	0.000000	0.000000	0.000000	-.185910	0.000000	0.000000	0.000000	0.000000	-M
				+M	.569008	0.000000	0.000000	0.000000	0.000000	0.000000	.801041	0.000000	+M
2.7153E+01	.8155	-.0254	.8159	-M	0.000000	0.000000	-.077523	0.000000	0.000000	0.000000	0.000000	0.000000	-M
				+M	0.000000	.970576	0.000000	0.000000	0.000000	0.000000	0.000000	.227975	+M
2.7153E+01	-.8155	-.0254	.8159	-M	0.000000	.970576	0.000000	0.000000	0.000000	0.000000	0.000000	.227975	-M
				+M	0.000000	0.000000	-.077523	0.000000	0.000000	0.000000	0.000000	0.000000	+M
9.4527E-01	-.1802	-.3890	.4287	-M	0.000000	0.000000	0.000000	0.000000	0.000000	.899220	0.000000	0.000000	-M
				+M	0.000000	0.000000	0.000000	0.000000	-.437496	0.000000	0.000000	0.000000	+M
9.4527E-01	.1802	-.3890	.4287	-M	0.000000	0.000000	0.000000	0.000000	-.437496	0.000000	0.000000	0.000000	-M
				+M	0.000000	0.000000	0.000000	0.000000	0.000000	.899220	0.000000	0.000000	+M
-5.5012E+01	-.7054	.0424	.7066	-M	0.000000	0.000000	.983027	0.000000	0.000000	0.000000	0.000000	0.000000	-M
				+M	0.000000	.036276	0.000000	0.000000	0.000000	0.000000	0.000000	.179837	+M
-5.5012E+01	.7054	.0424	.7066	-M	0.000000	.036276	0.000000	0.000000	0.000000	0.000000	0.000000	.179837	-M
				+M	0.000000	0.000000	.983027	0.000000	0.000000	0.000000	0.000000	0.000000	+M
-9.4065E+01	.3135	.3890	.4996	-M	0.000000	0.000000	0.000000	0.000000	0.000000	.437496	0.000000	0.000000	-M
				+M	0.000000	0.000000	0.000000	0.000000	.899220	0.000000	0.000000	0.000000	+M
-9.4065E+01	-.3135	.3890	.4996	-M	0.000000	0.000000	0.000000	0.000000	.899220	0.000000	0.000000	0.000000	-M
				+M	0.000000	0.000000	0.000000	0.000000	0.000000	.437496	0.000000	0.000000	+M
-1.0032E+02	.5616	0.0000	.5616	-M	.011342	0.000000	0.000000	0.000000	0.000000	0.000000	.218410	0.000000	-M
				+M	0.000000	0.000000	0.000000	.975791	0.000000	0.000000	0.000000	0.000000	+M
-1.0032E+02	-.5616	0.0000	.5616	-M	0.000000	0.000000	0.000000	.975791	0.000000	0.000000	0.000000	0.000000	-M
				+M	.011342	0.000000	0.000000	0.000000	0.000000	0.000000	.218410	0.000000	+M
ENERGY	HEPAL	HEPER	HE		M=15/2	M=13/2	M=11/2	M=9/2	M=7/2	M=5/2	M=3/2	M=1/2	

Y=-.80 X=-.20 OAS=2.0602E+02*W246 B4F4/B2F2= 8.00E-01 B4F4/B6F6=-2.50E-01 B2F2/B6F6=-3.12E-01

ENERGY	HEPAL	HEPER	HE		M=15/2	M=13/2	M=11/2	M=9/2	M=7/2	M=5/2	M=3/2	M=1/2
1.0035E+02	.2324	.4468	.5036	-M	0.000000	0.000000	-.260861	0.000000	0.000000	0.000000	0.000000	0.000000-M
				+M	0.000000	-.524667	0.000000	0.000000	0.000000	0.000000	0.000000	.810356+M
1.0035E+02	-.2324	.4468	.5036	-M	0.000000	-.524667	0.000000	0.000000	0.000000	0.000000	0.000000	.810356-M
				+M	0.000000	0.000000	-.260861	0.000000	0.000000	0.000000	0.000000	0.000000+M
5.2080E+01	-.1576	0.0000	.1576	-M	-.240071	0.000000	0.000000	0.000000	0.000000	0.000000	.911976	0.000000-M
				+M	0.000000	0.000000	0.000000	-.332666	0.000000	0.000000	0.000000	0.000000+M
5.2080E+01	.1576	0.0000	.1576	-M	0.000000	0.000000	0.000000	-.332666	0.000000	0.000000	0.000000	0.000000-M
				+M	-.240071	0.000000	0.000000	0.000000	0.000000	0.000000	.911976	0.000000+M
5.1864E+01	-.5817	-.0462	.5835	-M	0.000000	.845817	0.000000	0.000000	0.000000	0.000000	0.000000	.461461-M
				+M	0.000000	0.000000	-.267669	0.000000	0.000000	0.000000	0.000000	0.000000+M
5.1864E+01	.5817	-.0462	.5835	-M	0.000000	0.000000	-.267669	0.000000	0.000000	0.000000	0.000000	0.000000-M
				+M	0.000000	.845817	0.000000	0.000000	0.000000	0.000000	0.000000	.461461+M
6.6212E+00	-.9405	0.0000	.9405	-M	.969867	0.000000	0.000000	0.000000	0.000000	0.000000	.210673	0.000000-M
				+M	0.000000	0.000000	0.000000	-.122370	0.000000	0.000000	0.000000	0.000000+M
6.6212E+00	.9405	0.0000	.9405	-M	0.000000	0.000000	0.000000	-.122370	0.000000	0.000000	0.000000	0.000000-M
				+M	.969867	0.000000	0.000000	0.000000	0.000000	0.000000	.210673	0.000000+M
3.9080E+00	-.0995	-.4497	.4606	-M	0.000000	0.000000	0.000000	0.000000	0.000000	.841255	0.000000	0.000000-M
				+M	0.000000	0.000000	0.000000	0.000000	-.540639	0.000000	0.000000	0.000000+M
3.9080E+00	.0995	-.4497	.4606	-M	0.000000	0.000000	0.000000	0.000000	-.540639	0.000000	0.000000	0.000000-M
				+M	0.000000	0.000000	0.000000	0.000000	0.000000	.841255	0.000000	0.000000+M
-2.2253E+01	-.6141	.1327	.6283	-M	0.000000	0.000000	.927526	0.000000	0.000000	0.000000	0.000000	0.000000-M
				+M	0.000000	.096530	0.000000	0.000000	0.000000	0.000000	0.000000	.361077+M
-2.2253E+01	.6141	.1327	.6283	-M	0.000000	.096530	0.000000	0.000000	0.000000	0.000000	0.000000	.361077-M
				+M	0.000000	0.000000	.927526	0.000000	0.000000	0.000000	0.000000	0.000000+M
-8.6901E+01	.4981	0.0000	.4981	-M	.041515	0.000000	0.000000	0.000000	0.000000	0.000000	.352019	0.000000-M
				+M	0.000000	0.000000	0.000000	.935072	0.000000	0.000000	0.000000	0.000000+M
-8.6901E+01	-.4981	0.0000	.4981	-M	0.000000	0.000000	0.000000	.935072	0.000000	0.000000	0.000000	0.000000-M
				+M	.041515	0.000000	0.000000	0.000000	0.000000	0.000000	.352019	0.000000+M
-1.0567E+02	-.2328	.4497	.5064	-M	0.000000	0.000000	0.000000	0.000000	.841255	0.000000	0.000000	0.000000-M
				+M	0.000000	0.000000	0.000000	0.000000	0.000000	.540639	0.000000	0.000000+M
-1.0567E+02	.2328	.4497	.5064	-M	0.000000	0.000000	0.000000	0.000000	0.000000	.540639	0.000000	0.000000-M
				+M	0.000000	0.000000	0.000000	0.000000	.841255	0.000000	0.000000	0.000000+M
ENERGY	HEPAL	HEPER	HE		M=15/2	M=13/2	M=11/2	M=9/2	M=7/2	M=5/2	M=3/2	M=1/2

Y=-.80 X=0.00 OAS=2.3819E+02*W246 B4F4/B2F2= 0. B4F4/B6F6= 0. B2F2/B6F6=-2.50E-01

ENERGY	HEPAL	HEPER	HE		M=15/2	M=13/2	M=11/2	M=9/2	M=7/2	M=5/2	M=3/2	M=1/2
1.1862E+02	.4857	.3237	.5837	-M	0.000000	0.000000	.286166	0.000000	0.000000	0.000000	0.000000	0.000000-M
				+M	0.000000	.778281	0.000000	0.000000	0.000000	0.000000	0.000000	-.558916+M
1.1862E+02	-.4857	.3237	.5837	-M	0.000000	.778281	0.000000	0.000000	0.000000	0.000000	0.000000	-.558916-M
				+M	0.000000	0.000000	.286166	0.000000	0.000000	0.000000	0.000000	0.000000+M
7.2139E+01	.0717	-.1018	.1246	-M	0.000000	0.000000	-.596046	0.000000	0.000000	0.000000	0.000000	0.000000-M
				+M	0.000000	.601305	0.000000	0.000000	0.000000	0.000000	0.000000	.532130+M
7.2139E+01	-.0717	-.1018	.1246	-M	0.000000	.601305	0.000000	0.000000	0.000000	0.000000	0.000000	.532130-M
				+M	0.000000	0.000000	-.596046	0.000000	0.000000	0.000000	0.000000	0.000000+M
5.0135E+01	-.0195	0.0000	.0195	-M	-.128076	0.000000	0.000000	0.000000	0.000000	0.000000	.861150	0.000000-M
				+M	0.000000	0.000000	0.000000	-.491952	0.000000	0.000000	0.000000	0.000000+M
5.0135E+01	.0195	0.0000	.0195	-M	0.000000	0.000000	0.000000	-.491952	0.000000	0.000000	0.000000	0.000000-M
				+M	-.128076	0.000000	0.000000	0.000000	0.000000	0.000000	.861150	0.000000+M
9.1680E+00	.0340	-.4785	.4797	-M	0.000000	0.000000	0.000000	0.000000	-.611687	0.000000	0.000000	0.000000-M
				+M	0.000000	0.000000	0.000000	0.000000	0.000000	.791100	0.000000	0.000000+M
9.1680E+00	-.0340	-.4785	.4797	-M	0.000000	0.000000	0.000000	0.000000	0.000000	.791100	0.000000	0.000000-M
				+M	0.000000	0.000000	0.000000	0.000000	-.611687	0.000000	0.000000	0.000000+M
-4.1628E+00	-.3574	.3114	.4741	-M	0.000000	0.000000	.750226	0.000000	0.000000	0.000000	0.000000	0.000000-M
				+M	0.000000	.180862	0.000000	0.000000	0.000000	0.000000	0.000000	.635964+M
-4.1628E+00	.3574	.3114	.4741	-M	0.000000	.180862	0.000000	0.000000	0.000000	0.000000	0.000000	.635964-M
				+M	0.000000	0.000000	.750226	0.000000	0.000000	0.000000	0.000000	0.000000+M
-4.5140E+01	-.9107	0.0000	.9107	-M	.971659	0.000000	0.000000	0.000000	0.000000	0.000000	.009582	0.000000-M
				+M	0.000000	0.000000	0.000000	-.236192	0.000000	0.000000	0.000000	0.000000+M
-4.5140E+01	.9107	0.0000	.9107	-M	0.000000	0.000000	0.000000	-.236192	0.000000	0.000000	0.000000	0.000000-M
				+M	.971659	0.000000	0.000000	0.000000	0.000000	0.000000	.009582	0.000000+M
-8.1196E+01	-.3302	0.0000	.3302	-M	0.000000	0.000000	0.000000	.837972	0.000000	0.000000	0.000000	0.000000-M
				+M	.198683	0.000000	0.000000	0.000000	0.000000	0.000000	.508260	0.000000+M
-8.1196E+01	.3302	0.0000	.3302	-M	.198683	0.000000	0.000000	0.000000	0.000000	0.000000	.508260	0.000000-M
				+M	0.000000	0.000000	0.000000	.837972	0.000000	0.000000	0.000000	0.000000+M
-1.1957E+02	-.1673	.4785	.5069	-M	0.000000	0.000000	0.000000	0.000000	.791100	0.000000	0.000000	0.000000-M
				+M	0.000000	0.000000	0.000000	0.000000	0.000000	.611687	0.000000	0.000000+M
-1.1957E+02	.1673	.4785	.5069	-M	0.000000	0.000000	0.000000	0.000000	0.000000	.611687	0.000000	0.000000-M
				+M	0.000000	0.000000	0.000000	0.000000	.791100	0.000000	0.000000	0.000000+M
ENERGY	HEPAL	HEPER	HE		M=15/2	M=13/2	M=11/2	M=9/2	M=7/2	M=5/2	M=3/2	M=1/2

Y=-.80 X= .20 OAS=1.9109E+02*W246 B4F4/B2F2=-8.00E-01 B4F4/B6F6= 2.50E-01 B2F2/B6F6=-3.12E-01

ENERGY	HEPAL	HEPER	HE		M=15/2	M=13/2	M=11/2	M=9/2	M=7/2	M=5/2	M=3/2	M=1/2
1.0277E+02	-.6239	.2536	.6735	-M	0.000000	.887307	0.000000	0.000000	0.000000	0.000000	0.000000	-.349176-M
				+M	0.000000	0.000000	.301268	0.000000	0.000000	0.000000	0.000000	0.000000+M
1.0277E+02	.6239	.2536	.6735	-M	0.000000	0.000000	.301268	0.000000	0.000000	0.000000	0.000000	0.000000-M
				+M	0.000000	.887307	0.000000	0.000000	0.000000	0.000000	0.000000	-.349176+M
7.5534E+01	.3718	-.1931	.4189	-M	0.000000	-.416185	0.000000	0.000000	0.000000	0.000000	0.000000	-.324831-M
				+M	0.000000	0.000000	.849279	0.000000	0.000000	0.000000	0.000000	0.000000+M
7.5534E+01	-.3718	-.1931	.4189	-M	0.000000	0.000000	.849279	0.000000	0.000000	0.000000	0.000000	0.000000-M
				+M	0.000000	-.416185	0.000000	0.000000	0.000000	0.000000	0.000000	-.324831+M
3.8013E+01	-.2013	0.0000	.2013	-M	0.000000	0.000000	0.000000	.711526	0.000000	0.000000	0.000000	0.000000-M
				+M	.068072	0.000000	0.000000	0.000000	0.000000	0.000000	-.699355	0.000000+M
3.8013E+01	.2013	0.0000	.2013	-M	.068072	0.000000	0.000000	0.000000	0.000000	0.000000	-.699355	0.000000-M
				+M	0.000000	0.000000	0.000000	.711526	0.000000	0.000000	0.000000	0.000000+M
1.1522E+01	-.0433	-.4936	.4955	-M	0.000000	0.000000	0.000000	0.000000	-.686116	0.000000	0.000000	0.000000-M
				+M	0.000000	0.000000	0.000000	0.000000	0.000000	.727493	0.000000	0.000000+M
1.1522E+01	.0433	-.4936	.4955	-M	0.000000	0.000000	0.000000	0.000000	0.000000	.727493	0.000000	0.000000-M
				+M	0.000000	0.000000	0.000000	0.000000	-.686116	0.000000	0.000000	0.000000+M
-8.9831E+00	-.0521	.4728	.4757	-M	0.000000	0.000000	.433547	0.000000	0.000000	0.000000	0.000000	0.000000-M
				+M	0.000000	.198686	0.000000	0.000000	0.000000	0.000000	0.000000	.878955+M
-8.9831E+00	.0521	.4728	.4757	-M	0.000000	.198686	0.000000	0.000000	0.000000	0.000000	0.000000	.878955-M
				+M	0.000000	0.000000	.433547	0.000000	0.000000	0.000000	0.000000	0.000000+M
-4.9480E+01	-.1231	0.0000	.1231	-M	0.000000	0.000000	0.000000	.688662	0.000000	0.000000	0.000000	0.000000-M
				+M	-.265232	0.000000	0.000000	0.000000	0.000000	0.000000	.674831	0.000000+M
-4.9480E+01	.1231	0.0000	.1231	-M	-.265232	0.000000	0.000000	0.000000	0.000000	0.000000	.674831	0.000000-M
				+M	0.000000	0.000000	0.000000	.688662	0.000000	0.000000	0.000000	0.000000+M
-8.1053E+01	-.9244	0.0000	.9244	-M	.961779	0.000000	0.000000	0.000000	0.000000	0.000000	.235598	0.000000-M
				+M	0.000000	0.000000	0.000000	.139555	0.000000	0.000000	0.000000	0.000000+M
-8.1053E+01	.9244	0.0000	.9244	-M	0.000000	0.000000	0.000000	.139555	0.000000	0.000000	0.000000	0.000000-M
				+M	.961779	0.000000	0.000000	0.000000	0.000000	0.000000	.235598	0.000000+M
-8.8322E+01	-.0901	.4936	.5017	-M	0.000000	0.000000	0.000000	0.000000	.727493	0.000000	0.000000	0.000000-M
				+M	0.000000	0.000000	0.000000	0.000000	0.000000	.686116	0.000000	0.000000+M
-8.8322E+01	.0901	.4936	.5017	-M	0.000000	0.000000	0.000000	0.000000	0.000000	.686116	0.000000	0.000000-M
				+M	0.000000	0.000000	0.000000	0.000000	.727493	0.000000	0.000000	0.000000+M
ENERGY	HEPAL	HEPER	HE		M=15/2	M=13/2	M=11/2	M=9/2	M=7/2	M=5/2	M=3/2	M=1/2

Y=-.80 X= .40 OAS=2.1116E+02*W246 B4F4/B2F2=-1.60E+00 B4F4/B6F6= 6.67E-01 B2F2/B6F6=-4.17E-01

ENERGY	HEPAL	HEPER	HE		M=15/2	M=13/2	M=11/2	M=9/2	M=7/2	M=5/2	M=3/2	M=1/2
9.8595E+01	-.3531	.3069	.4678	-M	0.000000	0.000000	.856531	0.000000	0.000000	0.000000	0.000000	0.000000-M
				+M	0.000000	.457153	0.000000	0.000000	0.000000	0.000000	0.000000	-.239510+M
9.8595E+01	.3531	.3069	.4678	-M	0.000000	.457153	0.000000	0.000000	0.000000	0.000000	0.000000	-.239510-M
				+M	0.000000	0.000000	.856531	0.000000	0.000000	0.000000	0.000000	0.000000+M
8.9944E+01	.5040	-.2949	.5839	-M	0.000000	0.000000	-.475823	0.000000	0.000000	0.000000	0.000000	0.000000-M
				+M	0.000000	.879228	0.000000	0.000000	0.000000	0.000000	0.000000	-.023445+M
8.9944E+01	-.5040	-.2949	.5839	-M	0.000000	.879228	0.000000	0.000000	0.000000	0.000000	0.000000	-.023445-M
				+M	0.000000	0.000000	-.475823	0.000000	0.000000	0.000000	0.000000	0.000000+M
4.9421E+01	-.4855	0.0000	.4855	-M	0.000000	0.000000	0.000000	.925900	0.000000	0.000000	0.000000	0.000000-M
				+M	.019894	0.000000	0.000000	0.000000	0.000000	0.000000	-.377245	0.000000+M
4.9421E+01	.4855	0.0000	.4855	-M	.019894	0.000000	0.000000	0.000000	0.000000	0.000000	-.377245	0.000000-M
				+M	0.000000	0.000000	0.000000	.925900	0.000000	0.000000	0.000000	0.000000+M
1.7187E+01	.1735	-.4765	.5071	-M	0.000000	0.000000	0.000000	0.000000	0.000000	-.605351	0.000000	0.000000-M
				+M	0.000000	0.000000	0.000000	0.000000	.795959	0.000000	0.000000	0.000000+M
1.7187E+01	-.1735	-.4765	.5071	-M	0.000000	0.000000	0.000000	0.000000	.795959	0.000000	0.000000	0.000000-M
				+M	0.000000	0.000000	0.000000	0.000000	0.000000	-.605351	0.000000	0.000000+M
-3.6499E+01	.0491	.5213	.5237	-M	0.000000	0.000000	.199866	0.000000	0.000000	0.000000	0.000000	0.000000-M
				+M	0.000000	.134046	0.000000	0.000000	0.000000	0.000000	0.000000	.970611+M
-3.6499E+01	-.0491	.5213	.5237	-M	0.000000	.134046	0.000000	0.000000	0.000000	0.000000	0.000000	.970611-M
				+M	0.000000	0.000000	.199866	0.000000	0.000000	0.000000	0.000000	0.000000+M
-4.5700E+01	-.0976	0.0000	.0976	-M	-.118419	0.000000	0.000000	0.000000	0.000000	0.000000	.918675	0.000000-M
				+M	0.000000	0.000000	0.000000	.376846	0.000000	0.000000	0.000000	0.000000+M
-4.5700E+01	.0976	0.0000	.0976	-M	0.000000	0.000000	0.000000	.376846	0.000000	0.000000	0.000000	0.000000-M
				+M	-.118419	0.000000	0.000000	0.000000	0.000000	0.000000	.918675	0.000000+M
-6.0387E+01	-.0402	.4765	.4781	-M	0.000000	0.000000	0.000000	0.000000	0.000000	.795959	0.000000	0.000000-M
				+M	0.000000	0.000000	0.000000	0.000000	.605351	0.000000	0.000000	0.000000+M
-6.0387E+01	.0402	.4765	.4781	-M	0.000000	0.000000	0.000000	0.000000	.605351	0.000000	0.000000	0.000000-M
				+M	0.000000	0.000000	0.000000	0.000000	0.000000	.795959	0.000000	0.000000+M
-1.1256E+02	-.9879	0.0000	.9879	-M	.992764	0.000000	0.000000	0.000000	0.000000	0.000000	.117141	0.000000-
				+M	0.000000	0.000000	0.000000	.026396	0.000000	0.000000	0.000000	0.000000+M
-1.1256E+02	.9879	0.0000	.9879	-M	0.000000	0.000000	0.000000	.026396	0.000000	0.000000	0.000000	0.000000-
				+M	.992764	0.000000	0.000000	0.000000	0.000000	0.000000	.117141	0.000000+M
ENERGY	HEPAL	HEPER	HE		M=15/2	M=13/2	M=11/2	M=9/2	M=7/2	M=5/2	M=3/2	M=1/2

Y=-.80 X= .60 OAS=2.6720E+02*W246 B4F4/B2F2=-2.40E+00 B4F4/B6F6= 1.50E+00 B2F2/B6F6=-6.25E-01

ENERGY	HEPAL	HEPER	HE		M=15/2	M=13/2	M=11/2	M=9/2	M=7/2	M=5/2	M=3/2	M=1/2
1.2198E+02	-.7252	.0250	.7256	-M	0.000000	0.000000	.995310	0.000000	0.000000	0.000000	0.000000	0.000000-
				+M	0.000000	.029177	0.000000	0.000000	0.000000	0.000000	0.000000	-.092232+M
1.2198E+02	.7252	.0250	.7256	-M	0.000000	.029177	0.000000	0.000000	0.000000	0.000000	0.000000	-.092232-
				+M	0.000000	0.000000	.995310	0.000000	0.000000	0.000000	0.000000	0.000000+M
8.6130E+01	.8607	-.0225	.8610	-M	0.000000	0.000000	-.035679	0.000000	0.000000	0.000000	0.000000	0.000000-
				+M	0.000000	.996934	0.000000	0.000000	0.000000	0.000000	0.000000	-.069644+M
8.6130E+01	-.8607	-.0225	.8610	-M	0.000000	.996934	0.000000	0.000000	0.000000	0.000000	0.000000	-.069644-
				+M	0.000000	0.000000	-.035679	0.000000	0.000000	0.000000	0.000000	0.000000+M
8.0985E+01	.5798	0.0000	.5798	-M	.003980	0.000000	0.000000	0.000000	0.000000	0.000000	-.158806	0.000000-
				+M	0.000000	0.000000	0.000000	.987302	0.000000	0.000000	0.000000	0.000000+M
8.0985E+01	-.5798	0.0000	.5798	-M	0.000000	0.000000	0.000000	.987302	0.000000	0.000000	0.000000	0.000000-
				+M	.003980	0.000000	0.000000	0.000000	0.000000	0.000000	-.158806	0.000000+M
2.9548E+01	.3420	-.3587	.4956	-M	0.000000	0.000000	0.000000	0.000000	0.000000	-.394809	0.000000	0.000000-
				+M	0.000000	0.000000	0.000000	0.000000	.918763	0.000000	0.000000	0.000000+M
2.9548E+01	-.3420	-.3587	.4956	-M	0.000000	0.000000	0.000000	0.000000	.918763	0.000000	0.000000	0.000000-
				+M	0.000000	0.000000	0.000000	0.000000	0.000000	-.394809	0.000000	0.000000+M
-3.9148E+01	-.2086	.3587	.4149	-M	0.000000	0.000000	0.000000	0.000000	0.000000	.918763	0.000000	0.000000-
				+M	0.000000	0.000000	0.000000	0.000000	.394809	0.000000	0.000000	0.000000+M
-3.9148E+01	.2086	.3587	.4149	-M	0.000000	0.000000	0.000000	0.000000	.394809	0.000000	0.000000	0.000000
				+M	0.000000	0.000000	0.000000	0.000000	0.000000	.918763	0.000000	0.000000+
-6.0929E+01	-.1834	0.0000	.1834	-M	-.066443	0.000000	0.000000	0.000000	0.000000	0.000000	.985086	0.000000
				+M	0.000000	0.000000	0.000000	.158718	0.000000	0.000000	0.000000	0.000000+
-6.0929E+01	.1834	0.0000	.1834	-M	0.000000	0.000000	0.000000	.158718	0.000000	0.000000	0.000000	0.000000
				+M	-.066443	0.000000	0.000000	0.000000	0.000000	0.000000	.985086	0.000000+
-7.3355E+01	.0644	.5308	.5347	-M	0.000000	0.000000	.089917	0.000000	0.000000	0.000000	0.000000	0.000000
				+M	0.000000	.072608	0.000000	0.000000	0.000000	0.000000	0.000000	.993299+
-7.3355E+01	-.0644	.5308	.5347	-M	0.000000	.072608	0.000000	0.000000	0.000000	0.000000	0.000000	.993299
				+M	0.000000	0.000000	.089917	0.000000	0.000000	0.000000	0.000000	0.000000+
-1.4522E+02	-.9964	0.0000	.9964	-M	.997782	0.000000	0.000000	0.000000	0.000000	0.000000	.066231	0.000000
				+M	0.000000	0.000000	0.000000	.006631	0.000000	0.000000	0.000000	0.000000+
-1.4522E+02	.9964	0.0000	.9964	-M	0.000000	0.000000	0.000000	.006631	0.000000	0.000000	0.000000	0.000000
				+M	.997782	0.000000	0.000000	0.000000	0.000000	0.000000	.066231	0.000000+
ENERGY	HEPAL	HEPER	HE		M=15/2	M=13/2	M=11/2	M=9/2	M=7/2	M=5/2	M=3/2	M=1/2

Y=-.80 X= .80 OAS=3.2798E+02*W246 B4F4/B2F2=-3.20E+00 B4F4/B6F6= 4.00E+00 B2F2/B6F6=-1.25E+00

ENERGY	HEPAL	HEPER	HE		M=15/2	M=13/2	M=11/2	M=9/2	M=7/2	M=5/2	M=3/2	M=1/2
1.4977E+02	-.7324	.0026	.7324	-M	0.000000	0.000000	.999438	0.000000	0.000000	0.000000	0.000000	0.000000
				+M	0.000000	.002822	0.000000	0.000000	0.000000	0.000000	0.000000	-.033390+
1.4977E+02	.7324	.0026	.7324	-M	0.000000	.002822	0.000000	0.000000	0.000000	0.000000	0.000000	-.03339
				+M	0.000000	0.000000	.999438	0.000000	0.000000	0.000000	0.000000	0.000000+
1.1961E+02	.5976	0.0000	.5976	-M	.000524	0.000000	0.000000	0.000000	0.000000	0.000000	-.055098	0.00000
				+M	0.000000	0.000000	0.000000	.998481	0.000000	0.000000	0.000000	0.000000
1.1961E+02	-.5976	0.0000	.5976	-M	0.000000	0.000000	0.000000	.998481	0.000000	0.000000	0.000000	0.00000
				+M	.000524	0.000000	0.000000	0.000000	0.000000	0.000000	-.055098	0.000000
8.1332E+01	.8659	-.0022	.8659	-M	0.000000	0.000000	-.003813	0.000000	0.000000	0.000000	0.000000	0.00000
				+M	0.000000	.999554	0.000000	0.000000	0.000000	0.000000	0.000000	-.029633
8.1332E+01	-.8659	-.0022	.8659	-M	0.000000	.999554	0.000000	0.000000	0.000000	0.000000	0.000000	-.02963
				+M	0.000000	0.000000	-.003813	0.000000	0.000000	0.000000	0.000000	0.000000
5.0966E+01	.4457	-.1581	.4729	-M	0.000000	0.000000	0.000000	0.000000	0.000000	-.162015	0.000000	0.00000
				+M	0.000000	0.000000	0.000000	0.000000	.986788	0.000000	0.000000	0.00000
5.0966E+01	-.4457	-.1581	.4729	-M	0.000000	0.000000	0.000000	0.000000	.986788	0.000000	0.000000	0.00000
				+M	0.000000	0.000000	0.000000	0.000000	0.000000	-.162015	0.000000	0.00000
-2.6966E+01	-.3123	.1581	.3501	-M	0.000000	0.000000	0.000000	0.000000	0.000000	.986788	0.000000	0.00000
				+M	0.000000	0.000000	0.000000	0.000000	.162015	0.000000	0.000000	0.00000
-2.6966E+01	.3123	.1581	.3501	-M	0.000000	0.000000	0.000000	0.000000	.162015	0.000000	0.000000	0.00000
				+M	0.000000	0.000000	0.000000	0.000000	0.000000	.986788	0.000000	0.00000
-8.2891E+01	-.1983	0.0000	.1983	-M	-.029658	0.000000	0.000000	0.000000	0.000000	0.000000	.998041	0.00000
				+M	0.000000	0.000000	0.000000	.055089	0.000000	0.000000	0.000000	0.00000
-8.2891E+01	.1983	0.0000	.1983	-M	0.000000	0.000000	0.000000	.055089	0.000000	0.000000	0.000000	0.00000
				+M	-.029658	0.000000	0.000000	0.000000	0.000000	0.000000	.998041	0.00000
-1.1362E+02	.0665	.5330	.5371	-M	0.000000	0.000000	.033292	0.000000	0.000000	0.000000	0.000000	0.00000
				+M	0.000000	.029744	0.000000	0.000000	0.000000	0.000000	0.000000	.999003
-1.1362E+02	-.0665	.5330	.5371	-M	0.000000	.029744	0.000000	0.000000	0.000000	0.000000	0.000000	.99900
				+M	0.000000	0.000000	.033292	0.000000	0.000000	0.000000	0.000000	0.00000
-1.7820E+02	.9993	0.0000	.9993	-M	0.000000	0.000000	0.000000	.001111	0.000000	0.000000	0.000000	0.00000
				+M	.999560	0.000000	0.000000	0.000000	0.000000	0.000000	.029642	0.00000
-1.7820E+02	-.9993	0.0000	.9993	-M	.999560	0.000000	0.000000	0.000000	0.000000	0.000000	.029642	0.00000
				+M	0.000000	0.000000	0.000000	.001111	0.000000	0.000000	0.000000	0.00000
ENERGY	HEPAL	HEPER	HE		M=15/2	M=13/2	M=11/2	M=9/2	M=7/2	M=5/2	M=3/2	M=1/2

Y=-.60 X=-.80 OAS=2.4169E+02*W246 B4F4/B2F2= 1.20E+00 B4F4/B6F6=-4.00E+00 B2F2/B6F6=-3.33E+00

ENERGY	HEPAL	HEPER	HE		M=15/2	M=13/2	M=11/2	M=9/2	M=7/2	M=5/2	M=3/2	M=1/2
1.3730E+02	.9994	0.0000	.9994	-M	0.000000	0.000000	0.000000	.000922	0.000000	0.000000	0.000000	0.000000-M
				+M	.999645	0.000000	0.000000	0.000000	0.000000	0.000000	-.026625	0.000000+M
1.3730E+02	-.9994	0.0000	.9994	-M	.999645	0.000000	0.000000	0.000000	0.000000	0.000000	-.026625	0.000000-M
				+M	0.000000	0.000000	0.000000	.000922	0.000000	0.000000	0.000000	0.000000+M
9.1716E+01	.0669	.5328	.5370	-M	0.000000	0.000000	-.034652	0.000000	0.000000	0.000000	0.000000	0.000000-M
				+M	0.000000	-.038455	0.000000	0.000000	0.000000	0.000000	0.000000	.998659+M
9.1716E+01	-.0669	.5328	.5370	-M	0.000000	-.038455	0.000000	0.000000	0.000000	0.000000	0.000000	.998659-M
				+M	0.000000	0.000000	-.034652	0.000000	0.000000	0.000000	0.000000	0.000000+M
5.7695E+01	.1984	0.0000	.1984	-M	0.000000	0.000000	0.000000	-.051615	0.000000	0.000000	0.000000	0.000000-M
				+M	.026637	0.000000	0.000000	0.000000	0.000000	0.000000	.998312	0.000000+M
5.7695E+01	-.1984	0.0000	.1984	-M	.026637	0.000000	0.000000	0.000000	0.000000	0.000000	.998312	0.000000-M
				+M	0.000000	0.000000	0.000000	-.051615	0.000000	0.000000	0.000000	0.000000+M
1.0143E+00	.3164	-.1424	.3469	-M	0.000000	0.000000	0.000000	0.000000	-.145511	0.000000	0.000000	0.000000-M
				+M	0.000000	0.000000	0.000000	0.000000	0.000000	.989357	0.000000	0.000000+M
1.0143E+00	-.3164	-.1424	.3469	-M	0.000000	0.000000	0.000000	0.000000	0.000000	.989357	0.000000	0.000000-M
				+M	0.000000	0.000000	0.000000	0.000000	-.145511	0.000000	0.000000	0.000000+M
-2.1407E+01	.8655	-.0015	.8655	-M	0.000000	0.000000	-.003302	0.000000	0.000000	0.000000	0.000000	0.000000-M
				+M	0.000000	.999258	0.000000	0.000000	0.000000	0.000000	0.000000	.038364+M
-2.1407E+01	-.8655	-.0015	.8655	-M	0.000000	.999258	0.000000	0.000000	0.000000	0.000000	0.000000	.038364-M
				+M	0.000000	0.000000	-.003302	0.000000	0.000000	0.000000	0.000000	0.000000+M
-6.3894E+01	-.4497	.1424	.4717	-M	0.000000	0.000000	0.000000	0.000000	.989357	0.000000	0.000000	0.000000-M
				+M	0.000000	0.000000	0.000000	0.000000	0.000000	.145511	0.000000	0.000000+M
-6.3894E+01	.4497	.1424	.4717	-M	0.000000	0.000000	0.000000	0.000000	0.000000	.145511	0.000000	0.000000-M
				+M	0.000000	0.000000	0.000000	0.000000	.989357	0.000000	0.000000	0.000000+M
-9.8029E+01	-.7324	.0020	.7324	-M	0.000000	0.000000	.999394	0.000000	0.000000	0.000000	0.000000	0.000000-M
				+M	0.000000	.001969	0.000000	0.000000	0.000000	0.000000	0.000000	.034754+M
-9.8029E+01	.7324	.0020	.7324	-M	0.000000	.001969	0.000000	0.000000	0.000000	0.000000	0.000000	.034754-M
				+M	0.000000	0.000000	.999394	0.000000	0.000000	0.000000	0.000000	0.000000+M
-1.0439E+02	.5979	0.0000	.5979	-M	.000453	0.000000	0.000000	0.000000	0.000000	0.000000	.051621	0.000000-M
				+M	0.000000	0.000000	0.000000	.998667	0.000000	0.000000	0.000000	0.000000+M
-1.0439E+02	-.5979	0.0000	.5979	-M	0.000000	0.000000	0.000000	.998667	0.000000	0.000000	0.000000	0.000000-M
				+M	.000453	0.000000	0.000000	0.000000	0.000000	0.000000	.051621	0.000000+M
ENERGY	HEPAL	HEPER	HE		M=15/2	M=13/2	M=11/2	M=9/2	M=7/2	M=5/2	M=3/2	M=1/2

Y=-.60 X=-.60 OAS=1.8604E+02*W246 B4F4/B2F2= 9.00E-01 B4F4/B6F6=-1.50E+00 B2F2/B6F6=-1.67E+00

ENERGY	HEPAL	HEPER	HE		M=15/2	M=13/2	M=11/2	M=9/2	M=7/2	M=5/2	M=3/2	M=1/2
9.7036E+01	-.9943	0.0000	.9943	-M	.996482	0.000000	0.000000	0.000000	0.000000	0.000000	-.083467	0.000000-M
				+M	0.000000	0.000000	0.000000	.007585	0.000000	0.000000	0.000000	0.000000+M
9.7036E+01	.9943	0.0000	.9943	-M	0.000000	0.000000	0.000000	.007585	0.000000	0.000000	0.000000	0.000000-M
				+M	.996482	0.000000	0.000000	0.000000	0.000000	0.000000	-.083467	0.000000+M
7.9816E+01	.0706	.5294	.5341	-M	0.000000	0.000000	-.088891	0.000000	0.000000	0.000000	0.000000	0.000000-M
				+M	0.000000	-.113129	0.000000	0.000000	0.000000	0.000000	0.000000	.989596+M
7.9816E+01	-.0706	.5294	.5341	-M	0.000000	-.113129	0.000000	0.000000	0.000000	0.000000	0.000000	.989596-M
				+M	0.000000	0.000000	-.088891	0.000000	0.000000	0.000000	0.000000	0.000000+M
4.6574E+01	.1933	0.0000	.1933	-M	0.000000	0.000000	0.000000	-.123811	0.000000	0.000000	0.000000	0.000000-M
				+M	.083763	0.000000	0.000000	0.000000	0.000000	0.000000	.988764	0.000000+M
4.6574E+01	-.1933	0.0000	.1933	-M	.083763	0.000000	0.000000	0.000000	0.000000	0.000000	.988764	0.000000-M
				+M	0.000000	0.000000	0.000000	-.123811	0.000000	0.000000	0.000000	0.000000+M
2.7422E+00	-.8560	-.0082	.8560	-M	0.000000	.993519	0.000000	0.000000	0.000000	0.000000	0.000000	.111675-M
				+M	0.000000	0.000000	-.021181	0.000000	0.000000	0.000000	0.000000	0.000000+M
2.7422E+00	.8560	-.0082	.8560	-M	0.000000	0.000000	-.021181	0.000000	0.000000	0.000000	0.000000	0.000000-M
				+M	0.000000	.993519	0.000000	0.000000	0.000000	0.000000	0.000000	.111675+M
-2.5301E+00	-.2588	-.2874	.3868	-M	0.000000	0.000000	0.000000	0.000000	0.000000	.952284	0.000000	0.000000-M
				+M	0.000000	0.000000	0.000000	0.000000	-.305215	0.000000	0.000000	0.000000+M
-2.5301E+00	.2588	-.2874	.3868	-M	0.000000	0.000000	0.000000	0.000000	-.305215	0.000000	0.000000	0.000000-M
				+M	0.000000	0.000000	0.000000	0.000000	0.000000	.952284	0.000000	0.000000+M
-6.6830E+01	-.3921	.2874	.4862	-M	0.000000	0.000000	0.000000	0.000000	.952284	0.000000	0.000000	0.000000-M
				+M	0.000000	0.000000	0.000000	0.000000	0.000000	.305215	0.000000	0.000000+M
-6.6830E+01	.3921	.2874	.4862	-M	0.000000	0.000000	0.000000	0.000000	0.000000	.305215	0.000000	0.000000-M
				+M	0.000000	0.000000	0.000000	0.000000	.952284	0.000000	0.000000	0.000000+M
-6.7798E+01	.7266	.0121	.7267	-M	0.000000	.011034	0.000000	0.000000	0.000000	0.000000	0.000000	.090711-M
				+M	0.000000	0.000000	.995816	0.000000	0.000000	0.000000	0.000000	0.000000+M
-6.7798E+01	-.7266	.0121	.7267	-M	0.000000	0.000000	.995816	0.000000	0.000000	0.000000	0.000000	0.000000-M
				+M	0.000000	.011034	0.000000	0.000000	0.000000	0.000000	0.000000	.090711+M
-8.9009E+01	.5877	0.0000	.5877	-M	.002835	0.000000	0.000000	0.000000	0.000000	0.000000	.124011	0.000000-M
				+M	0.000000	0.000000	0.000000	.992277	0.000000	0.000000	0.000000	0.000000+M
-8.9009E+01	-.5877	0.0000	.5877	-M	0.000000	0.000000	0.000000	.992277	0.000000	0.000000	0.000000	0.000000-M
				+M	.002835	0.000000	0.000000	0.000000	0.000000	0.000000	.124011	0.000000+M
ENERGY	HEPAL	HEPER	HE		M=15/2	M=13/2	M=11/2	M=9/2	M=7/2	M=5/2	M=3/2	M=1/2

Y=-.60 X=-.40 OAS=1.4728E+02*W246 B4F4/B2F2= 6.00E-01 B4F4/B6F6=-6.67E-01 B2F2/B6F6=-1.11E+00

ENERGY	HEPAL	HEPER	HE		M=15/2	M=13/2	M=11/2	M=9/2	M=7/2	M=5/2	M=3/2	M=1/2
7.1568E+01	.1082	.5085	.5199	-M	0.000000	0.000000	-.173757	0.000000	0.000000	0.000000	0.000000	0.000000-M
				+M	0.000000	-.286647	0.000000	0.000000	0.000000	0.000000	0.000000	.942147+M
7.1568E+01	-.1082	.5085	.5199	-M	0.000000	-.286647	0.000000	0.000000	0.000000	0.000000	0.000000	.942147-M
				+M	0.000000	0.000000	-.173757	0.000000	0.000000	0.000000	0.000000	0.000000+M
5.8069E+01	.9264	0.0000	.9264	-M	0.000000	0.000000	0.000000	.057276	0.000000	0.000000	0.000000	0.000000-M
				+M	.954617	0.000000	0.000000	0.000000	0.000000	0.000000	-.292278	0.000000+M
5.8069E+01	-.9264	0.0000	.9264	-M	.954617	0.000000	0.000000	0.000000	0.000000	0.000000	-.292278	0.000000-M
				+M	0.000000	0.000000	0.000000	.057276	0.000000	0.000000	0.000000	0.000000+M
3.6244E+01	.2323	0.0000	.2323	-M	0.000000	0.000000	0.000000	-.219515	0.000000	0.000000	0.000000	0.000000-M
				+M	.297636	0.000000	0.000000	0.000000	0.000000	0.000000	.929099	0.000000+M
3.6244E+01	-.2323	0.0000	.2323	-M	.297636	0.000000	0.000000	0.000000	0.000000	0.000000	.929099	0.000000-M
				+M	0.000000	0.000000	0.000000	-.219515	0.000000	0.000000	0.000000	0.000000+M
2.4936E+01	.7934	-.0204	.7937	-M	0.000000	0.000000	-.089818	0.000000	0.000000	0.000000	0.000000	0.000000-M
				+M	0.000000	.957326	0.000000	0.000000	0.000000	0.000000	0.000000	.274700+M
2.4936E+01	-.7934	-.0204	.7937	-M	0.000000	.957326	0.000000	0.000000	0.000000	0.000000	0.000000	.274700-M
				+M	0.000000	0.000000	-.089818	0.000000	0.000000	0.000000	0.000000	0.000000+M
-2.7490E+00	-.1749	-.3941	.4311	-M	0.000000	0.000000	0.000000	0.000000	0.000000	.895518	0.000000	0.000000-M
				+M	0.000000	0.000000	0.000000	0.000000	-.445024	0.000000	0.000000	0.000000+M
-2.7490E+00	.1749	-.3941	.4311	-M	0.000000	0.000000	0.000000	0.000000	-.445024	0.000000	0.000000	0.000000-M
				+M	0.000000	0.000000	0.000000	0.000000	0.000000	.895518	0.000000	0.000000+M
-3.9264E+01	-.7016	.0452	.7031	-M	0.000000	0.000000	.980684	0.000000	0.000000	0.000000	0.000000	0.000000-M
				+M	0.000000	.036891	0.000000	0.000000	0.000000	0.000000	0.000000	.192089+M
-3.9264E+01	.7016	.0452	.7031	-M	0.000000	.036891	0.000000	0.000000	0.000000	0.000000	0.000000	.192089-M
				+M	0.000000	0.000000	.980684	0.000000	0.000000	0.000000	0.000000	0.000000+M
-7.3091E+01	.3082	.3941	.5003	-M	0.000000	0.000000	0.000000	0.000000	0.000000	.445024	0.000000	0.000000-M
				+M	0.000000	0.000000	0.000000	0.000000	.895518	0.000000	0.000000	0.000000+M
-7.3091E+01	-.3082	.3941	.5003	-M	0.000000	0.000000	0.000000	0.000000	.895518	0.000000	0.000000	0.000000-M
				+M	0.000000	0.000000	0.000000	0.000000	0.000000	.445024	0.000000	0.000000+M
-7.5714E+01	.5587	0.0000	.5587	-M	.010944	0.000000	0.000000	0.000000	0.000000	0.000000	.226601	0.000000-M
				+M	0.000000	0.000000	0.000000	.973926	0.000000	0.000000	0.000000	0.000000+M
-7.5714E+01	-.5587	0.0000	.5587	-M	0.000000	0.000000	0.000000	.973926	0.000000	0.000000	0.000000	0.000000-M
				+M	.010944	0.000000	0.000000	0.000000	0.000000	0.000000	.226601	0.000000+M
ENERGY	HEPAL	HEPER	HE		M=15/2	M=13/2	M=11/2	M=9/2	M=7/2	M=5/2	M=3/2	M=1/2

Y=-.60 X=-.20 OAS=1.5594E+02*W246 B4F4/B2F2= 3.00E-01 B4F4/B6F6=-2.50E-01 E2F2/B6F6=-8.33E-01

ENERGY	HEPAL	HEPER	HE		M=15/2	M=13/2	M=11/2	M=9/2	M=7/2	M=5/2	M=3/2	M=1/2
7.3997E+01	.3345	.2201	.4004	-M	0.000000	0.000000	-.247741	0.000000	0.000000	0.000000	0.000000	0.000000-M
				+M	0.000000	-.629442	0.000000	0.000000	0.000000	0.000000	0.000000	.736496+M
7.3997E+01	-.3345	-.2201	.4004	-M	0.000000	.629442	0.000000	0.000000	0.000000	0.000000	0.000000	-.736496-M
				+M	0.000000	0.000000	.247741	0.000000	0.000000	0.000000	0.000000	0.000000+M
4.1196E+01	-.4603	.2201	.5102	-M	0.000000	.770755	0.000000	0.000000	0.000000	0.000000	0.000000	.551275-M
				+M	0.000000	0.000000	-.319425	0.000000	0.000000	0.000000	0.000000	0.000000+M
4.1196E+01	.4603	-.2201	.5102	-M	0.000000	0.000000	-.319425	0.000000	0.000000	0.000000	0.000000	0.000000-M
				+M	0.000000	.770755	0.000000	0.000000	0.000000	0.000000	0.000000	.551275+M
3.5990E+01	.2189	0.0000	.2189	-M	0.000000	0.000000	0.000000	-.328282	0.000000	0.000000	0.000000	0.000000-M
				+M	-.362465	0.000000	0.000000	0.000000	0.000000	0.000000	.872267	0.000000+M
3.5990E+01	-.2189	0.0000	.2189	-M	-.362465	0.000000	0.000000	0.000000	0.000000	0.000000	.872267	0.000000-M
				+M	0.000000	0.000000	0.000000	-.328282	0.000000	0.000000	0.000000	0.000000+M
1.2601E+01	.8717	0.0000	.8717	-M	0.000000	0.000000	0.000000	-.165806	0.000000	0.000000	0.000000	0.000000-M
				+M	.931219	0.000000	0.000000	0.000000	0.000000	0.000000	.324560	0.000000+M
1.2601E+01	-.8717	0.0000	.8717	-M	.931219	0.000000	0.000000	0.000000	0.000000	0.000000	.324560	0.000000-M
				+M	0.000000	0.000000	0.000000	-.165806	0.000000	0.000000	0.000000	0.000000+M
-3.7483E-01	-.0934	-.4531	.4626	-M	0.000000	0.000000	0.000000	0.000000	0.000000	.836703	0.000000	0.000000-M
				+M	0.000000	0.000000	0.000000	0.000000	-.547656	0.000000	0.000000	0.000000+M
-3.7483E-01	.0934	-.4531	.4626	-M	0.000000	0.000000	0.000000	0.000000	-.547656	0.000000	0.000000	0.000000-M
				+M	0.000000	0.000000	0.000000	0.000000	0.000000	.836703	0.000000	0.000000+M
-1.5473E+01	-.5948	.1456	.6124	-M	0.000000	0.000000	.914654	0.000000	0.000000	0.000000	0.000000	0.000000-M
				+M	0.000000	.098682	0.000000	0.000000	0.000000	0.000000	0.000000	.392007+M
-1.5473E+01	.5948	.1456	.6124	-M	0.000000	.098682	0.000000	0.000000	0.000000	0.000000	0.000000	.392007-M
				+M	0.000000	0.000000	.914654	0.000000	0.000000	0.000000	0.000000	0.000000+M
-6.5991E+01	.4906	0.0000	.4906	-M	.038080	0.000000	0.000000	0.000000	0.000000	0.000000	.365802	0.000000-M
				+M	0.000000	0.000000	0.000000	.929913	0.000000	0.000000	0.000000	0.000000+M
-6.5991E+01	-.4906	0.0000	.4906	-M	0.000000	0.000000	0.000000	.929913	0.000000	0.000000	0.000000	0.000000-M
				+M	.038080	0.000000	0.000000	0.000000	0.000000	0.000000	.365802	0.000000+M
-8.1945E+01	-.2267	.4531	.5067	-M	0.000000	0.000000	0.000000	0.000000	.836703	0.000000	0.000000	0.000000-M
				+M	0.000000	0.000000	0.000000	0.000000	0.000000	.547656	0.000000	0.000000+M
-8.1945E+01	.2267	.4531	.5067	-M	0.000000	0.000000	0.000000	0.000000	0.000000	.547656	0.000000	0.000000-M
				+M	0.000000	0.000000	0.000000	0.000000	.836703	0.000000	0.000000	0.000000+M
ENERGY	HEPAL	HEPER	HE		M=15/2	M=13/2	M=11/2	M=9/2	M=7/2	M=5/2	M=3/2	M=1/2

Y=-.60 X=0.00 OAS=1.8364E+02*W246 B4F4/B2F2= 0. B4F4/B6F6= 0. B2F2/B6F6=-6.67E-01

ENERGY	HEPAL	HEPER	HE		M=15/2	M=13/2	M=11/2	M=9/2	M=7/2	M=5/2	M=3/2	M=1/2
9.1148E+01	.5807	.2719	.6412	-M	0.000000	0.000000	.248139	0.000000	0.000000	0.000000	0.000000	0.000000-M
				+M	0.000000	.839085	0.000000	0.000000	0.000000	0.000000	0.000000	-.484112+M
9.1148E+01	-.5807	.2719	.6412	-M	0.000000	.839085	0.000000	0.000000	0.000000	0.000000	0.000000	-.484112-M
				+M	0.000000	0.000000	.248139	0.000000	0.000000	0.000000	0.000000	0.000000+M
5.5052E+01	-.0625	-.0714	.0949	-M	0.000000	0.000000	-.652494	0.000000	0.000000	0.000000	0.000000	0.000000-M
				+M	0.000000	.514145	0.000000	0.000000	0.000000	0.000000	0.000000	.556693+M
5.5052E+01	.0625	-.0714	.0949	-M	0.000000	.514145	0.000000	0.000000	0.000000	0.000000	0.000000	.556693-M
				+M	0.000000	0.000000	-.652494	0.000000	0.000000	0.000000	0.000000	0.000000+M
3.4236E+01	.0133	0.0000	.0133	-M	0.000000	0.000000	0.000000	-.506442	0.000000	0.000000	0.000000	0.000000-M
				+M	-.152052	0.000000	0.000000	0.000000	0.000000	0.000000	.848762	0.000000+M
3.4236E+01	-.0133	0.0000	.0133	-M	-.152052	0.000000	0.000000	0.000000	0.000000	0.000000	.848762	0.000000-M
				+M	0.000000	0.000000	0.000000	-.506442	0.000000	0.000000	0.000000	0.000000+M
3.6927E+00	.0282	-.4803	.4811	-M	0.000000	0.000000	0.000000	0.000000	-.617640	0.000000	0.000000	0.000000-M
				+M	0.000000	0.000000	0.000000	0.000000	0.000000	.786461	0.000000	0.000000+M
3.6927E+00	-.0282	-.4803	.4811	-M	0.000000	0.000000	0.000000	0.000000	0.000000	.786461	0.000000	0.000000-M
				+M	0.000000	0.000000	0.000000	0.000000	-.617640	0.000000	0.000000	0.000000+M
-3.9998E+00	-.3182	.3328	.4605	-M	0.000000	0.000000	.716016	0.000000	0.000000	0.000000	0.000000	0.000000-M
				+M	0.000000	.177743	0.000000	0.000000	0.000000	0.000000	0.000000	.675077+M
-3.9998E+00	.3182	.3328	.4605	-M	0.000000	.177743	0.000000	0.000000	0.000000	0.000000	0.000000	.675077-M
				+M	0.000000	0.000000	.716016	0.000000	0.000000	0.000000	0.000000	0.000000+M
-2.5554E+01	-.9288	0.0000	.9288	-M	.976829	0.000000	0.000000	0.000000	0.000000	0.000000	.050966	0.000000-M
				+M	0.000000	0.000000	0.000000	-.207863	0.000000	0.000000	0.000000	0.000000+M
-2.5554E+01	.9288	0.0000	.9288	-M	0.000000	0.000000	0.000000	-.207863	0.000000	0.000000	0.000000	0.000000-M
				+M	.976829	0.000000	0.000000	0.000000	0.000000	0.000000	.050966	0.000000+M
-6.2083E+01	.3421	0.0000	.3421	-M	.150615	0.000000	0.000000	0.000000	0.000000	0.000000	.526314	0.000000-M
				+M	0.000000	0.000000	0.000000	.836845	0.000000	0.000000	0.000000	0.000000+M
-6.2083E+01	-.3421	0.0000	.3421	-M	0.000000	0.000000	0.000000	.836845	0.000000	0.000000	0.000000	0.000000-M
				+M	.150615	0.000000	0.000000	0.000000	0.000000	0.000000	.526314	0.000000+M
-9.2493E+01	-.1615	.4803	.5067	-M	0.000000	0.000000	0.000000	0.000000	.786461	0.000000	0.000000	0.000000-M
				+M	0.000000	0.000000	0.000000	0.000000	0.000000	.617640	0.000000	0.000000+M
-9.2493E+01	.1615	.4803	.5067	-M	0.000000	0.000000	0.000000	0.000000	0.000000	.617640	0.000000	0.000000-M
				+M	0.000000	0.000000	0.000000	0.000000	.786461	0.000000	0.000000	0.000000+M
ENERGY	HEPAL	HEPER	HE		M=15/2	M=13/2	M=11/2	M=9/2	M=7/2	M=5/2	M=3/2	M=1/2

Y=-.60 X= .20 OAS=1.5023E+02*W246 B4F4/B2F2=-3.00E-01 B4F4/B6F6= 2.50E-01 B2F2/B6F6=-8.33E-01

ENERGY	HEPAL	HEPER	HE		M=15/2	M=13/2	M=11/2	M=9/2	M=7/2	M=5/2	M=3/2	M=1/2
8.1023E+01	-.7044	.2034	.7332	-M	0.000000	.924764	0.000000	0.000000	0.000000	0.000000	0.000000	-.294649-M
				+M	0.000000	0.000000	.240819	0.000000	0.000000	0.000000	0.000000	0.000000+M
8.1023E+01	.7044	.2034	.7332	-M	0.000000	0.000000	.240819	0.000000	0.000000	0.000000	0.000000	0.000000-M
				+M	0.000000	.924764	0.000000	0.000000	0.000000	0.000000	0.000000	-.294649+M
5.8523E+01	.4699	-.1519	.4939	-M	0.000000	-.334084	0.000000	0.000000	0.000000	0.000000	0.000000	-.325609-M
				+M	0.000000	0.000000	.884515	0.000000	0.000000	0.000000	0.000000	0.000000+M
5.8523E+01	-.4699	-.1519	.4939	-M	0.000000	0.000000	.884515	0.000000	0.000000	0.000000	0.000000	0.000000-M
				+M	0.000000	-.334084	0.000000	0.000000	0.000000	0.000000	0.000000	-.325609+M
2.6179E+01	-.2260	0.0000	.2260	-M	0.000000	0.000000	0.000000	.733600	0.000000	0.000000	0.000000	0.000000-M
				+M	.075307	0.000000	0.000000	0.000000	0.000000	0.000000	-.675397	0.000000+M
2.6179E+01	.2260	0.0000	.2260	-M	.075307	0.000000	0.000000	0.000000	0.000000	0.000000	-.675397	0.000000-M
				+M	0.000000	0.000000	0.000000	.733600	0.000000	0.000000	0.000000	0.000000+M
5.6054E+00	-.0513	-.4940	.4967	-M	0.000000	0.000000	0.000000	0.000000	-.693362	0.000000	0.000000	0.000000-M
				+M	0.000000	0.000000	0.000000	0.000000	0.000000	.720589	0.000000	0.000000+M
5.6054E+00	.0513	-.4940	.4967	-M	0.000000	0.000000	0.000000	0.000000	0.000000	.720589	0.000000	0.000000-M
				+M	0.000000	0.000000	0.000000	0.000000	-.693362	0.000000	0.000000	0.000000+M
-1.0306E+01	-.0345	.4818	.4831	-M	0.000000	0.000000	.399549	0.000000	0.000000	0.000000	0.000000	0.000000-M
				+M	0.000000	.182209	0.000000	0.000000	0.000000	0.000000	0.000000	.898421+M
-1.0306E+01	.0345	.4818	.4831	-M	0.000000	.182209	0.000000	0.000000	0.000000	0.000000	0.000000	.898421-M
				+M	0.000000	0.000000	.399549	0.000000	0.000000	0.000000	0.000000	0.000000+M
-3.8206E+01	-.0509	0.0000	.0509	-M	-.455776	0.000000	0.000000	0.000000	0.000000	0.000000	.631028	0.000000-M
				+M	0.000000	0.000000	0.000000	.627751	0.000000	0.000000	0.000000	0.000000+M
-3.8206E+01	.0509	0.0000	.0509	-M	0.000000	0.000000	0.000000	.627751	0.000000	0.000000	0.000000	0.000000-M
				+M	-.455776	0.000000	0.000000	0.000000	0.000000	0.000000	.631028	0.000000+M
-5.3613E+01	-.7751	0.0000	.7751	-M	.886903	0.000000	0.000000	0.000000	0.000000	0.000000	.381631	0.000000-M
				+M	0.000000	0.000000	0.000000	.260309	0.000000	0.000000	0.000000	0.000000+M
-5.3613E+01	.7751	0.0000	.7751	-M	0.000000	0.000000	0.000000	.260309	0.000000	0.000000	0.000000	0.000000-M
				+M	.886903	0.000000	0.000000	0.000000	0.000000	0.000000	.381631	0.000000+M
-6.9205E+01	-.0821	.4940	.5008	-M	0.000000	0.000000	0.000000	0.000000	.720589	0.000000	0.000000	0.000000-M
				+M	0.000000	0.000000	0.000000	0.000000	0.000000	.693362	0.000000	0.000000+M
-6.9205E+01	.0821	.4940	.5008	-M	0.000000	0.000000	0.000000	0.000000	0.000000	.693362	0.000000	0.000000-M
				+M	0.000000	0.000000	0.000000	0.000000	.720589	0.000000	0.000000	0.000000+M
ENERGY	HEPAL	HEPER	HE		M=15/2	M=13/2	M=11/2	M=9/2	M=7/2	M=5/2	M=3/2	M=1/2

Y=-.60 X= .40 OAS=1.5265E+02*W246 B4F4/B2F2=-6.00E-01 B4F4/B6F6= 6.67E-01 E2F2/B6F6=-1.11E+00

ENERGY	HEPAL	HEPER	HE		M=15/2	M=13/2	M=11/2	M=9/2	M=7/2	M=5/2	M=3/2	M=1/2
7.6720E+01	-.0003	.3602	.3602	-M	0.000000	.658987	0.000000	0.000000	0.000000	0.000000	0.000000	-.220208-M
				+M	0.000000	0.000000	.719197	0.000000	0.000000	0.000000	0.000000	0.000000+M
7.6720E+01	.0003	.3602	.3602	-M	0.000000	0.000000	.719197	0.000000	0.000000	0.000000	0.000000	0.000000-M
				+M	0.000000	.658987	0.000000	0.000000	0.000000	0.000000	0.000000	-.220208+M
7.1732E+01	-.1492	-.3498	.3802	-M	0.000000	.742246	0.000000	0.000000	0.000000	0.000000	0.000000	.035698-M
				+M	0.000000	0.000000	-.669176	0.000000	0.000000	0.000000	0.000000	0.000000+M
7.1732E+01	.1492	-.3498	.3802	-M	0.000000	0.000000	-.669176	0.000000	0.000000	0.000000	0.000000	0.000000-M
				+M	0.000000	.742246	0.000000	0.000000	0.000000	0.000000	0.000000	.035698+M
3.6212E+01	.4969	0.0000	.4969	-M	.020504	0.000000	0.000000	0.000000	0.000000	0.000000	-.357873	0.000000-M
				+M	0.000000	0.000000	0.000000	.933545	0.000000	0.000000	0.000000	0.000000+M
3.6212E+01	-.4969	0.0000	.4969	-M	0.000000	0.000000	0.000000	.933545	0.000000	0.000000	0.000000	0.000000-M
				+M	.020504	0.000000	0.000000	0.000000	0.000000	0.000000	-.357873	0.000000+M
1.0099E+01	.1830	-.4730	.5072	-M	0.000000	0.000000	0.000000	0.000000	0.000000	-.595487	0.000000	0.000000-M
				+M	0.000000	0.000000	0.000000	0.000000	.803365	0.000000	0.000000	0.000000+M
1.0099E+01	-.1830	-.4730	.5072	-M	0.000000	0.000000	0.000000	0.000000	.803365	0.000000	0.000000	0.000000-M
				+M	0.000000	0.000000	0.000000	0.000000	0.000000	-.595487	0.000000	0.000000+M
-3.2173E+01	-.0505	.5228	.5253	-M	0.000000	.121684	0.000000	0.000000	0.000000	0.000000	0.000000	.974800-M
				+M	0.000000	0.000000	.186973	0.000000	0.000000	0.000000	0.000000	0.000000+M
-3.2173E+01	.0505	.5228	.5253	-M	0.000000	0.000000	.186973	0.000000	0.000000	0.000000	0.000000	0.000000-M
				+M	0.000000	.121684	0.000000	0.000000	0.000000	0.000000	0.000000	.974800+M
-3.8164E+01	.1187	0.0000	.1187	-M	0.000000	0.000000	0.000000	.356419	0.000000	0.000000	0.000000	0.000000-M
				+M	-.159477	0.000000	0.000000	0.000000	0.000000	0.000000	.920615	0.000000+M
-3.8164E+01	-.1187	0.0000	.1187	-M	-.159477	0.000000	0.000000	0.000000	0.000000	0.000000	.920615	0.000000-M
				+M	0.000000	0.000000	0.000000	.356419	0.000000	0.000000	0.000000	0.000000+M
-4.8499E+01	-.0496	.4730	.4756	-M	0.000000	0.000000	0.000000	0.000000	0.000000	.803365	0.000000	0.000000-M
				+M	0.000000	0.000000	0.000000	0.000000	.595487	0.000000	0.000000	0.000000+M
-4.8499E+01	.0496	.4730	.4756	-M	0.000000	0.000000	0.000000	0.000000	.595487	0.000000	0.000000	0.000000-M
				+M	0.000000	0.000000	0.000000	0.000000	0.000000	.803365	0.000000	0.000000+M
-7.5928E+01	-.9782	0.0000	.9782	-M	.986989	0.000000	0.000000	0.000000	0.000000	0.000000	.156187	0.000000-M
				+M	0.000000	0.000000	0.000000	.038196	0.000000	0.000000	0.000000	0.000000+M
-7.5928E+01	.9782	0.0000	.9782	-M	0.000000	0.000000	0.000000	.038196	0.000000	0.000000	0.000000	0.000000-M
				+M	.986989	0.000000	0.000000	0.000000	0.000000	0.000000	.156187	0.000000+M
ENERGY	HEPAL	HEPER	HE		M=15/2	M=13/2	M=11/2	M=9/2	M=7/2	M=5/2	M=3/2	M=1/2

Y=-.60 X= .60 OAS=1.9392E+02*W246 B4F4/B2F2=-9.00E-01 B4F4/B6F6= 1.50E+00 B2F2/B6F6=-1.67E+00

ENERGY	HEPAL	HEPER	HE		M=15/2	M=13/2	M=11/2	M=9/2	M=7/2	M=5/2	M=3/2	M=1/2
9.3681E+01	-.7256	.0261	.7261	-M	0.000000	0.000000	.995641	0.000000	0.000000	0.000000	0.000000	0.000000-M
				+M	0.000000	.031275	0.000000	0.000000	0.000000	0.000000	0.000000	-.087869+M
9.3681E+01	.7256	.0261	.7261	-M	0.000000	.031275	0.000000	0.000000	0.000000	0.000000	0.000000	-.087869-M
				+M	0.000000	0.000000	.995641	0.000000	0.000000	0.000000	0.000000	0.000000+M
6.9797E+01	.8612	-.0238	.8615	-M	0.000000	0.000000	-.036962	0.000000	0.000000	0.000000	0.000000	0.000000-M
				+M	0.000000	.997274	0.000000	0.000000	0.000000	0.000000	0.000000	-.063861+M
6.9797E+01	-.8612	-.0238	.8615	-M	0.000000	.997274	0.000000	0.000000	0.000000	0.000000	0.000000	-.063861-M
				+M	0.000000	0.000000	-.036962	0.000000	0.000000	0.000000	0.000000	0.000000+M
6.0380E+01	.5814	0.0000	.5814	-M	.004040	0.000000	0.000000	0.000000	0.000000	0.000000	-.152529	0.000000-M
				+M	0.000000	0.000000	0.000000	.988291	0.000000	0.000000	0.000000	0.000000+M
6.0380E+01	-.5814	0.0000	.5814	-M	0.000000	0.000000	0.000000	.988291	0.000000	0.000000	0.000000	0.000000-M
				+M	.004040	0.000000	0.000000	0.000000	0.000000	0.000000	-.152529	0.000000+M
1.9683E+01	.3479	-.3516	.4946	-M	0.000000	0.000000	0.000000	0.000000	0.000000	-.385277	0.000000	0.000000-M
				+M	0.000000	0.000000	0.000000	0.000000	.922801	0.000000	0.000000	0.000000+M
1.9683E+01	-.3479	-.3516	.4946	-M	0.000000	0.000000	0.000000	0.000000	.922801	0.000000	0.000000	0.000000-M
				+M	0.000000	0.000000	0.000000	0.000000	0.000000	-.385277	0.000000	0.000000+M
-3.2883E+01	-.2146	.3516	.4119	-M	0.000000	0.000000	0.000000	0.000000	0.000000	.922801	0.000000	0.000000-M
				+M	0.000000	0.000000	0.000000	0.000000	.385277	0.000000	0.000000	0.000000+M
-3.2883E+01	.2146	.3516	.4119	-M	0.000000	0.000000	0.000000	0.000000	.385277	0.000000	0.000000	0.000000-M
				+M	0.000000	0.000000	0.000000	0.000000	0.000000	.922801	0.000000	0.000000+M
-5.0263E+01	.1871	0.0000	.1871	-M	0.000000	0.000000	0.000000	.152325	0.000000	0.000000	0.000000	0.000000-M
				+M	-.084244	0.000000	0.000000	0.000000	0.000000	0.000000	.984734	0.000000+M
-5.0263E+01	-.1871	0.0000	.1871	-M	-.084244	0.000000	0.000000	0.000000	0.000000	0.000000	.984734	0.000000-M
				+M	0.000000	0.000000	0.000000	.152325	0.000000	0.000000	0.000000	0.000000+M
-6.0158E+01	-.0644	.5311	.5350	-M	0.000000	.066830	0.000000	0.000000	0.000000	0.000000	0.000000	.994083-M
				+M	0.000000	0.000000	.085632	0.000000	0.000000	0.000000	0.000000	0.000000+M
-6.0158E+01	.0644	.5311	.5350	-M	0.000000	0.000000	.085632	0.000000	0.000000	0.000000	0.000000	0.000000-M
				+M	0.000000	.066830	0.000000	0.000000	0.000000	0.000000	0.000000	.994083+M
-1.0024E+02	-.9942	0.0000	.9942	-M	.996437	0.000000	0.000000	0.000000	0.000000	0.000000	.083873	0.000000-M
				+M	0.000000	0.000000	0.000000	.008871	0.000000	0.000000	0.000000	0.000000+M
-1.0024E+02	.9942	0.0000	.9942	-M	0.000000	0.000000	0.000000	.008871	0.000000	0.000000	0.000000	0.000000-M
				+M	.996437	0.000000	0.000000	0.000000	0.000000	0.000000	.083873	0.000000+M
ENERGY	HEPAL	HEPER	HE		M=15/2	M=13/2	M=11/2	M=9/2	M=7/2	M=5/2	M=3/2	M=1/2

Y=-.60 X= .80 OAS=2.3949E+02*W246 B4F4/B2F2=-1.20E+00 B4F4/B6F6= 4.00E+00 E2F2/B6F6=-3.33E+00

ENERGY	HEPAL	HEPER	HE		M=15/2	M=13/2	M=11/2	M=9/2	M=7/2	M=5/2	M=3/2	M=1/2
1.1457E+02	.7325	.0026	.7325	-M	0.000000	.002888	0.000000	0.000000	0.000000	0.000000	0.000000	-.032171-M
				+M	0.000000	0.000000	.999478	0.000000	0.000000	0.000000	0.000000	0.000000+M
1.1457E+02	-.7325	.0026	.7325	-M	0.000000	0.000000	.999478	0.000000	0.000000	0.000000	0.000000	0.000000-M
				+M	0.000000	.002888	0.000000	0.000000	0.000000	0.000000	0.000000	-.032171+M
8.9448E+01	-.5977	0.0000	.5977	-M	0.000000	0.000000	0.000000	.998567	0.000000	0.000000	0.000000	0.000000-M
				+M	.000530	0.000000	0.000000	0.000000	0.000000	0.000000	-.053518	0.000000+M
8.9448E+01	.5977	0.0000	.5977	-M	.000530	0.000000	0.000000	0.000000	0.000000	0.000000	-.053518	0.000000-M
				+M	0.000000	0.000000	0.000000	.998567	0.000000	0.000000	0.000000	0.000000+M
6.6240E+01	.8660	-.0023	.8660	-M	0.000000	0.000000	-.003778	0.000000	0.000000	0.000000	0.000000	0.000000-M
				+M	0.000000	.999611	0.000000	0.000000	0.000000	0.000000	0.000000	-.027648+M
6.6240E+01	-.8660	-.0023	.8660	-M	0.000000	.999611	0.000000	0.000000	0.000000	0.000000	0.000000	-.027648-M
				+M	0.000000	0.000000	-.003778	0.000000	0.000000	0.000000	0.000000	0.000000+M
3.5936E+01	.4467	-.1543	.4726	-M	0.000000	0.000000	0.000000	0.000000	0.000000	-.158061	0.000000	0.000000-M
				+M	0.000000	0.000000	0.000000	0.000000	.987429	0.000000	0.000000	0.000000+M
3.5936E+01	-.4467	-.1543	.4726	-M	0.000000	0.000000	0.000000	0.000000	.987429	0.000000	0.000000	0.000000-M
				+M	0.000000	0.000000	0.000000	0.000000	0.000000	-.158061	0.000000	0.000000+M
-2.3936E+01	-.3133	.1543	.3493	-M	0.000000	0.000000	0.000000	0.000000	0.000000	.987429	0.000000	0.000000-M
				+M	0.000000	0.000000	0.000000	0.000000	.158061	0.000000	0.000000	0.000000+M
-2.3936E+01	.3133	.1543	.3493	-M	0.000000	0.000000	0.000000	0.000000	.158061	0.000000	0.000000	0.000000-M
				+M	0.000000	0.000000	0.000000	0.000000	0.000000	.987429	0.000000	0.000000+M
-6.6890E+01	.1988	0.0000	.1988	-M	0.000000	0.000000	0.000000	.053501	0.000000	0.000000	0.000000	0.000000-M
				+M	-.036548	0.000000	0.000000	0.000000	0.000000	0.000000	.997899	0.000000+M
-6.6890E+01	-.1988	0.0000	.1988	-M	-.036548	0.000000	0.000000	0.000000	0.000000	0.000000	.997899	0.000000-M
				+M	0.000000	0.000000	0.000000	.053501	0.000000	0.000000	0.000000	0.000000+M
-9.0452E+01	-.0665	.5330	.5371	-M	0.000000	.027755	0.000000	0.000000	0.000000	0.000000	0.000000	.999100-M
				+M	0.000000	0.000000	.032078	0.000000	0.000000	0.000000	0.000000	0.000000+M
-9.0452E+01	.0665	.5330	.5371	-M	0.000000	0.000000	.032078	0.000000	0.000000	0.000000	0.000000	0.000000-M
				+M	0.000000	.027755	0.000000	0.000000	0.000000	0.000000	0.000000	.999100+M
-1.2492E+02	-.9989	0.0000	.9989	-M	.999332	0.000000	0.000000	0.000000	0.000000	0.000000	.036524	0.000000-M
				+M	0.000000	0.000000	0.000000	.001427	0.000000	0.000000	0.000000	0.000000+M
-1.2492E+02	.9989	0.0000	.9989	-M	0.000000	0.000000	0.000000	.001427	0.000000	0.000000	0.000000	0.000000-M
				+M	.999332	0.000000	0.000000	0.000000	0.000000	0.000000	.036524	0.000000+M
ENERGY	HEPAL	HEPER	HE		M=15/2	M=13/2	M=11/2	M=9/2	M=7/2	M=5/2	M=3/2	M=1/2

Y=-.40 X=-.80 OAS=1.7313E+02*W246 B4F4/B2F2= 5.33E-01 B4F4/B6F6=-4.00E+00 B2F2/B6F6=-7.50E+00

ENERGY	HEPAL	HEPER	HE		M=15/2	M=13/2	M=11/2	M=9/2	M=7/2	M=5/2	M=3/2	M=1/2
1.0319E+02	.9997	0.0000	.9997	-M	0.000000	0.000000	0.000000	.000641	0.000000	0.000000	0.000000	0.000000-M
				+M	.999802	0.000000	0.000000	0.000000	0.000000	0.000000	-.019889	0.000000+M
1.0319E+02	-.9997	0.0000	.9997	-M	.999802	0.000000	0.000000	0.000000	0.000000	0.000000	-.019889	0.000000-M
				+M	0.000000	0.000000	0.000000	.000641	0.000000	0.000000	0.000000	0.000000+M
5.4182E+01	-.0673	.5326	.5369	-M	0.000000	-.047185	0.000000	0.000000	0.000000	0.000000	0.000000	.998178-M
				+M	0.000000	0.000000	-.037600	0.000000	0.000000	0.000000	0.000000	0.000000+M
5.4182E+01	.0673	.5326	.5369	-M	0.000000	0.000000	-.037600	0.000000	0.000000	0.000000	0.000000	0.000000-M
				+M	0.000000	-.047185	0.000000	0.000000	0.000000	0.000000	0.000000	.998178+M
3.2157E+01	.1979	0.0000	.1979	-M	0.000000	0.000000	0.000000	-.054643	0.000000	0.000000	0.000000	0.000000-M
				+M	.019894	0.000000	0.000000	0.000000	0.000000	0.000000	.998308	0.000000+M
3.2157E+01	-.1979	0.0000	.1979	-M	.019894	0.000000	0.000000	0.000000	0.000000	0.000000	.998308	0.000000-M
				+M	0.000000	0.000000	0.000000	-.054643	0.000000	0.000000	0.000000	0.000000+M
-4.2794E+00	.3148	-.1489	.3482	-M	0.000000	0.000000	0.000000	0.000000	-.152395	0.000000	0.000000	0.000000-M
				+M	0.000000	0.000000	0.000000	0.000000	0.000000	.988320	0.000000	0.000000+M
-4.2794E+00	-.3148	-.1489	.3482	-M	0.000000	0.000000	0.000000	0.000000	0.000000	.988320	0.000000	0.000000-M
				+M	0.000000	0.000000	0.000000	0.000000	-.152395	0.000000	0.000000	0.000000+M
-7.2967E+00	-.8649	-.0015	.8649	-M	0.000000	.998884	0.000000	0.000000	0.000000	0.000000	0.000000	.047077-M
				+M	0.000000	0.000000	-.003759	0.000000	0.000000	0.000000	0.000000	0.000000+M
-7.2967E+00	.8649	-.0015	.8649	-M	0.000000	0.000000	-.003759	0.000000	0.000000	0.000000	0.000000	0.000000-M
				+M	0.000000	.998884	0.000000	0.000000	0.000000	0.000000	0.000000	.047077+M
-4.5641E+01	-.4481	.1489	.4722	-M	0.000000	0.000000	0.000000	0.000000	.988320	0.000000	0.000000	0.000000-M
				+M	0.000000	0.000000	0.000000	0.000000	0.000000	.152395	0.000000	0.000000+M
-4.5641E+01	.4481	.1489	.4722	-M	0.000000	0.000000	0.000000	0.000000	0.000000	.152395	0.000000	0.000000-M
				+M	0.000000	0.000000	0.000000	0.000000	.988320	0.000000	0.000000	0.000000+M
-6.2366E+01	-.7322	.0022	.7322	-M	0.000000	0.000000	.999286	0.000000	0.000000	0.000000	0.000000	0.000000-M
				+M	0.000000	.001982	0.000000	0.000000	0.000000	0.000000	0.000000	.037736+M
-6.2366E+01	.7322	.0022	.7322	-M	0.000000	.001982	0.000000	0.000000	0.000000	0.000000	0.000000	.037736-M
				+M	0.000000	0.000000	.999286	0.000000	0.000000	0.000000	0.000000	0.000000+M
-6.9945E+01	.5976	0.0000	.5976	-M	.000447	0.000000	0.000000	0.000000	0.000000	0.000000	.054645	0.000000-M
				+M	0.000000	0.000000	0.000000	.998506	0.000000	0.000000	0.000000	0.000000+M
-6.9945E+01	-.5976	0.0000	.5976	-M	0.000000	0.000000	0.000000	.998506	0.000000	0.000000	0.000000	0.000000-M
				+M	.000447	0.000000	0.000000	0.000000	0.000000	0.000000	.054645	0.000000+M
ENERGY	HEPAL	HEPER	HE		M=15/2	M=13/2	M=11/2	M=9/2	M=7/2	M=5/2	M=3/2	M=1/2

Y=-.40 X=-.60 OAS=1.3605E+02*W246 B4F4/B2F2= 4.00E-01 B4F4/B6F6=-1.50E+00 B2F2/B6F6=-3.75E+00

ENERGY	HEPAL	HEPER	HE		M=15/2	M=13/2	M=11/2	M=9/2	M=7/2	M=5/2	M=3/2	M=1/2
7.6275E+01	.9976	0.0000	.9976	-M	0.000000	0.000000	0.000000	.004518	0.000000	0.000000	0.000000	0.000000-M
				+M	.998500	0.000000	0.000000	0.000000	0.000000	0.000000	-.054557	0.000000+M
7.6275E+01	-.9976	0.0000	.9976	-M	.998500	0.000000	0.000000	0.000000	0.000000	0.000000	-.054557	0.000000-M
				+M	0.000000	0.000000	0.000000	.004518	0.000000	0.000000	0.000000	0.000000+M
4.6542E+01	.0779	.5262	.5320	-M	0.000000	0.000000	-.098029	0.000000	0.000000	0.000000	0.000000	0.000000-M
				+M	0.000000	-.153855	0.000000	0.000000	0.000000	0.000000	0.000000	.983219+M
4.6542E+01	-.0779	.5262	.5320	-M	0.000000	-.153855	0.000000	0.000000	0.000000	0.000000	0.000000	.983219-M
				+M	0.000000	0.000000	-.098029	0.000000	0.000000	0.000000	0.000000	0.000000+M
2.4897E+01	.1883	0.0000	.1883	-M	0.000000	0.000000	0.000000	-.132593	0.000000	0.000000	0.000000	0.000000-M
				+M	.054674	0.000000	0.000000	0.000000	0.000000	0.000000	.989661	0.000000+M
2.4897E+01	-.1883	0.0000	.1883	-M	.054674	0.000000	0.000000	0.000000	0.000000	0.000000	.989661	0.000000-M
				+M	0.000000	0.000000	0.000000	-.132593	0.000000	0.000000	0.000000	0.000000+M
8.5881E+00	.8471	-.0063	.8471	-M	0.000000	0.000000	-.026652	0.000000	0.000000	0.000000	0.000000	0.000000-M
				+M	0.000000	.988029	0.000000	0.000000	0.000000	0.000000	0.000000	.151950+M
8.5881E+00	-.8471	-.0063	.8471	-M	0.000000	.988029	0.000000	0.000000	0.000000	0.000000	0.000000	.151950-M
				+M	0.000000	0.000000	-.026652	0.000000	0.000000	0.000000	0.000000	0.000000+M
-6.4922E+00	-.2521	-.2986	.3908	-M	0.000000	0.000000	0.000000	0.000000	0.000000	.947888	0.000000	0.000000-M
				+M	0.000000	0.000000	0.000000	0.000000	-.318604	0.000000	0.000000	0.000000+M
-6.4922E+00	.2521	-.2986	.3908	-M	0.000000	0.000000	0.000000	0.000000	-.318604	0.000000	0.000000	0.000000-M
				+M	0.000000	0.000000	0.000000	0.000000	0.000000	.947888	0.000000	0.000000+M
-4.2290E+01	.7250	.0134	.7251	-M	0.000000	.011309	0.000000	0.000000	0.000000	0.000000	0.000000	.100956-M
				+M	0.000000	0.000000	.994827	0.000000	0.000000	0.000000	0.000000	0.000000+M
-4.2290E+01	-.7250	.0134	.7251	-M	0.000000	0.000000	.994827	0.000000	0.000000	0.000000	0.000000	0.000000-M
				+M	0.000000	.011309	0.000000	0.000000	0.000000	0.000000	0.000000	.100956+M
-4.7748E+01	.3855	.2986	.4876	-M	0.000000	0.000000	0.000000	0.000000	0.000000	.318604	0.000000	0.000000-M
				+M	0.000000	0.000000	0.000000	0.000000	.947888	0.000000	0.000000	0.000000+M
-4.7748E+01	-.3855	.2986	.4876	-M	0.000000	0.000000	0.000000	0.000000	.947888	0.000000	0.000000	0.000000-M
				+M	0.000000	0.000000	0.000000	0.000000	0.000000	.318604	0.000000	0.000000+M
-5.9771E+01	-.5859	0.0000	.5859	-M	0.000000	0.000000	0.000000	.991160	0.000000	0.000000	0.000000	0.000000-M
				+M	.002762	0.000000	0.000000	0.000000	0.000000	0.000000	.132641	0.000000+M
-5.9771E+01	.5859	0.0000	.5859	-M	.002762	0.000000	0.000000	0.000000	0.000000	0.000000	.132641	0.000000-
				+M	0.000000	0.000000	0.000000	.991160	0.000000	0.000000	0.000000	0.000000+M
ENERGY	HEPAL	HEPER	HE		M=15/2	M=13/2	M=11/2	M=9/2	M=7/2	M=5/2	M=3/2	M=1/2

Y=-.40 X=-.40 OAS=1.0179E+02*W246 B4F4/B2F2= 2.67E-01 B4F4/B6F6=-6.67E-01 B2F2/B6F6=-2.50E+00

ENERGY	HEPAL	HEPER	HE		M=15/2	M=13/2	M=11/2	M=9/2	M=7/2	M=5/2	M=3/2	M=1/2
4.9652E+01	.9849	0.0000	.9849	-M	0.000000	0.000000	0.000000	.023097	0.000000	0.000000	0.000000	0.000000-
				+M	.990788	0.000000	0.000000	0.000000	0.000000	0.000000	-.133436	0.000000+M
4.9652E+01	-.9849	0.0000	.9849	-M	.990788	0.000000	0.000000	0.000000	0.000000	0.000000	-.133436	0.000000-
				+M	0.000000	0.000000	0.000000	.023097	0.000000	0.000000	0.000000	0.000000+M
4.2880E+01	.2047	.4642	.5074	-M	0.000000	0.000000	-.180855	0.000000	0.000000	0.000000	0.000000	0.000000-
				+M	0.000000	-.453086	0.000000	0.000000	0.000000	0.000000	0.000000	.872929+M
4.2880E+01	-.2047	.4642	.5074	-M	0.000000	-.453086	0.000000	0.000000	0.000000	0.000000	0.000000	.872929-
				+M	0.000000	0.000000	-.180855	0.000000	0.000000	0.000000	0.000000	0.000000+
2.1877E+01	.6868	.0161	.6869	-M	0.000000	0.000000	-.134502	0.000000	0.000000	0.000000	0.000000	0.000000-
				+M	0.000000	.890620	0.000000	0.000000	0.000000	0.000000	0.000000	.434402+M
2.1877E+01	-.6868	.0161	.6869	-M	0.000000	.890620	0.000000	0.000000	0.000000	0.000000	0.000000	.434402-
				+M	0.000000	0.000000	-.134502	0.000000	0.000000	0.000000	0.000000	0.000000+
1.8889E+01	-.1670	0.0000	.1670	-M	.135022	0.000000	0.000000	0.000000	0.000000	0.000000	.960347	0.000000-
				+M	0.000000	0.000000	0.000000	-.243932	0.000000	0.000000	0.000000	0.000000+
1.8889E+01	.1670	0.0000	.1670	-M	0.000000	0.000000	0.000000	-.243932	0.000000	0.000000	0.000000	0.000000-
				+M	.135022	0.000000	0.000000	0.000000	0.000000	0.000000	.960347	0.000000+M
-6.4227E+00	-.1636	-.4042	.4361	-M	0.000000	0.000000	0.000000	0.000000	0.000000	.887622	0.000000	0.000000-
				+M	0.000000	0.000000	0.000000	0.000000	-.460573	0.000000	0.000000	0.000000+
-6.4227E+00	.1636	-.4042	.4361	-M	0.000000	0.000000	0.000000	0.000000	-.460573	0.000000	0.000000	0.000000-
				+M	0.000000	0.000000	0.000000	0.000000	0.000000	.887622	0.000000	0.000000+
-2.3597E+01	.6915	.0530	.6935	-M	0.000000	.038847	0.000000	0.000000	0.000000	0.000000	0.000000	.222014-
				+M	0.000000	0.000000	.974269	0.000000	0.000000	0.000000	0.000000	0.000000+
-2.3597E+01	-.6915	.0530	.6935	-M	0.000000	0.000000	.974269	0.000000	0.000000	0.000000	0.000000	0.000000-
				+M	0.000000	.038847	0.000000	0.000000	0.000000	0.000000	0.000000	.222014+
-5.1141E+01	-.5519	0.0000	.5519	-M	0.000000	0.000000	0.000000	.969517	0.000000	0.000000	0.000000	0.000000-
				+M	.010368	0.000000	0.000000	0.000000	0.000000	0.000000	.244804	0.000000+
-5.1141E+01	.5519	0.0000	.5519	-M	.010368	0.000000	0.000000	0.000000	0.000000	0.000000	.244804	0.000000-
				+M	0.000000	0.000000	0.000000	.969517	0.000000	0.000000	0.000000	0.000000+
-5.2137E+01	-.2970	.4042	.5016	-M	0.000000	0.000000	0.000000	0.000000	.887622	0.000000	0.000000	0.000000-
				+M	0.000000	0.000000	0.000000	0.000000	0.000000	.460573	0.000000	0.000000+
-5.2137E+01	.2970	.4042	.5016	-M	0.000000	0.000000	0.000000	0.000000	0.000000	.460573	0.000000	0.000000
				+M	0.000000	0.000000	0.000000	0.000000	.887622	0.000000	0.000000	0.000000+
ENERGY	HEPAL	HEPER	HE		M=15/2	M=13/2	M=11/2	M=9/2	M=7/2	M=5/2	M=3/2	M=1/2

Y=-.40 X=-.20 OAS=1.0860E+02*W246 B4F4/B2F2= 1.33E-01 B4F4/B6F6=-2.50E-01 B2F2/B6F6=-1.87E+00

ENERGY	HEPAL	HEPER	HE		M=15/2	M=13/2	M=11/2	M=9/2	M=7/2	M=5/2	M=3/2	M=1/2
5.0358E+01	.5783	.2635	.6355	-M	0.000000	0.000000	.188150	0.000000	0.000000	0.000000	0.000000	0.000000-M
				+M	0.000000	.821570	0.000000	0.000000	0.000000	0.000000	0.000000	-.538166+M
5.0358E+01	-.5783	.2635	.6355	-M	0.000000	.821570	0.000000	0.000000	0.000000	0.000000	0.000000	-.538166-M
				+M	0.000000	0.000000	.188150	0.000000	0.000000	0.000000	0.000000	0.000000+M
2.8231E+01	-.1637	.0892	.1864	-M	0.000000	.560490	0.000000	0.000000	0.000000	0.000000	0.000000	.702092-M
				+M	0.000000	0.000000	-.439224	0.000000	0.000000	0.000000	0.000000	0.000000+M
2.8231E+01	.1637	.0892	.1864	-M	0.000000	0.000000	-.439224	0.000000	0.000000	0.000000	0.000000	0.000000-M
				+M	0.000000	.560490	0.000000	0.000000	0.000000	0.000000	0.000000	.702092+M
2.5207E+01	.7654	0.0000	.7654	-M	0.000000	0.000000	0.000000	.176313	0.000000	0.000000	0.000000	0.000000-M
				+M	.858966	0.000000	0.000000	0.000000	0.000000	0.000000	-.480718	0.000000+M
2.5207E+01	-.7654	0.0000	.7654	-M	.858966	0.000000	0.000000	0.000000	0.000000	0.000000	-.480718	0.000000-M
				+M	0.000000	0.000000	0.000000	.176313	0.000000	0.000000	0.000000	0.000000+M
1.3375E+01	.3073	0.0000	.3073	-M	0.000000	0.000000	0.000000	-.356303	0.000000	0.000000	0.000000	0.000000-M
				+M	.510944	0.000000	0.000000	0.000000	0.000000	0.000000	.782294	0.000000+M
1.3375E+01	-.3073	0.0000	.3073	-M	.510944	0.000000	0.000000	0.000000	0.000000	0.000000	.782294	0.000000-M
				+M	0.000000	0.000000	0.000000	-.356303	0.000000	0.000000	0.000000	0.000000+M
-4.6344E+00	-.0808	-.4596	.4666	-M	0.000000	0.000000	0.000000	0.000000	0.000000	.827218	0.000000	0.000000-M
				+M	0.000000	0.000000	0.000000	0.000000	-.561881	0.000000	0.000000	0.000000+M
-4.6344E+00	.0808	-.4596	.4666	-M	0.000000	0.000000	0.000000	0.000000	-.561881	0.000000	0.000000	0.000000-M
				+M	0.000000	0.000000	0.000000	0.000000	0.000000	.827218	0.000000	0.000000+M
-9.1086E+00	-.5420	.1806	.5713	-M	0.000000	0.000000	.878454	0.000000	0.000000	0.000000	0.000000	0.000000-M
				+M	0.000000	.104277	0.000000	0.000000	0.000000	0.000000	0.000000	.466310+M
-9.1086E+00	.5420	.1806	.5713	-M	0.000000	.104277	0.000000	0.000000	0.000000	0.000000	0.000000	.466310-M
				+M	0.000000	0.000000	.878454	0.000000	0.000000	0.000000	0.000000	0.000000+M
-4.5182E+01	.4727	0.0000	.4727	-M	.033352	0.000000	0.000000	0.000000	0.000000	0.000000	.396139	0.000000-M
				+M	0.000000	0.000000	0.000000	.917585	0.000000	0.000000	0.000000	0.000000+M
-4.5182E+01	-.4727	0.0000	.4727	-M	0.000000	0.000000	0.000000	.917585	0.000000	0.000000	0.000000	0.000000-M
				+M	.033352	0.000000	0.000000	0.000000	0.000000	0.000000	.396139	0.000000+M
-5.8246E+01	-.2141	.4596	.5070	-M	0.000000	0.000000	0.000000	0.000000	.827218	0.000000	0.000000	0.000000-M
				+M	0.000000	0.000000	0.000000	0.000000	0.000000	.561881	0.000000	0.000000+M
-5.8246E+01	.2141	.4596	.5070	-M	0.000000	0.000000	0.000000	0.000000	0.000000	.561881	0.000000	0.000000-M
				+M	0.000000	0.000000	0.000000	0.000000	.827218	0.000000	0.000000	0.000000+M
ENERGY	HEPAL	HEPER	HE		M=15/2	M=13/2	M=11/2	M=9/2	M=7/2	M=5/2	M=3/2	M=1/2

Y=-.40 X=0.00 OAS=1.3055E+02*W246 B4F4/B2F2= 0. B4F4/B6F6= 0. B2F2/B6F6=-1.50E+00

ENERGY	HEPAL	HEPER	HE		M=15/2	M=13/2	M=11/2	M=9/2	M=7/2	M=5/2	M=3/2	M=1/2
6.5110E+01	.7114	.1850	.7350	-M	0.000000	0.000000	.179141	0.000000	0.000000	0.000000	0.000000	0.000000-M
				+M	0.000000	.915399	0.000000	0.000000	0.000000	0.000000	0.000000	-.360489+M
6.5110E+01	-.7114	.1850	.7350	-M	0.000000	.915399	0.000000	0.000000	0.000000	0.000000	0.000000	-.360489-M
				+M	0.000000	0.000000	.179141	0.000000	0.000000	0.000000	0.000000	0.000000+M
3.7320E+01	.2728	-.0275	.2742	-M	0.000000	-.365079	0.000000	0.000000	0.000000	0.000000	0.000000	-.555974-M
				+M	0.000000	0.000000	.746733	0.000000	0.000000	0.000000	0.000000	0.000000+M
3.7320E+01	-.2728	-.0275	.2742	-M	0.000000	0.000000	.746733	0.000000	0.000000	0.000000	0.000000	0.000000-M
				+M	0.000000	-.365079	0.000000	0.000000	0.000000	0.000000	0.000000	-.555974+M
1.8769E+01	-.0241	0.0000	.0241	-M	-.242299	0.000000	0.000000	0.000000	0.000000	0.000000	.814066	0.000000-M
				+M	0.000000	0.000000	0.000000	-.527814	0.000000	0.000000	0.000000	0.000000+M
1.8769E+01	.0241	0.0000	.0241	-M	0.000000	0.000000	0.000000	-.527814	0.000000	0.000000	0.000000	0.000000-M
				+M	-.242299	0.000000	0.000000	0.000000	0.000000	0.000000	.814066	0.000000+M
-1.7604E+00	-.0162	-.4837	.4839	-M	0.000000	0.000000	0.000000	0.000000	0.000000	.776946	0.000000	0.000000-M
				+M	0.000000	0.000000	0.000000	0.000000	-.629568	0.000000	0.000000	0.000000+M
-1.7604E+00	.0162	-.4837	.4839	-M	0.000000	0.000000	0.000000	0.000000	-.629568	0.000000	0.000000	0.000000-M
				+M	0.000000	0.000000	0.000000	0.000000	0.000000	.776946	0.000000	0.000000+M
-4.6299E+00	.2386	.3758	.4451	-M	0.000000	.169591	0.000000	0.000000	0.000000	0.000000	0.000000	.748960-M
				+M	0.000000	0.000000	.640545	0.000000	0.000000	0.000000	0.000000	0.000000+M
-4.6299E+00	-.2386	.3758	.4451	-M	0.000000	0.000000	.640545	0.000000	0.000000	0.000000	0.000000	0.000000-M
				+M	0.000000	.169591	0.000000	0.000000	0.000000	0.000000	0.000000	.748960+M
-6.0552E+00	-.9054	0.0000	.9054	-M	.964623	0.000000	0.000000	0.000000	0.000000	0.000000	.143879	0.000000-M
				+M	0.000000	0.000000	0.000000	-.220912	0.000000	0.000000	0.000000	0.000000+M
-6.0552E+00	.9054	0.0000	.9054	-M	0.000000	0.000000	0.000000	-.220912	0.000000	0.000000	0.000000	0.000000-M
				+M	.964623	0.000000	0.000000	0.000000	0.000000	0.000000	.143879	0.000000+M
-4.3314E+01	.3295	0.0000	.3295	-M	.103896	0.000000	0.000000	0.000000	0.000000	0.000000	.562668	0.000000-M
				+M	0.000000	0.000000	0.000000	.820128	0.000000	0.000000	0.000000	0.000000+M
-4.3314E+01	-.3295	0.0000	.3295	-M	0.000000	0.000000	0.000000	.820128	0.000000	0.000000	0.000000	0.000000-M
				+M	.103896	0.000000	0.000000	0.000000	0.000000	0.000000	.562668	0.000000+M
-6.5440E+01	-.1496	.4837	.5063	-M	0.000000	0.000000	0.000000	0.000000	.776946	0.000000	0.000000	0.000000-M
				+M	0.000000	0.000000	0.000000	0.000000	0.000000	.629568	0.000000	0.000000+M
-6.5440E+01	.1496	.4837	.5063	-M	0.000000	0.000000	0.000000	0.000000	0.000000	.629568	0.000000	0.000000-M
				+M	0.000000	0.000000	0.000000	0.000000	.776946	0.000000	0.000000	0.000000+M
ENERGY	HEPAL	HEPER	HE		M=15/2	M=13/2	M=11/2	M=9/2	M=7/2	M=5/2	M=3/2	M=1/2

Y=-.40 X= .20 OAS=1.1007E+02*W246 B4F4/B2F2=-1.33E-01 B4F4/B6F6= 2.50E-01 B2F2/B6F6=-1.87E+00

ENERGY	HEPAL	HEPER	HE		M=15/2	M=13/2	M=11/2	M=9/2	M=7/2	M=5/2	M=3/2	M=1/2
5.9954E+01	.7888	.1323	.7999	-M	0.000000	0.000000	.157232	0.000000	0.000000	0.000000	0.000000	0.000000-M
				+M	0.000000	.963035	0.000000	0.000000	0.000000	0.000000	0.000000	-.218728+M
5.9954E+01	-.7888	.1323	.7999	-M	0.000000	.963035	0.000000	0.000000	0.000000	0.000000	0.000000	-.218728-M
				+M	0.000000	0.000000	.157232	0.000000	0.000000	0.000000	0.000000	0.000000+M
4.1314E+01	-.5816	-.0948	.5893	-M	0.000000	0.000000	.926790	0.000000	0.000000	0.000000	0.000000	0.000000-M
				+M	0.000000	-.220387	0.000000	0.000000	0.000000	0.000000	0.000000	-.304120+M
4.1314E+01	.5816	-.0948	.5893	-M	0.000000	-.220387	0.000000	0.000000	0.000000	0.000000	0.000000	-.304120-M
				+M	0.000000	0.000000	.926790	0.000000	0.000000	0.000000	0.000000	0.000000+M
1.4663E+01	-.2706	0.0000	.2706	-M	0.000000	0.000000	0.000000	.773179	0.000000	0.000000	0.000000	0.000000-M
				+M	.097691	0.000000	0.000000	0.000000	0.000000	0.000000	-.626619	0.000000+M
1.4663E+01	.2706	0.0000	.2706	-M	.097691	0.000000	0.000000	0.000000	0.000000	0.000000	-.626619	0.000000-M
				+M	0.000000	0.000000	0.000000	.773179	0.000000	0.000000	0.000000	0.000000+M
-2.8154E-01	.0673	-.4944	.4990	-M	0.000000	0.000000	0.000000	0.000000	0.000000	-.706539	0.000000	0.000000-M
				+M	0.000000	0.000000	0.000000	0.000000	.707674	0.000000	0.000000	0.000000+M
-2.8154E-01	-.0673	-.4944	.4990	-M	0.000000	0.000000	0.000000	0.000000	.707674	0.000000	0.000000	0.000000-M
				+M	0.000000	0.000000	0.000000	0.000000	0.000000	-.706539	0.000000	0.000000+M
-1.2109E+01	-.0072	.4958	.4958	-M	0.000000	0.000000	.341083	0.000000	0.000000	0.000000	0.000000	0.000000-M
				+M	0.000000	.154897	0.000000	0.000000	0.000000	0.000000	0.000000	.927183+M
-1.2109E+01	.0072	.4958	.4958	-M	0.000000	.154897	0.000000	0.000000	0.000000	0.000000	0.000000	.927183-M
				+M	0.000000	0.000000	.341083	0.000000	0.000000	0.000000	0.000000	0.000000+M
-2.0245E+01	-.8126	0.0000	.8126	-M	-.925656	0.000000	0.000000	0.000000	0.000000	0.000000	.228139	0.000000-M
				+M	0.000000	0.000000	0.000000	.301850	0.000000	0.000000	0.000000	0.000000+M
-2.0245E+01	.8126	0.0000	.8126	-M	0.000000	0.000000	0.000000	-.301850	0.000000	0.000000	0.000000	0.000000-M
				+M	.925656	0.000000	0.000000	0.000000	0.000000	0.000000	-.228139	0.000000+M
-3.3178E+01	-.0580	0.0000	.0580	-M	.365537	0.000000	0.000000	0.000000	0.000000	0.000000	.745186	0.000000-M
				+M	0.000000	0.000000	0.000000	.557746	0.000000	0.000000	0.000000	0.000000+M
-3.3178E+01	.0580	0.0000	.0580	-M	0.000000	0.000000	0.000000	.557746	0.000000	0.000000	0.000000	0.000000-M
				+M	.365537	0.000000	0.000000	0.000000	0.000000	0.000000	.745186	0.000000+M
-5.0118E+01	-.0660	.4944	.4988	-M	0.000000	0.000000	0.000000	0.000000	.706539	0.000000	0.000000	0.000000-M
				+M	0.000000	0.000000	0.000000	0.000000	0.000000	.707674	0.000000	0.000000+M
-5.0118E+01	.0660	.4944	.4988	-M	0.000000	0.000000	0.000000	0.000000	0.000000	.707674	0.000000	0.000000-M
				+M	0.000000	0.000000	0.000000	0.000000	.706539	0.000000	0.000000	0.000000+M
ENERGY	HEPAL	HEPER	HE		M=15/2	M=13/2	M=11/2	M=9/2	M=7/2	M=5/2	M=3/2	M=1/2

Y=-.40 X= .40 OAS=9.6832E+01*W246 B4F4/B2F2=-2.67E-01 B4F4/B6F6= 6.67E-01 B2F2/B6F6=-2.50E+00

ENERGY	HEPAL	HEPER	HE		M=15/2	M=13/2	M=11/2	M=9/2	M=7/2	M=5/2	M=3/2	M=1/2
5.6620E+01	-.7028	.2126	.7342	-M	0.000000	.941490	0.000000	0.000000	0.000000	0.000000	0.000000	-.149327-M
				+M	0.000000	0.000000	.302155	0.000000	0.000000	0.000000	0.000000	0.000000+M
5.6620E+01	.7028	.2126	.7342	-M	0.000000	0.000000	.302155	0.000000	0.000000	0.000000	0.000000	0.000000-M
				+M	0.000000	.941490	0.000000	0.000000	0.000000	0.000000	0.000000	-.149327+M
5.1865E+01	-.5560	-.2044	.5924	-M	0.000000	0.000000	.938832	0.000000	0.000000	0.000000	0.000000	0.000000-M
				+M	0.000000	-.321058	0.000000	0.000000	0.000000	0.000000	0.000000	-.124562+M
5.1865E+01	.5560	-.2044	.5924	-M	0.000000	-.321058	0.000000	0.000000	0.000000	0.000000	0.000000	-.124562-M
				+M	0.000000	0.000000	.938832	0.000000	0.000000	0.000000	0.000000	0.000000+M
2.3118E+01	.5154	0.0000	.5154	-M	.022371	0.000000	0.000000	0.000000	0.000000	0.000000	-.323588	0.000000-M
				+M	0.000000	0.000000	0.000000	.945934	0.000000	0.000000	0.000000	0.000000+M
2.3118E+01	-.5154	0.0000	.5154	-M	0.000000	0.000000	0.000000	.945934	0.000000	0.000000	0.000000	0.000000-M
				+M	.022371	0.000000	0.000000	0.000000	0.000000	0.000000	-.323588	0.000000+M
3.0468E+00	.2013	-.4656	.5072	-M	0.000000	0.000000	0.000000	0.000000	0.000000	-.575943	0.000000	0.000000-M
				+M	0.000000	0.000000	0.000000	0.000000	.817490	0.000000	0.000000	0.000000+M
3.0468E+00	-.2013	-.4656	.5072	-M	0.000000	0.000000	0.000000	0.000000	.817490	0.000000	0.000000	0.000000-M
				+M	0.000000	0.000000	0.000000	0.000000	0.000000	-.575943	0.000000	0.000000+M
-2.7965E+01	.0532	.5251	.5278	-M	0.000000	0.000000	.165216	0.000000	0.000000	0.000000	0.000000	0.000000-M
				+M	0.000000	.102556	0.000000	0.000000	0.000000	0.000000	0.000000	.980911+M
-2.7965E+01	-.0532	.5251	.5278	-M	0.000000	.102556	0.000000	0.000000	0.000000	0.000000	0.000000	.980911-M
				+M	0.000000	0.000000	.165216	0.000000	0.000000	0.000000	0.000000	0.000000+M
-2.9825E+01	-.2736	0.0000	.2736	-M	-.427801	0.000000	0.000000	0.000000	0.000000	0.000000	.852072	0.000000-M
				+M	0.000000	0.000000	0.000000	.301597	0.000000	0.000000	0.000000	0.000000+M
-2.9825E+01	.2736	0.0000	.2736	-M	0.000000	0.000000	0.000000	.301597	0.000000	0.000000	0.000000	0.000000-M
				+M	-.427801	0.000000	0.000000	0.000000	0.000000	0.000000	.852072	0.000000+M
-3.6647E+01	-.0680	.4656	.4705	-M	0.000000	0.000000	0.000000	0.000000	0.000000	.817490	0.000000	0.000000-M
				+M	0.000000	0.000000	0.000000	0.000000	.575943	0.000000	0.000000	0.000000+M
-3.6647E+01	.0680	.4656	.4705	-M	0.000000	0.000000	0.000000	0.000000	.575943	0.000000	0.000000	0.000000-M
				+M	0.000000	0.000000	0.000000	0.000000	0.000000	.817490	0.000000	0.000000+M
-4.0213E+01	-.8418	0.0000	.8418	-M	.903596	0.000000	0.000000	0.000000	0.000000	0.000000	.411418	0.000000-M
				+M	0.000000	0.000000	0.000000	.119369	0.000000	0.000000	0.000000	0.000000+M
-4.0213E+01	.8418	0.0000	.8418	-M	0.000000	0.000000	0.000000	.119369	0.000000	0.000000	0.000000	0.000000-M
				+M	.903596	0.000000	0.000000	0.000000	0.000000	0.000000	.411418	0.000000+M
ENERGY	HEPAL	HEPER	HE		M=15/2	M=13/2	M=11/2	M=9/2	M=7/2	M=5/2	M=3/2	M=1/2

Y=-.40 X= .60 OAS=1.2082E+02*W246 B4F4/B2F2=-4.00E-01 B4F4/B6F6= 1.50E+00 E2F2/B6F6=-3.75E+00

ENERGY	HEPAL	HEPER	HE		M=15/2	M=13/2	M=11/2	M=9/2	M=7/2	M=5/2	M=3/2	M=1/2
6.5388E+01	-.7258	.0303	.7265	-M	0.000000	0.000000	.996031	0.000000	0.000000	0.000000	0.000000	0.000000-M
				+M	0.000000	.038140	0.000000	0.000000	0.000000	0.000000	0.000000	-.080418+M
6.5388E+01	.7258	.0303	.7265	-M	0.000000	.038140	0.000000	0.000000	0.000000	0.000000	0.000000	-.080418-M
				+M	0.000000	0.000000	.996031	0.000000	0.000000	0.000000	0.000000	0.000000+M
5.3476E+01	.8614	-.0284	.8619	-M	0.000000	0.000000	-.042595	0.000000	0.000000	0.000000	0.000000	0.000000-M
				+M	0.000000	.997609	0.000000	0.000000	0.000000	0.000000	0.000000	-.054428+M
5.3476E+01	-.8614	-.0284	.8619	-M	0.000000	.997609	0.000000	0.000000	0.000000	0.000000	0.000000	-.054428-M
				+M	0.000000	0.000000	-.042595	0.000000	0.000000	0.000000	0.000000	0.000000+M
3.9791E+01	.5840	0.0000	.5840	-M	.004222	0.000000	0.000000	0.000000	0.000000	0.000000	-.141315	0.000000-M
				+M	0.000000	0.000000	0.000000	.989956	0.000000	0.000000	0.000000	0.000000+M
3.9791E+01	-.5840	0.0000	.5840	-M	0.000000	0.000000	0.000000	.989956	0.000000	0.000000	0.000000	0.000000-M
				+M	.004222	0.000000	0.000000	0.000000	0.000000	0.000000	-.141315	0.000000+M
9.8388E+00	.3588	-.3377	.4927	-M	0.000000	0.000000	0.000000	0.000000	0.000000	-.367212	0.000000	0.000000-M
				+M	0.000000	0.000000	0.000000	0.000000	.930137	0.000000	0.000000	0.000000+M
9.8388E+00	-.3588	-.3377	.4927	-M	0.000000	0.000000	0.000000	0.000000	.930137	0.000000	0.000000	0.000000-M
				+M	0.000000	0.000000	0.000000	0.000000	0.000000	-.367212	0.000000	0.000000+M
-2.6639E+01	-.2255	.3377	.4061	-M	0.000000	0.000000	0.000000	0.000000	0.000000	.930137	0.000000	0.000000-M
				+M	0.000000	0.000000	0.000000	0.000000	.367212	0.000000	0.000000	0.000000+M
-2.6639E+01	.2255	.3377	.4061	-M	0.000000	0.000000	0.000000	0.000000	.367212	0.000000	0.000000	0.000000-M
				+M	0.000000	0.000000	0.000000	0.000000	0.000000	.930137	0.000000	0.000000+M
-3.9442E+01	-.2097	0.0000	.2097	-M	-.178083	0.000000	0.000000	0.000000	0.000000	0.000000	.974034	0.000000-M
				+M	0.000000	0.000000	0.000000	.139802	0.000000	0.000000	0.000000	0.000000+M
-3.9442E+01	.2097	0.0000	.2097	-M	0.000000	0.000000	0.000000	.139802	0.000000	0.000000	0.000000	0.000000-M
				+M	-.178083	0.000000	0.000000	0.000000	0.000000	0.000000	.974034	0.000000+M
-4.6984E+01	.0644	.5315	.5354	-M	0.000000	0.000000	.078149	0.000000	0.000000	0.000000	0.000000	0.000000-M
				+M	0.000000	.057637	0.000000	0.000000	0.000000	0.000000	0.000000	.995274+M
-4.6984E+01	-.0644	.5315	.5354	-M	0.000000	.057637	0.000000	0.000000	0.000000	0.000000	0.000000	.995274-M
				+M	0.000000	0.000000	.078149	0.000000	0.000000	0.000000	0.000000	0.000000+M
-5.5429E+01	-.9743	0.0000	.9743	-M	.984006	0.000000	0.000000	0.000000	0.000000	0.000000	.176885	0.000000-M
				+M	0.000000	0.000000	0.000000	.021053	0.000000	0.000000	0.000000	0.000000+M
-5.5429E+01	.9743	0.0000	.9743	-M	0.000000	0.000000	0.000000	.021053	0.000000	0.000000	0.000000	0.000000-M
				+M	.984006	0.000000	0.000000	0.000000	0.000000	0.000000	.176885	0.000000+M
ENERGY	HEPAL	HEPER	HE		M=15/2	M=13/2	M=11/2	M=9/2	M=7/2	M=5/2	M=3/2	M=1/2

Y=-.40 X= .80 OAS=1.5103E+02*W246 B4F4/B2F2=-5.33E-01 B4F4/B6F6= 4.00E+00 E2F2/B6F6=-7.50E+00

ENERGY	HEPAL	HEPER	HE		M=15/2	M=13/2	M=11/2	M=9/2	M=7/2	M=5/2	M=3/2	M=1/2
7.9371E+01	.7326	.0026	.7326	-M	0.000000	.003073	0.000000	0.000000	0.000000	0.000000	0.000000	-.029982-M
				+M	0.000000	0.000000	.999546	0.000000	0.000000	0.000000	0.000000	0.000000+M
7.9371E+01	-.7326	.0026	.7326	-M	0.000000	0.000000	.999546	0.000000	0.000000	0.000000	0.000000	0.000000-M
				+M	0.000000	.003073	0.000000	0.000000	0.000000	0.000000	0.000000	-.029982+M
5.9282E+01	.5980	0.0000	.5980	-M	.000547	0.000000	0.000000	0.000000	0.000000	0.000000	-.050613	0.000000-M
				+M	0.000000	0.000000	0.000000	.998718	0.000000	0.000000	0.000000	0.000000+M
5.9282E+01	-.5980	0.0000	.5980	-M	0.000000	0.000000	0.000000	.998718	0.000000	0.000000	0.000000	0.000000-M
				+M	.000547	0.000000	0.000000	0.000000	0.000000	0.000000	-.050613	0.000000+M
5.1151E+01	.8662	-.0024	.8662	-M	0.000000	0.000000	-.003805	0.000000	0.000000	0.000000	0.000000	0.000000-M
				+M	0.000000	.999696	0.000000	0.000000	0.000000	0.000000	0.000000	-.024377+M
5.1151E+01	-.8662	-.0024	.8662	-M	0.000000	.999696	0.000000	0.000000	0.000000	0.000000	0.000000	-.024377-M
				+M	0.000000	0.000000	-.003805	0.000000	0.000000	0.000000	0.000000	0.000000+M
2.0910E+01	.4485	-.1473	.4721	-M	0.000000	0.000000	0.000000	0.000000	0.000000	-.150686	0.000000	0.000000-M
				+M	0.000000	0.000000	0.000000	0.000000	.988582	0.000000	0.000000	0.000000+M
2.0910E+01	-.4485	-.1473	.4721	-M	0.000000	0.000000	0.000000	0.000000	.988582	0.000000	0.000000	0.000000-M
				+M	0.000000	0.000000	0.000000	0.000000	0.000000	-.150686	0.000000	0.000000+M
-2.0910E+01	-.3152	.1473	.3479	-M	0.000000	0.000000	0.000000	0.000000	0.000000	.988582	0.000000	0.000000-M
				+M	0.000000	0.000000	0.000000	0.000000	.150686	0.000000	0.000000	0.000000+M
-2.0910E+01	.3152	.1473	.3479	-M	0.000000	0.000000	0.000000	0.000000	.150686	0.000000	0.000000	0.000000-M
				+M	0.000000	0.000000	0.000000	0.000000	0.000000	.988582	0.000000	0.000000+M
-5.0866E+01	.2017	0.0000	.2017	-M	0.000000	0.000000	0.000000	.050533	0.000000	0.000000	0.000000	0.000000-M
				+M	-.068127	0.000000	0.000000	0.000000	0.000000	0.000000	.996396	0.000000+M
-5.0866E+01	-.2017	0.0000	.2017	-M	-.068127	0.000000	0.000000	0.000000	0.000000	0.000000	.996396	0.000000-M
				+M	0.000000	0.000000	0.000000	.050533	0.000000	0.000000	0.000000	0.000000+M
-6.7282E+01	-.0664	.5331	.5372	-M	0.000000	.024480	0.000000	0.000000	0.000000	0.000000	0.000000	.999253-M
				+M	0.000000	0.000000	.029898	0.000000	0.000000	0.000000	0.000000	0.000000+M
-6.7282E+01	.0664	.5331	.5372	-M	0.000000	0.000000	.029898	0.000000	0.000000	0.000000	0.000000	0.000000-M
				+M	0.000000	.024480	0.000000	0.000000	0.000000	0.000000	0.000000	.999253+M
-7.1657E+01	-.9963	0.0000	.9963	-M	.997677	0.000000	0.000000	0.000000	0.000000	0.000000	.068067	0.000000-M
				+M	0.000000	0.000000	0.000000	.002903	0.000000	0.000000	0.000000	0.000000+M
-7.1657E+01	.9963	0.0000	.9963	-M	0.000000	0.000000	0.000000	.002903	0.000000	0.000000	0.000000	0.000000-M
				+M	.997677	0.000000	0.000000	0.000000	0.000000	0.000000	.068067	0.000000+M
ENERGY	HEPAL	HEPER	HE		M=15/2	M=13/2	M=11/2	M=9/2	M=7/2	M=5/2	M=3/2	M=1/2

Y=-.20 X=-.80 OAS=1.0459E+02*W246 B4F4/B2F2= 2.00E-01 B4F4/B6F6=-4.00E+00 E2F2/B6F6=-2.00E+01

ENERGY	HEPAL	HEPER	HE		M=15/2	M=13/2	M=11/2	M=9/2	M=7/2	M=5/2	M=3/2	M=1/2	
6.9088E+01	.9999	0.0000	.9999	-M	0.000000	0.000000	0.000000	.000302	0.000000	0.000000	0.000000	0.000000	-M
				+M	.999936	0.000000	0.000000	0.000000	0.000000	0.000000	-.011305	0.000000	+M
6.9088E+01	-.9999	0.0000	.9999	-M	.999936	0.000000	0.000000	0.000000	0.000000	0.000000	-.011305	0.000000	-M
				+M	0.000000	0.000000	0.000000	.000302	0.000000	0.000000	0.000000	0.000000	+M
1.6764E+01	.0816	.5258	.5321	-M	0.000000	0.000000	-.049946	0.000000	0.000000	0.000000	0.000000	0.000000	-M
				+M	0.000000	-.145525	0.000000	0.000000	0.000000	0.000000	0.000000	.988093	+M
1.6764E+01	-.0816	.5258	.5321	-M	0.000000	-.145525	0.000000	0.000000	0.000000	0.000000	0.000000	.988093	-M
				+M	0.000000	0.000000	-.049946	0.000000	0.000000	0.000000	0.000000	0.000000	+M
6.7072E+00	.8496	.0046	.8497	-M	0.000000	0.000000	-.009558	0.000000	0.000000	0.000000	0.000000	0.000000	-M
				+M	0.000000	.989352	0.000000	0.000000	0.000000	0.000000	0.000000	.145228	+M
6.7072E+00	-.8496	.0046	.8497	-M	0.000000	.989352	0.000000	0.000000	0.000000	0.000000	0.000000	.145228	-M
				+M	0.000000	0.000000	-.009558	0.000000	0.000000	0.000000	0.000000	0.000000	+M
6.6171E+00	-.1966	0.0000	.1966	-M	.011300	0.000000	0.000000	0.000000	0.000000	0.000000	.997737	0.000000	-M
				+M	0.000000	0.000000	0.000000	-.066282	0.000000	0.000000	0.000000	0.000000	+M
6.6171E+00	.1966	0.0000	.1966	-M	0.000000	0.000000	0.000000	-.066282	0.000000	0.000000	0.000000	0.000000	-M
				+M	.011300	0.000000	0.000000	0.000000	0.000000	0.000000	.997737	0.000000	+M
-9.5586E+00	-.3082	-.1726	.3532	-M	0.000000	0.000000	0.000000	0.000000	0.000000	.984142	0.000000	0.000000	-M
				+M	0.000000	0.000000	0.000000	0.000000	-.177382	0.000000	0.000000	0.000000	+M
-9.5586E+00	.3082	-.1726	.3532	-M	0.000000	0.000000	0.000000	0.000000	-.177382	0.000000	0.000000	0.000000	-M
				+M	0.000000	0.000000	0.000000	0.000000	0.000000	.984142	0.000000	0.000000	+M
-2.6712E+01	.7313	.0029	.7313	-M	0.000000	.002190	0.000000	0.000000	0.000000	0.000000	0.000000	.050805	-M
				+M	0.000000	0.000000	.998706	0.000000	0.000000	0.000000	0.000000	0.000000	+M
-2.6712E+01	-.7313	.0029	.7313	-M	0.000000	0.000000	.998706	0.000000	0.000000	0.000000	0.000000	0.000000	-M
				+M	0.000000	.002190	0.000000	0.000000	0.000000	0.000000	0.000000	.050805	+M
-2.7401E+01	.4415	.1726	.4740	-M	0.000000	0.000000	0.000000	0.000000	0.000000	.177382	0.000000	0.000000	-M
				+M	0.000000	0.000000	0.000000	0.000000	.984142	0.000000	0.000000	0.000000	+M
-2.7401E+01	-.4415	.1726	.4740	-M	0.000000	0.000000	0.000000	0.000000	.984142	0.000000	0.000000	0.000000	-M
				+M	0.000000	0.000000	0.000000	0.000000	0.000000	.177382	0.000000	0.000000	+M
-3.5505E+01	-.5965	0.0000	.5965	-M	0.000000	0.000000	0.000000	.997801	0.000000	0.000000	0.000000	0.000000	-M
				+M	.000448	0.000000	0.000000	0.000000	0.000000	0.000000	.066281	0.000000	+M
-3.5505E+01	.5965	0.0000	.5965	-M	.000448	0.000000	0.000000	0.000000	0.000000	0.000000	.066281	0.000000	-M
				+M	0.000000	0.000000	0.000000	.997801	0.000000	0.000000	0.000000	0.000000	+M
ENERGY	HEPAL	HEPER	HE		M=15/2	M=13/2	M=11/2	M=9/2	M=7/2	M=5/2	M=3/2	M=1/2	

Y=-.20 X=-.60 OAS=8.6183E+01*W246 B4F4/B2F2= 1.50E-01 B4F4/B6F6=-1.50E+00 B2F2/B6F6=-1.00E+01

ENERGY	HEPAL	HEPER	HE		M=15/2	M=13/2	M=11/2	M=9/2	M=7/2	M=5/2	M=3/2	M=1/2	
5.5598E+01	-.9994	0.0000	.9994	-M	.999642	0.000000	0.000000	0.000000	0.000000	0.000000	-.026710	0.000000	-M
				+M	0.000000	0.000000	0.000000	.001746	0.000000	0.000000	0.000000	0.000000	+M
5.5598E+01	.9994	0.0000	.9994	-M	0.000000	0.000000	0.000000	.001746	0.000000	0.000000	0.000000	0.000000	-M
				+M	.999642	0.000000	0.000000	0.000000	0.000000	0.000000	-.026710	0.000000	+M
1.7065E+01	.6330	.1923	.6615	-M	0.000000	0.000000	.070027	0.000000	0.000000	0.000000	0.000000	0.000000	-M
				+M	0.000000	.844250	0.000000	0.000000	0.000000	0.000000	0.000000	-.531354	+M
1.7065E+01	-.6330	.1923	.6615	-M	0.000000	.844250	0.000000	0.000000	0.000000	0.000000	0.000000	-.531354	-M
				+M	0.000000	0.000000	.070027	0.000000	0.000000	0.000000	0.000000	0.000000	+M
1.0730E+01	.2816	.3191	.4256	-M	0.000000	0.000000	-.135671	0.000000	0.000000	0.000000	0.000000	0.000000	-M
				+M	0.000000	.535773	0.000000	0.000000	0.000000	0.000000	0.000000	.833391	+M
1.0730E+01	-.2816	.3191	.4256	-M	0.000000	.535773	0.000000	0.000000	0.000000	0.000000	0.000000	.833391	-M
				+M	0.000000	0.000000	-.135671	0.000000	0.000000	0.000000	0.000000	0.000000	+M
3.1875E+00	-.1782	0.0000	.1782	-M	.026626	0.000000	0.000000	0.000000	0.000000	0.000000	.985550	0.000000	-M
				+M	0.000000	0.000000	0.000000	-.167281	0.000000	0.000000	0.000000	0.000000	+M
3.1875E+00	.1782	0.0000	.1782	-M	0.000000	0.000000	0.000000	-.167281	0.000000	0.000000	0.000000	0.000000	-M
				+M	.026626	0.000000	0.000000	0.000000	0.000000	0.000000	.985550	0.000000	+M
-1.0397E+01	-.2267	-.3361	.4054	-M	0.000000	0.000000	0.000000	0.000000	0.000000	.930956	0.000000	0.000000	-M
				+M	0.000000	0.000000	0.000000	0.000000	-.365131	0.000000	0.000000	0.000000	+M
-1.0397E+01	.2267	-.3361	.4054	-M	0.000000	0.000000	0.000000	0.000000	-.365131	0.000000	0.000000	0.000000	-M
				+M	0.000000	0.000000	0.000000	0.000000	0.000000	.930956	0.000000	0.000000	+M
-1.6875E+01	.7145	.0219	.7149	-M	0.000000	.013730	0.000000	0.000000	0.000000	0.000000	0.000000	.152059	-M
				+M	0.000000	0.000000	.988276	0.000000	0.000000	0.000000	0.000000	0.000000	+M
-1.6875E+01	-.7145	.0219	.7149	-M	0.000000	0.000000	.988276	0.000000	0.000000	0.000000	0.000000	0.000000	-M
				+M	0.000000	.013730	0.000000	0.000000	0.000000	0.000000	0.000000	.152059	+M
-2.8723E+01	.3600	.3361	.4925	-M	0.000000	0.000000	0.000000	0.000000	0.000000	.365131	0.000000	0.000000	-M
				+M	0.000000	0.000000	0.000000	0.000000	.930956	0.000000	0.000000	0.000000	+M
-2.8723E+01	-.3600	.3361	.4925	-M	0.000000	0.000000	0.000000	0.000000	.930956	0.000000	0.000000	0.000000	-M
				+M	0.000000	0.000000	0.000000	0.000000	0.000000	.365131	0.000000	0.000000	+M
-3.0585E+01	-.5776	0.0000	.5776	-M	0.000000	0.000000	0.000000	.985908	0.000000	0.000000	0.000000	0.000000	-M
				+M	.002747	0.000000	0.000000	0.000000	0.000000	0.000000	.167267	0.000000	+M
-3.0585E+01	.5776	0.0000	.5776	-M	.002747	0.000000	0.000000	0.000000	0.000000	0.000000	.167267	0.000000	-M
				+M	0.000000	0.000000	0.000000	.985908	0.000000	0.000000	0.000000	0.000000	+M
ENERGY	HEPAL	HEPER	HE		M=15/2	M=13/2	M=11/2	M=9/2	M=7/2	M=5/2	M=3/2	M=1/2	

Y=-.20 X=-.40 OAS=7.3421E+01*W246 B4F4/B2F2= 1.00E-01 B4F4/B6F6=-6.67E-01 E2F2/B6F6=-6.67E+00

ENERGY	HEPAL	HEPER	HE		M=15/2	M=13/2	M=11/2	M=9/2	M=7/2	M=5/2	M=3/2	M=1/2	
4.2145E+01	.9980	0.0000	.9980	-M	0.000000	0.000000	0.000000	.006237	0.000000	0.000000	0.000000	0.000000	-M
				+M	.998763	0.000000	0.000000	0.000000	0.000000	0.000000	-.049333	0.000000	+M
4.2145E+01	-.9980	0.0000	.9980	-M	.998763	0.000000	0.000000	0.000000	0.000000	0.000000	-.049333	0.000000	-M
				+M	0.000000	0.000000	0.000000	.006237	0.000000	0.000000	0.000000	0.000000	+M
2.4742E+01	-.8067	.0749	.8102	-M	0.000000	.963469	0.000000	0.000000	0.000000	0.000000	0.000000	-.261818	-M
				+M	0.000000	0.000000	.056372	0.000000	0.000000	0.000000	0.000000	0.000000	+M
2.4742E+01	.8067	.0749	.8102	-M	0.000000	0.000000	.056372	0.000000	0.000000	0.000000	0.000000	0.000000	-M
				+M	0.000000	.963469	0.000000	0.000000	0.000000	0.000000	0.000000	-.261818	+M
8.9553E+00	.0016	.3416	.3416	-M	0.000000	.262490	0.000000	0.000000	0.000000	0.000000	0.000000	.881368	-M
				+M	0.000000	0.000000	-.392797	0.000000	0.000000	0.000000	0.000000	0.000000	+M
8.9553E+00	-.0016	.3416	.3416	-M	0.000000	0.000000	-.392797	0.000000	0.000000	0.000000	0.000000	0.000000	-M
				+M	0.000000	.262490	0.000000	0.000000	0.000000	0.000000	0.000000	.881368	+M
8.2869E-01	.1204	0.0000	.1204	-M	0.000000	0.000000	0.000000	-.319114	0.000000	0.000000	0.000000	0.000000	-M
				+M	.048743	0.000000	0.000000	0.000000	0.000000	0.000000	.946462	0.000000	+M
8.2869E-01	-.1204	0.0000	.1204	-M	.048743	0.000000	0.000000	0.000000	0.000000	0.000000	.946462	0.000000	-M
				+M	0.000000	0.000000	0.000000	-.319114	0.000000	0.000000	0.000000	0.000000	+M
-8.6173E+00	-.6051	.1169	.6163	-M	0.000000	0.000000	.917896	0.000000	0.000000	0.000000	0.000000	0.000000	-M
				+M	0.000000	.053157	0.000000	0.000000	0.000000	0.000000	0.000000	.393245	+M
-8.6173E+00	.6051	.1169	.6163	-M	0.000000	.053157	0.000000	0.000000	0.000000	0.000000	0.000000	.393245	-M
				+M	0.000000	0.000000	.917896	0.000000	0.000000	0.000000	0.000000	0.000000	+M
-1.0004E+01	.1244	-.4344	.4518	-M	0.000000	0.000000	0.000000	0.000000	-.511066	0.000000	0.000000	0.000000	-M
				+M	0.000000	0.000000	0.000000	0.000000	0.000000	.859542	0.000000	0.000000	+M
-1.0004E+01	-.1244	-.4344	.4518	-M	0.000000	0.000000	0.000000	0.000000	0.000000	.859542	0.000000	0.000000	-M
				+M	0.000000	0.000000	0.000000	0.000000	-.511066	0.000000	0.000000	0.000000	+M
2.6774E+01	.5184	0.0000	.5184	-M	.009839	0.000000	0.000000	0.000000	0.000000	0.000000	.319024	0.000000	-M
				+M	0.000000	0.000000	0.000000	.947696	0.000000	0.000000	0.000000	0.000000	+M
-2.6774E+01	-.5184	0.0000	.5184	-M	0.000000	0.000000	0.000000	.947696	0.000000	0.000000	0.000000	0.000000	-M
				+M	.009839	0.000000	0.000000	0.000000	0.000000	0.000000	.319024	0.000000	+M
-3.1276E+01	-.2577	.4344	.5051	-M	0.000000	0.000000	0.000000	0.000000	.859542	0.000000	0.000000	0.000000	-M
				+M	0.000000	0.000000	0.000000	0.000000	0.000000	.511066	0.000000	0.000000	+M
3.1276E+01	.2577	.4344	.5051	-M	0.000000	0.000000	0.000000	0.000000	0.000000	.511066	0.000000	0.000000	-M
				+M	0.000000	0.000000	0.000000	0.000000	.859542	0.000000	0.000000	0.000000	+M
ENERGY	HEPAL	HEPER	HE		M=15/2	M=13/2	M=11/2	M=9/2	M=7/2	M=5/2	M=3/2	M=1/2	

Y=-.20 X=-.20 OAS=6.7742E+01*W246 B4F4/B2F2= 5.00E-02 B4F4/B6F6=-2.50E-01 B2F2/B6F6=-5.00E+00

ENERGY	HEPAL	HEPER	HE		M=15/2	M=13/2	M=11/2	M=9/2	M=7/2	M=5/2	M=3/2	M=1/2
3.3097E+01	.8267	.0659	.8293	-M	0.000000	0.000000	.063058	0.000000	0.000000	0.000000	0.000000	0.000000-M
				+M	0.000000	.976740	0.000000	0.000000	0.000000	0.000000	0.000000	-.204947+M
3.3097E+01	-.8267	.0659	.8293	-M	0.000000	.976740	0.000000	0.000000	0.000000	0.000000	0.000000	-.204947-M
				+M	0.000000	0.000000	.063058	0.000000	0.000000	0.000000	0.000000	0.000000+M
2.8769E+01	-.9932	0.0000	.9932	-M	.995939	0.000000	0.000000	0.000000	0.000000	0.000000	-.087610	0.000000-M
				+M	0.000000	0.000000	0.000000	.020751	0.000000	0.000000	0.000000	0.000000+M
2.8769E+01	.9932	0.0000	.9932	-M	0.000000	0.000000	0.000000	.020751	0.000000	0.000000	0.000000	0.000000-M
				+M	.995939	0.000000	0.000000	0.000000	0.000000	0.000000	-.087610	0.000000+M
1.1785E+01	.3705	.1107	.3867	-M	0.000000	-.179947	0.000000	0.000000	0.000000	0.000000	0.000000	-.623503-M
				+M	0.000000	0.000000	.760831	0.000000	0.000000	0.000000	0.000000	0.000000+M
1.1785E+01	-.3705	.1107	.3867	-M	0.000000	0.000000	.760831	0.000000	0.000000	0.000000	0.000000	0.000000-M
				+M	0.000000	-.179947	0.000000	0.000000	0.000000	0.000000	0.000000	-.623503+M
3.4387E-01	-.0035	0.0000	.0035	-M	0.000000	0.000000	0.000000	-.511567	0.000000	0.000000	0.000000	0.000000-M
				+M	.085865	0.000000	0.000000	0.000000	0.000000	0.000000	.854942	0.000000+M
3.4387E-01	.0035	0.0000	.0035	-M	.085865	0.000000	0.000000	0.000000	0.000000	0.000000	.854942	0.000000-M
				+M	0.000000	0.000000	0.000000	-.511567	0.000000	0.000000	0.000000	0.000000+M
-5.6422E+00	.2562	.3567	.4392	-M	0.000000	.116613	0.000000	0.000000	0.000000	0.000000	0.000000	.754481-M
				+M	0.000000	0.000000	.645879	0.000000	0.000000	0.000000	0.000000	0.000000+M
-5.6422E+00	-.2562	.3567	.4392	-M	0.000000	0.000000	.645879	0.000000	0.000000	0.000000	0.000000	0.000000-M
				+M	0.000000	.116613	0.000000	0.000000	0.000000	0.000000	0.000000	.754481+M
-8.7946E+00	.0398	-.4766	.4782	-M	0.000000	0.000000	0.000000	0.000000	-.605746	0.000000	0.000000	0.000000-M
				+M	0.000000	0.000000	0.000000	0.000000	0.000000	.795658	0.000000	0.000000+M
-8.7946E+00	-.0398	-.4766	.4782	-M	0.000000	0.000000	0.000000	0.000000	0.000000	.795658	0.000000	0.000000-M
				+M	0.000000	0.000000	0.000000	0.000000	-.605746	0.000000	0.000000	0.000000+M
-2.4913E+01	-.3897	0.0000	.3897	-M	0.000000	0.000000	0.000000	.858993	0.000000	0.000000	0.000000	0.000000-M
				+M	.027077	0.000000	0.000000	0.000000	0.000000	0.000000	.511271	0.000000+M
-2.4913E+01	.3897	0.0000	.3897	-M	.027077	0.000000	0.000000	0.000000	0.000000	0.000000	.511271	0.000000-M
				+M	0.000000	0.000000	0.000000	.858993	0.000000	0.000000	0.000000	0.000000+M
-3.4645E+01	-.1731	.4766	.5071	-M	0.000000	0.000000	0.000000	0.000000	.795658	0.000000	0.000000	0.000000-M
				+M	0.000000	0.000000	0.000000	0.000000	0.000000	.605746	0.000000	0.000000+M
-3.4645E+01	.1731	.4766	.5071	-M	0.000000	0.000000	0.000000	0.000000	0.000000	.605746	0.000000	0.000000-M
				+M	0.000000	0.000000	0.000000	0.000000	.795658	0.000000	0.000000	0.000000+M
ENERGY	HEPAL	HEPER	HE		M=15/2	M=13/2	M=11/2	M=9/2	M=7/2	M=5/2	M=3/2	M=1/2

Y=-.20 X=0.00 OAS=8.0031E+01*W246 B4F4/B2F2= 0. B4F4/B6F6= 0. B2F2/B6F6=-4.00E+00

ENERGY	HEPAL	HEPER	HE		M=15/2	M=13/2	M=11/2	M=9/2	M=7/2	M=5/2	M=3/2	M=1/2
4.1553E+01	-.8307	.0701	.8337	-M	0.000000	.980207	0.000000	0.000000	0.000000	0.000000	0.000000	-.183000-
				+M	0.000000	0.000000	.075533	0.000000	0.000000	0.000000	0.000000	0.000000+M
4.1553E+01	.8307	.0701	.8337	-M	0.000000	0.000000	.075533	0.000000	0.000000	0.000000	0.000000	0.000000-
				+M	0.000000	.980207	0.000000	0.000000	0.000000	0.000000	0.000000	-.183000+M
1.9937E+01	.5674	-.0045	.5674	-M	0.000000	-.145376	0.000000	0.000000	0.000000	0.000000	0.000000	-.406350-
				+M	0.000000	0.000000	.902078	0.000000	0.000000	0.000000	0.000000	0.000000+M
1.9937E+01	-.5674	-.0045	.5674	-M	0.000000	0.000000	.902078	0.000000	0.000000	0.000000	0.000000	0.000000-
				+M	0.000000	-.145376	0.000000	0.000000	0.000000	0.000000	0.000000	-.406350+M
1.5659E+01	-.9605	0.0000	.9605	-M	.979114	0.000000	0.000000	0.000000	0.000000	0.000000	-.182359	0.000000-
				+M	0.000000	0.000000	0.000000	.089892	0.000000	0.000000	0.000000	0.000000+M
1.5659E+01	.9605	0.0000	.9605	-M	0.000000	0.000000	0.000000	.089892	0.000000	0.000000	0.000000	0.000000-
				+M	.979114	0.000000	0.000000	0.000000	0.000000	0.000000	-.182359	0.000000+M
2.0646E+00	.1348	0.0000	.1348	-M	.194482	0.000000	0.000000	0.000000	0.000000	0.000000	.711193	0.000000-
				+M	0.000000	0.000000	0.000000	-.675560	0.000000	0.000000	0.000000	0.000000+M
2.0646E+00	-.1348	0.0000	.1348	-M	0.000000	0.000000	0.000000	-.675560	0.000000	0.000000	0.000000	0.000000-
				+M	.194482	0.000000	0.000000	0.000000	0.000000	0.000000	.711193	0.000000+M
-7.1223E+00	-.0207	-.4911	.4916	-M	0.000000	0.000000	0.000000	0.000000	-.665277	0.000000	0.000000	0.000000-
				+M	0.000000	0.000000	0.000000	0.000000	0.000000	.746597	0.000000	0.000000+M
-7.1223E+00	.0207	-.4911	.4916	-M	0.000000	0.000000	0.000000	0.000000	0.000000	.746597	0.000000	0.000000-
				+M	0.000000	0.000000	0.000000	0.000000	-.665277	0.000000	0.000000	0.000000+M
-8.0904E+00	-.0633	.4677	.4720	-M	0.000000	0.000000	.424911	0.000000	0.000000	0.000000	0.000000	0.000000-
				+M	0.000000	.134387	0.000000	0.000000	0.000000	0.000000	0.000000	.895204+M
-8.0904E+00	.0633	.4677	.4720	-M	0.000000	.134387	0.000000	0.000000	0.000000	0.000000	0.000000	.895204-
				+M	0.000000	0.000000	.424911	0.000000	0.000000	0.000000	0.000000	0.000000+M
-2.5523E+01	.2256	0.0000	.2256	-M	.059264	0.000000	0.000000	0.000000	0.000000	0.000000	.678932	0.000000-
				+M	0.000000	0.000000	0.000000	.731805	0.000000	0.000000	0.000000	0.000000+M
-2.5523E+01	-.2256	0.0000	.2256	-M	0.000000	0.000000	0.000000	.731805	0.000000	0.000000	0.000000	0.000000-
				+M	.059264	0.000000	0.000000	0.000000	0.000000	0.000000	.678932	0.000000+M
-3.8478E+01	-.1126	.4911	.5039	-M	0.000000	0.000000	0.000000	0.000000	.746597	0.000000	0.000000	0.000000-
				+M	0.000000	0.000000	0.000000	0.000000	0.000000	.665277	0.000000	0.000000+M
-3.8478E+01	.1126	.4911	.5039	-M	0.000000	0.000000	0.000000	0.000000	0.000000	.665277	0.000000	0.000000-
				+M	0.000000	0.000000	0.000000	0.000000	.746597	0.000000	0.000000	0.000000+M
ENERGY	HEPAL	HEPER	HE		M=15/2	M=13/2	M=11/2	M=9/2	M=7/2	M=5/2	M=3/2	M=1/2

Y=-.20 X= .20 OAS=7.1014E+01*W246 B4F4/B2F2=-5.00E-02 B4F4/B6F6= 2.50E-01 B2F2/B6F6=-5.00E+00

ENERGY	HEPAL	HEPER	HE		M=15/2	M=13/2	M=11/2	M=9/2	M=7/2	M=5/2	M=3/2	M=1/2
3.9863E+01	.8494	.0496	.8508	-M	0.000000	0.000000	.059891	0.000000	0.000000	0.000000	0.000000	0.000000
				+M	0.000000	.990953	0.000000	0.000000	0.000000	0.000000	0.000000	-.120103+
3.9863E+01	-.8494	.0496	.8508	-M	0.000000	.990953	0.000000	0.000000	0.000000	0.000000	0.000000	-.120103
				+M	0.000000	0.000000	.059891	0.000000	0.000000	0.000000	0.000000	0.000000+
2.4258E+01	.6832	-.0329	.6840	-M	0.000000	-.085310	0.000000	0.000000	0.000000	0.000000	0.000000	-.219216
				+M	0.000000	0.000000	.971940	0.000000	0.000000	0.000000	0.000000	0.000000+
2.4258E+01	-.6832	-.0329	.6840	-M	0.000000	0.000000	.971940	0.000000	0.000000	0.000000	0.000000	0.000000
				+M	0.000000	-.085310	0.000000	0.000000	0.000000	0.000000	0.000000	-.219216+
7.4171E+00	-.8274	0.0000	.8274	-M	.928898	0.000000	0.000000	0.000000	0.000000	0.000000	-.241956	0.000000
				+M	0.000000	0.000000	0.000000	.280367	0.000000	0.000000	0.000000	0.000000+
7.4171E+00	.8274	0.0000	.8274	-M	0.000000	0.000000	0.000000	.280367	0.000000	0.000000	0.000000	0.000000
				+M	.928898	0.000000	0.000000	0.000000	0.000000	0.000000	-.241956	0.000000+
3.3955E+00	-.2518	0.0000	.2518	-M	0.000000	0.000000	0.000000	.833542	0.000000	0.000000	0.000000	0.000000
				+M	-.360605	0.000000	0.000000	0.000000	0.000000	0.000000	-.418537	0.000000+
3.3955E+00	.2518	0.0000	.2518	-M	-.360605	0.000000	0.000000	0.000000	0.000000	0.000000	-.418537	0.000000
				+M	0.000000	0.000000	0.000000	.833542	0.000000	0.000000	0.000000	0.000000+
-6.0484E+00	-.1151	-.4908	.5041	-M	0.000000	0.000000	0.000000	0.000000	.748699	0.000000	0.000000	0.000000
				+M	0.000000	0.000000	0.000000	0.000000	0.000000	-.662910	0.000000	0.000000+
-6.0484E+00	.1151	-.4908	.5041	-M	0.000000	0.000000	0.000000	0.000000	0.000000	-.662910	0.000000	0.000000
				+M	0.000000	0.000000	0.000000	0.000000	.748699	0.000000	0.000000	0.000000+
-1.5040E+01	-.0339	.5166	.5177	-M	0.000000	.103603	0.000000	0.000000	0.000000	0.000000	0.000000	.968256
				+M	0.000000	0.000000	.227479	0.000000	0.000000	0.000000	0.000000	0.000000+
-1.5040E+01	.0339	.5166	.5177	-M	0.000000	0.000000	.227479	0.000000	0.000000	0.000000	0.000000	0.000000
				+M	0.000000	.103603	0.000000	0.000000	0.000000	0.000000	0.000000	.968256+
-2.2693E+01	.0244	0.0000	.0244	-M	0.000000	0.000000	0.000000	.476028	0.000000	0.000000	0.000000	0.000000
				+M	.084336	0.000000	0.000000	0.000000	0.000000	0.000000	.875377	0.000000+
-2.2693E+01	-.0244	0.0000	.0244	-M	.084336	0.000000	0.000000	0.000000	0.000000	0.000000	.875377	0.000000
				+M	0.000000	0.000000	0.000000	.476028	0.000000	0.000000	0.000000	0.000000+
-3.1152E+01	-.0182	.4908	.4911	-M	0.000000	0.000000	0.000000	0.000000	.662910	0.000000	0.000000	0.000000
				+M	0.000000	0.000000	0.000000	0.000000	0.000000	.748699	0.000000	0.000000+
-3.1152E+01	.0182	.4908	.4911	-M	0.000000	0.000000	0.000000	0.000000	0.000000	.748699	0.000000	0.000000
				+M	0.000000	0.000000	0.000000	0.000000	.662910	0.000000	0.000000	0.000000+
ENERGY	HEPAL	HEPER	HE		M=15/2	M=13/2	M=11/2	M=9/2	M=7/2	M=5/2	M=3/2	M=1/2

Y=-.20 X= .40 OAS=6.3378E+01*W246 B4F4/B2F2=-1.00E-01 B4F4/B6F6= 6.67E-01 B2F2/B6F6=-6.67E+00

ENERGY	HEPAL	HEPER	HE		M=15/2	M=13/2	M=11/2	M=9/2	M=7/2	M=5/2	M=3/2	M=1/2
3.8455E+01	-.8568	.0429	.8579	-M	0.000000	.995450	0.000000	0.000000	0.000000	0.000000	0.000000	-.076616-M
				+M	0.000000	0.000000	.056643	0.000000	0.000000	0.000000	0.000000	0.000000+M
3.8455E+01	.8568	.0429	.8579	-M	0.000000	0.000000	.056643	0.000000	0.000000	0.000000	0.000000	0.000000-M
				+M	0.000000	.995450	0.000000	0.000000	0.000000	0.000000	0.000000	-.076616+M
3.0337E+01	-.7155	-.0384	.7166	-M	0.000000	0.000000	.990974	0.000000	0.000000	0.000000	0.000000	0.000000-M
				+M	0.000000	-.065394	0.000000	0.000000	0.000000	0.000000	0.000000	-.117022+M
3.0337E+01	.7155	-.0384	.7166	-M	0.000000	-.065394	0.000000	0.000000	0.000000	0.000000	0.000000	-.117022-M
				+M	0.000000	0.000000	.990974	0.000000	0.000000	0.000000	0.000000	0.000000+M
1.0350E+01	.5473	0.0000	.5473	-M	.044053	0.000000	0.000000	0.000000	0.000000	0.000000	-.248863	0.000000-M
				+M	0.000000	0.000000	0.000000	.967536	0.000000	0.000000	0.000000	0.000000+M
1.0350E+01	-.5473	0.0000	.5473	-M	0.000000	0.000000	0.000000	.967536	0.000000	0.000000	0.000000	0.000000-M
				+M	.044053	0.000000	0.000000	0.000000	0.000000	0.000000	-.248863	0.000000+M
-1.4806E+00	-.9889	0.0000	.9889	-M	.995112	0.000000	0.000000	0.000000	0.000000	0.000000	-.074747	0.000000-M
				+M	0.000000	0.000000	0.000000	-.064534	0.000000	0.000000	0.000000	0.000000+M
-1.4806E+00	.9889	0.0000	.9889	-M	0.000000	0.000000	0.000000	-.064534	0.000000	0.000000	0.000000	0.000000-M
				+M	.995112	0.000000	0.000000	0.000000	0.000000	0.000000	-.074747	0.000000+M
-3.8765E+00	-.2506	-.4390	.5055	-M	0.000000	0.000000	0.000000	0.000000	.854378	0.000000	0.000000	0.000000-M
				+M	0.000000	0.000000	0.000000	0.000000	0.000000	-.519652	0.000000	0.000000+M
-3.8765E+00	.2506	-.4390	.5055	-M	0.000000	0.000000	0.000000	0.000000	0.000000	-.519652	0.000000	0.000000-M
				+M	0.000000	0.000000	0.000000	0.000000	.854378	0.000000	0.000000	0.000000+M
-2.4031E+01	-.0587	.5288	.5321	-M	0.000000	.069296	0.000000	0.000000	0.000000	0.000000	0.000000	.990170-M
				+M	0.000000	0.000000	.121499	0.000000	0.000000	0.000000	0.000000	0.000000+M
-2.4031E+01	.0587	.5288	.5321	-M	0.000000	0.000000	.121499	0.000000	0.000000	0.000000	0.000000	0.000000-M
				+M	0.000000	.069296	0.000000	0.000000	0.000000	0.000000	0.000000	.990170+M
-2.4829E+01	-.1585	0.0000	.1585	-M	.088381	0.000000	0.000000	0.000000	0.000000	0.000000	.965650	0.000000-M
				+M	0.000000	0.000000	0.000000	.244354	0.000000	0.000000	0.000000	0.000000+M
-2.4829E+01	.1585	0.0000	.1585	-M	0.000000	0.000000	0.000000	.244354	0.000000	0.000000	0.000000	0.000000-M
				+M	.088381	0.000000	0.000000	0.000000	0.000000	0.000000	.965650	0.000000+M
-2.4923E+01	.1173	.4390	.4544	-M	0.000000	0.000000	0.000000	0.000000	.519652	0.000000	0.000000	0.000000-M
				+M	0.000000	0.000000	0.000000	0.000000	0.000000	.854378	0.000000	0.000000+M
-2.4923E+01	-.1173	.4390	.4544	-M	0.000000	0.000000	0.000000	0.000000	0.000000	.854378	0.000000	0.000000-M
				+M	0.000000	0.000000	0.000000	0.000000	.519652	0.000000	0.000000	0.000000+M
ENERGY	HEPAL	HEPER	HE		M=15/2	M=13/2	M=11/2	M=9/2	M=7/2	M=5/2	M=3/2	M=1/2

Y=-.20 X= .60 OAS=7.1210E+01*W246 B4F4/B2F2=-1.50E-01 B4F4/B6F6= 1.50E+00 B2F2/B6F6=-1.00E+01

ENERGY	HEPAL	HEPER	HE		M=15/2	M=13/2	M=11/2	M=9/2	M=7/2	M=5/2	M=3/2	M=1/2
3.7340E+01	.2549	.3437	.4279	-M	0.000000	0.000000	.616369	0.000000	0.000000	0.000000	0.000000	0.000000-M
				+M	0.000000	.784316	0.000000	0.000000	0.000000	0.000000	0.000000	-.070269+M
3.7340E+01	-.2549	.3437	.4279	-M	0.000000	.784316	0.000000	0.000000	0.000000	0.000000	0.000000	-.070269-M
				+M	0.000000	0.000000	.616369	0.000000	0.000000	0.000000	0.000000	0.000000+M
3.6971E+01	.1198	-.3426	.3629	-M	0.000000	-.619021	0.000000	0.000000	0.000000	0.000000	0.000000	-.023372-M
				+M	0.000000	0.000000	.785027	0.000000	0.000000	0.000000	0.000000	0.000000+M
3.6971E+01	-.1198	-.3426	.3629	-M	0.000000	0.000000	.785027	0.000000	0.000000	0.000000	0.000000	0.000000-M
				+M	0.000000	-.619021	0.000000	0.000000	0.000000	0.000000	0.000000	-.023372+M
1.9249E+01	-.5893	0.0000	.5893	-M	0.000000	0.000000	0.000000	.993276	0.000000	0.000000	0.000000	0.000000-M
				+M	.005602	0.000000	0.000000	0.000000	0.000000	0.000000	-.115638	0.000000+M
1.9249E+01	.5893	0.0000	.5893	-M	.005602	0.000000	0.000000	0.000000	0.000000	0.000000	-.115638	0.000000-M
				+M	0.000000	0.000000	0.000000	.993276	0.000000	0.000000	0.000000	0.000000+M
6.6139E-02	.3846	-.3000	.4878	-M	0.000000	0.000000	0.000000	0.000000	0.000000	-.320276	0.000000	0.000000-M
				+M	0.000000	0.000000	0.000000	0.000000	.947324	0.000000	0.000000	0.000000+M
6.6139E-02	-.3846	-.3000	.4878	-M	0.000000	0.000000	0.000000	0.000000	.947324	0.000000	0.000000	0.000000-M
				+M	0.000000	0.000000	0.000000	0.000000	0.000000	-.320276	0.000000	0.000000+M
-9.8595E+00	-.9957	0.0000	.9957	-M	.997389	0.000000	0.000000	0.000000	0.000000	0.000000	-.070875	0.000000-M
				+M	0.000000	0.000000	0.000000	-.013876	0.000000	0.000000	0.000000	0.000000+M
-9.8595E+00	.9957	0.0000	.9957	-M	0.000000	0.000000	0.000000	-.013876	0.000000	0.000000	0.000000	0.000000-M
				+M	.997389	0.000000	0.000000	0.000000	0.000000	0.000000	-.070875	0.000000+M
-2.0466E+01	.2513	.3000	.3913	-M	0.000000	0.000000	0.000000	0.000000	.320276	0.000000	0.000000	0.000000-M
				+M	0.000000	0.000000	0.000000	0.000000	0.000000	.947324	0.000000	0.000000+M
-2.0466E+01	-.2513	.3000	.3913	-M	0.000000	0.000000	0.000000	0.000000	0.000000	.947324	0.000000	0.000000-M
				+M	0.000000	0.000000	0.000000	0.000000	.320276	0.000000	0.000000	0.000000+M
-2.9429E+01	-.1936	0.0000	.1936	-M	.072003	0.000000	0.000000	0.000000	0.000000	0.000000	.990760	0.000000-M
				+M	0.000000	0.000000	0.000000	.114939	0.000000	0.000000	0.000000	0.000000+M
-2.9429E+01	.1936	0.0000	.1936	-M	0.000000	0.000000	0.000000	.114939	0.000000	0.000000	0.000000	0.000000-M
				+M	.072003	0.000000	0.000000	0.000000	0.000000	0.000000	.990760	0.000000+M
-3.3870E+01	.0649	.5322	.5361	-M	0.000000	0.000000	.061829	0.000000	0.000000	0.000000	0.000000	0.000000-M
				+M	0.000000	.040758	0.000000	0.000000	0.000000	0.000000	0.000000	.997254+M
-3.3870E+01	-.0649	.5322	.5361	-M	0.000000	.040758	0.000000	0.000000	0.000000	0.000000	0.000000	.997254-M
				+M	0.000000	0.000000	.061829	0.000000	0.000000	0.000000	0.000000	0.000000+M
ENERGY	HEPAL	HEPER	HE		M=15/2	M=13/2	M=11/2	M=9/2	M=7/2	M=5/2	M=3/2	M=1/2

Y=-.20 X= .80 OAS=8.8295E+01*W246 B4F4/B2F2=-2.00E-01 B4F4/B6F6= 4.00E+00 B2F2/B6F6=-2.00E+01

ENERGY	HEPAL	HEPER	HE		M=15/2	M=13/2	M=11/2	M=9/2	M=7/2	M=5/2	M=3/2	M=1/2
4.4175E+01	-.7328	.0035	.7328	-M	0.000000	0.000000	.999680	0.000000	0.000000	0.000000	0.000000	0.000000-M
				+M	0.000000	.004440	0.000000	0.000000	0.000000	0.000000	0.000000	-.024913+M
4.4175E+01	.7328	.0035	.7328	-M	0.000000	.004440	0.000000	0.000000	0.000000	0.000000	0.000000	-.024913-M
				+M	0.000000	0.000000	.999680	0.000000	0.000000	0.000000	0.000000	0.000000+M
3.6066E+01	.8664	-.0033	.8664	-M	0.000000	0.000000	-.004889	0.000000	0.000000	0.000000	0.000000	0.000000-M
				+M	0.000000	.999827	0.000000	0.000000	0.000000	0.000000	0.000000	-.017962+M
3.6066E+01	-.8664	-.0033	.8664	-M	0.000000	.999827	0.000000	0.000000	0.000000	0.000000	0.000000	-.017962-M
				+M	0.000000	0.000000	-.004889	0.000000	0.000000	0.000000	0.000000	0.000000+M
2.9121E+01	.5985	0.0000	.5985	-M	.000650	0.000000	0.000000	0.000000	0.000000	0.000000	-.043523	0.000000-M
				+M	0.000000	0.000000	0.000000	.999052	0.000000	0.000000	0.000000	0.000000+M
2.9121E+01	-.5985	0.0000	.5985	-M	0.000000	0.000000	0.000000	.999052	0.000000	0.000000	0.000000	0.000000-M
				+M	.000650	0.000000	0.000000	0.000000	0.000000	0.000000	-.043523	0.000000+M
5.8951E+00	.4527	-.1295	.4709	-M	0.000000	0.000000	0.000000	0.000000	0.000000	-.132086	0.000000	0.000000-M
				+M	0.000000	0.000000	0.000000	0.000000	.991238	0.000000	0.000000	0.000000+M
5.8951E+00	-.4527	-.1295	.4709	-M	0.000000	0.000000	0.000000	0.000000	.991238	0.000000	0.000000	0.000000-M
				+M	0.000000	0.000000	0.000000	0.000000	0.000000	-.132086	0.000000	0.000000+M
-1.7895E+01	-.3194	.1295	.3446	-M	0.000000	0.000000	0.000000	0.000000	0.000000	.991238	0.000000	0.000000-M
				+M	0.000000	0.000000	0.000000	0.000000	.132086	0.000000	0.000000	0.000000+M
-1.7895E+01	.3194	.1295	.3446	-M	0.000000	0.000000	0.000000	0.000000	.132086	0.000000	0.000000	0.000000-M
				+M	0.000000	0.000000	0.000000	0.000000	0.000000	.991238	0.000000	0.000000+M
-1.8250E+01	.9986	0.0000	.9986	-M	0.000000	0.000000	0.000000	-.002488	0.000000	0.000000	0.000000	0.000000-M
				+M	.999107	0.000000	0.000000	0.000000	0.000000	0.000000	-.042188	0.000000+M
-1.8250E+01	-.9986	0.0000	.9986	-M	.999107	0.000000	0.000000	0.000000	0.000000	0.000000	-.042188	0.000000-M
				+M	0.000000	0.000000	0.000000	-.002488	0.000000	0.000000	0.000000	0.000000+M
-3.4991E+01	.1999	0.0000	.1999	-M	0.000000	0.000000	0.000000	.043457	0.000000	0.000000	0.000000	0.000000-M
				+M	.042257	0.000000	0.000000	0.000000	0.000000	0.000000	.998161	0.000000+M
-3.4991E+01	-.1999	0.0000	.1999	-M	.042257	0.000000	0.000000	0.000000	0.000000	0.000000	.998161	0.000000-M
				+M	0.000000	0.000000	0.000000	.043457	0.000000	0.000000	0.000000	0.000000+M
-4.4121E+01	.0664	.5331	.5373	-M	0.000000	0.000000	.024829	0.000000	0.000000	0.000000	0.000000	0.000000-M
				+M	0.000000	.018078	0.000000	0.000000	0.000000	0.000000	0.000000	.999528+M
-4.4121E+01	-.0664	.5331	.5373	-M	0.000000	.018078	0.000000	0.000000	0.000000	0.000000	0.000000	.999528-M
				+M	0.000000	0.000000	.024829	0.000000	0.000000	0.000000	0.000000	0.000000+M
ENERGY	HEPAL	HEPER	HE		M=15/2	M=13/2	M=11/2	M=9/2	M=7/2	M=5/2	M=3/2	M=1/2

Y= .20 X=-.80 OAS=9.1173E+01*W246 B4F4/B2F2=-2.00E-01 B4F4/B6F6=-4.00E+00 B2F2/B6F6= 2.00E+01

ENERGY	HEPAL	HEPER	HE		M=15/2	M=13/2	M=11/2	M=9/2	M=7/2	M=5/2	M=3/2	M=1/2
4.1053E+01	.7329	.0020	.7329	-M	0.000000	.002431	0.000000	0.000000	0.000000	0.000000	0.000000	.024095-M
				+M	0.000000	0.000000	.999707	0.000000	0.000000	0.000000	0.000000	0.000000+M
4.1053E+01	-.7329	.0020	.7329	-M	0.000000	0.000000	.999707	0.000000	0.000000	0.000000	0.000000	0.000000-M
				+M	0.000000	.002431	0.000000	0.000000	0.000000	0.000000	0.000000	.024095+M
3.3830E+01	-.5988	0.0000	.5988	-M	0.000000	0.000000	0.000000	.999223	0.000000	0.000000	0.000000	0.000000-M
				+M	.000594	0.000000	0.000000	0.000000	0.000000	0.000000	.039398	0.000000+M
3.3830E+01	.5988	0.0000	.5988	-M	.000594	0.000000	0.000000	0.000000	0.000000	0.000000	.039398	0.000000-M
				+M	0.000000	0.000000	0.000000	.999223	0.000000	0.000000	0.000000	0.000000+M
2.6707E+01	-.8664	-.0018	.8664	-M	0.000000	.999819	0.000000	0.000000	0.000000	0.000000	0.000000	.018806-M
				+M	0.000000	0.000000	-.002884	0.000000	0.000000	0.000000	0.000000	0.000000+M
2.6707E+01	.8664	-.0018	.8664	-M	0.000000	0.000000	-.002884	0.000000	0.000000	0.000000	0.000000	0.000000-M
				+M	0.000000	.999819	0.000000	0.000000	0.000000	0.000000	0.000000	.018806+M
1.2804E+01	.4559	.1139	.4699	-M	0.000000	0.000000	0.000000	0.000000	0.000000	.115944	0.000000	0.000000-M
				+M	0.000000	0.000000	0.000000	0.000000	.993256	0.000000	0.000000	0.000000+M
1.2804E+01	-.4559	.1139	.4699	-M	0.000000	0.000000	0.000000	0.000000	.993256	0.000000	0.000000	0.000000-M
				+M	0.000000	0.000000	0.000000	0.000000	0.000000	.115944	0.000000	0.000000+M
-1.3059E+01	-.9993	0.0000	.9993	-M	.999562	0.000000	0.000000	0.000000	0.000000	0.000000	.029530	0.000000-M
				+M	0.000000	0.000000	0.000000	-.001759	0.000000	0.000000	0.000000	0.000000+M
-1.3059E+01	.9993	0.0000	.9993	-M	0.000000	0.000000	0.000000	-.001759	0.000000	0.000000	0.000000	0.000000-M
				+M	.999562	0.000000	0.000000	0.000000	0.000000	0.000000	.029530	0.000000+M
-1.4244E+01	.3226	-.1139	.3421	-M	0.000000	0.000000	0.000000	0.000000	-.115944	0.000000	0.000000	0.000000-M
				+M	0.000000	0.000000	0.000000	0.000000	0.000000	.993256	0.000000	0.000000+M
-1.4244E+01	-.3226	-.1139	.3421	-M	0.000000	0.000000	0.000000	0.000000	0.000000	.993256	0.000000	0.000000-M
				+M	0.000000	0.000000	0.000000	0.000000	-.115944	0.000000	0.000000	0.000000+M
-3.6971E+01	-.1995	0.0000	.1995	-M	-.029576	0.000000	0.000000	0.000000	0.000000	0.000000	.998787	0.000000-M
				+M	0.000000	0.000000	0.000000	-.039363	0.000000	0.000000	0.000000	0.000000+M
-3.6971E+01	.1995	0.0000	.1995	-M	0.000000	0.000000	0.000000	-.039363	0.000000	0.000000	0.000000	0.000000-M
				+M	-.029576	0.000000	0.000000	0.000000	0.000000	0.000000	.998787	0.000000+M
-5.0120E+01	-.0665	.5332	.5373	-M	0.000000	-.018870	0.000000	0.000000	0.000000	0.000000	0.000000	.999533-M
				+M	0.000000	0.000000	-.024045	0.000000	0.000000	0.000000	0.000000	0.000000+M
-5.0120E+01	.0665	.5332	.5373	-M	0.000000	0.000000	-.024045	0.000000	0.000000	0.000000	0.000000	0.000000-M
				+M	0.000000	-.018870	0.000000	0.000000	0.000000	0.000000	0.000000	.999533+M
ENERGY	HEPAL	HEPER	HE		M=15/2	M=13/2	M=11/2	M=9/2	M=7/2	M=5/2	M=3/2	M=1/2

Y= .20 X=-.60 OAS=7.6716E+01*W246 B4F4/B2F2=-1.50E-01 B4F4/B6F6=-1.50E+00 B2F2/B6F6= 1.00E+01

ENERGY	HEPAL	HEPER	HE		M=15/2	M=13/2	M=11/2	M=9/2	M=7/2	M=5/2	M=3/2	M=1/2
3.0854E+01	.7304	.0112	.7305	-M	0.000000	.013424	0.000000	0.000000	0.000000	0.000000	0.000000	.057840-M
				+M	0.000000	0.000000	.998236	0.000000	0.000000	0.000000	0.000000	0.000000+M
3.0854E+01	-.7304	.0112	.7305	-M	0.000000	0.000000	.998236	0.000000	0.000000	0.000000	0.000000	0.000000-M
				+M	0.000000	.013424	0.000000	0.000000	0.000000	0.000000	0.000000	.057840+M
2.8548E+01	.5934	0.0000	.5934	-M	.004567	0.000000	0.000000	0.000000	0.000000	0.000000	.090733	0.000000-M
				+M	0.000000	0.000000	0.000000	.995865	0.000000	0.000000	0.000000	0.000000+M
2.8548E+01	-.5934	0.0000	.5934	-M	0.000000	0.000000	0.000000	.995865	0.000000	0.000000	0.000000	0.000000-M
				+M	.004567	0.000000	0.000000	0.000000	0.000000	0.000000	.090733	0.000000+M
1.8488E+01	.8647	-.0102	.8648	-M	0.000000	0.000000	-.015993	0.000000	0.000000	0.000000	0.000000	0.000000-M
				+M	0.000000	.998895	0.000000	0.000000	0.000000	0.000000	0.000000	.044185+M
1.8488E+01	-.8647	-.0102	.8648	-M	0.000000	.998895	0.000000	0.000000	0.000000	0.000000	0.000000	.044185-M
				+M	0.000000	0.000000	-.015993	0.000000	0.000000	0.000000	0.000000	0.000000+M
1.3457E+01	.4185	.2352	.4801	-M	0.000000	0.000000	0.000000	0.000000	0.000000	.245332	0.000000	0.000000-M
				+M	0.000000	0.000000	0.000000	0.000000	.969439	0.000000	0.000000	0.000000+M
1.3457E+01	-.4185	.2352	.4801	-M	0.000000	0.000000	0.000000	0.000000	.969439	0.000000	0.000000	0.000000-M
				+M	0.000000	0.000000	0.000000	0.000000	0.000000	.245332	0.000000	0.000000+M
4.9836E-01	.9985	0.0000	.9985	-M	0.000000	0.000000	0.000000	-.008337	0.000000	0.000000	0.000000	0.000000-M
				+M	.999116	0.000000	0.000000	0.000000	0.000000	0.000000	.041208	0.000000+M
4.9836E-01	-.9985	0.0000	.9985	-M	.999116	0.000000	0.000000	0.000000	0.000000	0.000000	.041208	0.000000-M
				+M	0.000000	0.000000	0.000000	-.008337	0.000000	0.000000	0.000000	0.000000+M
-1.2737E+01	.2852	-.2352	.3696	-M	0.000000	0.000000	0.000000	0.000000	-.245332	0.000000	0.000000	0.000000-M
				+M	0.000000	0.000000	0.000000	0.000000	0.000000	.969439	0.000000	0.000000+M
-1.2737E+01	-.2852	-.2352	.3696	-M	0.000000	0.000000	0.000000	0.000000	0.000000	.969439	0.000000	0.000000-M
				+M	0.000000	0.000000	0.000000	0.000000	-.245332	0.000000	0.000000	0.000000+M
-3.3246E+01	.1949	0.0000	.1949	-M	0.000000	0.000000	0.000000	-.090464	0.000000	0.000000	0.000000	0.000000-M
				+M	-.041794	0.000000	0.000000	0.000000	0.000000	0.000000	.995022	0.000000+M
-3.3246E+01	-.1949	0.0000	.1949	-M	-.041794	0.000000	0.000000	0.000000	0.000000	0.000000	.995022	0.000000-M
				+M	0.000000	0.000000	0.000000	-.090464	0.000000	0.000000	0.000000	0.000000+M
-4.5862E+01	-.0657	.5323	.5364	-M	0.000000	-.045032	0.000000	0.000000	0.000000	0.000000	0.000000	.997348-M
				+M	0.000000	0.000000	-.057183	0.000000	0.000000	0.000000	0.000000	0.000000+M
-4.5862E+01	.0657	.5323	.5364	-M	0.000000	0.000000	-.057183	0.000000	0.000000	0.000000	0.000000	0.000000-M
				+M	0.000000	-.045032	0.000000	0.000000	0.000000	0.000000	0.000000	.997348+M
ENERGY	HEPAL	HEPER	HE		M=15/2	M=13/2	M=11/2	M=9/2	M=7/2	M=5/2	M=3/2	M=1/2

Y= .20 X=-.40 OAS=6.5689E+01*W246 B4F4/B2F2=-1.00E-01 B4F4/B6F6=-6.67E-01 B2F2/B6F6= 6.67E+00

ENERGY	HEPAL	HEPER	HE		M=15/2	M=13/2	M=11/2	M=9/2	M=7/2	M=5/2	M=3/2	M=1/2
2.3709E+01	-.5778	0.0000	.5778	-M	0.000000	0.000000	0.000000	.986622	0.000000	0.000000	0.000000	0.000000-M
				+M	.034677	0.000000	0.000000	0.000000	0.000000	0.000000	.159291	0.000000+M
2.3709E+01	.5778	0.0000	.5778	-M	.034677	0.000000	0.000000	0.000000	0.000000	0.000000	.159291	0.000000-M
				+M	0.000000	0.000000	0.000000	.986622	0.000000	0.000000	0.000000	0.000000+M
2.0920E+01	-.7209	.0367	.7218	-M	0.000000	0.000000	.993129	0.000000	0.000000	0.000000	0.000000	0.000000-M
				+M	0.000000	.043443	0.000000	0.000000	0.000000	0.000000	0.000000	.108664+M
2.0920E+01	.7209	.0367	.7218	-M	0.000000	.043443	0.000000	0.000000	0.000000	0.000000	0.000000	.108664-M
				+M	0.000000	0.000000	.993129	0.000000	0.000000	0.000000	0.000000	0.000000+M
1.5039E+01	-.3573	.3397	.4930	-M	0.000000	0.000000	0.000000	0.000000	.929119	0.000000	0.000000	0.000000-M
				+M	0.000000	0.000000	0.000000	0.000000	0.000000	.369780	0.000000	0.000000+M
1.5039E+01	.3573	.3397	.4930	-M	0.000000	0.000000	0.000000	0.000000	0.000000	.369780	0.000000	0.000000-M
				+M	0.000000	0.000000	0.000000	0.000000	.929119	0.000000	0.000000	0.000000+M
1.4048E+01	.9958	0.0000	.9958	-M	0.000000	0.000000	0.000000	-.041796	0.000000	0.000000	0.000000	0.000000-M
				+M	.998261	0.000000	0.000000	0.000000	0.000000	0.000000	.041565	0.000000+M
1.4048E+01	-.9958	0.0000	.9958	-M	.998261	0.000000	0.000000	0.000000	0.000000	0.000000	.041565	0.000000-M
				+M	0.000000	0.000000	0.000000	-.041796	0.000000	0.000000	0.000000	0.000000+M
1.0380E+01	.8575	-.0333	.8581	-M	0.000000	0.000000	-.052058	0.000000	0.000000	0.000000	0.000000	0.000000-M
				+M	0.000000	.995613	0.000000	0.000000	0.000000	0.000000	0.000000	.077743+M
1.0380E+01	-.8575	-.0333	.8581	-M	0.000000	.995613	0.000000	0.000000	0.000000	0.000000	0.000000	.077743-M
				+M	0.000000	0.000000	-.052058	0.000000	0.000000	0.000000	0.000000	0.000000+M
-1.2159E+01	.2239	-.3397	.4069	-M	0.000000	0.000000	0.000000	0.000000	-.369780	0.000000	0.000000	0.000000-M
				+M	0.000000	0.000000	0.000000	0.000000	0.000000	.929119	0.000000	0.000000+M
-1.2159E+01	-.2239	-.3397	.4069	-M	0.000000	0.000000	0.000000	0.000000	0.000000	.929119	0.000000	0.000000-M
				+M	0.000000	0.000000	0.000000	0.000000	-.369780	0.000000	0.000000	0.000000+M
-2.9958E+01	-.1820	0.0000	.1820	-M	-.047667	0.000000	0.000000	0.000000	0.000000	0.000000	.986356	0.000000-M
				+M	0.000000	0.000000	0.000000	-.157573	0.000000	0.000000	0.000000	0.000000+M
-2.9958E+01	.1820	0.0000	.1820	-M	0.000000	0.000000	0.000000	-.157573	0.000000	0.000000	0.000000	0.000000-M
				+M	-.047667	0.000000	0.000000	0.000000	0.000000	0.000000	.986356	0.000000+M
-4.1979E+01	.0634	.5299	.5337	-M	0.000000	0.000000	-.104810	0.000000	0.000000	0.000000	0.000000	0.000000-M
				+M	0.000000	-.082865	0.000000	0.000000	0.000000	0.000000	0.000000	.991034+M
-4.1979E+01	-.0634	.5299	.5337	-M	0.000000	-.082865	0.000000	0.000000	0.000000	0.000000	0.000000	.991034-M
				+M	0.000000	0.000000	-.104810	0.000000	0.000000	0.000000	0.000000	0.000000+M
ENERGY	HEPAL	HEPER	HE		M=15/2	M=13/2	M=11/2	M=9/2	M=7/2	M=5/2	M=3/2	M=1/2

Y= .20 X=-.20 OAS=6.6450E+01*W246 B4F4/B2F2=-5.00E-02 B4F4/B6F6=-2.50E-01 E2F2/B6F6= 5.00E+00

ENERGY	HEPAL	HEPER	HE		M=15/2	M=13/2	M=11/2	M=9/2	M=7/2	M=5/2	M=3/2	M=1/2
2.7677E+01	-.9883	0.0000	.9883	-M	.995146	0.000000	0.000000	0.000000	0.000000	0.000000	.069118	0.000000-M
				+M	0.000000	0.000000	0.000000	.070044	0.000000	0.000000	0.000000	0.000000+M
2.7677E+01	.9883	0.0000	.9883	-M	0.000000	0.000000	0.000000	.070044	0.000000	0.000000	0.000000	0.000000-M
				+M	.995146	0.000000	0.000000	0.000000	0.000000	0.000000	.069118	0.000000+M
1.9451E+01	.5422	0.0000	.5422	-M	-.084762	0.000000	0.000000	0.000000	0.000000	0.000000	.240477	0.000000-M
				+M	0.000000	0.000000	0.000000	.966947	0.000000	0.000000	0.000000	0.000000+M
1.9451E+01	-.5422	0.0000	.5422	-M	0.000000	0.000000	0.000000	.966947	0.000000	0.000000	0.000000	0.000000-M
				+M	-.084762	0.000000	0.000000	0.000000	0.000000	0.000000	.240477	0.000000+M
1.7458E+01	-.2873	.4124	.5026	-M	0.000000	0.000000	0.000000	0.000000	.880804	0.000000	0.000000	0.000000-M
				+M	0.000000	0.000000	0.000000	0.000000	0.000000	.473481	0.000000	0.000000+M
1.7458E+01	.2873	.4124	.5026	-M	0.000000	0.000000	0.000000	0.000000	0.000000	.473481	0.000000	0.000000-M
				+M	0.000000	0.000000	0.000000	0.000000	.880804	0.000000	0.000000	0.000000+M
1.1529E+01	.6831	.0979	.6901	-M	0.000000	.113651	0.000000	0.000000	0.000000	0.000000	0.000000	.192228-M
				+M	0.000000	0.000000	.974747	0.000000	0.000000	0.000000	0.000000	0.000000+M
1.1529E+01	-.6831	.0979	.6901	-M	0.000000	0.000000	.974747	0.000000	0.000000	0.000000	0.000000	0.000000-M
				+M	0.000000	.113651	0.000000	0.000000	0.000000	0.000000	0.000000	.192228+M
2.4042E+00	-.8256	-.0883	.8303	-M	0.000000	.983669	0.000000	0.000000	0.000000	0.000000	0.000000	.116047-M
				+M	0.000000	0.000000	-.137577	0.000000	0.000000	0.000000	0.000000	0.000000+M
2.4042E+00	.8256	-.0883	.8303	-M	0.000000	0.000000	-.137577	0.000000	0.000000	0.000000	0.000000	0.000000-M
				+M	0.000000	.983669	0.000000	0.000000	0.000000	0.000000	0.000000	.116047+M
-1.2418E+01	.1540	-.4124	.4402	-M	0.000000	0.000000	0.000000	0.000000	-.473481	0.000000	0.000000	0.000000-M
				+M	0.000000	0.000000	0.000000	0.000000	0.000000	.880804	0.000000	0.000000+M
-1.2418E+01	-.1540	-.4124	.4402	-M	0.000000	0.000000	0.000000	0.000000	0.000000	.880804	0.000000	0.000000-M
				+M	0.000000	0.000000	0.000000	0.000000	-.473481	0.000000	0.000000	0.000000+M
-2.7328E+01	.1539	0.0000	.1539	-M	0.000000	0.000000	0.000000	-.245169	0.000000	0.000000	0.000000	0.000000-M
				+M	-.049990	0.000000	0.000000	0.000000	0.000000	0.000000	.968191	0.000000+M
-2.7328E+01	-.1539	0.0000	.1539	-M	-.049990	0.000000	0.000000	0.000000	0.000000	0.000000	.968191	0.000000-M
				+M	0.000000	0.000000	0.000000	-.245169	0.000000	0.000000	0.000000	0.000000+M
-3.8773E+01	-.0575	.5238	.5269	-M	0.000000	-.139563	0.000000	0.000000	0.000000	0.000000	0.000000	.974465-M
				+M	0.000000	0.000000	-.175900	0.000000	0.000000	0.000000	0.000000	0.000000+M
-3.8773E+01	.0575	.5238	.5269	-M	0.000000	0.000000	-.175900	0.000000	0.000000	0.000000	0.000000	0.000000-M
				+M	0.000000	-.139563	0.000000	0.000000	0.000000	0.000000	0.000000	.974465+M
ENERGY	HEPAL	HEPER	HE		M=15/2	M=13/2	M=11/2	M=9/2	M=7/2	M=5/2	M=3/2	M=1/2

Y= .20 X=0.00 OAS=7.8078E+01*W246 B4F4/B2F2= 0. B4F4/B6F6= 0. E2F2/B6F6= 4.00E+00

ENERGY	HEPAL	HEPER	HE		M=15/2	M=13/2	M=11/2	M=9/2	M=7/2	M=5/2	M=3/2	M=1/2
4.1227E+01	.9953	0.0000	.9953	-M	0.000000	0.000000	0.000000	.029560	0.000000	0.000000	0.000000	0.000000-M
				+M	.997503	0.000000	0.000000	0.000000	0.000000	0.000000	.064144	0.000000+M
4.1227E+01	-.9953	0.0000	.9953	-M	.997503	0.000000	0.000000	0.000000	0.000000	0.000000	.064144	0.000000-M
				+M	0.000000	0.000000	0.000000	.029560	0.000000	0.000000	0.000000	0.000000+M
2.0515E+01	.2227	.4552	.5068	-M	0.000000	0.000000	0.000000	0.000000	0.000000	.552182	0.000000	0.000000-M
				+M	0.000000	0.000000	0.000000	0.000000	.833724	0.000000	0.000000	0.000000+M
2.0515E+01	-.2227	.4552	.5068	-M	0.000000	0.000000	0.000000	0.000000	.833724	0.000000	0.000000	0.000000-M
				+M	0.000000	0.000000	0.000000	0.000000	0.000000	.552182	0.000000	0.000000+M
1.6251E+01	.4967	0.0000	.4967	-M	-.050346	0.000000	0.000000	0.000000	0.000000	0.000000	.352258	0.000000-M
				+M	0.000000	0.000000	0.000000	.934548	0.000000	0.000000	0.000000	0.000000+M
1.6251E+01	-.4967	0.0000	.4967	-M	0.000000	0.000000	0.000000	.934548	0.000000	0.000000	0.000000	0.000000-M
				+M	-.050346	0.000000	0.000000	0.000000	0.000000	0.000000	.352258	0.000000+M
3.4095E+00	.5511	.2145	.5914	-M	0.000000	.241460	0.000000	0.000000	0.000000	0.000000	0.000000	.333395-M
				+M	0.000000	0.000000	.911342	0.000000	0.000000	0.000000	0.000000	0.000000+M
3.4095E+00	-.5511	.2145	.5914	-M	0.000000	0.000000	.911342	0.000000	0.000000	0.000000	0.000000	0.000000-M
				+M	0.000000	.241460	0.000000	0.000000	0.000000	0.000000	0.000000	.333395+M
-5.5586E+00	-.7081	-.1899	.7332	-M	0.000000	.944304	0.000000	0.000000	0.000000	0.000000	0.000000	.135648-M
				+M	0.000000	0.000000	-.299817	0.000000	0.000000	0.000000	0.000000	0.000000+M
-5.5586E+00	.7081	-.1899	.7332	-M	0.000000	0.000000	-.299817	0.000000	0.000000	0.000000	0.000000	0.000000-M
				+M	0.000000	.944304	0.000000	0.000000	0.000000	0.000000	0.000000	.135648+M
-1.3315E+01	.0894	-.4552	.4639	-M	0.000000	0.000000	0.000000	0.000000	-.552182	0.000000	0.000000	0.000000-M
				+M	0.000000	0.000000	0.000000	0.000000	0.000000	.833724	0.000000	0.000000+M
-1.3315E+01	-.0894	-.4552	.4639	-M	0.000000	0.000000	0.000000	0.000000	0.000000	.833724	0.000000	0.000000-M
				+M	0.000000	0.000000	0.000000	0.000000	-.552182	0.000000	0.000000	0.000000+M
-2.5678E+01	.1014	0.0000	.1014	-M	0.000000	0.000000	0.000000	-.354608	0.000000	0.000000	0.000000	0.000000-M
				+M	-.049533	0.000000	0.000000	0.000000	0.000000	0.000000	.933702	0.000000+M
-2.5678E+01	-.1014	0.0000	.1014	-M	-.049533	0.000000	0.000000	0.000000	0.000000	0.000000	.933702	0.000000-M
				+M	0.000000	0.000000	0.000000	-.354608	0.000000	0.000000	0.000000	0.000000+M
-3.6851E+01	.0430	.5087	.5105	-M	0.000000	0.000000	-.282073	0.000000	0.000000	0.000000	0.000000	0.000000-M
				+M	0.000000	-.223579	0.000000	0.000000	0.000000	0.000000	0.000000	.932978+M
-3.6851E+01	-.0430	.5087	.5105	-M	0.000000	-.223579	0.000000	0.000000	0.000000	0.000000	0.000000	.932978-M
				+M	0.000000	0.000000	-.282073	0.000000	0.000000	0.000000	0.000000	0.000000+M
ENERGY	HEPAL	HEPER	HE		M=15/2	M=13/2	M=11/2	M=9/2	M=7/2	M=5/2	M=3/2	M=1/2

Y= .20 X= .20 OAS=7.6319E+01*W246 B4F4/B2F2= 5.00E-02 B4F4/B6F6= 2.50E-01 B2F2/B6F6= 5.00E+00

ENERGY	HEPAL	HEPER	HE		M=15/2	M=13/2	M=11/2	M=9/2	M=7/2	M=5/2	M=3/2	M=1/2
4.9451E+01	.9981	0.0000	.9981	-M	0.000000	0.000000	0.000000	.010579	0.000000	0.000000	0.000000	0.000000-M
				+M	.998868	0.000000	0.000000	0.000000	0.000000	0.000000	.046377	0.000000+M
4.9451E+01	-.9981	0.0000	.9981	-M	.998868	0.000000	0.000000	0.000000	0.000000	0.000000	.046377	0.000000-M
				+M	0.000000	0.000000	0.000000	.010579	0.000000	0.000000	0.000000	0.000000+M
1.2284E+01	.1685	.4781	.5069	-M	0.000000	0.000000	0.000000	0.000000	0.000000	.610498	0.000000	0.000000-M
				+M	0.000000	0.000000	0.000000	0.000000	.792018	0.000000	0.000000	0.000000+M
1.2284E+01	-.1685	.4781	.5069	-M	0.000000	0.000000	0.000000	0.000000	.792018	0.000000	0.000000	0.000000-M
				+M	0.000000	0.000000	0.000000	0.000000	0.000000	.610498	0.000000	0.000000+M
6.5710E+00	.4202	0.0000	.4202	-M	-.031248	0.000000	0.000000	0.000000	0.000000	0.000000	.472054	0.000000-M
				+M	0.000000	0.000000	0.000000	.881016	0.000000	0.000000	0.000000	0.000000+M
6.5710E+00	-.4202	0.0000	.4202	-M	0.000000	0.000000	0.000000	.881016	0.000000	0.000000	0.000000	0.000000-M
				+M	-.031248	0.000000	0.000000	0.000000	0.000000	0.000000	.472054	0.000000+M
-1.3772E+00	.0843	.0179	.0861	-M	0.000000	0.000000	.619285	0.000000	0.000000	0.000000	0.000000	0.000000-M
				+M	0.000000	.636812	0.000000	0.000000	0.000000	0.000000	0.000000	.459300+M
-1.3772E+00	-.0843	-.0179	.0861	-M	0.000000	.636812	0.000000	0.000000	0.000000	0.000000	0.000000	.459300-M
				+M	0.000000	0.000000	.619285	0.000000	0.000000	0.000000	0.000000	0.000000+M
-6.4355E+00	-.1355	.0179	.1367	-M	0.000000	-.732750	0.000000	0.000000	0.000000	0.000000	0.000000	.110619-M
				+M	0.000000	0.000000	.671446	0.000000	0.000000	0.000000	0.000000	0.000000+M
-6.4355E+00	.1355	-.0179	.1367	-M	0.000000	0.000000	-.671446	0.000000	0.000000	0.000000	0.000000	0.000000-M
				+M	0.000000	.732750	0.000000	0.000000	0.000000	0.000000	0.000000	-.110619+M
-1.3484E+01	.0352	-.4781	.4794	-M	0.000000	0.000000	0.000000	0.000000	-.610498	0.000000	0.000000	0.000000-M
				+M	0.000000	0.000000	0.000000	0.000000	0.000000	.792018	0.000000	0.000000+M
-1.3484E+01	-.0352	-.4781	.4794	-M	0.000000	0.000000	0.000000	0.000000	0.000000	.792018	0.000000	0.000000-M
				+M	0.000000	0.000000	0.000000	0.000000	-.610498	0.000000	0.000000	0.000000+M
-2.0142E+01	-.0221	0.0000	.0221	-M	-.035865	0.000000	0.000000	0.000000	0.000000	0.000000	.880349	0.000000-M
				+M	0.000000	0.000000	0.000000	-.472969	0.000000	0.000000	0.000000	0.000000+M
-2.0142E+01	.0221	0.0000	.0221	-M	0.000000	0.000000	0.000000	-.472969	0.000000	0.000000	0.000000	0.000000-M
				+M	-.035865	0.000000	0.000000	0.000000	0.000000	0.000000	.880349	0.000000+M
-2.6867E+01	-.0198	.4832	.4836	-M	0.000000	0.000000	-.406996	0.000000	0.000000	0.000000	0.000000	0.000000-M
				+M	0.000000	-.239891	0.000000	0.000000	0.000000	0.000000	0.000000	.881366+M
-2.6867E+01	.0198	.4832	.4836	-M	0.000000	-.239891	0.000000	0.000000	0.000000	0.000000	0.000000	.881366-M
				+M	0.000000	0.000000	-.406996	0.000000	0.000000	0.000000	0.000000	0.000000+M
ENERGY	HEPAL	HEPER	HE		M=15/2	M=13/2	M=11/2	M=9/2	M=7/2	M=5/2	M=3/2	M=1/2

Y= .20 X= .40 OAS=7.8007E+01*W246 B4F4/B2F2= 1.00E-01 B4F4/B6F6= 6.67E-01 B2F2/B6F6= 6.67E+00

ENERGY	HEPAL	HEPER	HE		M=15/2	M=13/2	M=11/2	M=9/2	M=7/2	M=5/2	M=3/2	M=1/2	
5.7710E+01	.9991	0.0000	.9991	-M	0.000000	0.000000	0.000000	.004068	0.000000	0.000000	0.000000	0.000000	-M
				+M	.999452	0.000000	0.000000	0.000000	0.000000	0.000000	.032859	0.000000	+M
5.7710E+01	-.9991	0.0000	.9991	-M	.999452	0.000000	0.000000	0.000000	0.000000	0.000000	.032859	0.000000	-M
				+M	0.000000	0.000000	0.000000	.004068	0.000000	0.000000	0.000000	0.000000	+M
4.5445E+00	-.0650	.4944	.4987	-M	0.000000	0.000000	0.000000	0.000000	.705592	0.000000	0.000000	0.000000	-M
				+M	0.000000	0.000000	0.000000	0.000000	0.000000	.708619	0.000000	0.000000	+M
4.5445E+00	.0650	.4944	.4987	-M	0.000000	0.000000	0.000000	0.000000	0.000000	.708619	0.000000	0.000000	-M
				+M	0.000000	0.000000	0.000000	0.000000	.705592	0.000000	0.000000	0.000000	+M
-4.7025E-01	-.1553	0.0000	.1553	-M	0.000000	0.000000	0.000000	.667001	0.000000	0.000000	0.000000	0.000000	-M
				+M	-.027194	0.000000	0.000000	0.000000	0.000000	0.000000	.744560	0.000000	+M
-4.7025E-01	.1553	0.0000	.1553	-M	-.027194	0.000000	0.000000	0.000000	0.000000	0.000000	.744560	0.000000	-M
				+M	0.000000	0.000000	0.000000	.667001	0.000000	0.000000	0.000000	0.000000	+M
-1.4231E+00	-.5023	.3147	.5927	-M	0.000000	.784919	0.000000	0.000000	0.000000	0.000000	0.000000	.558864	-M
				+M	0.000000	0.000000	.267532	0.000000	0.000000	0.000000	0.000000	0.000000	+M
-1.4231E+00	.5023	.3147	.5927	-M	0.000000	0.000000	.267532	0.000000	0.000000	0.000000	0.000000	0.000000	-M
				+M	0.000000	.784919	0.000000	0.000000	0.000000	0.000000	0.000000	.558864	+M
-8.6399E+00	-.0916	-.0696	.1150	-M	0.000000	-.596577	0.000000	0.000000	0.000000	0.000000	0.000000	.565069	-M
				+M	0.000000	0.000000	.569906	0.000000	0.000000	0.000000	0.000000	0.000000	+M
-8.6399E+00	.0916	-.0696	.1150	-M	0.000000	0.000000	.569906	0.000000	0.000000	0.000000	0.000000	0.000000	-M
				+M	0.000000	-.596577	0.000000	0.000000	0.000000	0.000000	0.000000	.565069	+M
-1.4144E+01	-.0684	-.4944	.4991	-M	0.000000	0.000000	0.000000	0.000000	.708619	0.000000	0.000000	0.000000	-M
				+M	0.000000	0.000000	0.000000	0.000000	0.000000	-.705592	0.000000	0.000000	+M
-1.4144E+01	.0684	-.4944	.4991	-M	0.000000	0.000000	0.000000	0.000000	0.000000	-.705592	0.000000	0.000000	-M
				+M	0.000000	0.000000	0.000000	0.000000	.708619	0.000000	0.000000	0.000000	+M
-1.7280E+01	-.2438	0.0000	.2438	-M	0.000000	0.000000	0.000000	.745046	0.000000	0.000000	0.000000	0.000000	-M
				+M	.018888	0.000000	0.000000	0.000000	0.000000	0.000000	-.666746	0.000000	+M
-1.7280E+01	.2438	0.0000	.2438	-M	.018888	0.000000	0.000000	0.000000	0.000000	0.000000	-.666746	0.000000	-M
				+M	0.000000	0.000000	0.000000	.745046	0.000000	0.000000	0.000000	0.000000	+M
-2.0297E+01	-.3938	.2882	.4880	-M	0.000000	0.000000	.776938	0.000000	0.000000	0.000000	0.000000	0.000000	-M
				+M	0.000000	.167326	0.000000	0.000000	0.000000	0.000000	0.000000	-.606934	+M
-2.0297E+01	.3938	.2882	.4880	-M	0.000000	.167326	0.000000	0.000000	0.000000	0.000000	0.000000	-.606934	-M
				+M	0.000000	0.000000	.776938	0.000000	0.000000	0.000000	0.000000	0.000000	+M
ENERGY	HEPAL	HEPER	HE		M=15/2	M=13/2	M=11/2	M=9/2	M=7/2	M=5/2	M=3/2	M=1/2	

Y= .20 X= .60 OAS=8.9275E+01*W246 B4F4/B2F2= 1.50E-01 B4F4/B6F6= 1.50E+00 B2F2/B6F6= 1.00E+01

ENERGY	HEPAL	HEPER	HE		M=15/2	M=13/2	M=11/2	M=9/2	M=7/2	M=5/2	M=3/2	M=1/2	
6.5990E+01	.9996	0.0000	.9996	-M	0.000000	0.000000	0.000000	.001358	0.000000	0.000000	0.000000	0.000000	-M
				+M	.999778	0.000000	0.000000	0.000000	0.000000	0.000000	.021004	0.000000	+M
6.5990E+01	-.9996	0.0000	.9996	-M	.999778	0.000000	0.000000	0.000000	0.000000	0.000000	.021004	0.000000	-M
				+M	0.000000	0.000000	0.000000	.001358	0.000000	0.000000	0.000000	0.000000	+M
2.1539E+00	.2105	.4612	.5070	-M	0.000000	0.000000	.156539	0.000000	0.000000	0.000000	0.000000	0.000000	-M
				+M	0.000000	.451991	0.000000	0.000000	0.000000	0.000000	0.000000	.878180	+M
2.1539E+00	-.2105	.4612	.5070	-M	0.000000	.451991	0.000000	0.000000	0.000000	0.000000	0.000000	.878180	-M
				+M	0.000000	0.000000	.156539	0.000000	0.000000	0.000000	0.000000	0.000000	+M
-1.9809E-01	-.1428	0.0000	.1428	-M	-.020598	0.000000	0.000000	0.000000	0.000000	0.000000	.963116	0.000000	-M
				+M	0.000000	0.000000	0.000000	.268295	0.000000	0.000000	0.000000	0.000000	+M
-1.9809E-01	.1428	0.0000	.1428	-M	0.000000	0.000000	0.000000	.268295	0.000000	0.000000	0.000000	0.000000	-M
				+M	-.020598	0.000000	0.000000	0.000000	0.000000	0.000000	.963116	0.000000	+M
-1.9220E+00	.1232	.4352	.4523	-M	0.000000	0.000000	0.000000	0.000000	.512488	0.000000	0.000000	0.000000	-M
				+M	0.000000	0.000000	0.000000	0.000000	0.000000	.858695	0.000000	0.000000	+M
-1.9220E+00	-.1232	.4352	.4523	-M	0.000000	0.000000	0.000000	0.000000	0.000000	.858695	0.000000	0.000000	-M
				+M	0.000000	0.000000	0.000000	0.000000	.512488	0.000000	0.000000	0.000000	+M
-4.9088E+00	-.6930	.0338	.6938	-M	0.000000	.891592	0.000000	0.000000	0.000000	0.000000	0.000000	-.439311	-M
				+M	0.000000	0.000000	-.109857	0.000000	0.000000	0.000000	0.000000	0.000000	+M
-4.9088E+00	.6930	.0338	.6938	-M	0.000000	0.000000	-.109857	0.000000	0.000000	0.000000	0.000000	0.000000	-M
				+M	0.000000	.891592	0.000000	0.000000	0.000000	0.000000	0.000000	-.439311	+M
-1.6078E+01	-.2566	-.4352	.5052	-M	0.000000	0.000000	0.000000	0.000000	.858695	0.000000	0.000000	0.000000	-M
				+M	0.000000	0.000000	0.000000	0.000000	0.000000	-.512488	0.000000	0.000000	+M
-1.6078E+01	.2566	-.4352	.5052	-M	0.000000	0.000000	0.000000	0.000000	0.000000	-.512488	0.000000	0.000000	-M
				+M	0.000000	0.000000	0.000000	0.000000	.858695	0.000000	0.000000	0.000000	+M
-2.1752E+01	-.5424	0.0000	.5424	-M	0.000000	0.000000	0.000000	.963336	0.000000	0.000000	0.000000	0.000000	-M
				+M	.004327	0.000000	0.000000	0.000000	0.000000	0.000000	-.268264	0.000000	+M
-2.1752E+01	.5424	0.0000	.5424	-M	.004327	0.000000	0.000000	0.000000	0.000000	0.000000	-.268264	0.000000	-M
				+M	0.000000	0.000000	0.000000	.963336	0.000000	0.000000	0.000000	0.000000	+M
-2.3285E+01	-.7035	.0383	.7045	-M	0.000000	0.000000	.981543	0.000000	0.000000	0.000000	0.000000	0.000000	-M
				+M	0.000000	.027705	0.000000	0.000000	0.000000	0.000000	0.000000	-.189224	+M
-2.3285E+01	.7035	.0383	.7045	-M	0.000000	.027705	0.000000	0.000000	0.000000	0.000000	0.000000	-.189224	-M
				+M	0.000000	0.000000	.981543	0.000000	0.000000	0.000000	0.000000	0.000000	+M
ENERGY	HEPAL	HEPER	HE		M=15/2	M=13/2	M=11/2	M=9/2	M=7/2	M=5/2	M=3/2	M=1/2	

Y= .20 X= .80 OAS=1.0511E+02*W246 B4F4/B2F2= 2.00E-01 B4F4/B6F6= 4.00E+00 E2F2/B6F6= 2.00E+01

ENERGY	HEPAL	HEPER	HE		M=15/2	M=13/2	M=11/2	M=9/2	M=7/2	M=5/2	M=3/2	M=1/2	
7.4287E+01	.9999	0.0000	.9999	-M	0.000000	0.000000	0.000000	.000269	0.000000	0.000000	0.000000	0.000000	-M
				+M	.999949	0.000000	0.000000	0.000000	0.000000	0.000000	.010139	0.000000	+M
7.4287E+01	-.9999	0.0000	.9999	-M	.999949	0.000000	0.000000	0.000000	0.000000	0.000000	.010139	0.000000	-M
				+M	0.000000	0.000000	0.000000	.000269	0.000000	0.000000	0.000000	0.000000	+M
1.0719E+01	.0739	.5296	.5347	-M	0.000000	0.000000	.053801	0.000000	0.000000	0.000000	0.000000	0.000000	-M
				+M	0.000000	.109362	0.000000	0.000000	0.000000	0.000000	0.000000	.992545	+M
1.0719E+01	-.0739	.5296	.5347	-M	0.000000	.109362	0.000000	0.000000	0.000000	0.000000	0.000000	.992545	-M
				+M	0.000000	0.000000	.053801	0.000000	0.000000	0.000000	0.000000	0.000000	+M
4.6530E+00	-.1951	0.0000	.1951	-M	-.010129	0.000000	0.000000	0.000000	0.000000	0.000000	.996841	0.000000	-M
				+M	0.000000	0.000000	0.000000	.078778	0.000000	0.000000	0.000000	0.000000	+M
4.6530E+00	.1951	0.0000	.1951	-M	0.000000	0.000000	0.000000	.078778	0.000000	0.000000	0.000000	0.000000	-M
				+M	-.010129	0.000000	0.000000	0.000000	0.000000	0.000000	.996841	0.000000	+M
-2.5991E+00	-.8570	.0002	.8570	-M	0.000000	.993998	0.000000	0.000000	0.000000	0.000000	0.000000	-.109045	-M
				+M	0.000000	0.000000	-.008813	0.000000	0.000000	0.000000	0.000000	0.000000	+M
-2.5991E+00	.8570	.0002	.8570	-M	0.000000	0.000000	-.008813	0.000000	0.000000	0.000000	0.000000	0.000000	-M
				+M	0.000000	.993998	0.000000	0.000000	0.000000	0.000000	0.000000	-.109045	+M
-5.8295E+00	-.2959	.2089	.3622	-M	0.000000	0.000000	0.000000	0.000000	0.000000	.976298	0.000000	0.000000	-M
				+M	0.000000	0.000000	0.000000	0.000000	.216432	0.000000	0.000000	0.000000	+M
-5.8295E+00	.2959	.2089	.3622	-M	0.000000	0.000000	0.000000	0.000000	.216432	0.000000	0.000000	0.000000	-M
				+M	0.000000	0.000000	0.000000	0.000000	0.000000	.976298	0.000000	0.000000	+M
-2.0571E+01	-.4292	-.2089	.4773	-M	0.000000	0.000000	0.000000	0.000000	.976298	0.000000	0.000000	0.000000	-M
				+M	0.000000	0.000000	0.000000	0.000000	0.000000	-.216432	0.000000	0.000000	+M
-2.0571E+01	.4292	-.2089	.4773	-M	0.000000	0.000000	0.000000	0.000000	0.000000	-.216432	0.000000	0.000000	-M
				+M	0.000000	0.000000	0.000000	0.000000	.976298	0.000000	0.000000	0.000000	+M
-2.9840E+01	-.7309	.0036	.7310	-M	0.000000	0.000000	.998513	0.000000	0.000000	0.000000	0.000000	0.000000	-M
				+M	0.000000	.002881	0.000000	0.000000	0.000000	0.000000	0.000000	-.054442	+M
-2.9840E+01	.7309	.0036	.7310	-M	0.000000	.002881	0.000000	0.000000	0.000000	0.000000	0.000000	-.054442	-M
				+M	0.000000	0.000000	.998513	0.000000	0.000000	0.000000	0.000000	0.000000	+M
-3.0820E+01	.5950	0.0000	.5950	-M	.000530	0.000000	0.000000	0.000000	0.000000	0.000000	-.078777	0.000000	-M
				+M	0.000000	0.000000	0.000000	.996892	0.000000	0.000000	0.000000	0.000000	+M
-3.0820E+01	-.5950	0.0000	.5950	-M	0.000000	0.000000	0.000000	.996892	0.000000	0.000000	0.000000	0.000000	-M
				+M	.000530	0.000000	0.000000	0.000000	0.000000	0.000000	-.078777	0.000000	+M
ENERGY	HEPAL	HEPER	HE		M=15/2	M=13/2	M=11/2	M=9/2	M=7/2	M=5/2	M=3/2	M=1/2	

Y= .40 X=-.80 OAS=1.5241E+02*W246 B4F4/B2F2=-5.33E-01 B4F4/B6F6=-4.00E+00 E2F2/B6F6= 7.50E+00

ENERGY	HEPAL	HEPER	HE		M=15/2	M=13/2	M=11/2	M=9/2	M=7/2	M=5/2	M=3/2	M=1/2
7.3126E+01	.7327	.0019	.7327	-M	0.000000	.002051	0.000000	0.000000	0.000000	0.000000	0.000000	.028830-M
				+M	0.000000	0.000000	.999582	0.000000	0.000000	0.000000	0.000000	0.000000+M
7.3126E+01	-.7327	.0019	.7327	-M	0.000000	0.000000	.999582	0.000000	0.000000	0.000000	0.000000	0.000000-M
				+M	0.000000	.002051	0.000000	0.000000	0.000000	0.000000	0.000000	.028830+M
6.8692E+01	-.5984	0.0000	.5984	-M	0.000000	0.000000	0.000000	.998981	0.000000	0.000000	0.000000	0.000000-M
				+M	.000492	0.000000	0.000000	0.000000	0.000000	0.000000	.045123	0.000000+M
6.8692E+01	.5984	0.0000	.5984	-M	.000492	0.000000	0.000000	0.000000	0.000000	0.000000	.045123	0.000000-M
				+M	0.000000	0.000000	0.000000	.998981	0.000000	0.000000	0.000000	0.000000+M
3.4698E+01	-.4531	.1276	.4707	-M	0.000000	0.000000	0.000000	0.000000	.991494	0.000000	0.000000	0.000000-M
				+M	0.000000	0.000000	0.000000	0.000000	0.000000	.130150	0.000000	0.000000+M
3.4698E+01	.4531	.1276	.4707	-M	0.000000	0.000000	0.000000	0.000000	0.000000	.130150	0.000000	0.000000-M
				+M	0.000000	0.000000	0.000000	0.000000	.991494	0.000000	0.000000	0.000000+M
3.2435E+01	-.8661	-.0016	.8661	-M	0.000000	.999661	0.000000	0.000000	0.000000	0.000000	0.000000	.025884-M
				+M	0.000000	0.000000	-.002797	0.000000	0.000000	0.000000	0.000000	0.000000+M
3.2435E+01	.8661	-.0016	.8661	-M	0.000000	0.000000	-.002797	0.000000	0.000000	0.000000	0.000000	0.000000-M
				+M	0.000000	.999661	0.000000	0.000000	0.000000	0.000000	0.000000	.025884+M
-1.3578E+01	-.3198	-.1276	.3443	-M	0.000000	0.000000	0.000000	0.000000	0.000000	.991494	0.000000	0.000000-M
				+M	0.000000	0.000000	0.000000	0.000000	-.130150	0.000000	0.000000	0.000000+M
-1.3578E+01	.3198	-.1276	.3443	-M	0.000000	0.000000	0.000000	0.000000	-.130150	0.000000	0.000000	0.000000-M
				+M	0.000000	0.000000	0.000000	0.000000	0.000000	.991494	0.000000	0.000000+M
-5.4626E+01	-.2342	0.0000	.2342	-M	.211431	0.000000	0.000000	0.000000	0.000000	0.000000	.976393	0.000000-M
				+M	0.000000	0.000000	0.000000	-.044207	0.000000	0.000000	0.000000	0.000000+M
-5.4626E+01	.2342	0.0000	.2342	-M	0.000000	0.000000	0.000000	-.044207	0.000000	0.000000	0.000000	0.000000-M
				+M	.211431	0.000000	0.000000	0.000000	0.000000	0.000000	.976393	0.000000+M
-6.1466E+01	-.9642	0.0000	.9642	-M	.977393	0.000000	0.000000	0.000000	0.000000	0.000000	-.211238	0.000000-M
				+M	0.000000	0.000000	0.000000	.009060	0.000000	0.000000	0.000000	0.000000+M
-6.1466E+01	.9642	0.0000	.9642	-M	0.000000	0.000000	0.000000	.009060	0.000000	0.000000	0.000000	0.000000-M
				+M	.977393	0.000000	0.000000	0.000000	0.000000	0.000000	-.211238	0.000000+M
-7.9282E+01	.0665	.5331	.5372	-M	0.000000	0.000000	-.028767	0.000000	0.000000	0.000000	0.000000	0.000000-M
				+M	0.000000	-.025954	0.000000	0.000000	0.000000	0.000000	0.000000	.999249+M
-7.9282E+01	-.0665	.5331	.5372	-M	0.000000	-.025954	0.000000	0.000000	0.000000	0.000000	0.000000	.999249-M
				+M	0.000000	0.000000	-.028767	0.000000	0.000000	0.000000	0.000000	0.000000+M
ENERGY	HEPAL	HEPER	HE		M=15/2	M=13/2	M=11/2	M=9/2	M=7/2	M=5/2	M=3/2	M=1/2

Y= .40 X=-.60 OAS=1.2924E+02*W246 B4F4/B2F2=-4.00E-01 B4F4/B6F6=-1.50E+00 B2F2/B6F6= 3.75E+00

ENERGY	HEPAL	HEPER	HE		M=15/2	M=13/2	M=11/2	M=9/2	M=7/2	M=5/2	M=3/2	M=1/2
5.8268E+01	-.5910	0.0000	.5910	-M	0.000000	0.000000	0.000000	.994361	0.000000	0.000000	0.000000	0.000000-M
				+M	.003247	0.000000	0.000000	0.000000	0.000000	0.000000	.105997	0.000000+M
5.8268E+01	.5910	0.0000	.5910	-M	.003247	0.000000	0.000000	0.000000	0.000000	0.000000	.105997	0.000000-M
				+M	0.000000	0.000000	0.000000	.994361	0.000000	0.000000	0.000000	0.000000+M
5.2831E+01	.7290	.0106	.7291	-M	0.000000	.011212	0.000000	0.000000	0.000000	0.000000	0.000000	.071747-M
				+M	0.000000	0.000000	.997360	0.000000	0.000000	0.000000	0.000000	0.000000+M
5.2831E+01	-.7290	.0106	.7291	-M	0.000000	0.000000	.997360	0.000000	0.000000	0.000000	0.000000	0.000000-M
				+M	0.000000	.011212	0.000000	0.000000	0.000000	0.000000	0.000000	.071747+M
3.6317E+01	-.4064	.2610	.4830	-M	0.000000	0.000000	0.000000	0.000000	.961574	0.000000	0.000000	0.000000-M
				+M	0.000000	0.000000	0.000000	0.000000	0.000000	.274546	0.000000	0.000000+M
3.6317E+01	.4064	.2610	.4830	-M	0.000000	0.000000	0.000000	0.000000	0.000000	.274546	0.000000	0.000000-M
				+M	0.000000	0.000000	0.000000	0.000000	.961574	0.000000	0.000000	0.000000+M
1.6101E+01	-.8628	-.0089	.8629	-M	0.000000	.997718	0.000000	0.000000	0.000000	0.000000	0.000000	.065610-M
				+M	0.000000	0.000000	-.015936	0.000000	0.000000	0.000000	0.000000	0.000000+M
1.6101E+01	.8628	-.0089	.8629	-M	0.000000	0.000000	-.015936	0.000000	0.000000	0.000000	0.000000	0.000000-M
				+M	0.000000	.997718	0.000000	0.000000	0.000000	0.000000	0.000000	.065610+M
-1.0877E+01	-.2730	-.2610	.3777	-M	0.000000	0.000000	0.000000	0.000000	0.000000	.961574	0.000000	0.000000-M
				+M	0.000000	0.000000	0.000000	0.000000	-.274546	0.000000	0.000000	0.000000+M
-1.0877E+01	.2730	-.2610	.3777	-M	0.000000	0.000000	0.000000	0.000000	-.274546	0.000000	0.000000	0.000000-M
				+M	0.000000	0.000000	0.000000	0.000000	0.000000	.961574	0.000000	0.000000+M
-3.3555E+01	-.9684	0.0000	.9684	-M	.980366	0.000000	0.000000	0.000000	0.000000	0.000000	.195715	0.000000-M
				+M	0.000000	0.000000	0.000000	-.024064	0.000000	0.000000	0.000000	0.000000+M
-3.3555E+01	.9684	0.0000	.9684	-M	0.000000	0.000000	0.000000	-.024064	0.000000	0.000000	0.000000	0.000000-M
				+M	.980366	0.000000	0.000000	0.000000	0.000000	0.000000	.195715	0.000000+M
-4.8113E+01	-.2226	0.0000	.2226	-M	-.197162	0.000000	0.000000	0.000000	0.000000	0.000000	.974916	0.000000-M
				+M	0.000000	0.000000	0.000000	-.103280	0.000000	0.000000	0.000000	0.000000+M
-4.8113E+01	.2226	0.0000	.2226	-M	0.000000	0.000000	0.000000	-.103280	0.000000	0.000000	0.000000	0.000000-M
				+M	-.197162	0.000000	0.000000	0.000000	0.000000	0.000000	.974916	0.000000+M
-7.0972E+01	-.0662	.5316	.5357	-M	0.000000	-.066580	0.000000	0.000000	0.000000	0.000000	0.000000	.995263-M
				+M	0.000000	0.000000	-.070847	0.000000	0.000000	0.000000	0.000000	0.000000+M
-7.0972E+01	.0662	.5316	.5357	-M	0.000000	0.000000	-.070847	0.000000	0.000000	0.000000	0.000000	0.000000-M
				+M	0.000000	-.066580	0.000000	0.000000	0.000000	0.000000	0.000000	.995263+M
ENERGY	HEPAL	HEPER	HE		M=15/2	M=13/2	M=11/2	M=9/2	M=7/2	M=5/2	M=3/2	M=1/2

Y= .40 X=-.40 OAS=1.1280E+02*W246 B4F4/B2F2=-2.67E-01 B4F4/B6F6=-6.67E-01 E2F2/B6F6= 2.50E+00

ENERGY	HEPAL	HEPER	HE		M=15/2	M=13/2	M=11/2	M=9/2	M=7/2	M=5/2	M=3/2	M=1/2
4.8944E+01	-.5710	0.0000	.5710	-M	0.000000	0.000000	0.000000	.981800	0.000000	0.000000	0.000000	0.000000-M
				+M	.014348	0.000000	0.000000	0.000000	0.000000	0.000000	.189373	0.000000+M
4.8944E+01	.5710	0.0000	.5710	-M	.014348	0.000000	0.000000	0.000000	0.000000	0.000000	.189373	0.000000-M
				+M	0.000000	0.000000	0.000000	.981800	0.000000	0.000000	0.000000	0.000000+M
3.9983E+01	.3337	.3681	.4968	-M	0.000000	0.000000	0.000000	0.000000	0.000000	.407649	0.000000	0.000000-M
				+M	0.000000	0.000000	0.000000	0.000000	.913139	0.000000	0.000000	0.000000+M
3.9983E+01	-.3337	.3681	.4968	-M	0.000000	0.000000	0.000000	0.000000	.913139	0.000000	0.000000	0.000000-M
				+M	0.000000	0.000000	0.000000	0.000000	0.000000	.407649	0.000000	0.000000+M
3.3282E+01	-.7152	.0358	.7161	-M	0.000000	0.000000	.989274	0.000000	0.000000	0.000000	0.000000	0.000000-M
				+M	0.000000	.036006	0.000000	0.000000	0.000000	0.000000	0.000000	.141562+M
3.3282E+01	.7152	.0358	.7161	-M	0.000000	.036006	0.000000	0.000000	0.000000	0.000000	0.000000	.141562-M
				+M	0.000000	0.000000	.989274	0.000000	0.000000	0.000000	0.000000	0.000000+M
2.1622E-01	-.8485	-.0292	.8490	-M	0.000000	.990102	0.000000	0.000000	0.000000	0.000000	0.000000	.129319-M
				+M	0.000000	0.000000	-.054541	0.000000	0.000000	0.000000	0.000000	0.000000+M
2.1622E-01	.8485	-.0292	.8490	-M	0.000000	0.000000	-.054541	0.000000	0.000000	0.000000	0.000000	0.000000-M
				+M	0.000000	.990102	0.000000	0.000000	0.000000	0.000000	0.000000	.129319+M
-6.5903E+00	-.9874	0.0000	.9874	-M	.992745	0.000000	0.000000	0.000000	0.000000	0.000000	.114537	0.000000-M
				+M	0.000000	0.000000	0.000000	-.036600	0.000000	0.000000	0.000000	0.000000+M
-6.5903E+00	.9874	0.0000	.9874	-M	0.000000	0.000000	0.000000	-.036600	0.000000	0.000000	0.000000	0.000000-M
				+M	.992745	0.000000	0.000000	0.000000	0.000000	0.000000	.114537	0.000000+M
-1.0223E+01	.2004	-.3681	.4191	-M	0.000000	0.000000	0.000000	0.000000	-.407649	0.000000	0.000000	0.000000-M
				+M	0.000000	0.000000	0.000000	0.000000	0.000000	.913139	0.000000	0.000000+M
-1.0223E+01	-.2004	-.3681	.4191	-M	0.000000	0.000000	0.000000	0.000000	0.000000	.913139	0.000000	0.000000-M
				+M	0.000000	0.000000	0.000000	0.000000	-.407649	0.000000	0.000000	0.000000+M
-4.1754E+01	-.1836	0.0000	.1836	-M	-.119384	0.000000	0.000000	0.000000	0.000000	0.000000	.975202	0.000000-M
				+M	0.000000	0.000000	0.000000	-.186356	0.000000	0.000000	0.000000	0.000000+M
-4.1754E+01	.1836	0.0000	.1836	-M	0.000000	0.000000	0.000000	-.186356	0.000000	0.000000	0.000000	0.000000-M
				+M	-.119384	0.000000	0.000000	0.000000	0.000000	0.000000	.975202	0.000000+M
-6.3858E+01	-.0667	.5267	.5309	-M	0.000000	-.135653	0.000000	0.000000	0.000000	0.000000	0.000000	.981446-M
				+M	0.000000	0.000000	-.135504	0.000000	0.000000	0.000000	0.000000	0.000000+M
-6.3858E+01	.0667	.5267	.5309	-M	0.000000	0.000000	-.135504	0.000000	0.000000	0.000000	0.000000	0.000000-M
				+M	0.000000	-.135653	0.000000	0.000000	0.000000	0.000000	0.000000	.981446+M
ENERGY	HEPAL	HEPER	HE		M=15/2	M=13/2	M=11/2	M=9/2	M=7/2	M=5/2	M=3/2	M=1/2

```
Y= .40   X=-.20    OAS=1.0471E+02*W246     B4F4/B2F2=-1.33E-01   B4F4/B6F6=-2.50E-01   B2F2/B6F6= 1.87E+00
  ENERGY      HEPAL   HEPER    HE          M=15/2      M=13/2      M=11/2     M=9/2      M=7/2      M=5/2      M=3/2      M=1/2
 4.5371E+01  -.2570   .4349   .5051   -M 0.000000   0.000000   0.000000   0.000000    .859013   0.000000   0.000000   0.000000-M
                                      +M 0.000000   0.000000   0.000000   0.000000   0.000000    .511954   0.000000   0.000000+M
 4.5371E+01   .2570   .4349   .5051   -M 0.000000   0.000000   0.000000   0.000000   0.000000    .511954   0.000000   0.000000-M
                                      +M 0.000000   0.000000   0.000000   0.000000    .859013   0.000000   0.000000   0.000000+M
 4.1485E+01  -.5161  0.0000   .5161   -M 0.000000   0.000000   0.000000    .949472   0.000000   0.000000   0.000000   0.000000-M
                                      +M  .079810   0.000000   0.000000   0.000000   0.000000   0.000000    .303535   0.000000+M
 4.1485E+01   .5161  0.0000   .5161   -M  .079810   0.000000   0.000000   0.000000   0.000000   0.000000    .303535   0.000000-M
                                      +M 0.000000   0.000000   0.000000    .949472   0.000000   0.000000   0.000000   0.000000+M
 2.0335E+01   .9791  0.0000   .9791   -M 0.000000   0.000000   0.000000   -.104422   0.000000   0.000000   0.000000   0.000000-M
                                      +M  .992360   0.000000   0.000000   0.000000   0.000000   0.000000    .065710   0.000000+M
 2.0335E+01  -.9791  0.0000   .9791   -M  .992360   0.000000   0.000000   0.000000   0.000000   0.000000    .065710   0.000000-M
                                      +M 0.000000   0.000000   0.000000   -.104422   0.000000   0.000000   0.000000   0.000000+M
 1.5472E+01  -.6625   .1012   .6702   -M 0.000000   0.000000    .959278   0.000000   0.000000   0.000000   0.000000   0.000000-M
                                      +M 0.000000    .093583   0.000000   0.000000   0.000000   0.000000   0.000000    .266509+M
 1.5472E+01   .6625   .1012   .6702   -M 0.000000    .093583   0.000000   0.000000   0.000000   0.000000   0.000000    .266509-M
                                      +M 0.000000   0.000000    .959278   0.000000   0.000000   0.000000   0.000000   0.000000+M
-1.1291E+01  -.1237  -.4349   .4521   -M 0.000000   0.000000   0.000000   0.000000   0.000000    .859013   0.000000   0.000000-M
                                      +M 0.000000   0.000000   0.000000   0.000000   -.511954   0.000000   0.000000   0.000000+M
-1.1291E+01   .1237  -.4349   .4521   -M 0.000000   0.000000   0.000000   0.000000   -.511954   0.000000   0.000000   0.000000-M
                                      +M 0.000000   0.000000   0.000000   0.000000   0.000000    .859013   0.000000   0.000000+M
-1.4815E+01  -.7859  -.0788   .7898   -M 0.000000    .961053   0.000000   0.000000   0.000000   0.000000   0.000000    .227470-M
                                      +M 0.000000   0.000000   -.156952   0.000000   0.000000   0.000000   0.000000   0.000000+M
-1.4815E+01   .7859  -.0788   .7898   -M 0.000000   0.000000   -.156952   0.000000   0.000000   0.000000   0.000000   0.000000-M
                                      +M 0.000000    .961053   0.000000   0.000000   0.000000   0.000000   0.000000    .227470+M
-3.7220E+01  -.1370  0.0000   .1370   -M -.094086   0.000000   0.000000   0.000000   0.000000   0.000000    .950552   0.000000-M
                                      +M 0.000000   0.000000   0.000000   -.295972   0.000000   0.000000   0.000000   0.000000+M
-3.7220E+01   .1370  0.0000   .1370   -M 0.000000   0.000000   0.000000   -.295972   0.000000   0.000000   0.000000   0.000000-M
                                      +M -.094086   0.000000   0.000000   0.000000   0.000000   0.000000    .950552   0.000000+M
-5.9337E+01  -.0766   .5109   .5167   -M 0.000000   -.260036   0.000000   0.000000   0.000000   0.000000   0.000000    .936606-M
                                      +M 0.000000   0.000000   -.234842   0.000000   0.000000   0.000000   0.000000   0.000000+M
-5.9337E+01   .0766   .5109   .5167   -M 0.000000   0.000000   -.234842   0.000000   0.000000   0.000000   0.000000   0.000000-M
                                      +M 0.000000   -.260036   0.000000   0.000000   0.000000   0.000000   0.000000    .936606+M
  ENERGY      HEPAL   HEPER    HE          M=15/2      M=13/2      M=11/2     M=9/2      M=7/2      M=5/2      M=3/2      M=1/2

Y= .40   X=0.00    OAS=1.1224E+02*W246     B4F4/B2F2= 0.          B4F4/B6F6= 0.          B2F2/B6F6= 1.50E+00
  ENERGY      HEPAL   HEPER    HE          M=15/2      M=13/2      M=11/2     M=9/2      M=7/2      M=5/2      M=3/2      M=1/2
 5.1976E+01  -.1911   .4699   .5073   -M 0.000000   0.000000   0.000000   0.000000    .809693   0.000000   0.000000   0.000000-M
                                      +M 0.000000   0.000000   0.000000   0.000000   0.000000    .586853   0.000000   0.000000+M
 5.1976E+01   .1911   .4699   .5073   -M 0.000000   0.000000   0.000000   0.000000   0.000000    .586853   0.000000   0.000000-M
                                      +M 0.000000   0.000000   0.000000   0.000000    .809693   0.000000   0.000000   0.000000+M
 4.8284E+01  -.9158  0.0000   .9158   -M  .965424   0.000000   0.000000   0.000000   0.000000   0.000000    .175191   0.000000-M
                                      +M 0.000000   0.000000   0.000000    .193039   0.000000   0.000000   0.000000   0.000000+M
 4.8284E+01   .9158  0.0000   .9158   -M 0.000000   0.000000   0.000000    .193039   0.000000   0.000000   0.000000   0.000000-M
                                      +M  .965424   0.000000   0.000000   0.000000   0.000000   0.000000    .175191   0.000000+M
 3.5633E+01   .3728  0.0000   .3728   -M -.248955   0.000000   0.000000   0.000000   0.000000   0.000000    .399987   0.000000-M
                                      +M 0.000000   0.000000   0.000000    .882061   0.000000   0.000000   0.000000   0.000000+M
 3.5633E+01  -.3728  0.0000   .3728   -M 0.000000   0.000000   0.000000    .882061   0.000000   0.000000   0.000000   0.000000-M
                                      +M -.248955   0.000000   0.000000   0.000000   0.000000   0.000000    .399987   0.000000+M
 2.1357E+00   .4890   .2394   .5445   -M 0.000000    .192050   0.000000   0.000000   0.000000   0.000000   0.000000    .481306-M
                                      +M 0.000000   0.000000    .855255   0.000000   0.000000   0.000000   0.000000   0.000000+M
 2.1357E+00  -.4890   .2394   .5445   -M 0.000000   0.000000    .855255   0.000000   0.000000   0.000000   0.000000   0.000000-M
                                      +M 0.000000    .192050   0.000000   0.000000   0.000000   0.000000   0.000000    .481306+M
-1.3576E+01  -.0578  -.4699   .4734   -M 0.000000   0.000000   0.000000   0.000000   0.000000    .809693   0.000000   0.000000-M
                                      +M 0.000000   0.000000   0.000000   0.000000   -.586853   0.000000   0.000000   0.000000+M
-1.3576E+01   .0578  -.4699   .4734   -M 0.000000   0.000000   0.000000   0.000000   -.586853   0.000000   0.000000   0.000000-M
                                      +M 0.000000   0.000000   0.000000   0.000000   0.000000    .809693   0.000000   0.000000+M
-2.8873E+01  -.5596  -.1759   .5866   -M 0.000000    .870040   0.000000   0.000000   0.000000   0.000000   0.000000    .319683-M
                                      +M 0.000000   0.000000   -.375276   0.000000   0.000000   0.000000   0.000000   0.000000+M
-2.8873E+01   .5596  -.1759   .5866   -M 0.000000   0.000000   -.375276   0.000000   0.000000   0.000000   0.000000   0.000000-M
                                      +M 0.000000    .870040   0.000000   0.000000   0.000000   0.000000   0.000000    .319683+M
-3.5317E+01  -.0570  0.0000   .0570   -M -.077316   0.000000   0.000000   0.000000   0.000000   0.000000    .899621   0.000000-M
                                      +M 0.000000   0.000000   0.000000   -.429772   0.000000   0.000000   0.000000   0.000000+M
-3.5317E+01   .0570  0.0000   .0570   -M 0.000000   0.000000   0.000000   -.429772   0.000000   0.000000   0.000000   0.000000-M
                                      +M -.077316   0.000000   0.000000   0.000000   0.000000   0.000000    .899621   0.000000+M
-6.0263E+01   .1294   .4698   .4873   -M 0.000000   0.000000   -.357360   0.000000   0.000000   0.000000   0.000000   0.000000-M
                                      +M 0.000000   -.454033   0.000000   0.000000   0.000000   0.000000   0.000000    .816179+M
-6.0263E+01  -.1294   .4698   .4873   -M 0.000000   -.454033   0.000000   0.000000   0.000000   0.000000   0.000000    .816179-M
                                      +M 0.000000   0.000000   -.357360   0.000000   0.000000   0.000000   0.000000   0.000000+M
  ENERGY      HEPAL   HEPER    HE          M=15/2      M=13/2      M=11/2     M=9/2      M=7/2      M=5/2      M=3/2      M=1/2

Y= .40   X= .20    OAS=1.0999E+02*W246     B4F4/B2F2= 1.33E-01   B4F4/B6F6= 2.50E-01   B2F2/B6F6= 1.87E+00
  ENERGY      HEPAL   HEPER    HE          M=15/2      M=13/2      M=11/2     M=9/2      M=7/2      M=5/2      M=3/2      M=1/2
 6.4131E+01   .9925  0.0000   .9925   -M 0.000000   0.000000   0.000000    .031112   0.000000   0.000000   0.000000   0.000000-M
                                      +M  .995770   0.000000   0.000000   0.000000   0.000000   0.000000    .086450   0.000000+M
 6.4131E+01  -.9925  0.0000   .9925   -M  .995770   0.000000   0.000000   0.000000   0.000000   0.000000    .086450   0.000000-M
                                      +M 0.000000   0.000000   0.000000    .031112   0.000000   0.000000   0.000000   0.000000+M
 3.5971E+01   .1232   .4894   .5047   -M 0.000000   0.000000   0.000000   0.000000   0.000000    .655198   0.000000   0.000000-M
                                      +M 0.000000   0.000000   0.000000   0.000000    .755457   0.000000   0.000000   0.000000+M
 3.5971E+01  -.1232   .4894   .5047   -M 0.000000   0.000000   0.000000   0.000000    .755457   0.000000   0.000000   0.000000-M
                                      +M 0.000000   0.000000   0.000000   0.000000   0.000000    .655198   0.000000   0.000000+M
 1.9280E+01   .2962  0.0000   .2962   -M -.077375   0.000000   0.000000   0.000000   0.000000   0.000000    .606455   0.000000-M
                                      +M 0.000000   0.000000   0.000000    .791344   0.000000   0.000000   0.000000   0.000000+M
 1.9280E+01  -.2962  0.0000   .2962   -M 0.000000   0.000000   0.000000    .791344   0.000000   0.000000   0.000000   0.000000-M
                                      +M -.077375   0.000000   0.000000   0.000000   0.000000   0.000000    .606455   0.000000+M
-2.8493E+00   .1485   .4212   .4466   -M 0.000000    .295557   0.000000   0.000000   0.000000   0.000000   0.000000    .745862-M
                                      +M 0.000000   0.000000    .596938   0.000000   0.000000   0.000000   0.000000   0.000000+M
-2.8493E+00  -.1485   .4212   .4466   -M 0.000000   0.000000    .596938   0.000000   0.000000   0.000000   0.000000   0.000000-M
                                      +M 0.000000    .295557   0.000000   0.000000   0.000000   0.000000   0.000000    .745862+M
-1.4371E+01  -.0101  -.4894   .4895   -M 0.000000   0.000000   0.000000   0.000000   -.655198   0.000000   0.000000   0.000000-M
                                      +M 0.000000   0.000000   0.000000   0.000000   0.000000    .755457   0.000000   0.000000+M
-1.4371E+01   .0101  -.4894   .4895   -M 0.000000   0.000000   0.000000   0.000000   0.000000    .755457   0.000000   0.000000-M
                                      +M 0.000000   0.000000   0.000000   0.000000   -.655198   0.000000   0.000000   0.000000+M
-2.6652E+01   .0963  0.0000   .0963   -M -.049544   0.000000   0.000000   0.000000   0.000000   0.000000    .790404   0.000000-M
                                      +M 0.000000   0.000000   0.000000   -.610579   0.000000   0.000000   0.000000   0.000000+M
-2.6652E+01  -.0963  0.0000   .0963   -M 0.000000   0.000000   0.000000   -.610579   0.000000   0.000000   0.000000   0.000000-M
                                      +M -.049544   0.000000   0.000000   0.000000   0.000000   0.000000    .790404   0.000000+M
-2.9657E+01  -.2745   .0236   .2755   -M 0.000000   -.784853   0.000000   0.000000   0.000000   0.000000   0.000000   -.166669-M
                                      +M 0.000000   0.000000    .596848   0.000000   0.000000   0.000000   0.000000   0.000000+M
-2.9657E+01   .2745  -.0236   .2755   -M 0.000000   0.000000   -.596848   0.000000   0.000000   0.000000   0.000000   0.000000-M
                                      +M 0.000000    .784853   0.000000   0.000000   0.000000   0.000000   0.000000    .166669+M
-4.5854E+01   .0740   .0236   .0777   -M 0.000000   0.000000    .536132   0.000000   0.000000   0.000000   0.000000   0.000000-M
                                      +M 0.000000    .544658   0.000000   0.000000   0.000000   0.000000   0.000000   -.644911+M
-4.5854E+01  -.0740  -.0236   .0777   -M 0.000000    .544658   0.000000   0.000000   0.000000   0.000000   0.000000   -.644911-M
                                      +M 0.000000   0.000000    .536132   0.000000   0.000000   0.000000   0.000000   0.000000+M
  ENERGY      HEPAL   HEPER    HE          M=15/2      M=13/2      M=11/2     M=9/2      M=7/2      M=5/2      M=3/2      M=1/2
```

Y= .40 X= .40 OAS=1.2439E+02*W246 B4F4/B2F2= 2.67E-01 B4F4/B6F6= 6.67E-01 B2F2/B6F6= 2.50E+00

ENERGY	HEPAL	HEPER	HE		M=15/2	M=13/2	M=11/2	M=9/2	M=7/2	M=5/2	M=3/2	M=1/2
8.0522E+01	-.9973	0.0000	.9973	-M	.998333	0.000000	0.000000	0.000000	0.000000	0.000000	.056910	0.000000-M
				+M	0.000000	0.000000	0.000000	.009597	0.000000	0.000000	0.000000	0.000000+M
8.0522E+01	.9973	0.0000	.9973	-M	0.000000	0.000000	0.000000	.009597	0.000000	0.000000	0.000000	0.000000-M
				+M	.998333	0.000000	0.000000	0.000000	0.000000	0.000000	.056910	0.000000+M
2.1341E+01	-.0016	.4878	.4878	-M	0.000000	0.000000	0.000000	0.000000	.647066	0.000000	0.000000	0.000000-M
				+M	0.000000	0.000000	0.000000	0.000000	0.000000	.762434	0.000000	0.000000+M
2.1341E+01	.0016	.4878	.4878	-M	0.000000	0.000000	0.000000	0.000000	0.000000	.762434	0.000000	0.000000-M
				+M	0.000000	0.000000	0.000000	0.000000	.647066	0.000000	0.000000	0.000000+M
1.2083E+01	.0198	0.0000	.0198	-M	0.000000	0.000000	0.000000	.477695	0.000000	0.000000	0.000000	0.000000-M
				+M	-.054576	0.000000	0.000000	0.000000	0.000000	0.000000	.876829	0.000000+M
1.2083E+01	-.0198	0.0000	.0198	-M	-.054576	0.000000	0.000000	0.000000	0.000000	0.000000	.876829	0.000000-M
				+M	0.000000	0.000000	0.000000	.477695	0.000000	0.000000	0.000000	0.000000+M
5.6122E+00	-.0484	.5097	.5120	-M	0.000000	.228007	0.000000	0.000000	0.000000	0.000000	0.000000	.934452-M
				+M	0.000000	0.000000	.273519	0.000000	0.000000	0.000000	0.000000	0.000000+M
5.6122E+00	.0484	.5097	.5120	-M	0.000000	0.000000	.273519	0.000000	0.000000	0.000000	0.000000	0.000000-M
				+M	0.000000	.228007	0.000000	0.000000	0.000000	0.000000	0.000000	.934452+M
-1.6541E+01	.1317	-.4878	.5053	-M	0.000000	0.000000	0.000000	0.000000	0.000000	-.647066	0.000000	0.000000-M
				+M	0.000000	0.000000	0.000000	0.000000	.762434	0.000000	0.000000	0.000000+M
-1.6541E+01	-.1317	-.4878	.5053	-M	0.000000	0.000000	0.000000	0.000000	.762434	0.000000	0.000000	0.000000-M
				+M	0.000000	0.000000	0.000000	0.000000	0.000000	-.647066	0.000000	0.000000+M
-2.7685E+01	-.4171	0.0000	.4171	-M	0.000000	0.000000	0.000000	.878473	0.000000	0.000000	0.000000	0.000000-M
				+M	.018770	0.000000	0.000000	0.000000	0.000000	0.000000	-.477422	0.000000+M
-2.7685E+01	.4171	0.0000	.4171	-M	.018770	0.000000	0.000000	0.000000	0.000000	0.000000	-.477422	0.000000-M
				+M	0.000000	0.000000	0.000000	.878473	0.000000	0.000000	0.000000	0.000000+M
-3.1465E+01	-.7381	-.1626	.7558	-M	0.000000	.952606	0.000000	0.000000	0.000000	0.000000	0.000000	-.155992-M
				+M	0.000000	0.000000	-.261166	0.000000	0.000000	0.000000	0.000000	0.000000+M
-3.1465E+01	.7381	-.1626	.7558	-M	0.000000	0.000000	-.261166	0.000000	0.000000	0.000000	0.000000	0.000000-M
				+M	0.000000	.952606	0.000000	0.000000	0.000000	0.000000	0.000000	-.155992+M
-4.3868E+01	-.5865	.1862	.6153	-M	0.000000	0.000000	.925732	0.000000	0.000000	0.000000	0.000000	0.000000-M
				+M	0.000000	.201381	0.000000	0.000000	0.000000	0.000000	0.000000	-.320104+M
-4.3868E+01	.5865	.1862	.6153	-M	0.000000	.201381	0.000000	0.000000	0.000000	0.000000	0.000000	-.320104-M
				+M	0.000000	0.000000	.925732	0.000000	0.000000	0.000000	0.000000	0.000000+M
ENERGY	HEPAL	HEPER	HE		M=15/2	M=13/2	M=11/2	M=9/2	M=7/2	M=5/2	M=3/2	M=1/2

Y= .40 X= .60 OAS=1.5193E+02*W246 B4F4/B2F2= 4.00E-01 B4F4/B6F6= 1.50E+00 B2F2/B6F6= 3.75E+00

ENERGY	HEPAL	HEPER	HE		M=15/2	M=13/2	M=11/2	M=9/2	M=7/2	M=5/2	M=3/2	M=1/2
9.7019E+01	.9990	0.0000	.9990	-M	0.000000	0.000000	0.000000	.002868	0.000000	0.000000	0.000000	0.000000-M
				+M	.999379	0.000000	0.000000	0.000000	0.000000	0.000000	.035110	0.000000+M
9.7019E+01	-.9990	0.0000	.9990	-M	.999379	0.000000	0.000000	0.000000	0.000000	0.000000	.035110	0.000000-M
				+M	0.000000	0.000000	0.000000	.002868	0.000000	0.000000	0.000000	0.000000+M
2.2435E+01	.0667	.5287	.5329	-M	0.000000	0.000000	.113402	0.000000	0.000000	0.000000	0.000000	0.000000-M
				+M	0.000000	.113588	0.000000	0.000000	0.000000	0.000000	0.000000	.987035+M
2.2435E+01	-.0667	.5287	.5329	-M	0.000000	.113588	0.000000	0.000000	0.000000	0.000000	0.000000	.987035-M
				+M	0.000000	0.000000	.113402	0.000000	0.000000	0.000000	0.000000	0.000000+M
1.7636E+01	-.1716	0.0000	.1716	-M	-.035009	0.000000	0.000000	0.000000	0.000000	0.000000	.980841	0.000000-M
				+M	0.000000	0.000000	0.000000	.191639	0.000000	0.000000	0.000000	0.000000+M
1.7636E+01	.1716	0.0000	.1716	-M	0.000000	0.000000	0.000000	.191639	0.000000	0.000000	0.000000	0.000000-M
				+M	-.035009	0.000000	0.000000	0.000000	0.000000	0.000000	.980841	0.000000+M
9.8022E+00	.1794	.3898	.4291	-M	0.000000	0.000000	0.000000	0.000000	.438689	0.000000	0.000000	0.000000-M
				+M	0.000000	0.000000	0.000000	0.000000	0.000000	.898639	0.000000	0.000000+M
9.8022E+00	-.1794	.3898	.4291	-M	0.000000	0.000000	0.000000	0.000000	0.000000	.898639	0.000000	0.000000-M
				+M	0.000000	0.000000	0.000000	0.000000	.438689	0.000000	0.000000	0.000000+M
-2.1802E+01	.3127	-.3898	.4997	-M	0.000000	0.000000	0.000000	0.000000	0.000000	-.438689	0.000000	0.000000-M
				+M	0.000000	0.000000	0.000000	0.000000	.898639	0.000000	0.000000	0.000000+M
-2.1802E+01	-.3127	-.3898	.4997	-M	0.000000	0.000000	0.000000	0.000000	.898639	0.000000	0.000000	0.000000-M
				+M	0.000000	0.000000	0.000000	0.000000	0.000000	-.438689	0.000000	0.000000+M
-2.8602E+01	-.8547	-.0203	.8549	-M	0.000000	.993209	0.000000	0.000000	0.000000	0.000000	0.000000	-.109918-M
				+M	0.000000	0.000000	-.038126	0.000000	0.000000	0.000000	0.000000	0.000000+M
-2.8602E+01	.8547	-.0203	.8549	-M	0.000000	0.000000	-.038126	0.000000	0.000000	0.000000	0.000000	0.000000-M
				+M	0.000000	.993209	0.000000	0.000000	0.000000	0.000000	0.000000	-.109918+M
-4.1576E+01	-.5706	0.0000	.5706	-M	0.000000	0.000000	0.000000	.981461	0.000000	0.000000	0.000000	0.000000-M
				+M	.003915	0.000000	0.000000	0.000000	0.000000	0.000000	-.191620	0.000000+M
-4.1576E+01	.5706	0.0000	.5706	-M	.003915	0.000000	0.000000	0.000000	0.000000	0.000000	-.191620	0.000000-M
				+M	0.000000	0.000000	0.000000	.981461	0.000000	0.000000	0.000000	0.000000+M
-5.4913E+01	.7214	.0249	.7218	-M	0.000000	.025167	0.000000	0.000000	0.000000	0.000000	0.000000	-.116963-M
				+M	0.000000	0.000000	.992817	0.000000	0.000000	0.000000	0.000000	0.000000+M
-5.4913E+01	-.7214	.0249	.7218	-M	0.000000	0.000000	.992817	0.000000	0.000000	0.000000	0.000000	0.000000-M
				+M	0.000000	.025167	0.000000	0.000000	0.000000	0.000000	0.000000	-.116963+M
ENERGY	HEPAL	HEPER	HE		M=15/2	M=13/2	M=11/2	M=9/2	M=7/2	M=5/2	M=3/2	M=1/2

Y= .40 X= .80 OAS=1.8220E+02*W246 B4F4/B2F2= 5.33E-01 B4F4/B6F6= 4.00E+00 B2F2/B6F6= 7.50E+00

ENERGY	HEPAL	HEPER	HE		M=15/2	M=13/2	M=11/2	M=9/2	M=7/2	M=5/2	M=3/2	M=1/2
1.1358E+02	.9998	0.0000	.9998	-M	0.000000	0.000000	0.000000	.000530	0.000000	0.000000	0.000000	0.000000-M
				+M	.999863	0.000000	0.000000	0.000000	0.000000	0.000000	.016542	0.000000+M
1.1358E+02	-.9998	0.0000	.9998	-M	.999863	0.000000	0.000000	0.000000	0.000000	0.000000	.016542	0.000000-M
				+M	0.000000	0.000000	0.000000	.000530	0.000000	0.000000	0.000000	0.000000+M
4.2177E+01	.0669	.5327	.5369	-M	0.000000	0.000000	.039565	0.000000	0.000000	0.000000	0.000000	0.000000-M
				+M	0.000000	.042535	0.000000	0.000000	0.000000	0.000000	0.000000	.998311+M
4.2177E+01	-.0669	.5327	.5369	-M	0.000000	.042535	0.000000	0.000000	0.000000	0.000000	0.000000	.998311-M
				+M	0.000000	0.000000	.039565	0.000000	0.000000	0.000000	0.000000	0.000000+M
2.8208E+01	.1971	0.0000	.1971	-M	0.000000	0.000000	0.000000	.062892	0.000000	0.000000	0.000000	0.000000-M
				+M	-.016542	0.000000	0.000000	0.000000	0.000000	0.000000	.997883	0.000000+M
2.8208E+01	-.1971	0.0000	.1971	-M	-.016542	0.000000	0.000000	0.000000	0.000000	0.000000	.997883	0.000000-M
				+M	0.000000	0.000000	0.000000	.062892	0.000000	0.000000	0.000000	0.000000+M
3.1059E+00	.3071	.1759	.3540	-M	0.000000	0.000000	0.000000	0.000000	.180914	0.000000	0.000000	0.000000-M
				+M	0.000000	0.000000	0.000000	0.000000	0.000000	.983499	0.000000	0.000000+M
3.1059E+00	-.3071	.1759	.3540	-M	0.000000	0.000000	0.000000	0.000000	0.000000	.983499	0.000000	0.000000-M
				+M	0.000000	0.000000	0.000000	0.000000	.180914	0.000000	0.000000	0.000000+M
-2.6003E+01	-.8652	-.0021	.8652	-M	0.000000	.999091	0.000000	0.000000	0.000000	0.000000	0.000000	-.042395-M
				+M	0.000000	0.000000	-.004380	0.000000	0.000000	0.000000	0.000000	0.000000+M
-2.6003E+01	.8652	-.0021	.8652	-M	0.000000	0.000000	-.004380	0.000000	0.000000	0.000000	0.000000	0.000000-M
				+M	0.000000	.999091	0.000000	0.000000	0.000000	0.000000	0.000000	-.042395+M
-3.1906E+01	-.4405	-.1759	.4743	-M	0.000000	0.000000	0.000000	0.000000	.983499	0.000000	0.000000	0.000000-M
				+M	0.000000	0.000000	0.000000	0.000000	0.000000	-.180914	0.000000	0.000000+M
-3.1906E+01	.4405	-.1759	.4743	-M	0.000000	0.000000	0.000000	0.000000	0.000000	-.180914	0.000000	0.000000-M
				+M	0.000000	0.000000	0.000000	0.000000	.983499	0.000000	0.000000	0.000000+M
-6.0551E+01	.5968	0.0000	.5968	-M	.000511	0.000000	0.000000	0.000000	0.000000	0.000000	-.062892	0.000000-M
				+M	0.000000	0.000000	0.000000	.998020	0.000000	0.000000	0.000000	0.000000+M
-6.0551E+01	-.5968	0.0000	.5968	-M	0.000000	0.000000	0.000000	.998020	0.000000	0.000000	0.000000	0.000000-M
				+M	.000511	0.000000	0.000000	0.000000	0.000000	0.000000	-.062892	0.000000+M
-6.8614E+01	-.7321	.0027	.7321	-M	0.000000	0.000000	.999207	0.000000	0.000000	0.000000	0.000000	0.000000-M
				+M	0.000000	.002695	0.000000	0.000000	0.000000	0.000000	0.000000	-.039715+M
-6.8614E+01	.7321	.0027	.7321	-M	0.000000	.002695	0.000000	0.000000	0.000000	0.000000	0.000000	-.039715-M
				+M	0.000000	0.000000	.999207	0.000000	0.000000	0.000000	0.000000	0.000000+M
ENERGY	HEPAL	HEPER	HE		M=15/2	M=13/2	M=11/2	M=9/2	M=7/2	M=5/2	M=3/2	M=1/2

Y= .60 X=-.80 OAS=2.1457E+02*W246 B4F4/B2F2=-1.20E+00 B4F4/B6F6=-4.00E+00 B2F2/B6F6= 3.33E+00

ENERGY	HEPAL	HEPER	HE		M=15/2	M=13/2	M=11/2	M=9/2	M=7/2	M=5/2	M=3/2	M=1/2
1.0520E+02	-.7326	.0019	.7326	-M	0.000000	0.000000	.999522	0.000000	0.000000	0.000000	0.000000	0.000000-M
				+M	0.000000	.001998	0.000000	0.000000	0.000000	0.000000	0.000000	.030852+M
1.0520E+02	.7326	.0019	.7326	-M	0.000000	.001998	0.000000	0.000000	0.000000	0.000000	0.000000	.030852-M
				+M	0.000000	0.000000	.999522	0.000000	0.000000	0.000000	0.000000	0.000000+M
1.0356E+02	.5982	0.0000	.5982	-M	.000473	0.000000	0.000000	0.000000	0.000000	0.000000	.047420	0.000000-M
				+M	0.000000	0.000000	0.000000	.998875	0.000000	0.000000	0.000000	0.000000+M
1.0356E+02	-.5982	0.0000	.5982	-M	0.000000	0.000000	0.000000	.998875	0.000000	0.000000	0.000000	0.000000-M
				+M	.000473	0.000000	0.000000	0.000000	0.000000	0.000000	.047420	0.000000+M
5.6600E+01	-.4519	.1329	.4711	-M	0.000000	0.000000	0.000000	0.000000	.990754	0.000000	0.000000	0.000000-M
				+M	0.000000	0.000000	0.000000	0.000000	0.000000	.135669	0.000000	0.000000+M
5.6600E+01	.4519	.1329	.4711	-M	0.000000	0.000000	0.000000	0.000000	0.000000	.135669	0.000000	0.000000-M
				+M	0.000000	0.000000	0.000000	0.000000	.990754	0.000000	0.000000	0.000000+M
3.8169E+01	.8660	-.0016	.8660	-M	0.000000	0.000000	-.002911	0.000000	0.000000	0.000000	0.000000	0.000000-M
				+M	0.000000	.999558	0.000000	0.000000	0.000000	0.000000	0.000000	.029589+M
3.8169E+01	-.8660	-.0016	.8660	-M	0.000000	.999558	0.000000	0.000000	0.000000	0.000000	0.000000	.029589-M
				+M	0.000000	0.000000	-.002911	0.000000	0.000000	0.000000	0.000000	0.000000+M
-1.2920E+01	-.3186	-.1329	.3452	-M	0.000000	0.000000	0.000000	0.000000	0.000000	.990754	0.000000	0.000000-M
				+M	0.000000	0.000000	0.000000	0.000000	-.135669	0.000000	0.000000	0.000000+M
-1.2920E+01	.3186	-.1329	.3452	-M	0.000000	0.000000	0.000000	0.000000	-.135669	0.000000	0.000000	0.000000-M
				+M	0.000000	0.000000	0.000000	0.000000	0.000000	.990754	0.000000	0.000000+M
-7.2793E+01	-.2009	0.0000	.2009	-M	.058069	0.000000	0.000000	0.000000	0.000000	0.000000	.997188	0.000000-M
				+M	0.000000	0.000000	0.000000	-.047367	0.000000	0.000000	0.000000	0.000000+M
-7.2793E+01	.2009	0.0000	.2009	-M	0.000000	0.000000	0.000000	-.047367	0.000000	0.000000	0.000000	0.000000-M
				+M	.058069	0.000000	0.000000	0.000000	0.000000	0.000000	.997188	0.000000+M
-1.0845E+02	-.0666	.5330	.5371	-M	0.000000	-.029665	0.000000	0.000000	0.000000	0.000000	0.000000	.999086-M
				+M	0.000000	0.000000	-.030780	0.000000	0.000000	0.000000	0.000000	0.000000+M
-1.0845E+02	.0666	.5330	.5371	-M	0.000000	0.000000	-.030780	0.000000	0.000000	0.000000	0.000000	0.000000-M
				+M	0.000000	-.029665	0.000000	0.000000	0.000000	0.000000	0.000000	.999086+M
-1.0936E+02	-.9973	0.0000	.9973	-M	.998312	0.000000	0.000000	0.000000	0.000000	0.000000	-.058026	0.000000-M
				+M	0.000000	0.000000	0.000000	.002282	0.000000	0.000000	0.000000	0.000000+M
-1.0936E+02	.9973	0.0000	.9973	-M	0.000000	0.000000	0.000000	.002282	0.000000	0.000000	0.000000	0.000000-M
				+M	.998312	0.000000	0.000000	0.000000	0.000000	0.000000	-.058026	0.000000+M
ENERGY	HEPAL	HEPER	HE		M=15/2	M=13/2	M=11/2	M=9/2	M=7/2	M=5/2	M=3/2	M=1/2

Y= .60 X=-.60 OAS=1.8416E+02*W246 B4F4/B2F2=-9.00E-01 B4F4/B6F6=-1.50E+00 B2F2/B6F6= 1.67E+00

ENERGY	HEPAL	HEPER	HE		M=15/2	M=13/2	M=11/2	M=9/2	M=7/2	M=5/2	M=3/2	M=1/2
8.8009E+01	-.5899	0.0000	.5899	-M	0.000000	0.000000	0.000000	.993672	0.000000	0.000000	0.000000	0.000000-M
				+M	.003042	0.000000	0.000000	0.000000	0.000000	0.000000	.112280	0.000000+M
8.8009E+01	.5899	0.0000	.5899	-M	.003042	0.000000	0.000000	0.000000	0.000000	0.000000	.112280	0.000000-M
				+M	0.000000	0.000000	0.000000	.993672	0.000000	0.000000	0.000000	0.000000+M
7.4829E+01	.7283	.0110	.7284	-M	0.000000	.010992	0.000000	0.000000	0.000000	0.000000	0.000000	.078021-M
				+M	0.000000	0.000000	.996891	0.000000	0.000000	0.000000	0.000000	0.000000+M
7.4829E+01	-.7283	.0110	.7284	-M	0.000000	0.000000	.996891	0.000000	0.000000	0.000000	0.000000	0.000000-M
				+M	0.000000	.010992	0.000000	0.000000	0.000000	0.000000	0.000000	.078021+M
5.9211E+01	-.4014	.2707	.4841	-M	0.000000	0.000000	0.000000	0.000000	.958324	0.000000	0.000000	0.000000-M
				+M	0.000000	0.000000	0.000000	0.000000	0.000000	.285685	0.000000	0.000000+M
5.9211E+01	.4014	.2707	.4841	-M	0.000000	0.000000	0.000000	0.000000	0.000000	.285685	0.000000	0.000000-M
				+M	0.000000	0.000000	0.000000	0.000000	.958324	0.000000	0.000000	0.000000+M
1.3761E+01	.8613	-.0088	.8614	-M	0.000000	0.000000	-.017100	0.000000	0.000000	0.000000	0.000000	0.000000-M
				+M	0.000000	.996802	0.000000	0.000000	0.000000	0.000000	0.000000	.078062+M
1.3761E+01	-.8613	-.0088	.8614	-M	0.000000	.996802	0.000000	0.000000	0.000000	0.000000	0.000000	.078062-M
				+M	0.000000	0.000000	-.017100	0.000000	0.000000	0.000000	0.000000	0.000000+M
-9.0513E+00	-.2680	-.2707	.3810	-M	0.000000	0.000000	0.000000	0.000000	0.000000	.958324	0.000000	0.000000-M
				+M	0.000000	0.000000	0.000000	0.000000	-.285685	0.000000	0.000000	0.000000+M
-9.0513E+00	.2680	-.2707	.3810	-M	0.000000	0.000000	0.000000	0.000000	-.285685	0.000000	0.000000	0.000000-M
				+M	0.000000	0.000000	0.000000	0.000000	0.000000	.958324	0.000000	0.000000+M
-5.9901E+01	.3416	0.0000	.3416	-M	0.000000	0.000000	0.000000	-.102517	0.000000	0.000000	0.000000	0.000000-M
				+M	.433011	0.000000	0.000000	0.000000	0.000000	0.000000	.895540	0.000000+M
-5.9901E+01	-.3416	0.0000	.3416	-M	.433011	0.000000	0.000000	0.000000	0.000000	0.000000	.895540	0.000000-M
				+M	0.000000	0.000000	0.000000	-.102517	0.000000	0.000000	0.000000	0.000000+M
-7.0708E+01	-.8483	0.0000	.8483	-M	.901383	0.000000	0.000000	0.000000	0.000000	0.000000	-.430583	0.000000-M
				+M	0.000000	0.000000	0.000000	.045895	0.000000	0.000000	0.000000	0.000000+M
-7.0708E+01	.8483	0.0000	.8483	-M	0.000000	0.000000	0.000000	.045895	0.000000	0.000000	0.000000	0.000000-M
				+M	.901383	0.000000	0.000000	0.000000	0.000000	0.000000	-.430583	0.000000+M
-9.6151E+01	-.0669	.5311	.5353	-M	0.000000	-.079154	0.000000	0.000000	0.000000	0.000000	0.000000	.993891-M
				+M	0.000000	0.000000	-.076914	0.000000	0.000000	0.000000	0.000000	0.000000+M
-9.6151E+01	.0669	.5311	.5353	-M	0.000000	0.000000	-.076914	0.000000	0.000000	0.000000	0.000000	0.000000-M
				+M	0.000000	-.079154	0.000000	0.000000	0.000000	0.000000	0.000000	.993891+M
ENERGY	HEPAL	HEPER	HE		M=15/2	M=13/2	M=11/2	M=9/2	M=7/2	M=5/2	M=3/2	M=1/2

Y= .60 X=-.40 OAS=1.6034E+02*W246 B4F4/B2F2=-6.00E-01 B4F4/B6F6=-6.67E-01 B2F2/B6F6= 1.11E+00

ENERGY	HEPAL	HEPER	HE		M=15/2	M=13/2	M=11/2	M=9/2	M=7/2	M=5/2	M=3/2	M=1/2
7.4258E+01	-.5670	0.0000	.5670	-M	0.000000	0.000000	0.000000	.979260	0.000000	0.000000	0.000000	0.000000-M
				+M	.012576	0.000000	0.000000	0.000000	0.000000	0.000000	.202218	0.000000+M
7.4258E+01	.5670	0.0000	.5670	-M	.012576	0.000000	0.000000	0.000000	0.000000	0.000000	.202218	0.000000-M
				+M	0.000000	0.000000	0.000000	.979260	0.000000	0.000000	0.000000	0.000000+M
6.4990E+01	.3245	.3780	.4982	-M	0.000000	0.000000	0.000000	0.000000	0.000000	.421507	0.000000	0.000000-M
				+M	0.000000	0.000000	0.000000	0.000000	.906825	0.000000	0.000000	0.000000+M
6.4990E+01	-.3245	.3780	.4982	-M	0.000000	0.000000	0.000000	0.000000	.906825	0.000000	0.000000	0.000000-M
				+M	0.000000	0.000000	0.000000	0.000000	0.000000	.421507	0.000000	0.000000+M
4.5750E+01	-.7114	.0381	.7125	-M	0.000000	0.000000	.986863	0.000000	0.000000	0.000000	0.000000	0.000000-M
				+M	0.000000	.035678	0.000000	0.000000	0.000000	0.000000	0.000000	.157571+M
4.5750E+01	.7114	.0381	.7125	-M	0.000000	.035678	0.000000	0.000000	0.000000	0.000000	0.000000	.157571-M
				+M	0.000000	0.000000	.986863	0.000000	0.000000	0.000000	0.000000	0.000000+M
-8.3503E+00	.1912	-.3780	.4236	-M	0.000000	0.000000	0.000000	0.000000	-.421507	0.000000	0.000000	0.000000-M
				+M	0.000000	0.000000	0.000000	0.000000	0.000000	.906825	0.000000	0.000000+M
-8.3503E+00	-.1912	-.3780	.4236	-M	0.000000	0.000000	0.000000	0.000000	0.000000	.906825	0.000000	0.000000-M
				+M	0.000000	0.000000	0.000000	0.000000	-.421507	0.000000	0.000000	0.000000+M
-9.7107E+00	-.8391	-.0286	.8396	-M	0.000000	.984579	0.000000	0.000000	0.000000	0.000000	0.000000	.163685-M
				+M	0.000000	0.000000	-.061730	0.000000	0.000000	0.000000	0.000000	0.000000+M
-9.7107E+00	.8391	-.0286	.8396	-M	0.000000	0.000000	-.061730	0.000000	0.000000	0.000000	0.000000	0.000000-M
				+M	0.000000	.984579	0.000000	0.000000	0.000000	0.000000	0.000000	.163685+M
-2.6645E+01	-.9538	0.0000	.9538	-M	.972499	0.000000	0.000000	0.000000	0.000000	0.000000	.225306	0.000000-M
				+M	0.000000	0.000000	0.000000	-.059016	0.000000	0.000000	0.000000	0.000000+M
-2.6645E+01	.9538	0.0000	.9538	-M	0.000000	0.000000	0.000000	-.059016	0.000000	0.000000	0.000000	0.000000-M
				+M	.972499	0.000000	0.000000	0.000000	0.000000	0.000000	.225306	0.000000+M
-5.4213E+01	-.2132	0.0000	.2132	-M	-.232567	0.000000	0.000000	0.000000	0.000000	0.000000	.953071	0.000000-M
				+M	0.000000	0.000000	0.000000	-.193823	0.000000	0.000000	0.000000	0.000000+M
-5.4213E+01	.2132	0.0000	.2132	-M	0.000000	0.000000	0.000000	-.193823	0.000000	0.000000	0.000000	0.000000-M
				+M	-.232567	0.000000	0.000000	0.000000	0.000000	0.000000	.953071	0.000000+M
-8.6079E+01	-.0723	.5238	.5288	-M	0.000000	-.171262	0.000000	0.000000	0.000000	0.000000	0.000000	.973847-M
				+M	0.000000	0.000000	-.149301	0.000000	0.000000	0.000000	0.000000	0.000000+M
-8.6079E+01	.0723	.5238	.5288	-M	0.000000	0.000000	-.149301	0.000000	0.000000	0.000000	0.000000	0.000000-M
				+M	0.000000	-.171262	0.000000	0.000000	0.000000	0.000000	0.000000	.973847+M
ENERGY	HEPAL	HEPER	HE		M=15/2	M=13/2	M=11/2	M=9/2	M=7/2	M=5/2	M=3/2	M=1/2

Y= .60 X=-.20 OAS=1.5471E+02*W246 B4F4/B2F2=-3.00E-01 B4F4/B6F6=-2.50E-01 B2F2/B6F6= 8.33E-01

ENERGY	HEPAL	HEPER	HE		M=15/2	M=13/2	M=11/2	M=9/2	M=7/2	M=5/2	M=3/2	M=1/2
7.3363E+01	-.2458	.4421	.5058	-M	0.000000	0.000000	0.000000	0.000000	.850826	0.000000	0.000000	0.000000-M
				+M	0.000000	0.000000	0.000000	0.000000	0.000000	.525447	0.000000	0.000000+M
7.3363E+01	.2458	.4421	.5058	-M	0.000000	0.000000	0.000000	0.000000	0.000000	.525447	0.000000	0.000000-M
				+M	0.000000	0.000000	0.000000	0.000000	.850826	0.000000	0.000000	0.000000+M
6.3538E+01	-.5110	0.0000	.5110	-M	0.000000	0.000000	0.000000	.944260	0.000000	0.000000	0.000000	0.000000-M
				+M	.053955	0.000000	0.000000	0.000000	0.000000	0.000000	.324750	0.000000+M
6.3538E+01	.5110	0.0000	.5110	-M	.053955	0.000000	0.000000	0.000000	0.000000	0.000000	.324750	0.000000-M
				+M	0.000000	0.000000	0.000000	.944260	0.000000	0.000000	0.000000	0.000000+M
1.9877E+01	-.6447	.1124	.6544	-M	0.000000	0.000000	.947609	0.000000	0.000000	0.000000	0.000000	0.000000-M
				+M	0.000000	.093680	0.000000	0.000000	0.000000	0.000000	0.000000	.305388+M
1.9877E+01	.6447	.1124	.6544	-M	0.000000	.093680	0.000000	0.000000	0.000000	0.000000	0.000000	.305388-M
				+M	0.000000	0.000000	.947609	0.000000	0.000000	0.000000	0.000000	0.000000+M
1.3372E+01	.9765	0.0000	.9765	-M	0.000000	0.000000	0.000000	-.093901	0.000000	0.000000	0.000000	0.000000-M
				+M	.989640	0.000000	0.000000	0.000000	0.000000	0.000000	.108610	0.000000+M
1.3372E+01	-.9765	0.0000	.9765	-M	.989640	0.000000	0.000000	0.000000	0.000000	0.000000	.108610	0.000000-M
				+M	0.000000	0.000000	0.000000	-.093901	0.000000	0.000000	0.000000	0.000000+M
-1.0243E+01	-.1125	-.4421	.4561	-M	0.000000	0.000000	0.000000	0.000000	0.000000	.850826	0.000000	0.000000-M
				+M	0.000000	0.000000	0.000000	0.000000	-.525447	0.000000	0.000000	0.000000+M
-1.0243E+01	.1125	-.4421	.4561	-M	0.000000	0.000000	0.000000	0.000000	-.525447	0.000000	0.000000	0.000000-M
				+M	0.000000	0.000000	0.000000	0.000000	0.000000	.850826	0.000000	0.000000+M
-3.1055E+01	.7310	-.0749	.7349	-M	0.000000	0.000000	-.191865	0.000000	0.000000	0.000000	0.000000	0.000000-M
				+M	0.000000	.931285	0.000000	0.000000	0.000000	0.000000	0.000000	.309672+M
-3.1055E+01	-.7310	-.0749	.7349	-M	0.000000	.931285	0.000000	0.000000	0.000000	0.000000	0.000000	.309672-M
				+M	0.000000	0.000000	-.191865	0.000000	0.000000	0.000000	0.000000	0.000000+M
-4.7509E+01	-.1345	0.0000	.1345	-M	-.133050	0.000000	0.000000	0.000000	0.000000	0.000000	.939543	0.000000-M
				+M	0.000000	0.000000	0.000000	-.315526	0.000000	0.000000	0.000000	0.000000+M
-4.7509E+01	.1345	0.0000	.1345	-M	0.000000	0.000000	0.000000	-.315526	0.000000	0.000000	0.000000	0.000000-M
				+M	-.133050	0.000000	0.000000	0.000000	0.000000	0.000000	.939543	0.000000+M
-8.1342E+01	-.1136	.4959	.5087	-M	0.000000	-.352041	0.000000	0.000000	0.000000	0.000000	0.000000	.900467-M
				+M	0.000000	0.000000	-.255393	0.000000	0.000000	0.000000	0.000000	0.000000+M
-8.1342E+01	.1136	.4959	.5087	-M	0.000000	0.000000	-.255393	0.000000	0.000000	0.000000	0.000000	0.000000-M
				+M	0.000000	-.352041	0.000000	0.000000	0.000000	0.000000	0.000000	.900467+M
ENERGY	HEPAL	HEPER	HE		M=15/2	M=13/2	M=11/2	M=9/2	M=7/2	M=5/2	M=3/2	M=1/2

Y= .60 X=0.00 OAS=1.7116E+02*W246 B4F4/B2F2= 0. B4F4/B6F6= 0. B2F2/B6F6= 6.67E-01

ENERGY	HEPAL	HEPER	HE		M=15/2	M=13/2	M=11/2	M=9/2	M=7/2	M=5/2	M=3/2	M=1/2
8.3517E+01	-.1800	.4742	.5072	-M	0.000000	0.000000	0.000000	0.000000	.801020	0.000000	0.000000	0.000000-M
				+M	0.000000	0.000000	0.000000	0.000000	0.000000	.598638	0.000000	0.000000+M
8.3517E+01	.1800	.4742	.5072	-M	0.000000	0.000000	0.000000	0.000000	0.000000	.598638	0.000000	0.000000-M
				+M	0.000000	0.000000	0.000000	0.000000	.801020	0.000000	0.000000	0.000000+M
6.0571E+01	.0998	0.0000	.0998	-M	0.000000	0.000000	0.000000	.688418	0.000000	0.000000	0.000000	0.000000-M
				+M	.590458	0.000000	0.000000	0.000000	0.000000	0.000000	.421237	0.000000+M
6.0571E+01	-.0998	0.0000	.0998	-M	.590458	0.000000	0.000000	0.000000	0.000000	0.000000	.421237	0.000000-M
				+M	0.000000	0.000000	0.000000	.688418	0.000000	0.000000	0.000000	0.000000+M
5.0280E+01	-.4614	0.0000	.4614	-M	.801432	0.000000	0.000000	0.000000	0.000000	0.000000	-.205391	0.000000-M
				+M	0.000000	0.000000	0.000000	-.561713	0.000000	0.000000	0.000000	0.000000+M
5.0280E+01	.4614	0.0000	.4614	-M	0.000000	0.000000	0.000000	-.561713	0.000000	0.000000	0.000000	0.000000-M
				+M	.801432	0.000000	0.000000	0.000000	0.000000	0.000000	-.205391	0.000000+M
2.4801E+00	.4347	.2690	.5112	-M	0.000000	.186463	0.000000	0.000000	0.000000	0.000000	0.000000	.551126-M
				+M	0.000000	0.000000	.813321	0.000000	0.000000	0.000000	0.000000	0.000000+M
2.4801E+00	-.4347	.2690	.5112	-M	0.000000	0.000000	.813321	0.000000	0.000000	0.000000	0.000000	0.000000-M
				+M	0.000000	.186463	0.000000	0.000000	0.000000	0.000000	0.000000	.551126+M
-1.3917E+01	-.0466	-.4742	.4765	-M	0.000000	0.000000	0.000000	0.000000	0.000000	.801020	0.000000	0.000000-M
				+M	0.000000	0.000000	0.000000	0.000000	-.598638	0.000000	0.000000	0.000000+M
-1.3917E+01	.0466	-.4742	.4765	-M	0.000000	0.000000	0.000000	0.000000	-.598638	0.000000	0.000000	0.000000-M
				+M	0.000000	0.000000	0.000000	0.000000	0.000000	.801020	0.000000	0.000000+M
-4.5451E+01	.0388	0.0000	.0388	-M	0.000000	0.000000	0.000000	-.458867	0.000000	0.000000	0.000000	0.000000-M
				+M	-.095219	0.000000	0.000000	0.000000	0.000000	0.000000	.883388	0.000000+M
-4.5451E+01	-.0388	0.0000	.0388	-M	-.095219	0.000000	0.000000	0.000000	0.000000	0.000000	.883388	0.000000-M
				+M	0.000000	0.000000	0.000000	-.458867	0.000000	0.000000	0.000000	0.000000+M
-4.9840E+01	-.3731	.1674	.4089	-M	0.000000	.775054	0.000000	0.000000	0.000000	0.000000	0.000000	.426219-M
				+M	0.000000	0.000000	-.466506	0.000000	0.000000	0.000000	0.000000	0.000000+M
-4.9840E+01	.3731	-.1674	.4089	-M	0.000000	0.000000	-.466506	0.000000	0.000000	0.000000	0.000000	0.000000-M
				+M	0.000000	.775054	0.000000	0.000000	0.000000	0.000000	0.000000	.426219+M
-8.7640E+01	.2616	.1674	.3105	-M	0.000000	0.000000	-.347679	0.000000	0.000000	0.000000	0.000000	0.000000-M
				+M	0.000000	-.603757	0.000000	0.000000	0.000000	0.000000	0.000000	.717354+M
-8.7640E+01	-.2616	-.1674	.3105	-M	0.000000	.603757	0.000000	0.000000	0.000000	0.000000	0.000000	-.717354-M
				+M	0.000000	0.000000	.347679	0.000000	0.000000	0.000000	0.000000	0.000000+M
ENERGY	HEPAL	HEPER	HE		M=15/2	M=13/2	M=11/2	M=9/2	M=7/2	M=5/2	M=3/2	M=1/2

Y= .60 X= .20 OAS=1.4972E+02*W246 B4F4/B2F2= 3.00E-01 B4F4/B6F6= 2.50E-01 B2F2/B6F6= 8.33E-01

ENERGY	HEPAL	HEPER	HE		M=15/2	M=13/2	M=11/2	M=9/2	M=7/2	M=5/2	M=3/2	M=1/2
7.9016E+01	.9831	0.0000	.9831	-M	0.000000	0.000000	0.000000	.054762	0.000000	0.000000	0.000000	0.000000-M
				+M	.990869	0.000000	0.000000	0.000000	0.000000	0.000000	.123206	0.000000+M
7.9016E+01	-.9831	0.0000	.9831	-M	.990869	0.000000	0.000000	0.000000	0.000000	0.000000	.123206	0.000000-M
				+M	0.000000	0.000000	0.000000	.054762	0.000000	0.000000	0.000000	0.000000+M
5.9774E+01	.1075	.4918	.5034	-M	0.000000	0.000000	0.000000	0.000000	0.000000	.670001	0.000000	0.000000-M
				+M	0.000000	0.000000	0.000000	0.000000	.742360	0.000000	0.000000	0.000000+M
5.9774E+01	-.1075	.4918	.5034	-M	0.000000	0.000000	0.000000	0.000000	.742360	0.000000	0.000000	0.000000-M
				+M	0.000000	0.000000	0.000000	0.000000	0.000000	.670001	0.000000	0.000000+M
3.2810E+01	-.2376	0.0000	.2376	-M	0.000000	0.000000	0.000000	.749612	0.000000	0.000000	0.000000	0.000000-M
				+M	-.122310	0.000000	0.000000	0.000000	0.000000	0.000000	.650478	0.000000+M
3.2810E+01	.2376	0.0000	.2376	-M	-.122310	0.000000	0.000000	0.000000	0.000000	0.000000	.650478	0.000000-M
				+M	0.000000	0.000000	0.000000	.749612	0.000000	0.000000	0.000000	0.000000+M
3.6696E-02	-.1005	.4477	.4588	-M	0.000000	0.000000	.517721	0.000000	0.000000	0.000000	0.000000	0.000000-M
				+M	0.000000	.243062	0.000000	0.000000	0.000000	0.000000	0.000000	.820296+M
3.6696E-02	.1005	.4477	.4588	-M	0.000000	.243062	0.000000	0.000000	0.000000	0.000000	0.000000	.820296-M
				+M	0.000000	0.000000	.517721	0.000000	0.000000	0.000000	0.000000	0.000000+M
-1.5374E+01	-.0258	-.4918	.4925	-M	0.000000	0.000000	0.000000	0.000000	-.670001	0.000000	0.000000	0.000000-M
				+M	0.000000	0.000000	0.000000	0.000000	0.000000	.742360	0.000000	0.000000+M
-1.5374E+01	.0258	-.4918	.4925	-M	0.000000	0.000000	0.000000	0.000000	0.000000	.742360	0.000000	0.000000-M
				+M	0.000000	0.000000	0.000000	0.000000	-.670001	0.000000	0.000000	0.000000+M
-3.4186E+01	.1455	0.0000	.1455	-M	-.056735	0.000000	0.000000	0.000000	0.000000	0.000000	.749466	0.000000-M
				+M	0.000000	0.000000	0.000000	-.659608	0.000000	0.000000	0.000000	0.000000+M
-3.4186E+01	-.1455	0.0000	.1455	-M	0.000000	0.000000	0.000000	-.659608	0.000000	0.000000	0.000000	0.000000-M
				+M	-.056735	0.000000	0.000000	0.000000	0.000000	0.000000	.749466	0.000000+M
-5.1374E+01	-.0268	-.1559	.1582	-M	0.000000	0.000000	-.722936	0.000000	0.000000	0.000000	0.000000	0.000000-M
				+M	0.000000	.637022	0.000000	0.000000	0.000000	0.000000	0.000000	.267518+M
-5.1374E+01	.0268	.1559	.1582	-M	0.000000	-.637022	0.000000	0.000000	0.000000	0.000000	0.000000	-.267518-M
				+M	0.000000	0.000000	.722936	0.000000	0.000000	0.000000	0.000000	0.000000+M
-7.0703E+01	-.3273	-.1559	.3625	-M	0.000000	.731521	0.000000	0.000000	0.000000	0.000000	0.000000	-.505518-M
				+M	0.000000	0.000000	.457523	0.000000	0.000000	0.000000	0.000000	0.000000+M
-7.0703E+01	.3273	.1559	.3625	-M	0.000000	0.000000	.457523	0.000000	0.000000	0.000000	0.000000	0.000000-M
				+M	0.000000	.731521	0.000000	0.000000	0.000000	0.000000	0.000000	-.505518+M
ENERGY	HEPAL	HEPER	HE		M=15/2	M=13/2	M=11/2	M=9/2	M=7/2	M=5/2	M=3/2	M=1/2

= .60 X= .40 OAS=1.7270E+02*W246 B4F4/B2F2= 6.00E-01 B4F4/B6F6= 6.67E-01 B2F2/B6F6= 1.11E+00

ENERGY	HEPAL	HEPER	HE		M=15/2	M=13/2	M=11/2	M=9/2	M=7/2	M=5/2	M=3/2	M=1/2
1.0340E+02	.9951	0.0000	.9951	-M	0.000000	0.000000	0.000000	.014447	0.000000	0.000000	0.000000	0.000000-M
				+M	.997051	0.000000	0.000000	0.000000	0.000000	0.000000	.075365	0.000000+M
1.0340E+02	-.9951	0.0000	.9951	-M	.997051	0.000000	0.000000	0.000000	0.000000	0.000000	.075365	0.000000-M
				+M	0.000000	0.000000	0.000000	.014447	0.000000	0.000000	0.000000	0.000000+M
3.8293E+01	.0186	.4830	.4834	-M	0.000000	0.000000	0.000000	0.000000	.627181	0.000000	0.000000	0.000000-M
				+M	0.000000	0.000000	0.000000	0.000000	0.000000	.778874	0.000000	0.000000+M
3.8293E+01	-.0186	.4830	.4834	-M	0.000000	0.000000	0.000000	0.000000	0.000000	.778874	0.000000	0.000000-M
				+M	0.000000	0.000000	0.000000	0.000000	.627181	0.000000	0.000000	0.000000+M
2.5679E+01	.0593	0.0000	.0593	-M	0.000000	0.000000	0.000000	.425942	0.000000	0.000000	0.000000	0.000000-M
				+M	-.074329	0.000000	0.000000	0.000000	0.000000	0.000000	.901692	0.000000+M
2.5679E+01	-.0593	0.0000	.0593	-M	-.074329	0.000000	0.000000	0.000000	0.000000	0.000000	.901692	0.000000-M
				+M	0.000000	0.000000	0.000000	.425942	0.000000	0.000000	0.000000	0.000000+M
1.7130E+01	.0464	.5167	.5188	-M	0.000000	0.000000	.234182	0.000000	0.000000	0.000000	0.000000	0.000000-M
				+M	0.000000	.171774	0.000000	0.000000	0.000000	0.000000	0.000000	.956897+M
1.7130E+01	-.0464	.5167	.5188	-M	0.000000	.171774	0.000000	0.000000	0.000000	0.000000	0.000000	.956897-M
				+M	0.000000	0.000000	.234182	0.000000	0.000000	0.000000	0.000000	0.000000+M
-1.9093E+01	.1520	-.4830	.5064	-M	0.000000	0.000000	0.000000	0.000000	0.000000	-.627181	0.000000	0.000000-M
				+M	0.000000	0.000000	0.000000	0.000000	.778874	0.000000	0.000000	0.000000+M
-1.9093E+01	-.1520	-.4830	.5064	-M	0.000000	0.000000	0.000000	0.000000	.778874	0.000000	0.000000	0.000000-M
				+M	0.000000	0.000000	0.000000	0.000000	0.000000	-.627181	0.000000	0.000000+M
-3.9201E+01	.4544	0.0000	.4544	-M	.019074	0.000000	0.000000	0.000000	0.000000	0.000000	-.425759	0.000000-M
				+M	0.000000	0.000000	0.000000	.904635	0.000000	0.000000	0.000000	0.000000+M
-3.9201E+01	-.4544	0.0000	.4544	-M	0.000000	0.000000	0.000000	.904635	0.000000	0.000000	0.000000	0.000000-M
				+M	.019074	0.000000	0.000000	0.000000	0.000000	0.000000	-.425759	0.000000+M
-5.6910E+01	-.7194	-.1928	.7448	-M	0.000000	.950338	0.000000	0.000000	0.000000	0.000000	0.000000	-.098333-M
				+M	0.000000	0.000000	-.295276	0.000000	0.000000	0.000000	0.000000	0.000000+M
-5.6910E+01	.7194	-.1928	.7448	-M	0.000000	0.000000	-.295276	0.000000	0.000000	0.000000	0.000000	0.000000-M
				+M	0.000000	.950338	0.000000	0.000000	0.000000	0.000000	0.000000	-.098333+M
-6.9300E+01	-.5658	.2094	.6033	-M	0.000000	0.000000	.926267	0.000000	0.000000	0.000000	0.000000	0.000000-M
				+M	0.000000	.259521	0.000000	0.000000	0.000000	0.000000	0.000000	-.273273+M
-6.9300E+01	.5658	.2094	.6033	-M	0.000000	.259521	0.000000	0.000000	0.000000	0.000000	0.000000	-.273273-M
				+M	0.000000	0.000000	.926267	0.000000	0.000000	0.000000	0.000000	0.000000+M
ENERGY	HEPAL	HEPER	HE		M=15/2	M=13/2	M=11/2	M=9/2	M=7/2	M=5/2	M=3/2	M=1/2

Y= .60 X= .60 OAS=2.1476E+02*W246 B4F4/B2F2= 9.00E-01 B4F4/B6F6= 1.50E+00 B2F2/B6F6= 1.67E+00

ENERGY	HEPAL	HEPER	HE		M=15/2	M=13/2	M=11/2	M=9/2	M=7/2	M=5/2	M=3/2	M=1/2
1.2807E+02	-.9983	0.0000	.9983	-M	.998968	0.000000	0.000000	0.000000	0.000000	0.000000	.045238	0.000000-M
				+M	0.000000	0.000000	0.000000	.004051	0.000000	0.000000	0.000000	0.000000+M
1.2807E+02	.9983	0.0000	.9983	-M	0.000000	0.000000	0.000000	.004051	0.000000	0.000000	0.000000	0.000000-M
				+M	.998968	0.000000	0.000000	0.000000	0.000000	0.000000	.045238	0.000000+M
4.3767E+01	.0649	.5300	.5340	-M	0.000000	0.000000	.100952	0.000000	0.000000	0.000000	0.000000	0.000000-M
				+M	0.000000	.089594	0.000000	0.000000	0.000000	0.000000	0.000000	.990849+M
4.3767E+01	-.0649	.5300	.5340	-M	0.000000	.089594	0.000000	0.000000	0.000000	0.000000	0.000000	.990849-M
				+M	0.000000	0.000000	.100952	0.000000	0.000000	0.000000	0.000000	0.000000+M
3.5611E+01	.1773	0.0000	.1773	-M	0.000000	0.000000	0.000000	.174522	0.000000	0.000000	0.000000	0.000000-M
				+M	-.045250	0.000000	0.000000	0.000000	0.000000	0.000000	.983613	0.000000+M
3.5611E+01	-.1773	0.0000	.1773	-M	-.045250	0.000000	0.000000	0.000000	0.000000	0.000000	.983613	0.000000-M
				+M	0.000000	0.000000	0.000000	.174522	0.000000	0.000000	0.000000	0.000000+M
2.1654E+01	.1942	.3748	.4221	-M	0.000000	0.000000	0.000000	0.000000	.417003	0.000000	0.000000	0.000000-M
				+M	0.000000	0.000000	0.000000	0.000000	0.000000	.908905	0.000000	0.000000+M
2.1654E+01	-.1942	.3748	.4221	-M	0.000000	0.000000	0.000000	0.000000	0.000000	.908905	0.000000	0.000000-M
				+M	0.000000	0.000000	0.000000	0.000000	.417003	0.000000	0.000000	0.000000+M
-2.7654E+01	.3276	-.3748	.4977	-M	0.000000	0.000000	0.000000	0.000000	0.000000	-.417003	0.000000	0.000000-M
				+M	0.000000	0.000000	0.000000	0.000000	.908905	0.000000	0.000000	0.000000+M
-2.7654E+01	-.3276	-.3748	.4977	-M	0.000000	0.000000	0.000000	0.000000	.908905	0.000000	0.000000	0.000000-M
				+M	0.000000	0.000000	0.000000	0.000000	0.000000	-.417003	0.000000	0.000000+M
-5.3195E+01	.8587	-.0209	.8589	-M	0.000000	0.000000	-.035381	0.000000	0.000000	0.000000	0.000000	0.000000-M
				+M	0.000000	.995630	0.000000	0.000000	0.000000	0.000000	0.000000	-.086422+M
-5.3195E+01	-.8587	-.0209	.8589	-M	0.000000	.995630	0.000000	0.000000	0.000000	0.000000	0.000000	-.086422-M
				+M	0.000000	0.000000	-.035381	0.000000	0.000000	0.000000	0.000000	0.000000+M
-6.1563E+01	-.5756	0.0000	.5756	-M	0.000000	0.000000	0.000000	.984645	0.000000	0.000000	0.000000	0.000000-M
				+M	.003911	0.000000	0.000000	0.000000	0.000000	0.000000	-.174526	0.000000+M
-6.1563E+01	.5756	0.0000	.5756	-M	.003911	0.000000	0.000000	0.000000	0.000000	0.000000	-.174526	0.000000-M
				+M	0.000000	0.000000	0.000000	.984645	0.000000	0.000000	0.000000	0.000000+M
-8.6691E+01	.7236	.0242	.7240	-M	0.000000	.026333	0.000000	0.000000	0.000000	0.000000	0.000000	-.103681-M
				+M	0.000000	0.000000	.994262	0.000000	0.000000	0.000000	0.000000	0.000000+M
-8.6691E+01	-.7236	.0242	.7240	-M	0.000000	0.000000	.994262	0.000000	0.000000	0.000000	0.000000	0.000000-M
				+M	0.000000	.026333	0.000000	0.000000	0.000000	0.000000	0.000000	-.103681+M
ENERGY	HEPAL	HEPER	HE		M=15/2	M=13/2	M=11/2	M=9/2	M=7/2	M=5/2	M=3/2	M=1/2

Y= .60 X= .80 OAS=2.6028E+02*W246 B4F4/B2F2= 1.20E+00 B4F4/B6F6= 4.00E+00 B2F2/B6F6= 3.33E+00

ENERGY	HEPAL	HEPER	HE		M=15/2	M=13/2	M=11/2	M=9/2	M=7/2	M=5/2	M=3/2	M=1/2
1.5288E+02	.9996	0.0000	.9996	-M	0.000000	0.000000	0.000000	.000722	0.000000	0.000000	0.000000	0.000000-M
				+M	.999780	0.000000	0.000000	0.000000	0.000000	0.000000	.020952	0.000000+M
1.5288E+02	-.9996	0.0000	.9996	-M	.999780	0.000000	0.000000	0.000000	0.000000	0.000000	.020952	0.000000-M
				+M	0.000000	0.000000	0.000000	.000722	0.000000	0.000000	0.000000	0.000000+M
7.3713E+01	.0666	.5329	.5370	-M	0.000000	0.000000	.036310	0.000000	0.000000	0.000000	0.000000	0.000000-M
				+M	0.000000	.035308	0.000000	0.000000	0.000000	0.000000	0.000000	.998717+M
7.3713E+01	-.0666	.5329	.5370	-M	0.000000	.035308	0.000000	0.000000	0.000000	0.000000	0.000000	.998717-M
				+M	0.000000	0.000000	.036310	0.000000	0.000000	0.000000	0.000000	0.000000+M
5.1769E+01	.1976	0.0000	.1976	-M	0.000000	0.000000	0.000000	.058922	0.000000	0.000000	0.000000	0.000000-M
				+M	-.020958	0.000000	0.000000	0.000000	0.000000	0.000000	.998043	0.000000+M
5.1769E+01	-.1976	0.0000	.1976	-M	-.020958	0.000000	0.000000	0.000000	0.000000	0.000000	.998043	0.000000-M
				+M	0.000000	0.000000	0.000000	.058922	0.000000	0.000000	0.000000	0.000000+M
1.2066E+01	.3098	.1670	.3520	-M	0.000000	0.000000	0.000000	0.000000	.171417	0.000000	0.000000	0.000000-M
				+M	0.000000	0.000000	0.000000	0.000000	0.000000	.985199	0.000000	0.000000+M
1.2066E+01	-.3098	.1670	.3520	-M	0.000000	0.000000	0.000000	0.000000	0.000000	.985199	0.000000	0.000000-M
				+M	0.000000	0.000000	0.000000	0.000000	.171417	0.000000	0.000000	0.000000+M
-4.3266E+01	.4432	-.1670	.4736	-M	0.000000	0.000000	0.000000	0.000000	0.000000	-.171417	0.000000	0.000000-M
				+M	0.000000	0.000000	0.000000	0.000000	.985199	0.000000	0.000000	0.000000+M
-4.3266E+01	-.4432	-.1670	.4736	-M	0.000000	0.000000	0.000000	0.000000	.985199	0.000000	0.000000	0.000000-M
				+M	0.000000	0.000000	0.000000	0.000000	0.000000	-.171417	0.000000	0.000000+M
-4.9473E+01	-.8657	-.0022	.8657	-M	0.000000	.999373	0.000000	0.000000	0.000000	0.000000	0.000000	-.035185-M
				+M	0.000000	0.000000	-.004011	0.000000	0.000000	0.000000	0.000000	0.000000+M
-4.9473E+01	.8657	-.0022	.8657	-M	0.000000	0.000000	-.004011	0.000000	0.000000	0.000000	0.000000	0.000000-M
				+M	0.000000	.999373	0.000000	0.000000	0.000000	0.000000	0.000000	-.035185+M
-9.0293E+01	-.5972	0.0000	.5972	-M	0.000000	0.000000	0.000000	.998262	0.000000	0.000000	0.000000	0.000000-M
				+M	.000514	0.000000	0.000000	0.000000	0.000000	0.000000	-.058924	0.000000+M
-9.0293E+01	.5972	0.0000	.5972	-M	.000514	0.000000	0.000000	0.000000	0.000000	0.000000	-.058924	0.000000-M
				+M	0.000000	0.000000	0.000000	.998262	0.000000	0.000000	0.000000	0.000000+M
-1.0740E+02	-.7323	.0026	.7323	-M	0.000000	0.000000	.999333	0.000000	0.000000	0.000000	0.000000	0.000000-M
				+M	0.000000	.002728	0.000000	0.000000	0.000000	0.000000	0.000000	-.036429+M
-1.0740E+02	.7323	.0026	.7323	-M	0.000000	.002728	0.000000	0.000000	0.000000	0.000000	0.000000	-.036429-M
				+M	0.000000	0.000000	.999333	0.000000	0.000000	0.000000	0.000000	0.000000+M
ENERGY	HEPAL	HEPER	HE		M=15/2	M=13/2	M=11/2	M=9/2	M=7/2	M=5/2	M=3/2	M=1/2

Y= .80 X=-.80 OAS=2.9586E+02*W246 B4F4/B2F2=-3.20E+00 B4F4/B6F6=-4.00E+00 B2F2/B6F6= 1.25E+00

ENERGY	HEPAL	HEPER	HE		M=15/2	M=13/2	M=11/2	M=9/2	M=7/2	M=5/2	M=3/2	M=1/2
1.3842E+02	.5981	0.0000	.5981	-M	.000466	0.000000	0.000000	0.000000	0.000000	0.000000	.048657	0.000000-M
				+M	0.000000	0.000000	0.000000	.998815	0.000000	0.000000	0.000000	0.000000+M
1.3842E+02	-.5981	0.0000	.5981	-M	0.000000	0.000000	0.000000	.998815	0.000000	0.000000	0.000000	0.000000-M
				+M	.000466	0.000000	0.000000	0.000000	0.000000	0.000000	.048657	0.000000+M
1.3728E+02	-.7325	.0019	.7325	-M	0.000000	0.000000	.999487	0.000000	0.000000	0.000000	0.000000	0.000000-M
				+M	0.000000	.001982	0.000000	0.000000	0.000000	0.000000	0.000000	.031974+M
1.3728E+02	.7325	.0019	.7325	-M	0.000000	.001982	0.000000	0.000000	0.000000	0.000000	0.000000	.031974-M
				+M	0.000000	0.000000	.999487	0.000000	0.000000	0.000000	0.000000	0.000000+M
7.8504E+01	-.4513	.1357	.4713	-M	0.000000	0.000000	0.000000	0.000000	.990348	0.000000	0.000000	0.000000-M
				+M	0.000000	0.000000	0.000000	0.000000	0.000000	.138603	0.000000	0.000000+M
7.8504E+01	.4513	.1357	.4713	-M	0.000000	0.000000	0.000000	0.000000	0.000000	.138603	0.000000	0.000000-M
				+M	0.000000	0.000000	0.000000	0.000000	.990348	0.000000	0.000000	0.000000+M
4.3905E+01	.8658	-.0016	.8658	-M	0.000000	0.000000	-.003001	0.000000	0.000000	0.000000	0.000000	0.000000-M
				+M	0.000000	.999488	0.000000	0.000000	0.000000	0.000000	0.000000	.031868+M
4.3905E+01	-.8658	-.0016	.8658	-M	0.000000	.999488	0.000000	0.000000	0.000000	0.000000	0.000000	.031868-M
				+M	0.000000	0.000000	-.003001	0.000000	0.000000	0.000000	0.000000	0.000000+M
-1.2264E+01	-.3180	-.1357	.3457	-M	0.000000	0.000000	0.000000	0.000000	0.000000	.990348	0.000000	0.000000-M
				+M	0.000000	0.000000	0.000000	0.000000	-.138603	0.000000	0.000000	0.000000+M
-1.2264E+01	.3180	-.1357	.3457	-M	0.000000	0.000000	0.000000	0.000000	-.138603	0.000000	0.000000	0.000000-M
				+M	0.000000	0.000000	0.000000	0.000000	0.000000	.990348	0.000000	0.000000+M
-9.0783E+01	-.1995	0.0000	.1995	-M	.042441	0.000000	0.000000	0.000000	0.000000	0.000000	.997915	0.000000-M
				+M	0.000000	0.000000	0.000000	-.048633	0.000000	0.000000	0.000000	0.000000+M
-9.0783E+01	.1995	0.0000	.1995	-M	0.000000	0.000000	0.000000	-.048633	0.000000	0.000000	0.000000	0.000000-M
				+M	.042441	0.000000	0.000000	0.000000	0.000000	0.000000	.997915	0.000000+M
-1.3763E+02	-.0667	.5330	.5371	-M	0.000000	-.031948	0.000000	0.000000	0.000000	0.000000	0.000000	.998981-M
				+M	0.000000	0.000000	-.031894	0.000000	0.000000	0.000000	0.000000	0.000000+M
-1.3763E+02	.0667	.5330	.5371	-M	0.000000	0.000000	-.031894	0.000000	0.000000	0.000000	0.000000	0.000000-M
				+M	0.000000	-.031948	0.000000	0.000000	0.000000	0.000000	0.000000	.998981+M
-1.5744E+02	.9986	0.0000	.9986	-M	0.000000	0.000000	0.000000	.001600	0.000000	0.000000	0.000000	0.000000-M
				+M	.999099	0.000000	0.000000	0.000000	0.000000	0.000000	-.042414	0.000000+M
-1.5744E+02	-.9986	0.0000	.9986	-M	.999099	0.000000	0.000000	0.000000	0.000000	0.000000	-.042414	0.000000-M
				+M	0.000000	0.000000	0.000000	.001600	0.000000	0.000000	0.000000	0.000000+M
ENERGY	HEPAL	HEPER	HE		M=15/2	M=13/2	M=11/2	M=9/2	M=7/2	M=5/2	M=3/2	M=1/2

Y= .80 X=-.60 OAS=2.3912E+02*W246 B4F4/B2F2=-2.40E+00 B4F4/B6F6=-1.50E+00 B2F2/B6F6= 6.25E-01

ENERGY	HEPAL	HEPER	HE		M=15/2	M=13/2	M=11/2	M=9/2	M=7/2	M=5/2	M=3/2	M=1/2
1.1776E+02	-.5893	0.0000	.5893	-M	0.000000	0.000000	0.000000	.993279	0.000000	0.000000	0.000000	0.000000-M
				+M	.002963	0.000000	0.000000	0.000000	0.000000	0.000000	.115705	0.000000+M
1.1776E+02	.5893	0.0000	.5893	-M	.002963	0.000000	0.000000	0.000000	0.000000	0.000000	.115705	0.000000-M
				+M	0.000000	0.000000	0.000000	.993279	0.000000	0.000000	0.000000	0.000000+M
9.6836E+01	-.7278	.0113	.7279	-M	0.000000	0.000000	.996606	0.000000	0.000000	0.000000	0.000000	0.000000-M
				+M	0.000000	.010954	0.000000	0.000000	0.000000	0.000000	0.000000	.081590+M
9.6836E+01	.7278	.0113	.7279	-M	0.000000	.010954	0.000000	0.000000	0.000000	0.000000	0.000000	.081590-M
				+M	0.000000	0.000000	.996606	0.000000	0.000000	0.000000	0.000000	0.000000+M
8.2115E+01	-.3987	.2758	.4847	-M	0.000000	0.000000	0.000000	0.000000	.956555	0.000000	0.000000	0.000000-M
				+M	0.000000	0.000000	0.000000	0.000000	0.000000	.291552	0.000000	0.000000+M
8.2115E+01	.3987	.2758	.4847	-M	0.000000	0.000000	0.000000	0.000000	0.000000	.291552	0.000000	0.000000-M
				+M	0.000000	0.000000	0.000000	0.000000	.956555	0.000000	0.000000	0.000000+M
1.1444E+01	-.8602	-.0087	.8602	-M	0.000000	.996115	0.000000	0.000000	0.000000	0.000000	0.000000	.086206-M
				+M	0.000000	0.000000	-.018006	0.000000	0.000000	0.000000	0.000000	0.000000+M
1.1444E+01	.8602	-.0087	.8602	-M	0.000000	0.000000	-.018006	0.000000	0.000000	0.000000	0.000000	0.000000-M
				+M	0.000000	.996115	0.000000	0.000000	0.000000	0.000000	0.000000	.086206+M
-7.2350E+00	-.2653	-.2758	.3827	-M	0.000000	0.000000	0.000000	0.000000	0.000000	.956555	0.000000	0.000000-M
				+M	0.000000	0.000000	0.000000	0.000000	-.291552	0.000000	0.000000	0.000000+M
-7.2350E+00	.2653	-.2758	.3827	-M	0.000000	0.000000	0.000000	0.000000	-.291552	0.000000	0.000000	0.000000-M
				+M	0.000000	0.000000	0.000000	0.000000	0.000000	.956555	0.000000	0.000000+M
-7.5188E+01	-.2205	0.0000	.2205	-M	.196480	0.000000	0.000000	0.000000	0.000000	0.000000	.973855	0.000000-M
				+M	0.000000	0.000000	0.000000	-.114028	0.000000	0.000000	0.000000	0.000000+M
-7.5188E+01	.2205	0.0000	.2205	-M	0.000000	0.000000	0.000000	-.114028	0.000000	0.000000	0.000000	0.000000-M
				+M	.196480	0.000000	0.000000	0.000000	0.000000	0.000000	.973855	0.000000+M
-1.0437E+02	-.9688	0.0000	.9688	-M	.980503	0.000000	0.000000	0.000000	0.000000	0.000000	-.195498	0.000000-M
				+M	0.000000	0.000000	0.000000	.019848	0.000000	0.000000	0.000000	0.000000+M
-1.0437E+02	.9688	0.0000	.9688	-M	0.000000	0.000000	0.000000	.019848	0.000000	0.000000	0.000000	0.000000-M
				+M	.980503	0.000000	0.000000	0.000000	0.000000	0.000000	-.195498	0.000000+M
-1.2136E+02	.0676	.5308	.5351	-M	0.000000	0.000000	-.080328	0.000000	0.000000	0.000000	0.000000	0.000000-M
				+M	0.000000	-.087382	0.000000	0.000000	0.000000	0.000000	0.000000	.992931+M
-1.2136E+02	-.0676	.5308	.5351	-M	0.000000	-.087382	0.000000	0.000000	0.000000	0.000000	0.000000	.992931-M
				+M	0.000000	0.000000	-.080328	0.000000	0.000000	0.000000	0.000000	0.000000+M
ENERGY	HEPAL	HEPER	HE		M=15/2	M=13/2	M=11/2	M=9/2	M=7/2	M=5/2	M=3/2	M=1/2

Y= .80 X=-.40 OAS=2.0807E+02*W246 B4F4/B2F2=-1.60E+00 B4F4/B6F6=-6.67E-01 B2F2/B6F6= 4.17E-01

ENERGY	HEPAL	HEPER	HE		M=15/2	M=13/2	M=11/2	M=9/2	M=7/2	M=5/2	M=3/2	M=1/2
9.9596E+01	-.5647	0.0000	.5647	-M	0.000000	0.000000	0.000000	.977781	0.000000	0.000000	0.000000	0.000000-M
				+M	.011944	0.000000	0.000000	0.000000	0.000000	0.000000	.209287	0.000000+M
9.9596E+01	.5647	0.0000	.5647	-M	.011944	0.000000	0.000000	0.000000	0.000000	0.000000	.209287	0.000000-M
				+M	0.000000	0.000000	0.000000	.977781	0.000000	0.000000	0.000000	0.000000+M
9.0015E+01	.3197	.3830	.4988	-M	0.000000	0.000000	0.000000	0.000000	0.000000	.428677	0.000000	0.000000-M
				+M	0.000000	0.000000	0.000000	0.000000	.903458	0.000000	0.000000	0.000000+M
9.0015E+01	-.3197	.3830	.4988	-M	0.000000	0.000000	0.000000	0.000000	.903458	0.000000	0.000000	0.000000-M
				+M	0.000000	0.000000	0.000000	0.000000	0.000000	.428677	0.000000	0.000000+M
5.8258E+01	-.7090	.0398	.7101	-M	0.000000	0.000000	.985308	0.000000	0.000000	0.000000	0.000000	0.000000-M
				+M	0.000000	.035825	0.000000	0.000000	0.000000	0.000000	0.000000	.166988+M
5.8258E+01	.7090	.0398	.7101	-M	0.000000	.035825	0.000000	0.000000	0.000000	0.000000	0.000000	.166988-M
				+M	0.000000	0.000000	.985308	0.000000	0.000000	0.000000	0.000000	0.000000+M
-6.4951E+00	.1863	-.3830	.4259	-M	0.000000	0.000000	0.000000	0.000000	-.428677	0.000000	0.000000	0.000000-M
				+M	0.000000	0.000000	0.000000	0.000000	0.000000	.903458	0.000000	0.000000+M
-6.4951E+00	-.1863	-.3830	.4259	-M	0.000000	0.000000	0.000000	0.000000	0.000000	.903458	0.000000	0.000000-M
				+M	0.000000	0.000000	0.000000	0.000000	-.428677	0.000000	0.000000	0.000000+M
-1.9500E+01	-.8311	-.0278	.8315	-M	0.000000	.979823	0.000000	0.000000	0.000000	0.000000	0.000000	.188119-M
				+M	0.000000	0.000000	-.067507	0.000000	0.000000	0.000000	0.000000	0.000000+M
-1.9500E+01	.8311	-.0278	.8315	-M	0.000000	0.000000	-.067507	0.000000	0.000000	0.000000	0.000000	0.000000-M
				+M	0.000000	.979823	0.000000	0.000000	0.000000	0.000000	0.000000	.188119+M
-4.5499E+01	-.8611	0.0000	.8611	-M	.914039	0.000000	0.000000	0.000000	0.000000	0.000000	.394214	0.000000-M
				+M	0.000000	0.000000	0.000000	-.095544	0.000000	0.000000	0.000000	0.000000+M
-4.5499E+01	.8611	0.0000	.8611	-M	0.000000	0.000000	0.000000	-.095544	0.000000	0.000000	0.000000	0.000000-M
				+M	.914039	0.000000	0.000000	0.000000	0.000000	0.000000	.394214	0.000000+M
-6.7897E+01	.3037	0.0000	.3037	-M	0.000000	0.000000	0.000000	-.186588	0.000000	0.000000	0.000000	0.000000-M
				+M	-.405451	0.000000	0.000000	0.000000	0.000000	0.000000	.894871	0.000000+M
-6.7897E+01	-.3037	0.0000	.3037	-M	-.405451	0.000000	0.000000	0.000000	0.000000	0.000000	.894871	0.000000-M
				+M	0.000000	0.000000	0.000000	-.186588	0.000000	0.000000	0.000000	0.000000+M
-1.0848E+02	-.0779	.5214	.5271	-M	0.000000	-.196628	0.000000	0.000000	0.000000	0.000000	0.000000	.967846-M
				+M	0.000000	0.000000	-.156879	0.000000	0.000000	0.000000	0.000000	0.000000+M
-1.0848E+02	.0779	.5214	.5271	-M	0.000000	0.000000	-.156879	0.000000	0.000000	0.000000	0.000000	0.000000-M
				+M	0.000000	-.196628	0.000000	0.000000	0.000000	0.000000	0.000000	.967846+M
ENERGY	HEPAL	HEPER	HE		M=15/2	M=13/2	M=11/2	M=9/2	M=7/2	M=5/2	M=3/2	M=1/2

Y= .80 X=-.20 OAS=2.0558E+02*W246 B4F4/B2F2=-8.00E-01 B4F4/B6F6=-2.50E-01 E2F2/B6F6= 3.12E-01

ENERGY	HEPAL	HEPER	HE		M=15/2	M=13/2	M=11/2	M=9/2	M=7/2	M=5/2	M=3/2	M=1/2
1.0138E+02	-.2400	.4456	.5061	-M	0.000000	0.000000	0.000000	0.000000	.846550	0.000000	0.000000	0.000000-M
				+M	0.000000	0.000000	0.000000	0.000000	0.000000	.532309	0.000000	0.000000+M
1.0138E+02	.2400	.4456	.5061	-M	0.000000	0.000000	0.000000	0.000000	0.000000	.532309	0.000000	0.000000-M
				+M	0.000000	0.000000	0.000000	0.000000	.846550	0.000000	0.000000	0.000000+M
8.5676E+01	.5058	0.0000	.5058	-M	.047186	0.000000	0.000000	0.000000	0.000000	0.000000	.336636	0.000000-M
				+M	0.000000	0.000000	0.000000	.940452	0.000000	0.000000	0.000000	0.000000+M
8.5676E+01	-.5058	0.0000	.5058	-M	0.000000	0.000000	0.000000	.940452	0.000000	0.000000	0.000000	0.000000-M
				+M	.047186	0.000000	0.000000	0.000000	0.000000	0.000000	.336636	0.000000+M
2.4472E+01	-.6325	.1204	.6439	-M	0.000000	0.000000	.939652	0.000000	0.000000	0.000000	0.000000	0.000000-M
				+M	0.000000	.094638	0.000000	0.000000	0.000000	0.000000	0.000000	.328783+M
2.4472E+01	.6325	.1204	.6439	-M	0.000000	.094638	0.000000	0.000000	0.000000	0.000000	0.000000	.328783-M
				+M	0.000000	0.000000	.939652	0.000000	0.000000	0.000000	0.000000	0.000000+M
6.5533E+00	.9678	0.0000	.9678	-M	0.000000	0.000000	0.000000	-.100270	0.000000	0.000000	0.000000	0.000000-M
				+M	.984762	0.000000	0.000000	0.000000	0.000000	0.000000	.142090	0.000000+M
6.5533E+00	-.9678	0.0000	.9678	-M	.984762	0.000000	0.000000	0.000000	0.000000	0.000000	.142090	0.000000-M
				+M	0.000000	0.000000	0.000000	-.100270	0.000000	0.000000	0.000000	0.000000+M
-9.2173E+00	-.1067	-.4456	.4582	-M	0.000000	0.000000	0.000000	0.000000	0.000000	.846550	0.000000	0.000000-M
				+M	0.000000	0.000000	0.000000	0.000000	-.532309	0.000000	0.000000	0.000000+M
-9.2173E+00	.1067	-.4456	.4582	-M	0.000000	0.000000	0.000000	0.000000	-.532309	0.000000	0.000000	0.000000-M
				+M	0.000000	0.000000	0.000000	0.000000	0.000000	.846550	0.000000	0.000000+M
-4.6632E+01	-.6800	-.0675	.6834	-M	0.000000	.902864	0.000000	0.000000	0.000000	0.000000	0.000000	.369287-M
				+M	0.000000	0.000000	-.220146	0.000000	0.000000	0.000000	0.000000	0.000000+M
-4.6632E+01	.6800	-.0675	.6834	-M	0.000000	0.000000	-.220146	0.000000	0.000000	0.000000	0.000000	0.000000-M
				+M	0.000000	.902864	0.000000	0.000000	0.000000	0.000000	0.000000	.369287+M
-5.8029E+01	-.1380	0.0000	.1380	-M	-.167384	0.000000	0.000000	0.000000	0.000000	0.000000	.930853	0.000000-M
				+M	0.000000	0.000000	0.000000	-.324802	0.000000	0.000000	0.000000	0.000000+M
-5.8029E+01	.1380	0.0000	.1380	-M	0.000000	0.000000	0.000000	-.324802	0.000000	0.000000	0.000000	0.000000-M
				+M	-.167384	0.000000	0.000000	0.000000	0.000000	0.000000	.930853	0.000000+M
-1.0420E+02	.1525	.4804	.5041	-M	0.000000	0.000000	-.261897	0.000000	0.000000	0.000000	0.000000	0.000000-M
				+M	0.000000	-.419381	0.000000	0.000000	0.000000	0.000000	0.000000	.869212+M
-1.0420E+02	-.1525	.4804	.5041	-M	0.000000	-.419381	0.000000	0.000000	0.000000	0.000000	0.000000	.869212-M
				+M	0.000000	0.000000	-.261897	0.000000	0.000000	0.000000	0.000000	0.000000+M
ENERGY	HEPAL	HEPER	HE		M=15/2	M=13/2	M=11/2	M=9/2	M=7/2	M=5/2	M=3/2	M=1/2

Y= .80 X=0.00 OAS=2.3191E+02*W246 B4F4/B2F2= 0. B4F4/B6F6= 0. E2F2/B6F6= 2.50E-01

ENERGY	HEPAL	HEPER	HE		M=15/2	M=13/2	M=11/2	M=9/2	M=7/2	M=5/2	M=3/2	M=1/2
1.1508E+02	-.1743	.4762	.5071	-M	0.000000	0.000000	0.000000	0.000000	.796560	0.000000	0.000000	0.000000-M
				+M	0.000000	0.000000	0.000000	0.000000	0.000000	.604560	0.000000	0.000000+M
1.1508E+02	.1743	.4762	.5071	-M	0.000000	0.000000	0.000000	0.000000	0.000000	.604560	0.000000	0.000000-M
				+M	0.000000	0.000000	0.000000	0.000000	.796560	0.000000	0.000000	0.000000+M
7.9957E+01	-.2455	0.0000	.2455	-M	0.000000	0.000000	0.000000	.813972	0.000000	0.000000	0.000000	0.000000-M
				+M	.325055	0.000000	0.000000	0.000000	0.000000	0.000000	.481445	0.000000+M
7.9957E+01	.2455	0.0000	.2455	-M	.325055	0.000000	0.000000	0.000000	0.000000	0.000000	.481445	0.000000-M
				+M	0.000000	0.000000	0.000000	.813972	0.000000	0.000000	0.000000	0.000000+M
5.8001E+01	-.8159	0.0000	.8159	-M	.939536	0.000000	0.000000	0.000000	0.000000	0.000000	-.066332	0.000000-M
				+M	0.000000	0.000000	0.000000	-.335964	0.000000	0.000000	0.000000	0.000000+M
5.8001E+01	.8159	0.0000	.8159	-M	0.000000	0.000000	0.000000	-.335964	0.000000	0.000000	0.000000	0.000000-M
				+M	.939536	0.000000	0.000000	0.000000	0.000000	0.000000	-.066332	0.000000+M
3.4194E+00	.4016	.2872	.4937	-M	0.000000	.184025	0.000000	0.000000	0.000000	0.000000	0.000000	.589059-M
				+M	0.000000	0.000000	.786857	0.000000	0.000000	0.000000	0.000000	0.000000+M
3.4194E+00	-.4016	.2872	.4937	-M	0.000000	0.000000	.786857	0.000000	0.000000	0.000000	0.000000	0.000000-M
				+M	0.000000	.184025	0.000000	0.000000	0.000000	0.000000	0.000000	.589059+M
-1.4280E+01	-.0409	-.4762	.4779	-M	0.000000	0.000000	0.000000	0.000000	0.000000	.796560	0.000000	0.000000-M
				+M	0.000000	0.000000	0.000000	0.000000	-.604560	0.000000	0.000000	0.000000+M
-1.4280E+01	.0409	-.4762	.4779	-M	0.000000	0.000000	0.000000	0.000000	-.604560	0.000000	0.000000	0.000000-M
				+M	0.000000	0.000000	0.000000	0.000000	0.000000	.796560	0.000000	0.000000+M
-5.5758E+01	.0296	0.0000	.0296	-M	0.000000	0.000000	0.000000	-.473897	0.000000	0.000000	0.000000	0.000000-M
				+M	-.107756	0.000000	0.000000	0.000000	0.000000	0.000000	.873962	0.000000+M
-5.5758E+01	-.0296	0.0000	.0296	-M	-.107756	0.000000	0.000000	0.000000	0.000000	0.000000	.873962	0.000000-M
				+M	0.000000	0.000000	0.000000	-.473897	0.000000	0.000000	0.000000	0.000000+M
-6.9587E+01	-.2424	.2127	.3225	-M	0.000000	.703165	0.000000	0.000000	0.000000	0.000000	0.000000	.480460-M
				+M	0.000000	0.000000	-.524135	0.000000	0.000000	0.000000	0.000000	0.000000+M
-6.9587E+01	.2424	-.2127	.3225	-M	0.000000	0.000000	-.524135	0.000000	0.000000	0.000000	0.000000	0.000000-M
				+M	0.000000	.703165	0.000000	0.000000	0.000000	0.000000	0.000000	.480460+M
-1.1683E+02	.3591	.2127	.4174	-M	0.000000	0.000000	-.325789	0.000000	0.000000	0.000000	0.000000	0.000000-M
				+M	0.000000	-.686800	0.000000	0.000000	0.000000	0.000000	0.000000	.649744+M
-1.1683E+02	-.3591	-.2127	.4174	-M	0.000000	.686800	0.000000	0.000000	0.000000	0.000000	0.000000	-.649744-M
				+M	0.000000	0.000000	.325789	0.000000	0.000000	0.000000	0.000000	0.000000+M
ENERGY	HEPAL	HEPER	HE		M=15/2	M=13/2	M=11/2	M=9/2	M=7/2	M=5/2	M=3/2	M=1/2

Y= .80 X= .20 OAS=1.9157E+02*W246 B4F4/B2F2= 8.00E-01 B4F4/B6F6= 2.50E-01 E2F2/B6F6= 3.12E-01

ENERGY	HEPAL	HEPER	HE		M=15/2	M=13/2	M=11/2	M=9/2	M=7/2	M=5/2	M=3/2	M=1/2
9.4089E+01	.9702	0.0000	.9702	-M	0.000000	0.000000	0.000000	.079051	0.000000	0.000000	0.000000	0.000000-M
				+M	.984385	0.000000	0.000000	0.000000	0.000000	0.000000	.157282	0.000000+M
9.4089E+01	-.9702	0.0000	.9702	-M	.984385	0.000000	0.000000	0.000000	0.000000	0.000000	.157282	0.000000-M
				+M	0.000000	0.000000	0.000000	.079051	0.000000	0.000000	0.000000	0.000000+M
8.3607E+01	.0996	.4927	.5027	-M	0.000000	0.000000	0.000000	0.000000	0.000000	.677352	0.000000	0.000000-M
				+M	0.000000	0.000000	0.000000	0.000000	.735659	0.000000	0.000000	0.000000+M
8.3607E+01	-.0996	.4927	.5027	-M	0.000000	0.000000	0.000000	0.000000	.735659	0.000000	0.000000	0.000000-M
				+M	0.000000	0.000000	0.000000	0.000000	0.000000	.677352	0.000000	0.000000+M
4.6449E+01	-.1993	0.0000	.1993	-M	0.000000	0.000000	0.000000	.725520	0.000000	0.000000	0.000000	0.000000-M
				+M	-.165014	0.000000	0.000000	0.000000	0.000000	0.000000	.668125	0.000000+M
4.6449E+01	.1993	0.0000	.1993	-M	-.165014	0.000000	0.000000	0.000000	0.000000	0.000000	.668125	0.000000-M
				+M	0.000000	0.000000	0.000000	.725520	0.000000	0.000000	0.000000	0.000000+M
3.8152E+00	.0770	.4599	.4663	-M	0.000000	.221389	0.000000	0.000000	0.000000	0.000000	0.000000	.849900-M
				+M	0.000000	0.000000	.478181	0.000000	0.000000	0.000000	0.000000	0.000000+M
3.8152E+00	-.0770	.4599	.4663	-M	0.000000	0.000000	.478181	0.000000	0.000000	0.000000	0.000000	0.000000-M
				+M	0.000000	.221389	0.000000	0.000000	0.000000	0.000000	0.000000	.849900+M
-1.6407E+01	-.0337	-.4927	.4939	-M	0.000000	0.000000	0.000000	0.000000	-.677352	0.000000	0.000000	0.000000-M
				+M	0.000000	0.000000	0.000000	0.000000	0.000000	.735659	0.000000	0.000000+M
-1.6407E+01	.0337	-.4927	.4939	-M	0.000000	0.000000	0.000000	0.000000	0.000000	.735659	0.000000	0.000000-M
				+M	0.000000	0.000000	0.000000	0.000000	-.677352	0.000000	0.000000	0.000000+M
-4.2018E+01	.1709	0.0000	.1709	-M	-.061296	0.000000	0.000000	0.000000	0.000000	0.000000	.727236	0.000000-M
				+M	0.000000	0.000000	0.000000	-.683645	0.000000	0.000000	0.000000	0.000000+M
-4.2018E+01	-.1709	0.0000	.1709	-M	0.000000	0.000000	0.000000	-.683645	0.000000	0.000000	0.000000	0.000000-M
				+M	-.061296	0.000000	0.000000	0.000000	0.000000	0.000000	.727236	0.000000+M
-7.2052E+01	.2014	-.2482	.3196	-M	0.000000	-.534971	0.000000	0.000000	0.000000	0.000000	0.000000	-.304131-M
				+M	0.000000	0.000000	.788232	0.000000	0.000000	0.000000	0.000000	0.000000+M
-7.2052E+01	-.2014	-.2482	.3196	-M	0.000000	0.000000	.788232	0.000000	0.000000	0.000000	0.000000	0.000000-M
				+M	0.000000	-.534971	0.000000	0.000000	0.000000	0.000000	0.000000	-.304131+M
-9.7483E+01	-.4785	.3216	.5765	-M	0.000000	.815348	0.000000	0.000000	0.000000	0.000000	0.000000	-.430319-M
				+M	0.000000	0.000000	.387341	0.000000	0.000000	0.000000	0.000000	0.000000+M
-9.7483E+01	.4785	.3216	.5765	-M	0.000000	0.000000	.387341	0.000000	0.000000	0.000000	0.000000	0.000000-M
				+M	0.000000	.815348	0.000000	0.000000	0.000000	0.000000	0.000000	-.430319+M
ENERGY	HEPAL	HEPER	HE		M=15/2	M=13/2	M=11/2	M=9/2	M=7/2	M=5/2	M=3/2	M=1/2

Y= .80 X= .40 OAS=2.2136E+02*W246 B4F4/B2F2= 1.60E+00 B4F4/B6F6= 6.67E-01 B2F2/B6F6= 4.17E-01

ENERGY	HEPAL	HEPER	HE		M=15/2	M=13/2	M=11/2	M=9/2	M=7/2	M=5/2	M=3/2	M=1/2
1.2633E+02	.9930	0.0000	.9930	-M	0.000000	0.000000	0.000000	.018505	0.000000	0.000000	0.000000	0.000000-M
				+M	.995773	0.000000	0.000000	0.000000	0.000000	0.000000	.089961	0.000000+M
1.2633E+02	-.9930	0.0000	.9930	-M	.995773	0.000000	0.000000	0.000000	0.000000	0.000000	.089961	0.000000-M
				+M	0.000000	0.000000	0.000000	.018505	0.000000	0.000000	0.000000	0.000000+M
5.5284E+01	.0285	.4802	.4811	-M	0.000000	0.000000	0.000000	0.000000	.617243	0.000000	0.000000	0.000000-M
				+M	0.000000	0.000000	0.000000	0.000000	0.000000	.786773	0.000000	0.000000+M
5.5284E+01	-.0285	.4802	.4811	-M	0.000000	0.000000	0.000000	0.000000	0.000000	.786773	0.000000	0.000000-M
				+M	0.000000	0.000000	0.000000	0.000000	.617243	0.000000	0.000000	0.000000+M
3.9423E+01	.0767	0.0000	.0767	-M	0.000000	0.000000	0.000000	.402701	0.000000	0.000000	0.000000	0.000000-M
				+M	-.089778	0.000000	0.000000	0.000000	0.000000	0.000000	.910918	0.000000+M
3.9423E+01	-.0767	0.0000	.0767	-M	-.089778	0.000000	0.000000	0.000000	0.000000	0.000000	.910918	0.000000-M
				+M	0.000000	0.000000	0.000000	.402701	0.000000	0.000000	0.000000	0.000000+M
2.8964E+01	-.0474	.5191	.5213	-M	0.000000	.152425	0.000000	0.000000	0.000000	0.000000	0.000000	.964089-M
				+M	0.000000	0.000000	.217481	0.000000	0.000000	0.000000	0.000000	0.000000+M
2.8964E+01	.0474	.5191	.5213	-M	0.000000	0.000000	.217481	0.000000	0.000000	0.000000	0.000000	0.000000-M
				+M	0.000000	.152425	0.000000	0.000000	0.000000	0.000000	0.000000	.964089+M
-2.1684E+01	.1619	-.4802	.5068	-M	0.000000	0.000000	0.000000	0.000000	0.000000	-.617243	0.000000	0.000000-M
				+M	0.000000	0.000000	0.000000	0.000000	.786773	0.000000	0.000000	0.000000+M
-2.1684E+01	-.1619	-.4802	.5068	-M	0.000000	0.000000	0.000000	0.000000	.786773	0.000000	0.000000	0.000000-M
				+M	0.000000	0.000000	0.000000	0.000000	0.000000	-.617243	0.000000	0.000000+M
-5.0910E+01	.4697	0.0000	.4697	-M	.019371	0.000000	0.000000	0.000000	0.000000	0.000000	-.402661	0.000000-M
				+M	0.000000	0.000000	0.000000	.915144	0.000000	0.000000	0.000000	0.000000+M
-5.0910E+01	-.4697	0.0000	.4697	-M	0.000000	0.000000	0.000000	.915144	0.000000	0.000000	0.000000	0.000000-M
				+M	.019371	0.000000	0.000000	0.000000	0.000000	0.000000	-.402661	0.000000+M
-8.2373E+01	-.6699	-.2265	.7071	-M	0.000000	.935194	0.000000	0.000000	0.000000	0.000000	0.000000	-.069525-M
				+M	0.000000	0.000000	-.347244	0.000000	0.000000	0.000000	0.000000	0.000000+M
-8.2373E+01	.6699	-.2265	.7071	-M	0.000000	0.000000	-.347244	0.000000	0.000000	0.000000	0.000000	0.000000-M
				+M	0.000000	.935194	0.000000	0.000000	0.000000	0.000000	0.000000	-.069525+M
-9.5031E+01	-.5173	.2408	.5706	-M	0.000000	0.000000	.912208	0.000000	0.000000	0.000000	0.000000	0.000000-M
				+M	0.000000	.319653	0.000000	0.000000	0.000000	0.000000	0.000000	-.256316+M
-9.5031E+01	.5173	.2408	.5706	-M	0.000000	.319653	0.000000	0.000000	0.000000	0.000000	0.000000	-.256316-M
				+M	0.000000	0.000000	.912208	0.000000	0.000000	0.000000	0.000000	0.000000+M
ENERGY	HEPAL	HEPER	HE		M=15/2	M=13/2	M=11/2	M=9/2	M=7/2	M=5/2	M=3/2	M=1/2

Y= .80 X= .60 OAS=2.7763E+02*W246 B4F4/B2F2= 2.40E+00 B4F4/B6F6= 1.50E+00 B2F2/B6F6= 6.25E-01

ENERGY	HEPAL	HEPER	HE		M=15/2	M=13/2	M=11/2	M=9/2	M=7/2	M=5/2	M=3/2	M=1/2
1.5914E+02	.9977	0.0000	.9977	-M	0.000000	0.000000	0.000000	.004972	0.000000	0.000000	0.000000	0.000000-M
				+M	.998590	0.000000	0.000000	0.000000	0.000000	0.000000	.052858	0.000000+M
1.5914E+02	-.9977	0.0000	.9977	-M	.998590	0.000000	0.000000	0.000000	0.000000	0.000000	.052858	0.000000-M
				+M	0.000000	0.000000	0.000000	.004972	0.000000	0.000000	0.000000	0.000000+M
6.5157E+01	.0646	.5304	.5343	-M	0.000000	0.000000	.095635	0.000000	0.000000	0.000000	0.000000	0.000000-M
				+M	0.000000	.080994	0.000000	0.000000	0.000000	0.000000	0.000000	.992116+M
6.5157E+01	-.0646	.5304	.5343	-M	0.000000	.080994	0.000000	0.000000	0.000000	0.000000	0.000000	.992116-M
				+M	0.000000	0.000000	.095635	0.000000	0.000000	0.000000	0.000000	0.000000+M
5.3596E+01	-.1799	0.0000	.1799	-M	-.052945	0.000000	0.000000	0.000000	0.000000	0.000000	.984534	0.000000-M
				+M	0.000000	0.000000	0.000000	.166999	0.000000	0.000000	0.000000	0.000000+M
5.3596E+01	.1799	0.0000	.1799	-M	0.000000	0.000000	0.000000	.166999	0.000000	0.000000	0.000000	0.000000-M
				+M	-.052945	0.000000	0.000000	0.000000	0.000000	0.000000	.984534	0.000000+M
3.3533E+01	.2010	.3674	.4188	-M	0.000000	0.000000	0.000000	0.000000	.406703	0.000000	0.000000	0.000000-M
				+M	0.000000	0.000000	0.000000	0.000000	0.000000	.913561	0.000000	0.000000+M
3.3533E+01	-.2010	.3674	.4188	-M	0.000000	0.000000	0.000000	0.000000	0.000000	.913561	0.000000	0.000000-M
				+M	0.000000	0.000000	0.000000	0.000000	.406703	0.000000	0.000000	0.000000+M
-3.3533E+01	.3343	-.3674	.4968	-M	0.000000	0.000000	0.000000	0.000000	0.000000	-.406703	0.000000	0.000000-M
				+M	0.000000	0.000000	0.000000	0.000000	.913561	0.000000	0.000000	0.000000+M
-3.3533E+01	-.3343	-.3674	.4968	-M	0.000000	0.000000	0.000000	0.000000	.913561	0.000000	0.000000	0.000000-M
				+M	0.000000	0.000000	0.000000	0.000000	0.000000	-.406703	0.000000	0.000000+M
-7.7828E+01	.8598	-.0214	.8601	-M	0.000000	0.000000	-.035112	0.000000	0.000000	0.000000	0.000000	0.000000-M
				+M	0.000000	.996338	0.000000	0.000000	0.000000	0.000000	0.000000	-.077954+M
-7.7828E+01	-.8598	-.0214	.8601	-M	0.000000	.996338	0.000000	0.000000	0.000000	0.000000	0.000000	-.077954-M
				+M	0.000000	0.000000	-.035112	0.000000	0.000000	0.000000	0.000000	0.000000+M
-8.1576E+01	-.5777	0.0000	.5777	-M	0.000000	0.000000	0.000000	.985945	0.000000	0.000000	0.000000	0.000000-M
				+M	.003932	0.000000	0.000000	0.000000	0.000000	0.000000	-.167027	0.000000+M
-8.1576E+01	.5777	0.0000	.5777	-M	.003932	0.000000	0.000000	0.000000	0.000000	0.000000	-.167027	0.000000-M
				+M	0.000000	0.000000	0.000000	.985945	0.000000	0.000000	0.000000	0.000000+M
-1.1849E+02	-.7244	.0244	.7248	-M	0.000000	0.000000	.994797	0.000000	0.000000	0.000000	0.000000	0.000000-M
				+M	0.000000	.027380	0.000000	0.000000	0.000000	0.000000	0.000000	-.098129+M
-1.1849E+02	.7244	.0244	.7248	-M	0.000000	.027380	0.000000	0.000000	0.000000	0.000000	0.000000	-.098129-M
				+M	0.000000	0.000000	.994797	0.000000	0.000000	0.000000	0.000000	0.000000+M
ENERGY	HEPAL	HEPER	HE		M=15/2	M=13/2	M=11/2	M=9/2	M=7/2	M=5/2	M=3/2	M=1/2

Y= .80 X= .80 OAS=3.3838E+02*W246 B4F4/B2F2= 3.20E+00 B4F4/B6F6= 4.00E+00 B2F2/B6F6= 1.25E+00

ENERGY	HEPAL	HEPER	HE		M=15/2	M=13/2	M=11/2	M=9/2	M=7/2	M=5/2	M=3/2	M=1/2
1.9219E+02	.9995	0.0000	.9995	-M	0.000000	0.000000	0.000000	.000865	0.000000	0.000000	0.000000	0.000000-M
				+M	.999707	0.000000	0.000000	0.000000	0.000000	0.000000	.024173	0.000000+M
1.9219E+02	-.9995	0.0000	.9995	-M	.999707	0.000000	0.000000	0.000000	0.000000	0.000000	.024173	0.000000-M
				+M	0.000000	0.000000	0.000000	.000865	0.000000	0.000000	0.000000	0.000000+M
1.0526E+02	.0665	.5329	.5371	-M	0.000000	0.000000	.034874	0.000000	0.000000	0.000000	0.000000	0.000000-M
				+M	0.000000	.032541	0.000000	0.000000	0.000000	0.000000	0.000000	.998862+M
1.0526E+02	-.0665	.5329	.5371	-M	0.000000	.032541	0.000000	0.000000	0.000000	0.000000	0.000000	.998862-M
				+M	0.000000	0.000000	.034874	0.000000	0.000000	0.000000	0.000000	0.000000+M
7.5329E+01	.1979	0.0000	.1979	-M	0.000000	0.000000	0.000000	.057117	0.000000	0.000000	0.000000	0.000000-M
				+M	-.024183	0.000000	0.000000	0.000000	0.000000	0.000000	.998075	0.000000+M
7.5329E+01	-.1979	0.0000	.1979	-M	-.024183	0.000000	0.000000	0.000000	0.000000	0.000000	.998075	0.000000-M
				+M	0.000000	0.000000	0.000000	.057117	0.000000	0.000000	0.000000	0.000000+M
2.1031E+01	.3110	.1628	.3511	-M	0.000000	0.000000	0.000000	0.000000	.167017	0.000000	0.000000	0.000000-M
				+M	0.000000	0.000000	0.000000	0.000000	0.000000	.985954	0.000000	0.000000+M
2.1031E+01	-.3110	.1628	.3511	-M	0.000000	0.000000	0.000000	0.000000	0.000000	.985954	0.000000	0.000000-M
				+M	0.000000	0.000000	0.000000	0.000000	.167017	0.000000	0.000000	0.000000+M
-5.4631E+01	.4444	-.1628	.4732	-M	0.000000	0.000000	0.000000	0.000000	0.000000	-.167017	0.000000	0.000000-M
				+M	0.000000	0.000000	0.000000	0.000000	.985954	0.000000	0.000000	0.000000+M
-5.4631E+01	-.4444	-.1628	.4732	-M	0.000000	0.000000	0.000000	0.000000	.985954	0.000000	0.000000	0.000000-M
				+M	0.000000	0.000000	0.000000	0.000000	0.000000	-.167017	0.000000	0.000000+M
-7.2948E+01	-.8658	-.0022	.8658	-M	0.000000	.999467	0.000000	0.000000	0.000000	0.000000	0.000000	-.032425-M
				+M	0.000000	0.000000	-.003898	0.000000	0.000000	0.000000	0.000000	0.000000+M
-7.2948E+01	.8658	-.0022	.8658	-M	0.000000	0.000000	-.003898	0.000000	0.000000	0.000000	0.000000	0.000000-M
				+M	0.000000	.999467	0.000000	0.000000	0.000000	0.000000	0.000000	-.032425+M
-1.2004E+02	-.5974	0.0000	.5974	-M	0.000000	0.000000	0.000000	.998367	0.000000	0.000000	0.000000	0.000000-M
				+M	.000518	0.000000	0.000000	0.000000	0.000000	0.000000	-.057121	0.000000+M
-1.2004E+02	.5974	0.0000	.5974	-M	.000518	0.000000	0.000000	0.000000	0.000000	0.000000	-.057121	0.000000-M
				+M	0.000000	0.000000	0.000000	.998367	0.000000	0.000000	0.000000	0.000000+M
-1.4619E+02	-.7323	.0026	.7323	-M	0.000000	0.000000	.999384	0.000000	0.000000	0.000000	0.000000	0.000000-M
				+M	0.000000	.002763	0.000000	0.000000	0.000000	0.000000	0.000000	-.034982+M
-1.4619E+02	.7323	.0026	.7323	-M	0.000000	.002763	0.000000	0.000000	0.000000	0.000000	0.000000	-.034982-M
				+M	0.000000	0.000000	.999384	0.000000	0.000000	0.000000	0.000000	0.000000+M
ENERGY	HEPAL	HEPER	HE		M=15/2	M=13/2	M=11/2	M=9/2	M=7/2	M=5/2	M=3/2	M=1/2

AUTHOR INDEX

Numbers in parentheses are reference numbers and indicate that an author's work is referred to although his name is not cited in the text. Numbers in italics show the page on which the complete reference is listed.

V

W

Y

Z